Polyploidy and Hybridization for Crop Improvement

Polyploidy and Hybridization for Crop Improvement

Editor
Annaliese S. Mason
Department of Plant Breeding
Justus Liebig University Giessen
Giessen, Germany

CRC Press
Taylor & Francis Group
Boca Raton London New York

CRC Press is an imprint of the
Taylor & Francis Group, an **informa** business
A SCIENCE PUBLISHERS BOOK

CRC Press
Taylor & Francis Group
6000 Broken Sound Parkway NW, Suite 300
Boca Raton, FL 33487-2742

Printed on acid-free paper
Version Date: 20160225

International Standard Book Number-13: 978-1-4987-4066-1 (Hardback)

Library of Congress Cataloging-in-Publication Data

Names: Mason, Annaliese, editor.
Title: Polyploidy and hybridizaton for crop improvement / editor: Annaliese Mason.
Description: Boca Raton, FL : CRC Press, Taylor & Francis Group, [2016] | Includes bibliographical references and index.
Identifiers: LCCN 2016007064 | ISBN 9781498740661 (hardcover)
Subjects: LCSH: Crop improvement. | Polyploidy. | Plant hybridization.
Classification: LCC SB106.I47 P59 2016 | DDC 631.5/23--dc23
LC record available at http://lccn.loc.gov/2016007064

Visit the Taylor & Francis Web site at
http://www.taylorandfrancis.com

and the CRC Press Web site at
http://www.crcpress.com

Preface

Polyploidy and interspecific hybridization are critical processes in plant evolution and speciation. Polyploidy refers to the presence of more than two sets of chromosomes or the presence of more than one genome within a single organism, and interspecific hybridization refers to the process by which two different species come together to form a new organism with genetic information from both parents. When two different species each contribute a complete set of chromosomes to make a new, stable species with all the genetic information from both parents, this is referred to as "allopolyploid" formation, while "autopolyploids" have three or more sets of chromosomes from the same species. However, categorization of individuals or species as "auto" vs "allo" polyploid or as "intra" or "inter" specific hybrids is on some level arbitrary, depending on species definitions and genetic divergence between the parent genomes. In the middle are hybrids between subspecies or genetically distinct populations of individuals, or between geographically isolated taxonomic species with conserved genome structure.

Many of our current agricultural crops are either natural or agricultural hybrids or polyploids, including potato, oats, cotton, oilseed rape, wheat (and triticale), strawberries, kiwifruit, banana and many others. There is also a great deal of potential to utilise these natural evolutionary processes for targeted crop improvement, for example through introgression of desired traits from wild species into crops, production of seedless fruits, or even creation of entirely new crop types.

Polyploidy and interspecific hybridization are increasingly being revealed as both complex and common phenomena in plants. Characterization of crop genomes with modern genetics and genomics technologies has revealed that polyploidy and interspecific hybridization processes shaped many of our extant crop species. In parallel, breeding approaches utilizing polyploidy and interspecific hybridization as tools for crop improvement are becoming more and more common. Some crop genera have hundreds of years of manipulation of interspecific hybridization and polyploidy processes behind them, while in others use of these processes for crop improvement is still at the theoretical stage. As our knowledge grows and technology advances, it is increasingly important to bring together expert researchers and plant breeders to form a cohesive picture of how best to utilize these evolutionary processes for crop improvement across diverse genera. Which processes and problems are crop-specific, and which are widely applicable? Answering these questions has potential to not only aid in our understanding of plant evolutionary processes, but to add additional fundamental tools to the plant breeding tool kit, helping the agricultural processes on which we all rely for life. This book will provide a comprehensive summary of how the processes of polyploidy and hybridisation have shaped the foods we eat, and how these processes have been and can be used for crop improvement. This book is targeted to both researchers and breeders, facilitating sharing of knowledge and stories across the wide range of crops where polyploidy and hybridization processes are relevant and potentially useful for crop improvement.

Contents

Chapter 1

Interspecific Hybridization for Upland Cotton Improvement

Peng W. Chee[1,*], Andrew H. Paterson[2], Joshua A. Udall[3] and Jonathan F. Wendel[4]

ABSTRACT

Interspecific hybridization has been central to the evolution, domestication, and improvement of Upland cotton, the cultivated form of *Gossypium hirsutum*. As the world's most important fiber crop species, *Gossypium hirsutum* belongs to the allotetraploid *Gossypium* clade that consists of six additional species. The lint fiber evolved only once in the history of *Gossypium*, in the ancestor diploid A-genome species, and this trait was passed on to the allopolyploid species when the A-genome united in a common nucleus with a D-genome from the other ancestor that produced no lint fibers. The domestication history of *G. hirsutum* involved the collection and use of lint fibers by indigenous people for the purpose of making strings and other textile products; hence, spinnable lint fibers were likely to have evolved under domestication. The geographical distribution of *G. hirsutum* overlaps with *G. barbadense* and *G. mustelinum*, and gene flow among these species has been documented. Therefore, the introgression of novel alleles into *G. hirsutum* possibly contributed to greater ecological adaptation in colonizing new habitats and providing important sources of genetic variation for artificial selection in the early domestication efforts. In modern Upland cotton, numerous germplasm lines have been developed from crossing with *G. barbadense*. However, reproductive barriers such as reduced fertility, segregation distortion, and hybrid breakdown are often observed in later

[1] Institute of Plant Breeding, Genetics, and Genomics, University of Georgia – Tifton Campus, Tifton, GA 31794, USA, E-mail: pwchee@uga.edu
[2] Plant Genome Mapping Laboratory, University of Georgia, Athens, GA 30602, USA, E-mail: paterson@uga.edu
[3] Department of Plant and Wildlife Science, Brigham Young University, Provo, Utah, USA, E-mail: jaudall@byu.edu
[4] Department of Ecology, Evolution, and Organismal Biology, Iowa State University, Ames, IA 50011, USA, E-mail: jfw@iastate.edu
* Corresponding author

generation hybrids between *G. hirsutum* and the other polyploid species, complicating the task of introgressing new, stably inherited allelic variation from inter specific hybridization. Recent efforts in molecular genetic research have provided insights into the location and effects of QTLs from wild species that are associated with traits important to cotton production. These and future research efforts will undoubtedly provide the tools that can be utilized by plant breeders to access novel genes from wild and domesticated allotetraploid *Gossypium* for Upland cotton improvement.

The Cotton Crop

The cotton plant is a source of both food and fiber, contributing to two basic needs of humanity. Cotton fiber in the form of textile products has contributed greatly to the comfort, style, and culture of human society. Although not commonly viewed as a food source, cotton is an important source of vegetable oil used extensively in foodstuffs such as baking and frying fats, mayonnaise, margarine, and snack food. Furthermore, after oil extraction, the seed by-product is used as raw material in livestock feed, fertilizer, and paper. This versatility has made cotton one of the most important field crops in the world.

According to the International Cotton Advisory Committee (ICAC), which collects statistics on world cotton production, consumption and trade, about 36 million hectares of cotton are planted in over 100 countries from latitudes 45°N in Ukraine to 32°S in Australia (ICAC, 2015). The top five cotton producing countries in the 2014-15 season include, in order of importance, China, India, the United States, Pakistan and Brazil, which collectively account for nearly two-thirds of the world's cotton production. Total cotton fiber production has now reached 106 million bales, and contributes about 40% of the world fiber market (ICAC, 2015), thus making cotton the single most important natural fiber in the textile industries and a vital agricultural commodity in the global economy. The aggregate value of the world's cotton crop is estimated to be about US$30 billion/yr, with 90% of its value residing in lint fiber. More than 350 million people are engaged in jobs related to the production and processing of cotton. The economic importance of cotton as a natural fiber for the global textile industry has fueled considerable interest in improving the inherent genetic potential of the crop through breeding for cultivars with higher levels of biotic and abiotic tolerance as well as higher lint yield and the further enhancement of fiber quality.

Interspecific hybridization has been central to the evolution of cotton, and to its improvement. As a crop, cotton is unique in that four different species in the genus *Gossypium* (Malvaceae) were domesticated independently on two separate continents for lint fiber production (Wendel and Cronn 2003; Wendel and Grover 2015). Therefore, the word "cotton" in the textile industry can apply equally to the two allotetraploid species *G. hirsutum* L. and *G. barbadense* L., endemic to the Americas, and the two diploid species

G. arboreum L. and *G. herbaceum* L., endemic to Africa and Asia. Allotetraploid *Gossypium* trace to a single polyploidization, joining progenitors resembling *G. herbaceum* or *G. arboreum* (A genome) and *G. raimondii* (D genome) (Wendel and Grover, 2015). Currently, the two allotetraploids supply an overwhelming majority of the world's textile fiber, with *G. hirsutum* responsible for over 90% of the total cotton production. This chapter will therefore focus on the cultivated allotetraploid species *G. hirsutum*, commonly referred to as "Upland" cotton. However, we will reference other domesticated and non-domesticated allotetraploid species, as they represent a vast reservoir of untapped genetic resources to sustain continued genetic improvement.

Origin and Taxonomy of *Gossypium hirsutum*

The genus *Gossypium* includes over 50 species of shrubs and small trees indigenous to the arid and semiarid tropics (Fryxell 1979; Wendel and Grover 2015). Diploid *Gossypium* species have 13 pairs of chromosomes ($2n = 2x = 26$) and are grouped into eight different genome types (A-G, K) based on chromosome pairing affinities (et al. 1984). Collectively, these species have a widespread distribution, although several primary centers of diversity are recognized. Specifically, A-genome species are found in Africa/Asia, B- and F-genome species in Africa, E-genome species in Arabia, C-, G- and K-genome species in Australia and D-genome species in Central and South America. Molecular phylogenetic analyses (Cronn et al. 2002; reviewed in Wendel and Grover 2015) have provided a phylogenetic framework for the genus and its different genome types, although the complete picture of the evolutionary relationships between each species is still not fully clear. The taxonomy of the diploid *Gossypium* species is described in detail by Fryxell (1979).

In comparison to the diploids, the phylogenetic history of polyploid cottons is better established (Grover et al. 2015), although new species continue to be discovered (Grover et al. 2014, and Wendel unpublished). Classically, five allotetraploid *Gossypium* species have been widely recognized that include the two domesticated species, *G, hirsutum* and *G. barbadense*, and three wild species, *G. tomentosum* Nutt ex Seem, *G. darwinii* Watt, and *G. mustelinum* Miers ex G. Watt. A sixth species, *G. ekmanianum* Wittmack, was recently resurrected and a seventh identified (see below). The aggregate geographic range of polyploid *Gossypium* encompasses many seasonally arid subtropical and tropical regions of the North and South American continents, mostly near coastlines, and extends to many islands in the Caribbean and the Pacific. Hence, these species collectively are often referred to as the New World cottons. The island endemic nature of *G. darwinii* (Galapagos Islands), *G. ekmanianum* (Hispaniola) and *G. tomentosum* (Hawaiian Islands) indicates that these species likely originated following long-distance dispersal events (Wendel and Cronn 2003; Wendel and Grover 2015). *Gossypium mustelinum* is indigenous to a small region of northeast Brazil (Wendel et al. 2009). As for the two domesticated species, *G. hirsutum* is indigenous to Central America and *G. barbadense* to South America but their ranges overlap, particularly in northwest South America and extensively in the Caribbean.

Recently, two additional species have been added to the allopolyploid cotton clade. The species *G. ekmanianum* Wittm. was recently resurrected (Krapovickas and Seijo 2008; Grover et al. 2014), based on accessions found in the Dominican Republic. In addition, a seventh species has been identified from the Wake Atoll in the Pacific Ocean (Wake, Peale, Wilkes Islands), *Gossypium* sp. Nov. (Wendel et al. unpublished data). Therefore, the most current iteration of the allopolyploid cotton phylogeny consists of seven lineages in three clades, with *G. hirsutum, G. tomentosum,* G. *ekmanianum* and the newly discovered species forming one clade, *G. barbadense* and *G. darwinii* forming a second clade, and *G. mustelinum* remaining the basal clade of the allopolyploid phylogeny (Wendel and Grover 2015).

All seven polyploid species have two sets of 13 homoeologous chromosomes ($2n = 4x = 52$) and exhibit strict disomic chromosome pairing (Kimber 1961). They all contain A-genome cytoplasm due to their monophyletic origin, with polyploidization occurring about 1-2 million years ago by transoceanic migration of an Old World (A-genome) progenitor followed by hybridization with a native New World (D-genome) progenitor (Wendel 1989; reviewed in Wendel and Grover 2015). Both meiotic chromosome pairing and comparative genome analyses have shown that the At- and Dt-subgenomes of the tetraploids were contributed by diploid progenitors that resemble *G. arboreum* or *G. herbaceum*, and *G. raimondii*, respectively. DNA sequence data indicates the progenitors of these species diverged from a common ancestor 5-7 million years ago (Senchina et al. 2003; Wendel and Grover 2015).

The A- and D-genome progenitors of cotton differ by at least 9 chromosomal rearrangements (Reinisch et al. 1994). Additionally, the A-genome has about twice the gametic DNA content as the D-genome (Hendrix and Stewart 2004), with the larger genome size mostly due to the repetitive DNA fraction, as the amount of single-copy DNA is almost the same in both genomes (Greever et al. 1989). The corresponding homoeologous chromosomes of the At- and Dt-subgenomes have been established by genetic mapping (Reinisch et al. 1994; Rong et al. 2004; Wang et al. 2014), and more recently verified by two draft genome sequences (Li et al. 2015; Zhang et al. 2015). Direct comparisons of gene order and synteny between the two subgenomes showed two reciprocal translocations in the At subgenome between chromosomes 02/03 and between chromosomes 04/05 as well as several possible inversions (Rong et al. 2004; Wang et al. 2014). The tetraploid chromosomes have also been aligned with those of their diploid progenitors, revealing that additional rearrangement has occurred since polyploid formation (Brubaker et al. 1999; Rong et al. 2004; Desai et al. 2006). Nonetheless, these results collectively indicate that gene colinearity is high between the two subgenomes and that chromosomal structural rearrangement has been modest following allopolyploid formation.

Domestication and "Upland" Cotton

The epidermal layer of cotton seed contains 'linters' or fuzz, which are short fibers that tightly adhere to the seed coat, and longer 'lint' fibers, which loosely adhere to the seed. The fuzz fibers represent an important source of raw material for the manufacture of paper and other industrial cellulose products. However, it is the longer, spinnable lint fibers that make cotton the world's most important fiber crop, and these novel single-celled seed epidermal trichomes may have first attracted ancient peoples to the cotton plant. As mentioned earlier, four different *Gossypium* species were domesticated independently by four geographically different civilizations on two separate continents; A genome diploids *G. herbaceum* and *G. arboreum* were domesticated in Africa and Asia, and allopolyploids *G. hirsutum* and *G. barbadense* were domesticated in Central and South America (Fryxell 1979; Brubaker et al. 1999).

The domestication history of *G. hirsutum* is not unlike those of the other three cultivated cotton species; it is likely to have involved the collection and use of lint fibers by indigenous people for the purpose of making strings and other textile products (Brubaker et al. 1999). The long lint fiber, which has a flat convoluted ribbon structure that permits the fiber to be spun into yarn, evolved under domestication (Brubaker et al. 1999). Accordingly, wild cottons have short and coarse lint fibers that are far different from those found in the cottonseed of modern cultivars. The biological function of lint fiber has not been identified although the elongated seed trichomes may act as a dispersal agent in some ecological settings, and/or function in maintaining an appropriate microbial and hydration context for seed germination or early seedling development (Wendel et al. 2009). Primitive lint fiber likely attracted the attention of early domesticators, with the long lint fiber of modern germplasm arising through continual human selection of genotypes with improved fiber quality characteristics as well as increased lint yield and other agronomic properties (Stephens 1967; Fryxell 1979; Brubaker et al. 1999). Interestingly, the evolution of long spinnable lint fiber is strongly tied to the elongation of fiber cells during development; specifically, the developing fiber cells in domesticated diploid and tetraploid cottons all have a prolonged fiber-cell development program. This developmental innovation arose only in the F-genome/A-genome lineages, which may have facilitated domestication of the A-genome cottons (Applequist et al. 2001; Hovav et al. 2008). This propensity for the development of elongated lint fiber was also passed on to the allopolyploids when the A-genome united with the D-genome during polyploidization.

The domestication of *G. hirsutum* may have taken place in the Yucatan peninsula in Mesoamerica. Brubaker and Wendel (1994) suggested that race 'punctatum', which has agronomic characteristics that are immediate between the truly wild race 'yucatanense' and races with more advanced properties such as 'latifolium' and 'palmeri', represents the earliest domesticated form of

G. hirsutum. Early domestication efforts possibly involved selection of a more desirable plant from the wild population to form door-yard cultigens, which led to large-scale field production as civilization became more industrialized and agriculture became specialized. Traits that may have been favored by early agriculturalists include decreasing seed dormancy to accelerate crop maturity and better control of plant density in the fields, increasing lint yield by selecting for larger and more bolls per plant, reducing plant size from the shrubby/small treelike habit of wild cottons to a scale that humans can utilize, and selecting for a more annual life cycle. Later selections were made to improve the quality of the fiber, initially focusing on finer, stronger fibers that were uniformly longer. Additional fiber qualities such as elongation and short fiber content have subsequently also become important with increased mechanization in the textile industry. As agriculture converges with agribusiness, the industrialization of production will continue to define traits that increase production efficiency. Examples include the increased mechanization that has required selection for appropriate plant stature, and the susceptibility of monoculture to pests and diseases that require increased selection for resistance to these threats.

While it is difficult to establish with certainty the geographic location of *G. hirsutum* domestication, the origin and rise of Upland germplasm, the modern cultivated form of *G. hirsutum*, took place in the southern United States of America. Ware (1951) noted that: *"American Upland cotton (Gossypium hirsutum) is a Dixie product. Although the stocks of the species were brought from elsewhere, new types, through a series of adaptational changes, formed this distinctive group the final characteristics of which are a product of the Cotton Belt of the United States".*

Ware (1951) has provided a comprehensive historical account of the development of *G. hirsutum* as Upland cotton cultivars from the time of immigration of European settlers to the middle of the 20th Century. Most genetic improvement of Upland cotton arose within the 'Cotton Belt' of the United States. Accordingly, the cotton crop was introduced into the eastern coastal areas of North America as Europeans immigrated into this area. While all four domesticated species of *Gossypium* were planted as crops early in the history of the United States, the allotetraploids were found to be more productive than the A genome diploids. The domesticated form of *G. barbadense* known as Sea Island cotton was introduced in the late 18th century and a small but flourishing industry was established in coastal areas of South Carolina, Georgia, and Florida by the 19th century (Percy 2009). However, while prized for its exceptionally long and fine fiber, Sea Island cotton has limited range of adaptation. As the population began to migrate inland from the coastal states, *G. hirsutum* was found to yield significantly better than Sea Island cotton in the upland areas of the country, which led to its common name of "Upland cotton".

The development of Upland cotton germplasm involved continual adaptation to new geographic ranges by artificial selection, transforming a day-length-sensitive, tropical perennial species to an annualized crop plant with a day-neutral reproductive habit and bigger bolls, thinner seed coat, and higher lint to seed ratio (lint percent) than in its natural habitat. Modern Upland cotton germplasm is thought to have been derived from multiple introductions, admixtures and introgressions, but it is thought that material from the Mexican highlands, introduced to the cotton belt in the first part of the 19th century via Texas, is prominent in the current gene pool. Stephens (1967) observed that of the seven botanical races of wild and domesticated *G. hirsutum*, only 'latifolium', found in southern Mexico/Guatemala, has members that are day-length neutral and therefore likely progenitors of today's modern cultivated Upland cotton germplasm. All except one of the *G. hirsutum* races are associated with humans and contain traits that could be considered valuable in a domesticated form of cotton (Hutchinson et al. 1947; Hutchinson 1951). Race 'yucatanense' is considered to exist solely as a wild form and has the characteristics of truly wild species; small seeds with hard seed coats and delayed germination; and small bolls with brown, short, and scanty lint (Wendel et al. 1992; Brubaker and Wendel 1994; d'Eeckenbrugge and Lacape 2014).

The major thrust to increase cotton production in the United States was brought on by growing demand for cotton, initially in England, and later domestically when cottage industry was replaced by large-scale mechanization to produce cotton textiles. As production expanded, early farmers/agronomists began reselecting from germplasm introduced from Mexico and elsewhere or from existing cultivars, which at the time often consisted of genetically diverse populations. Goals of these early selection efforts were to improve productivity or to propagate unique phenotypes, such as big bolls to facilitate hand picking or exceptionally long lint fiber (Calhoun et al. 1994). By the middle to late 1800s, distinctive cultivar "types" had been developed with adaptation to specific local environments across the US Cotton Belt. Duggar (1907) and Tyler (1910) described nine types at the height of differentiation of Upland cultivars, including the Eastern Big Boll type, Western Big Boll type, Semi-cluster type, Cluster type, Rio Grande type, Early type, Long Limb type, Upland Long Staple type, and Intermediate or Miscellaneous type. The arrival of the boll weevil caused a devastating loss of important germplasm. Many cultivars that either survived the boll weevil or were developed for combatting the insect came from the Western Big Boll type. An exception was Acala, which was introduced from Guatemala and Mexico to combat boll weevil. As the 20th century continued, a new grouping based on geographical regions and production practices emerged and persists to this day; the Acala type, the Plains type, the Delta type, and the Eastern type (Niles and Feaster 1984).

Prior to the development of the US cotton industry, cotton producing countries such as India, China, and Russia were growing the Asiatic diploid species *G. arboreum* and *G. herbaceum*, with staple length that was much shorter than that of Upland or Sea Island cottons (Ware 1936). The fiber produced by *G. arboreum* and *G. herbaceum* could not be effectively utilized by the then-new spinning technology (May 2000). While breeding for longer fiber length within the diploid species has met with some success (Kulkarni et al. 2009), new Upland germplasm was introduced to meet the demand of new varieties with improved fiber quality. By the 1920s, most international breeding efforts were working with the allopolyploid *G. hirsutum* and had almost completely abandoned the short-fiber Asiatic species (Ware 1951).

Cotton Improvement

The current approach utilized by the vast majority of cotton breeding programs has not changed much since the summary of Niles and Feaster (1984). Since cotton is predominantly a self-pollinated species, most cotton breeders practice selection based on modifications of a pedigree breeding system aimed at developing pure line cultivars. Basically, parents with different attributes or traits of interest are selected for cross hybridization, segregating populations are evaluated in the field to identify individual plants with the desired trait combinations, seed from the selected plants are evaluated in progeny rows, and inbred lines that perform better than "check" cultivars are evaluated in replicated tests over multiple locations and years. Borland and Myers (2015) provided a detailed description of the operation of a modern cotton breeding program, highlighting changes that have occurred since 1980. Current breeding programs take advantage of significantly greater mechanization, thus handling a greater number of progeny rows with fewer personnel and with modern fertilization, plant growth control, and pest management.

The traits that are desired in current cultivars are not much different from those of the earlier years of cotton production, although the relative importance of these traits may have changed (Borland and Myers 2015). Lint yield is still the first priority in any commercial cotton breeding program. Lint percent, which is a component of lint yield, was selected early in the domestication process and may have been among the first traits to be selected (Fryxell 1979). Other agronomic traits that have gained in importance include environmental stability and early maturity. Although the introduction of transgenic *Bt* cultivars has relaxed the focus on insect resistance, host-plant resistance continues to be important for a number of pathogens and nematodes. Similarly, selection for cotton genotypes that do not have excessive vegetative growth and upright stature was a major consideration in breeding programs, but characteristics such as plant height and compact growth habit can now be easily managed by the use of plant growth regulators such as Mepiquat (Ren et al. 2013).

Fiber quality was the second most important priority in early breeding efforts (Fig. 1) and remains so in commercial cotton breeding programs. The impetus for this effort was the replacement of hand spinning methods with machine spinning and weaving technology that required sufficient fiber length and strength to operate most efficiently (May 2000). Cotton fiber quality is defined by the physical properties of the individual fibers. Lint fiber is generally spun into yarn, which is then weaved or knitted into different types of fabrics as dictated by quality and the desired properties of the end product. Fiber quality is therefore the combination of fiber properties that affect the efficiency of yarn spinning, weaving, and other fabric manufacturing processes as well as the quality associated with cotton fabrics. The main fiber properties highly correlated with spinning performance and end product quality include those related to length, strength, elongation, and fineness/maturity (measured as micronaire) of the fiber. Chee and Campbell (2009) summarized the importance of fiber properties and how they are measured.

Figure 1. Phenotypic variation in cotton fiber length.

Both lint yield and fiber quality traits are quantitatively inherited. Campbell and Myers (2015) summarized the mean broad-sense and narrow-sense heritability for yield and fiber quality traits. Although yield components and fiber quality traits are heritable and show additive genetic variance, they often show a negative correlation. Meredith (1984) indicated that this is likely due to linkage rather than to pleiotropy; however, Campbell et al. (2012) reported that the negative relationship persists in the Pee Dee germplasm even after nearly 80 years of breeding. Therefore, the greatest challenge in cotton breeding is simultaneous improvement of yield and fiber quality.

Polyploidization and the Evolution of Spinnable Fiber

With respect to fiber quality improvement, it is noteworthy that spinnable fiber evolved only once in the history of *Gossypium*, in an ancestor of the two domesticated diploid A-genome species, after divergence from the F-genome lineage. This trait was passed on to the allopolyploid cottons when the A genome united in a common nucleus with a D-genome from an ancestor which produced no lint fibers (Applequist et al. 2001; Hovav et al. 2008; Jiang et al. 1998; Paterson et al. 2012). As mentioned earlier, the evolution of long spinnable lint fiber is strongly tied to a prolonged phase of elongation during fiber cell development. Applequist et al. (2001) measured the elongation of lint fiber on developing ovules using accessions from AD-genome allopolyploids *G. hirsutum* and *G. tomentosum*, and from the diploid species *G. herbaceum* (A-genome), *G. arboreum* (A-genome), *G. raimondii* (D-genome), *G. davidsonii* (D-genome), *G. anomalum* (B-genome), *G. sturtianum* (C-genome), and *G. longicalyx* (F-genome). Comparison of the growth curves across species showed that accessions from the AD-genome allopolyploids and the A-genome diploids have a significantly greater rate of fiber elongation than the other diploids. A similar conclusion was reached by Hovav et al. (2008) studying comparative gene expression profile across a developmental time-course of fiber from *G. herbaceum* and *G. longicalyx*. Their results showed major differences in the expression of genes related to stress responses and cell elongation as well as a prolonged developmental profile in domesticated A-genome *G. herbaceum*. Thus, the development of lint fiber involved a prolongation of an ancestral developmental program, which evolved in the ancestral A-genome predating polyploidization.

In addition to facilitating the evolution of a prolonged phase of fiber elongation, polyploidization has another equally important implication for the evolution of spinnable fiber. The union of the A-genome with the D-genome in a common nucleus may have allowed the recruitment of D-genome alleles (genes) into fiber development, which contributed to greater fiber quality and yield of polyploid cottons. Since only A-genome diploid species produce spinnable fiber, the role of the D-genome in genetic determination of fiber quality in polyploid cotton has long been a question. Evidence of the extent to which genetic variation in fiber quality properties is determined by loci on the Dt-subgenome was first provided by Jiang et al. (1998), who showed a majority of QTLs for fiber quality mapping to the Dt-subgenome. Numerous genetic mapping analyses (Chee et al. 2005a, b; Rong et al. 2007; Draye et al. 2005), summarized by Chee and Campbell (2009), have now supported the observation that the Dt-subgenome, from the ancestor that did not have spinnable fiber, plays a large role in the genetic control of fiber growth and development in polyploid cotton. Collectively, these data demonstrate that Dt-subgenome genes have been recruited to the genetic control of fiber quality properties, contributing to the transgressive fiber quality and yield of polyploid cottons relative to their diploid progenitors. Jiang et al. (1998) speculated that since the At-subgenome has a much longer history of selection for fiber development, many favorable alleles at major loci for fiber properties might have already been fixed as a result of natural selection. On the other hand, loci for fiber development on the Dt-subgenome may not have experienced strong selection until after polyploid formation, and therefore, mutations that enhanced this trait may have become favorable only after polyploidization. Hence, when artificial selection was recently imposed via domestication and breeding, the Dt-subgenome may have had more 'room for improvement' of fiber properties. Recruitment of Dt-subgenome loci may confer additional flexibility of polyploid cotton for artificial selection through breeding and thereby explain the superiority of the fiber properties of polyploid cottons relative to cultivated A-genome diploids.

A secondary consequence of polyploidization for spinnable fiber and a host of other phenotypes that geneticists are only beginning to comprehend by analyses of genome sequences is the differential rate of evolution in the polyploid AD-subgenome. By definition, in formation of allopolyploid genomes most if not all loci are duplicated. Lynch and Conery (2000, 2003) discussed the various fates of duplicated genes, suggesting that most copies of duplicated genes experience a brief period of relaxed selection but ultimately are silenced or pseudogenized. However, a small fraction of duplicated genes that diverge to some degree via mutations, which are somehow favorable, survive in duplicate and contribute to the evolution of phenotypic complexity by natural selection. In recent allopolyploids such as cotton, where the two subgenomes contain duplicated but slightly divergent copies of most genes, a null hypothesis is that homoeologous genes would evolve independently and at equal rates following polyploid formation. For some genes, allopolyploid cottons have experienced bi-directional concerted evolution that shows different directional biases in different parts of the genome. Clues to this phenomenon were first observed in repetitive elements using *in situ* hybridization, which showed that dispersed repetitive sequences that are A-genome specific at the diploid level have spread to the Dt-subgenome in the allopolyploids (Hanson et al. 1998; Wendel et al. 1995; Zhao et al. 1998). Also, numerous studies have observed allelic diversity of both homoeologous genes (Small et al. 1999; Small and Wendel 2002) and loci controlling quantitative characters to be significantly more abundant in the Dt-subgenome than At-subgenome (Chee et al. 2005a, b; Jiang et al. 1998; Rong et al. 2007).

Recent comparative analyses of genome sequences have now confirmed that non-reciprocal DNA conversion favors genes in the Dt-subgenome over the At-subgenome (Guo et al. 2014; Page et al. 2013; Paterson et al. 2012; Zhang et al. 2015). For example, about 40% of At and Dt genes from an elite cotton cultivar differ in sequence from their diploid progenitors. Most of these mutations are convergent, with At genes converted to the Dt state at more than twice the rate as the reciprocal (Paterson et al. 2012). While At to Dt-conversion is the opposite of what would be expected to contribute to fiber evolution, this type of conversion is enriched in heterochromatin, while conversions in gene rich euchromatin are predominantly of Dt to At alleles, thus potentially doubling the copy numbers of At alleles that may contribute to fiber development (Guo et al. 2014). In addition, the rate of gene loss was higher in the AD-allopolyploid than in the A- and D-diploids, and within the polyploid genome, more gene loss was observed in the At-subgenome than in the Dt-subgenome (Paterson and Wendel 2015; Zhang et al. 2015). Mutation in the At-subgenome appears to be enriched in genomic regions which prior QTL mapping has identified as hotspots for fiber quality traits, suggesting that the improved fiber quality of domesticated allopolyploid cottons involved recursive changes in the At and new changes in the Dt-genes (Paterson et al. 2012). Finally, expression of homoeologous genes seems to be biased toward the Dt-subgenome copies, although this expression varied between different tissues and developmental stages (Small and Wendel 2002; Zhang et al. 2015). These observations collectively suggest that polyploidization allowed D-genome genes to adopt new functions in the allopolyploid genome, possibly explaining the underlying agronomic and fiber quality superiority of domesticated allopolyploid cotton over the modern descendants of their diploid progenitors.

Interspecific Hybridization and Gene Introgression

The steps in cotton domestication, from wild *G. hirsutum* through feral cultigens to the rise of Upland germplasm and finally to modern improved cultivars have each imposed severe genetic bottlenecks, reducing allelic diversity. The levels and patterns of genetic erosion that were associated with the development of early cultigens, landraces, and modern cultivars have been documented by morphological (Hutchinson 1951, 1959), allozyme (Wendel et al. 1989, 1992), and molecular characterization (Brubaker et al. 1999; Brubaker and Wendel 1994, 2001; Fang et al. 2013; Tyagi et al. 2014). For example, the average genetic distance among 378 Upland accessions from the United States surveyed with 120 SSR markers was found to be only 0.195, indicating that Upland cotton has a very narrow germplasm base (Tyagi et al. 2014). Lubbers and Chee (2009) surveyed 320 Upland cultivars/germplasm from the United States National Plant Germplasm Collection with 250 RFLPs and showed cotton to have a lower level of genetic variation than most major crops. The level of genetic diversity was not improved when elite germplasm from a broad geographical origin was surveyed. Indeed, in a survey of 157 elite cultivars from China, United States, Africa, Former Soviet Union, and Australia using 146 SSR loci, the average number of alleles detected per locus was a meager 2.3 (Zhao et al. 2015). Interestingly, these genetic bottlenecks were also accompanied by steady improvement in many traits important to cotton production, especially fiber quality. Given the extremely low levels of allelic diversity within the Upland cotton gene pool, one can speculate that the number of favorable alleles for fiber quality (such as fiber length and strength) that have yet to reach fixation may be low, because these traits have been under intense selection pressure since the early stages of domestication. Therefore, it is not surprising that interspecific introgression has long been of interest to improve Upland cotton (Bowman and Gutierrez 2003; Campbell and Myers 2015; Campbell et al. 2009; Zhang et al. 2014).

Breeding of Upland cotton has focused on maximum yield and broad adaptation, while breeding of domesticated forms of *G. barbadense*, also known as Pima, Egyptian, or Sea Island cotton, has emphasized superior fiber quality. Consequently, the cultivated forms of *G. barbadense* have significantly longer, finer, and stronger fiber than the more widely grown Upland cotton. However, of the *G. barbadense* that is currently still in cultivation, both Pima and Egyptian cottons have a narrow range of environmental adaptation limited to irrigated regions in arid zones of the Western United States and Lower Egypt, respectively. Nevertheless, the unique fiber properties of this species offer an ideal candidate for providing new genetic variation to improve the fiber quality of Upland cotton. Therefore, it is of little surprise that both classical genetic and recent molecular quantitative genetic studies have evaluated the genetic basis and heritability of specific fiber properties by studying populations derived from interspecific hybridization between wild and domesticated forms of *G. barbadense* and Upland cotton (Campbell and Myers 2015; Chee and Campbell 2009). The bulk of these genetic studies have documented the heritable nature of fiber properties, thus indicating the ability to improve specific fiber properties of Upland cotton through interspecific introgression.

In addition to fiber quality, wild and domesticated allotetraploid *Gossypium* is an important source of diversity for disease and pest resistance genes that can be introgressed into Upland cotton. Pathogens, nematodes, and insects cause significant crop losses wherever cotton is grown, and crop protection accounts for much of the high unit cost of cotton production, explaining the popularity of transgenic pest-management cultivars. Meredith (1980) noted that breeding for disease resistance is more important than breeding for insect resistance in most breeding programs. This is even more the case after the adoption of *Bt* cotton cultivars, which are resistant to many insects. Some resistance phenotypes from wild species are simply inherited, and breeders have taken advantage of these traits because of their ease of selection. However, many of the resistance traits are quantitatively inherited, and their genetic manipulation has recently become much more effective with the aid of DNA markers. Table 1 provides a summary of disease resistance genes identified in wild and domesticated allopolyploid *Gossypium*.

It is noteworthy that *G. hirsutum*, *G. barbadense*, and the wild allopolyploid species diverged from a common ancestor after polyploid formation during the Mid-Pleistocene, about 1-2 Mya (Wendel and Cronn 2003). Mutations that have accumulated in different allopolyploid lineages may be beneficial or benign in their native genetic background, but may or may not interact favorably when introgressed into a divergent genetic background (Orr 1995). While allopolyploid species are sexually compatible, partial reproductive barriers such as reduced fertility, segregation distortion (non-Mendelian inheritance), and hybrid breakdown are often observed in later generation hybrids (Stephen 1946). The effects of these barriers to gene introgression

Table 1. List of disease and insect resistance genes in polyploid Gossypium.

Disease	**Causal agent**	**Source**[1]	**Genetic control**	**Ref.**
Bacterial Blight	*Xanthomonas campestris pv. malvacearum*	Gb	quantitative	Knight 1953
Fusarium Wilt	*Fusarium oxysporum f. sp. vasinfectum*	Gb	quantitative	Wang and Roberts 2006
Leaf Curl Virus	*Begomoviruses*	Gh	quantitative	Ali 1997
Verticillium Wilt Root Knot	*Verticillium dahliae*	Gb	quantitative	Bolek et al. 2005
Nematode Reniform	*Meloidogyne incognita Rotylenchulus*	Gh, Gd	quantitative	He et al. 2014, Turcotte et al. 1963
Nematodes Cotton Blue	*Reniformis*	Gb	quantitative	Bell et al. 2015
Disease	*Polerovirus*	Gh	quantitative	Fang et al. 2010
Spider mites	*Tetranychus urticae*	Gb	quantitative	Zhang et al. 1992
Thrips	*Frankliniella occidentalis*	Gb, Gm Gd, Gt	quantitative	Zhang et al. 2013
Root rot	*Rhizoctonia bataticola*	Gh	n/a	Monga and Raj 2000

[1]Gb = *G. barbadense,* Gd = *G. darwinii,* Gh = *G. hirsutum,* Gm = *G. mustelinum,* Gt = *G. tomentosum*

between allotetraploid cottons were demonstrated by Jiang et al. (2000), who used DNA markers to study the transmission genetics of an advanced-backcross generation interspecific hybrid population between *G. hirsutum* and *G. barbadense.* The transmission patterns of individual alleles generally favor the elimination of the donor genotype, thus preserving the integrity of the recurrent genotype. For example, segregation distortion were widespread, and under-representation of donor alleles in early generations following hybridization resulted in complete elimination of some donor alleles as early as the BC_3 generation. Interestingly, the segregating ratios at identical loci introgressed into different independently-derived BC_3F_2 families were highly variable, with some families favoring recurrent parent alleles while others favored those from the donor parent, suggesting that hybrid incompatibility is best accounted for by multi-locus epistatic interactions affecting gamete success and genotype fecundity.

These results highlight the difficulty in working with interspecific populations to pyramid multiple favorable alleles for quantitative characters such as lint yield, fiber length, fiber strength, and fiber fineness in a single genotype to develop superior Upland cotton cultivars. As noted by Brown and Ware (1958), most attempts to directly combine Upland cotton with Sea Island (*G. barbadense*) types showed that while the F_1 generation is attractive, the F_2 and F_3 generations were considered a "mess" and were almost always discarded (Fig. 2). Therefore, while pedigree analyses indicated that many cultivars were developed through interspecific hybridization, molecular analysis based on isozymes and DNA markers indicate that the Upland cotton gene pool is largely homogeneous. Rare alleles, which have likely arisen through introgression, are restricted to only a few closely related cultivars within a germplasm group (Brubaker and Wendel 1994; Chee et al. 2004; Wendel et al. 1992). Further, none of these introgressed alleles were found in modern cotton cultivars, suggesting that the potential benefits of introgression from *G. barbadense* into Upland cotton are still largely unrealized. Conversely, the introgression of genes from Upland cotton into Pima cotton, a cultivated form of *G. barbadense*, has greatly contributed to the productivity and broader adaptation of the modern Pima cultivars (Percy 2009; Wang et al. 1995).

Figure 2. Floral abnormality in later generation hybrids between *Gossypium hirsutum* and *G. barbadense*.

Gene Introgression in Feral and Landrace Populations

Three (possibly four) of the seven allotetraploid species are island endemics, however the geographic ranges of *G. barbadense* and *G. mustelinum* overlap with wild and feral *G. hirsutum* (Wendel and Grover 2015; Fryxell 1979). *Gossypium barbadense* and *G. hirsutum* have the largest indigenous geographic distribution of all polyploid species, and sympatric populations can be found throughout the Caribbean and Central America. Both allozymes (Percy and

Wendel 1990) and DNA markers (Brubaker et al. 1993) have shown extensive gene flow in sympatric populations of *G. hirsutum* and *G. barbadense,* with a significantly higher proportion of *G. barbadense* alleles introgressed into *G. hirsutum* than in the reciprocal direction. These results are consistent with the observation of Stephens (1967, 1974) who noted that within sympatric populations there is a shift in the morphology of Caribbean and Central American *G. hirsutum* toward *G. barbadense,* and proposed that the race 'marie-galante' arose via introgression. In Brazil, a *G. hirsutum* botanical form called 'moco' cotton has a history of introgression from *G. barbadense* and *G. mustelinum*. Gene flow from other allopolyploids into *G. hirsutum* has possibly contributed to greater ecological adaptation and the success of this species in colonizing new habitats.

In Upland cotton germplasm, there is anecdotal evidence that a number of Extra Long Staple (ELS) varieties cultivated in the late 1800s resulted from unintended Sea Island (*G. barbadense*) introgression. This is a period that predates scientific plant breeding, but agriculturists at that time were known to have actively practiced re-selection within existing varieties, taking advantage of genetic variability from heterogeneous populations along with chance outcrosses. For example, Ware (1951) reported a number of Upland varieties such as 'Allen Yellow Bloom' and 'Coxe Yellow Bloom' that he described as having "a sign of Sea Island blood." These and other varieties developed during this period were abandoned in favor of shorter season varieties that also had shorter fiber during the boll weevil (*Anthonomus grandis* Boheman) era. Unfortunately, of the ELS varieties mentioned by Ware (1951), many of which contained novel alleles for fiber quality likely to have derived from *G. barbadense* introgression, most did not survive the boll weevil invasion and were lost.

Several ELS genotypes developed by modern plant breeding are mentioned in the literature as having *G. barbadense* in their background. Most notable are the "Sealand" varieties developed by W.H. Jenkins, who led the cotton breeding program at the Pee Dee Experiment Station in Florence, South Carolina in the mid-1930s (Ware 1951). Due to the collapse of the Sea Island cotton industry in coastal Georgia and South Carolina in the early 1900s, cotton breeding programs diverted interest from improving Sea Island cotton to utilizing it to improve Upland cotton. The varieties 'Sealand 542' and 'Sealand 883' were developed by crossing the Upland variety 'Coker Wilds' with the Sea Island variety 'Bleak Hall' followed by multiple backcrosses to Coker Wilds (Bowman et al. 2006). Both cultivars have exceptionally long fiber approaching that of *G. barbadense,* and 'Sealand-542' was grown commercially on a small scale in the coastal plains of Georgia and South Carolina in the mid-1940s. The Sealand germplasm from the USDA-ARS Pee Dee program at Florence, SC has been further utilized by private breeding firms across the United States Cotton Belt (Bowman et al. 1996). Other introgression lines have since been developed (Levy et al. 2009; Stelly et al. 2005; Zhang et al. 2014), including several populations that have also been subjected to molecular analysis. For these populations, the individual *G. barbadense* alleles introgressed into each line can be monitored via linked DNA markers (Chee et al. 2005a; Wang et al. 2012).

Developmental Introgressive Breeding

Interspecific populations from crosses between *G. hirsutum* and *G. barbadense* solved the quintessential problem of low genetic diversity faced in the early days of DNA marker based genetic and QTL mapping in Upland cotton. In addition to providing the DNA-level polymorphism needed to expedite genetic map construction, these two cultivated species are each prized for somewhat different characteristics. As mentioned above, breeding of *G. hirsutum* has focused on maximum yield and broad adaptation, while breeding for *G. barbadense* has emphasized fiber quality. Therefore, genetic mapping (Reinisch et al. 1994; Rong et al. 2004; Wang et al. 2015) and molecular quantitative genetic studies of fiber properties have mostly utilized populations derived from interspecific hybridization involving wild and domesticated forms of *G. barbadense* crossed with Upland cotton. Chee and Campbell (2009) summarized QTLs for various fiber quality traits mapped in cotton, and this list was recently updated by Fang (2015). Numerous major QTLs for fiber length, strength, and fineness have now been identified, with several of these having been validated (Cao et al. 2015; Kumar et al. 2012; Shen et al. 2011; Zhang et al. 2013). This information provides cotton breeders with additional tools to improve specific fiber properties of Upland cotton through the introgression of genes from *G. barbadense* with minimum disruption of the favorable allelic combinations created during more than a century of selective breeding.

A consortium effort to reduce the genetic vulnerability of Upland cotton has been undertaken by developing inbred backcross (also called Advanced Backcross, Fig. 3) populations derived from crossing Upland varieties with the allotetraploid species *G. barbadense, G. tomentosum,* and *G. mustelinum* (Paterson et al. 2004). The inbred backcross design was utilized to mitigate reproductive barriers associated with interspecific introgression between these species (Tanksley and Nelson 1996; Wang and Chee 2010). By developing an extensive set of Near Isogenic Introgression lines from BC_2 or BC_3 families, one can evaluate relatively small segments of introgressed DNA for agronomic or fiber quality performance and analyze these for QTLs. This allows the action of individual genetic loci to be more clearly resolved than in early generations, because recombination and segregation have broken the donor genome into smaller components. The number of QTLs discovered in each of the inbred backcross populations is listed in Table 2. Additional information on the genetic dissection of each of the fiber traits is presented by Rong et al. (2007), who reported the alignment of 432 fiber QTLs mapped in 10 interspecific *G. hirsutum* by *G. barbadense* populations onto a consensus map.

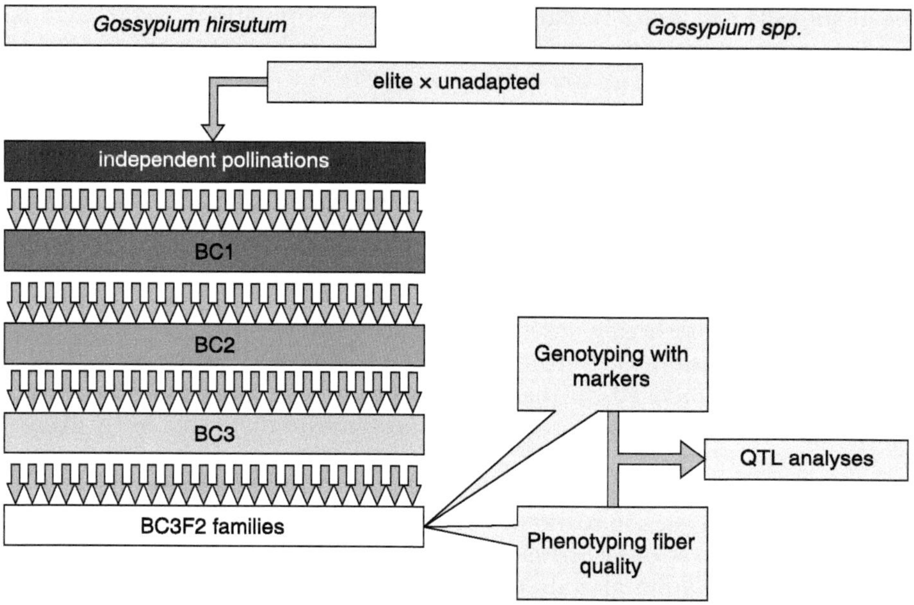

Figure 3. Inbred backcross design for cotton improvement via interspecific hybridization.

Table 2. Summary of QTL mapped in interspecific *Gossypium* species.

	G. barbadense[1]			*G. tomentosum*[2]			*G. mustelinum*[3]		
	Gh	Gb	**Total**	Gh	Gt	**Total**	Gh	Gm	**Total**
Fiber length	17	11	**28**	4	0	**4**	12	2	**14**
Fiber strength	—	—	—	4	0	**4**	6	5	**11**
Micronair	1	8	**9**	1	3	**4**	12	6	**18**
Fiber elongation	14	8	**22**	0	4	**4**	10	5	**15**

Favorable alleles: Gh = *G. hirsutum*, Gb = *G. barbadense*, Gt = *G. tomentosum*, Gm = *G. mustelinum*
[1]Chee et al. (2005a, b), Draye et al. (2005), [2]Zhang et al. (2011), [3]Unpublished data.

The numbers of QTLs detected for each fiber quality trait, their distributions in the genome, and the sources of the favorable alleles from the respective polyploid progenitors have revealed several intriguing observations with regard to the genetic control of cotton fiber properties and the potential benefit of interspecific introgression. Introgressed alleles can enhance the fiber length, strength, elongation, and fineness of Upland cotton, but this also includes alleles from donor parents of non-domesticated species that have inferior fiber quality (Table 2). Nearly half of the QTLs detected were located in a small number of genomic regions or 'fiber QTL hotspots', and most of the loci have predominantly additive gene action, supporting the long held notion that the polyploid *Gossypium* species are a source of valuable alleles for fiber quality improvement in Upland cotton. However, the generally large number of loci with small effects, a low level of correspondence between loci found in different populations with the same pedigree, and the lack of homoeologous association reflects the general complexity of manipulating these quantitatively-inherited fiber traits. Collectively, these results suggest that lint fiber development may involve a complex gene network, and that the evolution of spinnable fiber may have involved coordinated changes in expression of functionally diverse cotton genes. Interestingly, as noted above, the Dt-subgenome, from the ancestor that does not produce spinnable fiber, plays a large role in the genetic control of fiber growth and development, as indicated by the large number of QTLs mapped to this subgenome.

Despite the complications in manipulating alien alleles during introgressive breeding, a number of QTLs for fiber strength and length have now been stably introgressed into Upland cotton and their authenticity and improvement value validated using molecular markers: these QTLs represent attractive candidates for marker-assisted breeding. For example, scientists from Nanjing Agricultural University in China have reported a major fiber strength QTL (QTL_{FS1}) located on chromosome 24 in the germplasm line 'Suyuan 7235' which is thought to have been introgressed from the diploid species *G. anomalum* (Shen et al. 2005). This QTL was identified using F_2, $F_{2:3}$, backcross, and recombinant inbred mapping populations derived from the cross of line 'Suyuan 7235' × 'TM-1' (Shen et al. 2005, 2006; Zhang et al. 2003). QTL_{FS1} was found to be stably expressed in four test locations in China and the United States over two growing seasons (Shen et al. 2007), suggesting that it may be a good candidate for improving the fiber strength of Upland cotton germplasm. To test this hypothesis, Kumar et al. (2012) introgressed this QTL region into two Upland genotypes of United States origin ('Sealand 542' and 'Sealand 883') with different fiber strength and determined that the fiber strength of segregating progenies carrying this QTL was improved. Interestingly, by using a high marker density flanking the 10 cM interval within the QTL region, this locus has now been independently shown to harbor at least three QTL clusters with the allele from 'Suyuan 7235' conferring increased fiber strength (Chen et al. 2009; Kumar et al. 2012). Similarly, a fiber length QTL on Chromosome 1 (*qFL-chr1*) that was initially introgressed from *G. barbadense* via the inbred backcross approach was confirmed by Shen et al. (2011). In the initial population, the *G. barbadense* allele improved fiber length in three independently derived BC_3 families, explaining up to 24% of phenotypic variation (Chee et al. 2005b). Three BC_3F_2 heterozygote plants for the QTL region were self-pollinated to develop three independent populations of near-isogenic introgression lines and the effects of *qFL-chr1* was detected in all three populations when planted in Nanjing, China and Georgia, United States over two years.

The magnitudes of the genetic effects of *qFL-chr1* and QTL_{FS1} are only modest, as the donor alleles increase fiber length and strength by only a maximum of 1.45 mm (Kumar et al. 2012) and 22.8 kN m kg^{-1} (Shen et al. 2011), respectively. However, the genetic resources and DNA marker toolkits developed make two significant contributions to fiber quality improvement of Upland cotton. First, near-isogenic lines carrying the QTLs represent a new genetic source for improving fiber length and strength in Upland germplasm. This is important because, as mentioned earlier, Upland cotton has a narrow gene pool resulting from its evolutionary history, domestication, and modern plant breeding practices. A high degree of relatedness within cultivated germplasm suggests that many favorable genes, especially those related to yield and fiber quality, might have reached fixation in the elite gene pool. Therefore, while some degree of transgressive segregation in fiber properties will continue to be discovered from crossing among elite parents, the use of interspecific gene combinations such as these QTLs offers an important source of new genetic variation to ensure continued genetic gain in Upland cotton fiber improvement.

Summary

Interspecific hybridization between an A-genome diploid species which produces spinnable lint fiber and a D-genome diploid species that lacked this trait led to the development of allopolyploid cottons, including the two species that supply most of the world's natural textile fiber. Polyploid formation conferred advantages to the tetraploid genome over its diploid progenitors by allowing the recruitment of alleles from the D-genome to increase fiber productivity and quality of allopolyploid cottons. New genomic tools, such as whole genome sequencing and re-sequencing, have greatly advanced our understanding of the origin, diversification, genome structure, and impact of human-directed evolution on the *Gossypium* species. More importantly, the merger of these genomic tools with classical cotton breeding will allow better utilization of interspecific hybridization and gene introgression to provide the needed useful genetic diversity for continued improvement of the genetically depauperate Upland cotton gene pool. The use of interspecific gene combinations may provide the foundation for Upland cotton to meet current and future challenges in cotton production, such as those imposed by environmental change, disease, and the need for yield stability and specific combinations of desirable fiber traits to compete favorably with synthetic textiles created from oil-based products.

References

Ali, M. 1997. Breeding of cotton varieties for resistance to cotton leaf curl virus. Pakistan J. Phytopathology 9: 1-7.

Applequist, W.L., R. Cronn and J.F. Wendel. 2001. Comparative development of fiber in wild and cultivated cotton. Evol. Development 3: 3-17.

Bell, A.A., A. Robinson, J. Quintana, S.E. Duke, J.L. Starr, D.M. Stelly and Xiuting Zheng. 2015. Registration of BARBREN-713 Germplasm line of Upland cotton resistant to reniform and root-knot nematodes. *J. Plant Registrations* 9: 89-93.

Bolek, Y., K.M. El-Zik, A.E. Pepper, A.A. Bell, C.W. Magill, P.M. Thaxton, et al. 2005. Mapping of verticillium wilt resistance genes in cotton. Plant Sci. 168: 1581-1590. doi: 10.1016/j.plantsci.2005.02.008.

Borland, F. and G.O. Myers. 2015. Conventional cotton breeding. pp. 205-228. *In*: Fang D.D. and R.G. Percy (eds.). Cotton. American Society of Agronomy, Madison, WI.

Bowman, D.T. and O.A. Gutierrez. 2003. Sources of fiber strength in the U.S. Upland cotton crop from 1980-2000. *J. Cotton Sci.* 7: 164-169.

Bowman, D.T., O.A. Gutierrez, R.G. Percy, D.S. Calhoun and O.L. May. 2006. Pedigrees of upland and pima cotton cultivars released between 1970 and 2005. Mississippi Agricultural and Forestry Experiment Station Technical Bulletin 1155: 57.

Bowman, D.T., O.L. May and D.S. Calhoun. 1996. Genetic base of upland cotton cultivars released between 1970 and 1990. Crop Science 36: 577-581.

Brown, H.B. and J.O. Ware. 1958. Cotton. McGraw-Hill. New York.

Brubaker, C.L. and J.F. Wendel. 1994. Reevaluating the origin of domesticated cotton (*Gossypium hirsutum* L.) using nuclear restriction fragment length polymorphisms (RFLPs). Am. J. Bot. 81: 1309-1326. doi: 10.2307/2445407.

Brubaker, C.L. and J.F. Wendel. 2001. RFLP diversity in cotton. pp. 81-102. *In*: Jenkins J.N. and S. Saha (eds.). Genetic improvement of cotton: Emerging technologies. Science Publishers, Inc., Enfield, NH.

Brubaker, C.L., J.A. Koontz and J.F. Wendel. 1993. Bidirectional cytoplasmic and nuclear introgression in the New World cottons, *Gossypium barbadense* and *G. hirsutum* (Malavaceae). *Am. J. Bot.* 80:1203-1208.

Brubaker, C.L., F.M. Borland and J.F. Wendel. 1999. The origin and domestication of cotton. pp. 3-31. *In*: Smith C.W. and J.T. Cothren (eds.). Cotton: origin, history, technology and production. John Willey & Sons, New York.

Calhoun, D.S., D.T. Bowman and O.L. May. 1994. Pedigree of Upland and Pima cotton cultivars released between 1970 and 1990. pp. 1-53. Mississippi Agricultural and Forestry Experiment Station Technical Bulletin 1017. Mississippi State, MS.

Campbell, B.T. and G.O. Myers. 2015. Quantitative genetics. pp. 187-204. *In*: Fang D.D. and R.G. Percy (eds.). Cotton. American Society of Agronomy, Madison, WI.

Campbell, B.T., V. Williams and W. Park. 2009. Using molecular markers and field performance data to characterized the Pee Dee cotton germplasm resources. Euphytica 169: 285-301. doi: 10.1007/s10681-009-9917-4.

Campbell, B.T., P.W. Chee, E. Lubbers, D.T. Bowman, W.R. Meredith, J. Johnson, F. Fraser, W. Bridges and D.C. Jones. 2012. Dissecting genotype *x* environment interactions and trait correlations present in the Pee Dee cotton germplasm collection following seventy years of plant breeding. Crop Sci. 52: 690-699. doi: 10.2135/cropsci2011.07.0380.

Cao, Z., X. Zhu, H. Chen and T. Zhang. 2015. Fine mapping of clustered quantitative trait loci for fiebr quality on chromosome 7 using a *Gossypium barbadense* introgressed line. Mol. Breed. 35: 215-220.

Chee, P.W. and T. Campbell. 2009. Bridging classical and molecular genetics of cotton fiber quality and development. pp. 283-311. *In:* Paterson, A.H. (ed.). Genetics and Genomics of Cotton. Springer, New York.

Chee, P., E.L. Lubbers, O.L. May, J. Gannaway and A.H. Paterson. 2004. Secondary gene pool contributions in domesticated cotton. *In*: Beltwide Cotton Conference. National Cotton Council, San Antonio, TX.

Chee, P., X. Draye, C. Jiang, L. Decanini, T. Delmonte, R. Bredhauer, C.W. Smith and A.H. Paterson. 2005a. Molecular dissection of phenotypic variation between *Gossypium hirsutum* and *G. barbadense* (cotton) by a backcross-self approach: I Fiber Elongation. Theor. Appl. Genet. 111: 757-763.

Chee, P., X. Draye, C. Jiang, L. Decanini, T. Delmonte, R. Bredhauer, C.W. Smith and A.H. Paterson. 2005b. Molecular dissection of phenotypic variation between *Gossypium hirsutum* and G. *barbadense* (cotton) by a backcross-self approach. III. Fiber Length. Theor. Appl. Genet. 111: 772-781.

Chen, H., N. Qian, W.Z. Guo, Q.P. Song, B.C. Li, F.J. Deng, C.G. Dong and T.Z. Zhang. 2009. Using three overlapped RILs to dissect genetically clustered QTL for fiber strength on Chro. D8 in Upland cotton. Theor. Appl. Genet. 119: 605-612. doi: 10.1007/s00122-009-1070-x.

Cronn, R.C., R.L. Small, T. Haselkorn and J.F. Wendel. 2002. Rapid diversification of the cotton genus (*Gossypium*: Malvaceae) revealed by analysis of sixteen nuclear and chloroplast genes. Am. J. Bot. 89: 707-725.

Desai, A., P. Chee, J. Rong, L. May and A.H. Paterson. 2006. Chromosome structural changes in diploid and tetraploid A genomes of *Gossypium*. Genome 49: 336-345.

d'Eeckenbrugge, G.C. and J.M. Lacape. 2014. Distribution and differentiation of wild, feral, and cultivated populations of perennial Upland cotton (*Gossypium hirsutum* L.) in mesoamerica and the Caribbean. PLoS One. doi: 10.137/journal.pone.0107458.

Draye, X., P. Chee, C. Jiang, L. Decanini, T. Delmonte, R. Bredhauer, C.W. Smith and A.H. Paterson. 2005. Molecular dissection of phenotypic variation between *Gossypium hirsutum* and *G. barbadense* (cotton) by a backcross-self approach. II Fiber Fineness. Theor. Appl. Genet. 111: 764-771.

Duggar, J.F. 1907. Descriptions and classification of varieties of American Upland cotton. Agricultural Experiment Station of the Alabama Polytechnic Institute Bull. Auburn, AL.

Endrizzi, J.E., E.L. Turcotte and R.J. Kohel. 1984. Genetics, cytogenetics, and evolution of *Gossypium*. Adv. Genet. 23: 271-375.

Fang, D.D. 2015. Molecular breeding. pp. 255-289. *In*: Fang, D.D. and R.G. Percy. Cotton. Amer. Soc. Agron. Madison, WI.

Fang, D.D., L.L. Hinze, R.G. Percy, P. Li, D. Deng and G. Thyssen. 2013. A microsatellite based genome-wide analysis of genetic diversity and linkage disequilibrium in Upland cotton (*Gossypium hirsutum* L.) cultivars from major cotton-growing countries. Euphytica 191: 391-401. doi: 10.1007/s10681-013-0886-2.

Fang, D.D., J.H. Xiao, P.C. Canci and R.G. Cantrell. 2010. A new SNP haplotype associated with blue disease resistance gene in cotton (*Gossypium hirsutum* L.). Theor. Appl. Genet. 120: 943-953. doi: 10.1007/s00122-009-1223-y.

Fryxell, P.A. 1979. The natural history of the cotton tribe. College Station, TX: Texas A&M University Press.

Greever, R.F., F.R.H. Katterman and J.E. . 1989. DNA hybridization analyses of a *Gossypium* allotetraploid and two closely related diploid species. Theor. Appl. Genet. 77: 553-559.

Grover, C.E., J.P. Gallagher, J.J. Jareczek, J.T. Page, J.A. Udall, M.A. Gore and J.F. Wendel. 2015. Re-evaluating the phylogeny of allopolyploid *Gossypium* L. Mol. Phylogenet. and Evol. 92: 45-52.

Grover, C.E., X. Zhu, K.K. Grupp, J.J. Kareczek, J.P. Gallagher, E. Szadkowski, J.G. Seijo, and J.F. Wendel. 2014. Molecular confirmation of species status for the allotetraploid cotton species, *Gossypium ekmanianum* Wittm. Genet. Resour. Crop Evol. 1-12.

Guo, H., X. Wang, H. Gundlach, K.F. Mayer, D.G. Peterson, B.E. Scheffler, P.W. Chee and A.H. Paterson. 2014. Extensive and biased inter-genomic non-reciprocal DNA exchanges shaped a nascent polyploid genome, *Gossypium* (cotton). Genetics 197: 1153-1163.

Hanson, R.E., X. Zhao, M.N. Islam-Faridi, A.H. Paterson, M.S. Zwick, C.F. Crane, T.D. Mcknight, D.M. Stelly and H.J. Price. 1998. Evolution of interspersed repetitive elements in *Gossypium* (Malvaceae). Am. J. Bot. 85: 1364-1368.

He, Y.J., P. Kumar, X.L. Shen, R.F. Davis, G. Van Becelaere, O.L. May, et al. 2014. Re-evaluation of the inheritance for root-knot nematode resistance in the Upland cotton germplasm line M-120 RNR revealed two epistatic QTLs conferring resistance. Theor. Appl. Genet. 127: 1343-1351. doi: 10.1007/s00122-014-2302-2.

Hovav, R., J.A. Udall, B. Chaudhary, E. Hovav, L. Flagel, G. Hu and J.F. Wendel. 2008. The evolution of spinnable cotton fiber entailed prolonged development and a novel metabolism. PLOS Genet. 4: E25. Doi: 10.1371/journal.pgen.0040025.

Hendrix, B. and J.M. Stewart. 2004. Estimation of the Nuclear DNA content of *Gossypium* species. Annals of Botany 95: 789-797. doi: 10.1093/aob/mci078.

Hutchinson, J.B. 1951. Intra-specific differentiation in *Gossypium hirsutum*. Heredity 5: 161-193.

Hutchinson, J.B. 1959. The application of genetics to cotton improvement. Cambridge Univ. Press, Cambridge, US.

Hutchinson, J.B., R.A. Silow and S.G. Stephens. 1947. The evolution of *Gossypium* and the differentiation of the cultivated cottons. Oxford Univ. Press. London.

Jiang C.X., R.J. Wright, K.M. El-Zik and A.H. Paterson. 1998. Polyploid formation created unique avenues for response to selection in *Gossypium* (cotton). Proc. Natl. Acad. Sci. USA 95: 4419-4424.

Jiang, C.X., P.W. Chee, X. Draye, P.L. Morrell, C.W. Smith and A.H. Paterson. 2000. Multi-locus interactions restrict gene introgression in interspecific populations of polyploid *Gossypium* (cotton). Evolution 54: 798-814.

Kimber, G. 1961. Basics of the diploid-like meiotic behavior of polyploid cotton. Nature 191: 98-100.

Knight, R.L. 1953. The genetics of blackarm resistance. IX. The gene B_{6M} from *Gossypium arboreum*. J. Genet. 51: 270-275.

Krapovickas, A. and G. Seijo. 2008. *Gossypium ekmanianum* (malvaceae), algodon Silvestre de la Repoblico Dominican. Bonplandia 17: 55-63.

Kulkarni, V.N., B.M. Khadi, M.S. Maralappanavar, L.D. Deshapande and S.S. Narayanan. 2009. The worldwide gene pools of *Gossypium arboreum* and *G. herbaceum* and their improvement. pp. 69-97. *In:* Paterson, A.H. (ed.). Genomics of Cotton. Springer, New York.

Kumar, K., R. Singh, E.L. Lubbers, X. Shen, A.H. Paterson, B.T. Campbell, D.C. Jones and P.W. Chee. 2012. Mapping and validation of fiber strength quantitative trait loci on Chromosome 24 in Upland cotton. Crop Science 52: 1115-1122.

Levi, A., A.H. Paterson, V. Barak, D. Yakir, B. Wang, P.W. Chee and Y. Saranga. 2009. Field evaluation of cotton near-isogenic lines introgressed with QTLs for productivity and drought related traits. Mol. Breeding 23: 179-195. doi.10.1007/s11032-008-9224-0.

Li, F. et al. 2015. Genome sequence of cultivated upland cotton (*Gossypium hirsutum* TM-1) provides insights into genome evolution. Nature Biotechnology 33: 524-530. doi: 10.1038/nbt.3208.

Lubbers, E.L. and P.W. Chee. 2009. The worldwide gene pool of *G. hirsutum* and its improvement. pp. 23-52. *In:* Paterson, A.H. (ed.). Genomics of Cotton. Springer, New York.

Lynch, M. and J.S. Conery. 2000. The evolution fate and consequences of duplicated genes. Science 290(2494): 1151-1155.

Lynch, M. and J.S. Conery. 2003. The origin of genome complexity. Science 302(5649): 1401-1404.

May, L. 2000. Genetic variation in fiber quality. pp. 183-230. *In*: A.S. Basra (ed.), Cotton fibers, development biology, quality improvement and textile processing. Food Products Press, New York.

Meredith, W.R. 1980. Use of insect resistance germplasm in reducing the cost of production in the 1980s. Proc. Beltwide Cotton Prod. Res. Conf., St. Louis, MO.

Meredith, W.R. 1984. Genotype × environment interactions. pp. 138-141. *In*: Kohel, J.K. and C.F. Lewis (eds.), Cotton. Vol. 24. Amer. Soc. Agr., Madison.

Monga, D., and S. Raj. 2000: Screening of germplasm lines against root rot of cotton. Adv. In Plant Sciences. 13(2): 603-607.

Niles, G.A. and C.V. Feaster. 1984. Breeding. pp. 202-229. *In*: Cotton, Kohel R.J. and C.F. Lewis (eds.). ASA/CSSA/SSSA. Madison, WI.

Orr, H.A. 1995. The population genetics of speciation: the evolution of hybrid incompatibilities. Genetics 139: 1805-1813.

Page J.T., M.D. Huynh, Z.S. Liechty, K. Grupp, D. Stelly, A.M. Hules, H. Ashrafi, A. Van Deynze, J.F. Wendel and J.A. Udall. 2013. Insights into the evolution of cotton diploids and polyploids from whole-genome resequencing. Genes Genomes Genetics. 3(10): 1809-1818. doi: 10.1534/g3.113.007229.

Paterson, A.H., R.K. Bowman, S.M. Brown, P.W. Chee, J.R. Gannaway, A.R. Gingle, O.L. May and C.W. Smith. 2004. Reducing the genetic vulnerability of cotton. Crop Science 44: 1900-1901.

Paterson, A.H. et al. 2012. Repeated polyploidization of *Gossypium* genomes and the evolution of spinnable cotton fibres. Nature 492: 423-428. doi: 10.1038/nature11798.

Paterson, A.H. and J.F Wendel. 2015. Unraveling the fabric of polyploidy. Nature Biotechnol. 33: 491-493. Doi: 10.1038/nbt.

Percy, R.G. 2009. The worldwide gene pool of *Gossypium barbadense* L. and its improvement. pp. 53-68. *In:* Paterson, A.H. (ed.). Genetics and Genomics of Cotton. Springer, New York.

Percy, R.G. and J.F. Wendel. 1990. Allozyme evidence for the origin and diversification of *Gossypium barbadense* L. Theor. Appl. Genet. 79: 529-542.

Reinisch, A.J., J.M. Dong, C. Brubaker, D. Stelly, J.F. Wendel and A.P. Paterson. 1994. A detailed RFLP map of cotton (*Gossypium hirsutum* × *Gossypium barbadense*): Chromosome organization and evolution in a disomic polyploid genome. Genetics 138: 829-847.

Ren, X., L. Zhang, M. Du, J.B Evers, W. Werf, X. Tian and Z. Li. 2013. Managing mepiquat chloride and plant density for optimal yield and quality of cotton. Field Crop Res. 149: 1-10. doi: 10.1016/j.fcr.2013.04.014.

Rong, J., C. Abbey, J.E. Bowers, C.L. Brubaker, C. Chang, P.W. Chee, T.A. Delmonte, X. Ding, J.J. Garza, B.S. Marler, C. Park, G.J. Pierce, K.M. Rainey, V.K. Rastogi, S.R. Schulze, N.L. Trolinder, J.F. Wendel, T.A. Wilkins, D. Williams-Coplin, R.A. Wing, R.J. Wright, X. Zhao, L. Zhu and A.H. Paterson. 2004. A 3347-locus genetic recombination map of sequence-tagged sites reveals features of genome organization, transmission and evolution of cotton (*Gossypium*). Genetics 166: 389-417.

Rong, J., R.J. Wright, Y. Saranga, O.L. May, T.A. Wilkins, X. Draye, V.N. Waghmare, F.A. Feltus, P.W. Chee, G.J. Pierce and A.H. Paterson. 2007. Meta-analysis of polyploid cotton QTL shows unequal contributions of subgenomes to a complex network of genes and gene clusters implicated in lint fiber development. Genetics 176: 2577-2588. doi: 10.1534/genetics.107.074518.

Senchina, D.S. et al. 2003. Rate variation among nuclear genes and the age of polyploidy in *Gossypium*. Molecular Biology and Evolution 20, 633-643. doi: 10.1093/molbev/msg065.

Shen, X.L., W.Z. Guo, X.F. Zhu, Y.L. Yuan, J.Z. Yu, R.J. Kohel and T.Z. Zhang. 2005. Molecular mapping of QTLs for fiber qualities in three diverse lines in Upland cotton using SSR markers. Mol. Breed. 15: 169-181.

Shen, X., W. Guo, Q. Lu, X. Zhu, Y. Yuan and T.Z. Zhang. 2007. Genetic mapping of quantitative trait loci for fiber quality and yield trait by RIL approach in Upland cotton. Euphytica 155: 371-380.

Shen, X.L., T.Z. Zhang, W.Z. Guo, X.F. Zhu and X.Y. Zhang. 2006. Mapping Fiber and Yield QTLs with Main, Epistatic, and QTL by Environment Interaction Effects in Recombinant Inbred Lines of Upland Cotton. Crop Sci. 46: 61-66.

Shen, X., Z. Cao, R. Singh, E.L. Lubbers, P. Xu, C.W. Smith, A.H. Paterson and P.W. Chee. 2011. Efficacy of *qFL-chr1*, a quantitative trait locus for fiber length in cotton (*Gossypium* spp.). Crop Sci. 51: 2005-2010.

Small, R.L. and J.F. Wendel. 2002. Differential evolutionary dynamics of duplicated paralogous *Adh* loci in allotetraploid cotton (*Gossypium*). Mol. Biol. Evol. 19: 597-607.

Small, R.L., J.A. Ryburn and J.F. Wendel. 1999. Low levels of nucleotide diversity at homoeologous *Adh* loci in allotetraploid cotton (*Gossypium* L.). Mol. Biol. Evol. 16: 491-501.

Stelly, D.M., S. Saha, D.A. Raska, J.N. Jenkins, J.C. McCarty and O.A. Gutierrez. 2005. Registration of 17 Upland (*Gossypium hirsutum*) Cotton Germplasm Lines Disomic for Different *G. barbadense* Chromosome or Arm Substitutions. Crop Sci 45: 2663-2665.

Stephen, S.G. 1946. The genetics of "Corky". The new world alleles and their possible role as an interspecific isolating mechanism. J. Genet. 47: 150-161.

Stephen, S.G. 1967. Evolution under domestication of the New World cotton (*Gossypium* spp.). Cienciae Cultura 19: 118-134.

Stephen, S.G. 1974. Geographical and taxonomic distribution of anthocyanin genes in New World cottons. J. Genetics 61: 128-141.

Tanksley, S.D. and C.J. Nelson. 1996. Advanced backcross QTL analysis: a method for the simultaneous discovery and transfer of valuable QTLs from unadapted germplasm into elite breeding lines. Theor. Appl. Genet. 92: 191-203.

Turcotte, E.L., W.R. Harold, J.H. O'Bannon and C.V. Feaster. 1963. Evaluation of cotton root-knot nematodes resistance of a strain of *G. barbadense* var. *darwinii*. Cotton Improve Conf Proc 15: 36-44.

Tyler, F.J. 1910. Varieties of American Upland cotton. USDA-ARS, BPI, Bulletin no. 163.

Tyagi, P., M.A. Gore, D.T. Bowman, B.T. Campbell, J.A. Udall and V. Kuraparthy. 2014. Genetic diversity and population structure in the US Upland cotton (*Gossypium hirsutum* L.). Theor. Appl. Genet. 127: 283-295. doi: 10.1007/s00122-013-2217-3.

Wang, B. and P.W. Chee. 2010. Application of advanced backcross QTL analysis in crop improvement. J. Plant Breed. and Crop Sci. 22(3): 1-12.

Wang, S., J. Chen, W. Zhang, Y. Hu, L. Chang, L. Fang, Q. Wang, F. Lv, H. Wu, Z. Si, S. Chen, C. Cai, X. Zhu, B. Zhou, W. Gou and T. Zhang. 2015. Sequence-based ultra-dense genetic and physical maps reveal structural variations of allopolyploid cotton genomes. Genome Biol. 16: 108-114. doi: 10.1186/s13059-015-0678-1.

Wang, G.L., J.M. Dong and A.H. Paterson. 1995. The distribution of *Gossypium hirsutum* chromatin in *G. barbadense* germplasm: molecular analysis of introgressive breeding. Theor. Appl. Genet. 91: 1153-1161.

Wang, C. and P.A. Roberts. 2006. A fusarium wilt resistance gene in *Gossypium barbadense* and its effect on root-knot nematode-wilt disease complex. Phytopathology 96: 727-734. doi: 10.1094/Phyto-96-0727.

Wang, P., Y.J. Zhu, X.L. Song, Z.B. Cao, Y.Z. Ding, B.L. Liu, X.F. Zhu, S. Wang, W.Z. Guo and T.Z. Zhang. 2012. Genetic dissection of long staple fiber qualities in *Gossypium barbadense* using inter-specific chromosome segment introgression lines. Theor. Appl. Genet. 124: 1415-1428.

Wang, Z., D. Zhang, Z. Wang, Z. Tan, H, Guo and A.H. Paterson. 2014. A whole genome DNA marker map of cotton based on the D-genome sequence of *Gossypium raimondii* L. G3: Genes, Genomes, Genetics 1759-1767. doi: 10.1534/3g.113.006890.

Ware, J.O. 1951. Origin, rise, and development of American Upland cotton varieties and their status at present. Univ. of Arkansas, College of Agriculture, Agricultural Experiment Station. Fayetteville, AR.

Ware, J.O. 1936. Plant breeding and the cotton industry. pp. 657-744. *In*: Yearbook of Agriculture. United State of Department of Agriculture. Washington, DC.

Wendel, J.F. 1989. New world tetraploid cottons contain old world cytoplasm. Proc. Natl. Acad. Sci. USA 86: 4132-4136.

Wendel, J.F., A. Schnabel and T. Seelanan. 1995. Bidirectional interlocus concerted evolution following allopolyploid speciation in cotton (*Gossypium*)." Proc. Natl. Acad. Sci. 92: 280-284.

Wendel, J.F., C.L. Brubaker, J.P. Alvarez, R.C. Cronn and J.M. Stewart. 2009. Evolution and natural history of the cotton genus. pp. 3-22. *In*: Paterson, A.H. (ed). Genetics and genomics of cotton. (ed.). Springer, New York.

Wendel, J.F., C.L. Brubaker and A.E. Percival. 1992. Genetic diversity in *Gossypium hirsutum* and the origin of Upland cotton. Am. J. Bot. 79: 1291-1310. doi.10.2307/2445058.

Wendel, J.F. and R.C. Cronn. 2003. Polyploidy and the evolutionary history of cotton. Adv. Agron. 78: 139-186.

Wendel, J.F. and C.E. Grover. 2015. Taxonomy and evolution of the cotton genus, *Gossypium*. 25-44. *In*: Fang, D.D. and R.G. Percy. Cotton. Amer. Soc. Agron. Madison, WI.

Wendel, J.F., P.D. Olson and J.M. Stewart. 1989. Genetic diversity, introgression and independent domestication of Old World cultivated cottons. Am. J. Bot. 76: 1795-1806.

Zhang, T. et al. 2015. Sequencing of allotetraploid cotton (*Gossypium hirsutum* L. acc. TM-1) provides a resource for fibre improvement. Nature Biotechnol. 33: 531-537. doi: 10.1038/nbt.3207.

Zhang, J., R.G. Percy and J.C. McCarty. 2014. Introgression genetics and breeding between Upland and Pima cotton: a review. Euphytica 198: 1-12. doi: 10.1007/s10681-014-1094-4.

Zhang, T., Y. Yuan, J. Yu, W. Guo and R.J Kohel. 2013. Molecular tagging of a major QTL for fiber strength in Upland cotton and its marker-assisted selection. Theor. Appl. Genet. 106: 262-268. doi: 10.1007/s01122-022-1101-3.

Zhang, J., H. Fang, H. Zhou, S.E. Hughs and D.C. Jones. 2013. Inheritance and transfer of thrips resistance from Pima cotton to Upland cotton. J. Cotton Science 17: 163-169.

Zhang, J.F., J.Z. Sun, J.L. Liu and Z.B. Wu. 1992. Genetic analysis of cotton resistance to spider mites. J. Huazhong Agric. Univ. 11: 127-133.

Zhang, Z., J. Rong, V. Waghmare, P.W. Chee, O.L. May, R.J. Wright, J.R. Gannaway and A.H. Paterson. 2011. QTL alleles for improved fiber quality from a wild Hawaiian cotton, *Gossypium tomentosum*. Theor. Appl. Genet. 123:1075-1088.

Zhang, T.Z., Y.L. Yuan, J. Yu, W.Z. Guo and R.J. Kohel. 2003. Molecular tagging of a major QTL for fiber strength in Upland cotton and its marker-asssisted selection. Theor. Appl. Genet. 106: 262-268.

Zhao, X.P., Y. Si, R.E. Hanson, C.F. Crane, H.J. Price, D.M. Stelly, J.F. Wendel and A.H. Paterson. 1998. Dispersed repetitive DNA has spread to new genomes since polyploid formation in cotton. *Genome Res.* 8:479-492.

Zhao, Y., H. Wang, W. Chen, Y. Li, H. Gong, X. Sang, F. Huo and F. Zeng. 2015. Genetic diversity and population structure of elite cotton (*Gossypium hirsutum* L.) germplasm revealed by SSR markers. *Plant Sys. and Evol.* 301: 327-336.

Chapter 2

Allopolyploidy and Interspecific Hybridization for Wheat Improvement

Dengcai Liu[1,*], Ming Hao[2], Aili Li[3], Lianquan Zhang[4], Youliang Zheng[5] and Long Mao[6]

ABSTRACT

Triticeae is a big tribe with a large number of species, including important cereal crops common wheat, durum wheat, barley, rye, and triticale. Interspecific hybridization and allopolyploidy have played important roles in Triticeae speciation and evolution. Common wheat has a highly heterogeneous and plastic genome structure that is the product of two allopolyploidization events involving the A, B, and D genomes. The ancestral lineages of the three genomes are also thought to contain ancestral hybridization events between diploid species. Hence, common wheat is an important model system for investigation of biological mechanisms and genetic systems related to interspecific hybridization. As well, artificial allopolyploids and interspecific hybrids have been widely used incommon wheat improvement. The donor species of common wheat have been successfully used to enhance genetic diversity of common wheat through homologous recombination. However, most species in the Triticeae are wild species that do not share

[1] Triticeae Research Institute, Sichuan Agricultural University, Chengdu, Sichuan 611130, China; Key Laboratory of Adaptation and Evolution of Plateau Biota, Northwest Institute of Plateau Biology, Chinese Academy of Sciences, Xining 810001, China, E-mail: dcliu7@yahoo.com

[2] Triticeae Research Institute, Sichuan Agricultural University, Chengdu, Sichuan 611130, China, E-mail: haomingluo@foxmail.com

[3] National Key Facility for Crop Gene Resources and Genetic Improvement, Institute of Crop Science, Chinese Academy of Agricultural Sciences, Beijing 100081, China, E-mail: liaili@caas.cn

[4] Triticeae Research Institute, Sichuan Agricultural University, Chengdu, Sichuan 611130, China, E-mail: zhanglianquan1977@126.com

[5] Triticeae Research Institute, Sichuan Agricultural University, Chengdu, Sichuan 611130, China, E-mail: ylzheng@sicau.edu.cn

[6] National Key Facility for Crop Gene Resources and Genetic Improvement, Institute of Crop Science, Chinese Academy of Agricultural Sciences, Beijing 100081, China, E-mail: maolong@caas.cn

* Corresponding author

homologous genomes with wheat. A beneficial trait from a wild species can be used in wheat breeding via a collinear translocation between homoeologous chromosomes. However, it is cumbersome to develop a collinear translocation that contains target genes but not undesirable linked genes. To date, only a few wheat-alien homoeologous translocations have been successfully used in commercial cultivars. The utilization of high-throughput technologies in identification of alien introgressions will accelerate identification and tracking of tiny introgressions. In future, more and more small introgressions without linkage drag are expected to be used as operational modules in wheat breeding.

Introduction

Interspecific hybridization and allopolyploidization have played important roles in plant evolution (Grant 1981) and are common in natural ecosystems. Many important crop species, such as bread and durum wheat, oat, cotton, sugarcane, canola, coffee, and tobacco, have resulted from these processes. Common wheat belongs to the Triticeae tribe, which is part of the Pooideae grass subfamily. Triticeae is relatively large and contains over 500 species (Wang and Lu 2014). Based on genomic constitutions, these species fall in 32 genera (Table 1), which includes nine annual and 20 perennial genera, and three genera containing species with both perennial and annual growing habits. Triticeae contains the world's most agriculturally important cereals: common wheat (*Triticum aestivum* L., $2n = 6x = 42$, AABBDD), durum wheat (*T. turgidum* L. ssp. *durum*, $2n = 4x = 28$, AABB), barley (*Hordeum vulgare* L., $2n = 2x = 14$, HH), rye (*Secale cereale* L., $2n = 2x = 14$, RR), and the first man-made crop, triticale (× *Triticosecale*, $2n = 6x = 42$, AABBRR).

The Triticeae group demonstrates the impact of interspecific hybridization and allopolyploidy on speciation and evolution. Triticeae combines a wide variety of biological mechanisms and genetic systems that affect the outcome of interspecific hybridization, which make it an excellent model system for plant evolution studies (Bothmer and Salomon 1994, Feldman et al. 2012, De Storme and Mason 2014, Li et al. 2015). The large number of Triticeae species form one of most ideal experimental systems for conducting genetic improvement of cultivated and fodder species through interspecific hybridization.

Interspecific Hybridization and Allopolyploidy in Triticeae

Allopolyploidy and Genome Clusters

Speciation by allopolyploidy has long been recognized as a common phenomenon in Triticeae (Kihara 1930). Allopolyploids originate by interspecific hybridization followed by polyploidization via the union of unreduced gametes (Hao et al. 2014). There are more perennial than annual

Table 1. Triticeae genera.

Genus name*	**Basic genome****	**Ploidy**	**Comments*****
Triticum L.	A, B/G**, D	2*x*-6*x*	Annual; A-cluster
Aegilops L.	C, D, M, N, S, U	2*x*-6*x*	Annual; D/U-cluster
Eremopyrum (Ledeb.) Jaub. et Spach	F, Xe	2*x*/4*x*	Annual
Triticosecale Wittmack ex Yen et J.L. Yang	A, B, D, R	4*x*/6*x*	Annual; Man-made
Amblyopyrum Eig.	T	2*x*	Annual
Heteranthelium Hochst.	Q	2*x*	Annual
Crithopsis Jaub. et Spach	K	2*x*	Annual
Henrardia C.E. Hubbard	O	2*x*	Annual
Taeniatherum Nevski	Ta	2*x*	Annual
Secale L.	R	2*x*	Annual/perennial
Hordeum L.	I, H, Xa, Xu	2*x*-6*x*	Annual/perennial
Pseudosecale (Godron) Degen.	V, Xv	2*x*/4*x*	Annual/perennial
Festucopsis (C.E. Hubbard) Melseris	L	2*x*	Perennial
Peridictyon O. Seberg, S. Fredriksen et C. Baden	Xp	2*x*	Perennial
Australopyrum (Tzvelev) A. Love	W	2*x*	Perennial
Psathyrostachys Nevski	Ns	2*x*/4*x*	Perennial
Agropyron J. Gaertn.	P	2*x*-6*x*	Perennial
Lophopyrum A. Love	E/J**	2*x*-10*x*	Perennial
Stenostachys Turcz.	H, W	4*x*	Perennial
Hordelymus (Jensen) C. Hart, Carl (Karl) Otto	Xo, Xr	4*x*	Perennial
Psammopyrum A. Love	L, E	4*x*-8*x*	Perennial
Leymus Hochst.	Ns, Xm	4*x*-12*x*	Perennial
Pseudoroegneria (Nevski) A. Love	St	2*x*/4*x*	Perennial; St-cluster donor
Trichopyrum A. Love	St, E/J**	4*x*-12*x*	Perennial
Roegneria C. Koch.	St, Y	4*x*/6*x*	Perennial
Elymus L.	St, H	4*x*-8*x*	Perennial
Douglasdewey C. Yen, J.L. Yang et B.R. Baum	St, P	4*x*	Perennial
Campeiostachys Drobov	St, H, Y	6*x*	Perennial
Anthosachne Steudel	St, W, Y	6*x*	Perennial
Kengyilia C. Yen et J.L. Yang	St, P, Y	6*x*	Perennial
Connorochloa S.W.L. Jacobs, M.R. Barkworth et H.Q. Zhang	St, H, W, Y	8*x*	Perennial;
Pascopyrum A. Love	St, H, Ns, Xm	8*x*	Perennial

* Genomic constitution was used as the standard of generic classification, as suggested by Dewey (1984) and Löve (1984); except for the traditional generic treatment of *Triticum, Aegilops*, and *Hordeum*. *Connorochloa* designation was based on Barkworth et al. (2009), and other genera were adapted from Yen et al. (2005, 2013e) and Yen and Yang (2009, 2013a, b, c, d). Genomic relationships were mainly determined by genome analysis, which is the study of meiotic chromosome pairing in artificial interspecific F1 hybrids (Kihara 1930).

** The 26 basic genomes (haplomes) represented by symbols A to W and Ta, St, and Ns have a known diploid donor, and some genomes with an unknown donor are indicated by X with a lower-case letter that indicates species name, and the unknown donor of the Y genome is unique to the St-clustered allopolyploid groups (Wang et al. 1994). However, differentiated genomes from a basic genome may exist in some species, thus producing difficulty in genome recognition. For example, it is now generally accepted that J and E are different versions of the same basic genome, although some authors prefer to use J, whereas others use E (Wang and Lu 2014). Similarly, B, G, and S are recognized as different modifiers of the common genome S, which is derived from an *Ae. speltoides*-like species.

*** Based on the concept of pivotal-differential genomes (Kimber and Yen 1988), *Triticum* species were in the A-genome cluster. All of the allopolyploid *Aegilops* species were in either the U- or D-genome clusters. Most perennial species of allopolyploids, as denoted by *Pseudoroegneria*, were in the St-genome cluster.

allopolyploid Triticeae species (Dewey 1984; Table 1). According to the concept of pivotal-differential genomes (Kimber and Yen 1988), Triticeae species can be grouped into four clusters: A, D, and U in the annual *Triticum-Aegilops* complex and the St genome cluster in perennial *Elymus* and related genera (Yen et al. 2005). Species in each cluster share a common, often slightly modified nuclear genome, referred to as a pivotal genome, and additional genomes that are more extensively modified (Kimber and Yen 1988). Morphologically, species inside each cluster resemble the diploid donor of their pivotal genome more, particularly in spike and seed characteristics. *Triticum* species are formed around the pivotal A genome contributed by *T. urartu*. All allopolyploid *Aegilops* species belong to either U- or D-genome clusters, which were derived from *Ae. umbellula* and *Ae. tauschii*, respectively. The A-, D-, and U-genome clusters have a common basic genome, S (also designated as the B genome), in the *Triticum–Aegilops* complex. Out of the 13 perennial allopolyploid genera (Table 1), nine are clustered with the pivotal St genome, which was contributed by *Pseudoroegneria*. Therefore, the A, D, U, and St genomes played important roles in Triticeae allopolyploidy.

Hybridization and Introgression

Introgression, the movement of genetic material from one species into the gene pool of another by repeated backcrossing of an interspecific hybrid with one of its parental species, is an important source of genetic variation in natural populations that may contribute to adaptation and even adaptive radiation. With the utilization of molecular approaches, introgression was shown to have occurred frequently during Triticeae evolution, both in diploid and allopolyploid species. Introgression can lead to a reticulate evolutionary pattern, in which evolutionary relationships do not fit a simple bifurcate tree but instead a network structure, causing taxonomic uncertainties in Triticeae, especially in *Elymus* and related genera (Kellogg et al. 1996, Mason-Gamer 2004, 2008, Sun 2014). For example, *Elymusrepens* has introgressed sequences from the H genome of *Hordeum* and Ta genome of *Taeniatherum*-like.

Introgression also played a role in speciation and genome differentiation during the evolution of cultivated species in Triticeae. Rye and wheat diverged 7 million years ago (mya), and both rye and wheat lineages and the barley lineage (*Hordeum*) diverged from a common Triticeae ancestor around 11 mya (Huang et al. 2002). Global sequence analysis revealed a heterogeneous composition of the rye genome, with the genome containing a series of translocations. This heterogeneous composition is thought to be the result of introgression between rye species, which was facilitated by the outbreeding status of rye, and thus provided an important prerequisite for the formation of the modern rye genome (Escobar et al. 2011, Martis et al. 2013). Rye may have also introgressed with *Trichopyrum* (*Thinopyrum*), as the presence of R genome sequences has been detected in the J/E genome of *Trichopyrum* (Mahelka et al. 2011, Tang et al. 2011). In wheat, the B genome is thought to have differentiated from the S genome (Petersen et al. 2006). DNA sequence analysis indicates that the S-genome species probably originated through introgressive speciation between ancestral species of *Triticum–Aegilops* and *Hordeum* (Nakamura et al. 2009). *Horedum* also substantially contributed to introgression; in addition to the above-mentioned introgressions with *Elymus* and *Triticum–Aegilops*, *Horedum* was also probably involved in introgressions with *Trichopyrum* (Yang et al. 2006) and *Dasypyrum* (*Pseudosecale*) (Nakamura et al. 2009).

Hybridization and Genome Rearrangement

The rich diversity of Triticeae genomes suggests features inducing variability, such as propensity to differentiation. Genome sequence comparison between Triticeae species and other crops such as rice, maize, and sorghum of the Pooideae family reveals greatly accelerated genome evolution in the large Triticeae genomes, and modern Triticeae genomes were derived from numerous genomic reconstructions (Luo et al. 2009, Murat et al. 2014). Spontaneous genome rearrangement is not an accidental phenomenon in natural populations of Triticeae species. For example, chromosomal aberrations were frequently observed in diploid *Ae. speltoides* populations in marginal environments (Belyayev and Raskina 2013). Genome rearrangement is more often observed in individuals or populations of allopolyploid species, such as individuals of the annual *Triticum* species (e.g., Liu et al. 1992, Qi et al. 2006, Badaeva et al. 2007, Kawshara and Taketa 2000, Ma et al. 2014), perennial *Elymus* and related genera (Dou et al. 2009, Scoles et al. 2010, Tomas et al. 2012, Wang et al. 2012).

The universality of genome rearrangements in allopolyploids could be attributed to the ability of allopolyploids to tolerate genome rearrangements.

In fact, allopolyploidization can also contribute to genome rearrangements through genome shock (McClintock 1984), which means that their genomes are not simply additive with respect to their parental genomes. In Triticeae, obvious genome structural changes were observed in some newly formed allopolyploids, such as Triticale, which has a higher frequency of variations in the R genome (from rye) than wheat genomes (Ma and Gustafson 2008, Bento et al. 2011, Hao et al. 2013), and in some synthetic tetraploid wheat (Zhang et al. 2013).

Natural hybridization brings the dispersedly rearranged chromosomes among individual plants back together, enhancing genome differentiation relative to its original genomes. Chromosome rearrangements, such as translocations, provide important driving forces in this evolutionary process because they can immediately change the gene linkages, and thus probably result in substantial changes in gene expression, epigenetic effects, and, most importantly, possible post-zygotic reproductive barriers that can lead to speciation.

The Contribution of Interspecific Hybridization and Allopolyploidy to Common Wheat Genome Formation

Common Wheat Resulted from Multiple Rounds of Interspecific Hybrid Speciation

Common wheat (*T. aestivum*, AABBDD) is a young but very important species, providing nearly 20% of the calories and proteins consumed by the world's population and comprising 95% of the global wheat production. One key factor in the success of bread wheat as a global food crop is its adaptability to a wide range of climates, which is believed to result from its highly heterogeneous and plastic genome structures derived from two allopolyploidization events (Dubcovsky and Dvorak 2007, Feldman et al. 2012). Tetraploid *T. turgidum* (AABB) was generated by the first allopolyploidization event (<0.5 mya) between the wild diploid species *T. urartu* (AA) (Dvorak and Zhang 1990) and an unknown close relative of *Ae. speltoides* (BB) (Petersen et al. 2006). Durum wheat used for pasta belongs to *T. turgidum*. Common wheat was then produced by the second allopolyploidization event (7,000–12,000 ya) between the cultivated *T. turgidum* and the wild diploid species *Ae. tauschii* Cosson (DD) (Kihara 1944, McFadden and Sears 1944, Huang et al. 2002).

Homoploid and introgressive hybridization also contributed to the evolution of diploid donor species of common wheat. Recent work showed that *Ae. tauschii* is a product of homoploid hybrid speciation (IWGSC 2014; Marcussen et al. 2014). It is thought that the A and B genomes diverged from a common ancestor approximately 7 mya and that these genomes gave rise to the D genome by hybridization approximately 1–2 mya later. The hybrid D lineage originated from a nearly equal contribution of the A and B lineages, which makes the gene contents of A and B subgenomes more similar to that of the D subgenome than to each other at the whole chromosome level based on gene sequence relatedness. The B genome of wheat is thought to have differentiated from the S genome and is probably a product of introgressive speciation between ancestral species of *Triticum–Aegilops* and *Hordeum* (Nakamura et al. 2009).

Some factors may favor the evolutionary success of 'young' common wheat species. Although it is impossible to artificially re-synthesize *T. turgidum* because of the uncertainty of the B-genome donor, synthetic hexaploid wheats (SHWs) produced by *T. turgidum* and *Ae. tauschii* have been created based on pioneering work by Kihara (1944) and McFadden and Sears (1944) more than half a century ago. Similar work on common wheat origination indicated that some factors may favor hexaploidization (Li et al. 2015). For example, *T. turgidum* can be crossed with *Ae. tauschii*, which produces triploid F1 plants without special treatment after pollination (Zhang et al. 2008), and most of the resultant hybrids can spontaneously generate hexaploid wheat through unreduced gametes (Zhang et al. 2010). The gene and chromosomal structure are relatively conserved in some nascent SHWs (Mestiri et al. 2010, Zhao et al. 2011, Luo et al. 2012), although this may be related to particular hybrid combinations or the DNA markers used (Feldman et al. 2012). Accordingly, gene expression at the genome-wide level indicated that SHWs mainly exhibited gene additivity, in which the expression in SHWs equals the average 'mid-parent value' of parents (e.g., Chagué et al. 2010, Li et al. 2014a), which indicates relative compatibility of genomes from two parents. Based on similar gene expression patterns observed between natural bread wheat and SHWs, it is thought that regulation of gene expression was established immediately after allohexaploidization and maintained over generations (Chagué et al. 2010). Partial subgenome dominance can also contribute to the success of wheat (see the following section). These factors also promote the utilization of these donor species in common wheat breeding through re-synthesis of hexaploid wheat.

Regional Subgenome Expression Level Dominance in Common Wheat

Subgenomic asymmetry or subgenome dominance may be important for the evolutionary success of allopolyploids. Subgenome dominance results from the biased deletion of duplicate gene redundancy during diploidization after polyploidization, in which only one of the duplicated blocks or gene families retain the majority of ancestral copies of the duplicates (Akhunov et al. 2013, Pont et al. 2013).

Common wheat is a species with two pivotal (dominant) genomes (A and D) (Kimber and Yen 1988). A pivotal genome is often conserved after allopolyploidization. However, common wheat has traditionally been included in the A-genome cluster, although classical genome analysis indicates that its D genome seems to experience less change than its A genome

(Yen et al. 1996). The existence of pivotal-differential genomes reflects the asymmetric differentiation in chromosomal structure between subgenomes after allopolyploidization (Feldman et al. 2012), with relative conservation of pivotal genomes and easier changes for other genomes (Kimber and Yen 1988). Pont et al. (2013) proposed that an allotetraploidization event produced the dominant (i.e., stable) A subgenome and sensitive (i.e., plastic) B subgenome. Following the hexaploidization event, a supra-dominance where the tetraploid became sensitive (subgenomes A and B) and the D subgenome supra-dominant (i.e., pivotal) was formed (Pont et al. 2013, Murat et al. 2014). Wheat subgenome dominance was also shown by the differential genome control of morphological and physiological traits. For example, in natural hexaploid wheat, the A genome dominantly controls morphological traits, whereas the B and D genomes control the reaction to biotic and abiotic factors (Feldman et al. 2012). On the transcriptome level, parental expression-level dominance (ELD), in which the genes in polyploid progeny have equal expression levels to one parent and not the other, exists in nascent allohexaploid wheats derived from *T. turgidum* and *Ae. tauschii* (Li et al. 2014a). ELD genes similar to the *T. turgidum* parent are more likely to be associated with development, whereas ELDs similar to *Ae. tauschii* tend to be associated with adaptive traits, such as stress response and photoperiod adaptability. ELD thus seems to ensure the combination of the growth vigor of *T. turgidum* with the adaptation of *Ae. tauschii*, which may confer nascent allohexaploid wheat with better adaptability to a wide range of climates relative to its maternal parent, *T. turgidum*. Therefore, ELD probably plays a role in the success of bread wheat as a global food crop. Transcriptome analysis also indicates that ELD in nascent polyploids may precede and give rise to genome dominance in later stages of evolution (Eckardt 2014). Small RNA-mediated regulation may be one mechanism of ELD formation and thus produces gene ELD (Li et al. 2014a).

Natural Introgression Between Common Wheat and Other Species

Natural introgression contributed to wheat evolution. The introgression of *T. turgidum* into common wheat enriched the genetic diversity of the A and B genomes of current common wheat populations (Dvorak et al. 2006). 'Barbela,' a collective name for wheat landraces from northern Portugal, is an excellent example of introgression of rye into natural common wheat populations (Ribeiro-Carvalho et al. 2004). Natural introgression played a role in increasing the genetic diversity after wheat originated.

Alternatively, introgression of genes from wheat into wild species can also occur naturally. This is viewed as a possible threat to the environment and agriculture because it may increase the capability of the wild species to adapt to agricultural environments and compete with wheat (David et al. 2004, Hegde and Waines 2004, Zaharieva and Monneveux 2006). Wheat sympatrically grows with some wild relatives, and spontaneous interspecific hybridization occurs frequently, especially between wheat and *Aegilops*. Because of the existence of genes for unreduced gamete production (Hao et al. 2014), many wheat-*Aegilops* hybrids can produce amphidiploids that provide a bridge for introgression. The unstable cytology of newly formed amphidiploids further enhances the chance of introgression (Yang et al. 2011). Gene transfer risk via homologous recombination has been analyzed, for example, between D chromosomes of wheat and *Ae. cylindrica* ($2n = 28$, CCDD) (Caldwell et al. 2004). Although gene flow between homoeologous chromosomes is usually prevented by *Ph* genes (Sears, 1976), wheat DNA is detected in natural wild populations of *Ae. peregrina* ($2n = 28$, UUSS) (Weissmann et al. 2005).

Interspecific Manipulation of Genetic Recombination for Wheat Improvement

Gene Transfer from Donor Species of Wheat by Homologous Recombination

In the past years, common wheat breeders have mainly utilized natural variations of the original hybridization between *T. turgidum* and *Ae. tauschii*, subsequent mutations, and introgression. Classical breeding combines variations of the wheat parents into the offspring. However, only a limited number of individuals of donor species were involved in the origin and evolution of common wheat. This evolutionary bottleneck lead to a large amount of genetic variation in the two donor species that was excluded from common wheat, and subsequent domestication and modern breeding further reduced genetic diversity. The lack of genetic variation impedes further genetic improvement to meet increasing worldwide demand. One strategy for solving this problem is to transfer genes from the donor species into modern wheat.

The genetic diversity of *T. turgidum* and *Ae. tauschii* can be simultaneously introduced into common wheat by the 'bridge' of synthetic hexaploid wheat (SHW) derived from artificial synthesis of hexaploid wheat (*T. turgidum* × *Ae. tauschii*) in a manner analogous to the origination of hexaploid wheat (Mujeeb-Kazi et al. 1996). SHW is easy to cross with common wheat and provides a convenient way to simultaneously transfer multiple *T. turgidum* and *Ae. tauschii* genes that are dispersed in different chromosomes into a wheat line by genetic recombination between homologous chromosomes. Undesirable gene linkages can be mostly broken by repeated backcrossing to common wheat. SHW has been preferentially used for enhancing genetic diversity of common wheat by numerous groups (e.g., Mujeeb-Kazi et al. 1996, Yang et al. 2009, Ogbonnaya et al. 2013).

The great potential of SHWs in wheat breeding to improve yield has been demonstrated by different groups. SHWs alone are of poor agronomic value, because they have wild traits and produce low yields. However, when crossed with adapted wheat, some elite synthetic-derived lines have been shown to yield up to 35% more grain than the best local checks (Yang et al. 2009, Ogbonnaya et al. 2013, Li et al. 2014b). Successful high-yield breeding of commercial wheat cultivars has been demonstrated by scientists in Sichuan Province, China. For example, the SHW-derived cultivar Chuanmai 42, released in 2003, created a yield record greater than 6 t/ha in the regional trial in Sichuan Province and outperformed the commercial check Chuanmai 107 by 20% on average in Sichuan Province in 2002 and 2003 (Yang et al. 2009). This cultivar was produced by crossing CIMMYT SHW line Syn769 (*T. turgidum* ssp. *durum* Decoy 1/*Ae. tauschii* 188) with the local elite line SW3243 and then top-crossed with the elite line Chuan 6415. The high yield makes Chuanmai 42 a leading cultivar that can increases grain yield by 0.5–0.8 t/ha in farmers' fields (Li et al. 2014b). Chuanmai 42 has also become a key parent for yield improvement in Southwestern China, and about 20 commercial cultivars were derived from Chuanmai 42. Some cultivars, such as Chuanmai 104, Chuanmai 64, and Shumai 969, exhibit similar grain yield to Chuanmai 42. Large grains and high tiller numbers are the two key components of the high yield of SHW-derived cultivars. Compared with non-SHW-derived cultivars, SHW-derived cultivars have greater early growth vigor, more dry matter accumulation, good chlorophyll retention capacity, and higher canopy photosynthesis capacity (Tang et al. 2014).

The synthetic wheat SHW-L1 was also successfully used in wheat breeding in Sichuan Province (Fig. 1A). SHW-L1 was synthesized from interspecific hybridization between Chinese *T. turgidum* ssp. *turgidum* line AS2255 and Middle Eastern *Ae. tauschii* accession AS60. SHW-L1 has many poor agronomic traits, including wild characters from the diploid grass parent *Ae. tauschii*. A single cross or backcross using a single wheat parent is typically unable to overcome all of the poor agronomic traits. Therefore, the hybrids between SHW-L1 and Chuanmai 32 were further top-crossed to Chuanyu 16 and then Chuanmai 42. In 2013, a new cultivar, Shumai 969, was released (Fig. 1A). Because of its excellent performance, Shumai 969 quickly became a leading cultivar in Sichuan Province. It is the only cultivar that had grain yield greater than 6 t/ha in the production test of Sichuan Province (PTSP) during the past 10 years, and it out performed the commercial check Mianmai 37 by 10% in the PTSP in 2013. More importantly, this cultivar matures early and produces high yield and good quality, strong gluten. In Sichuan Province, only a few cultivars meet the quality standard for strong gluten because of unfavorable climate conditions. Shumai 969 contains genetic material from SHW-L1, including large chromosome fragments from the wild *Ae. tauschii* (Fig. 1B; unpublished data).

However, to date, only a few SHWs have been successfully used to develop commercial cultivars. SHW as a bridge is expected to remain an important high-yield breeding strategy for the foreseeable future, and an increasing amount of SHWs will be exploited. Consequently, it should be considered which *T. turgidum* and *Ae. tauschii* should be preferentially

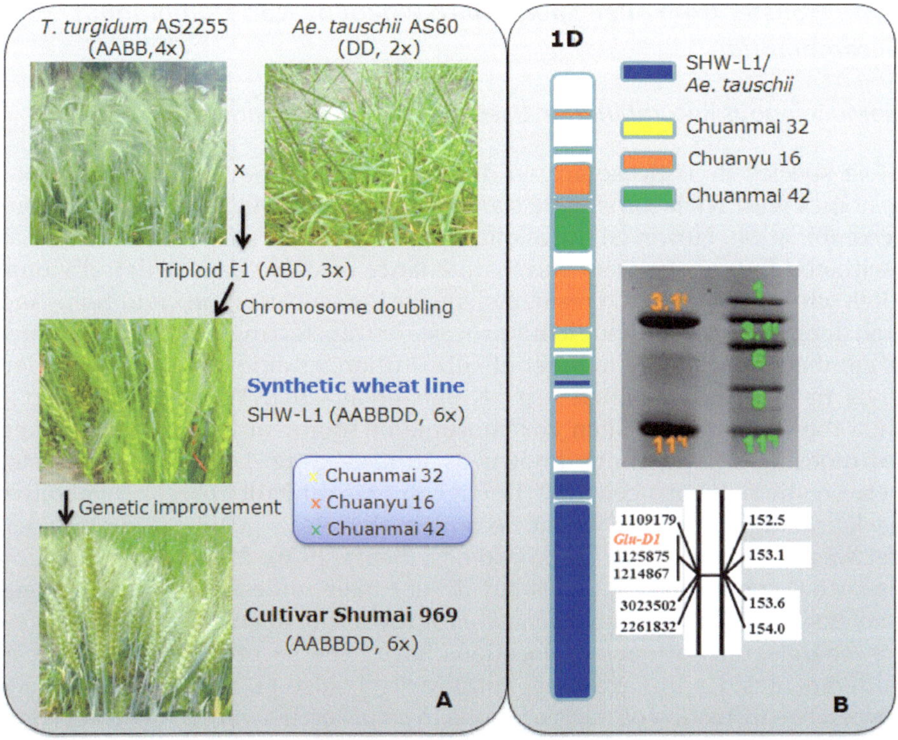

Figure 1. Transferring a chromosomal segment harboring the Glu-D1 locus from wild diploid species *Aegilops tauschii* into commercial wheat Shumai 969. (A) Transferring procedure via the artificially synthesized allohexaploid wheat as a bridge. The development of the new cultivar involves two major phases: the synthesis of nascent allohexaploid wheat SHW-L1 and genetic improvement through top-crosses involving three elite wheat lines. The total process takes 14 years and 16 generations; (B) the new cultivar with the chromosomal segment containing the high-molecular-weight gluten subunits (HMW-GS) $3.1^{t} + 11^{*t}$ at Glu-D1 locus on chromosome 1D, derived from synthetic wheat SHW-L1, donated by *Ae. tauschii* (Chen et al. 2012). The middle figure shows visualization of HMW-GS and the bottom figure shows the Glu-D1 locus and closely linked markers (left) and their genetic distance (right) on chromosome 1D, as determined by DArTseq. Wheat HMW-GS affects the flour processing quality.

used in interspecific hybridization to synthesize SHWs for further yield increasing. However, the genetic basis for using SHWs in yield breeding is largely unknown. SHW is different from common wheat since its newly synthetic process may introduce transcriptome shock which probably affects plant growth. Recent studies on the relationship between gene expression patterns and growth vigor or traits of newly formed SHWs provides new clues for further research on these issues (e.g., Li et al. 2014a, 2015).

Gene Transfer from Alien Species into Wheat by Non-Homologous Recombination

Homoeologous Recombination Used in Commercial Cultivars

Most species in Triticeae are wild species that do not share homologous genomes with wheat, and gene transfer cannot be achieved by homologous recombination. However, the allohexaploid structure of common wheat has facilitated the construction and maintenance of stable lines with individual alien chromosomes or fragments, including substitution, addition, and translocation lines. Although this process for transferring alien chromosomes is cumbersome, a large number of substitution, addition, and translocation lines have been generated (e.g., Kilian et al. 2011, Wang 2011). Despite this, the incorporated alien chromosome or fragment is frequently linked to undesirable wild traits, known as linkage drag. Unlike gene transfer between homologous genomes, linkage drag is difficult to break by iterative backcrossing to common wheat, because of the lack of genetic recombination between homoeologous chromosomes (Sears 1976). Hence, translocation lines that contain target genes but do not have undesirable gene linkages may be a more promising alternative for the use of alien germplasm.

To date, only a few translocations have been successful in commercial cultivars. Besides linkage drag, collinearity is also an important factor for successful utilization of translocations in commercial wheat. A chromosome translocation usually involves the replacement of a wheat chromosome fragment by an alien fragment. If the incorporated alien fragment and the corresponding replaced wheat fragment are not collinear, the non-collinear translocation event will lead to gene deletion or gene redundancy (i.e., genetically non-compensating translocations). Although this kind of translocation may be used in genetic analysis, gene deletion and redundancy is harmful for wheat breeding because of gene dosage imbalance. The gene balance hypothesis posits that altering the stoichiometry of members of multisubunit complexes will affect the function of the whole gene complex as a result of the kinetics and mode of assembly (Birchler and Veitia 2012).

The 1BL·1RS wheat/rye translocation is a collinear translocation between homoeologous chromosomes. This translocation is the most widespread alien translocation that is distributed in a large number of commercial wheat cultivars worldwide. Most cultivars with the 1BL·1RS translocation contain 1RS from "Petkus" rye. This chromosome arm possesses important disease resistance genes, including *Lr26*, *Sr31*, *Yr9*, and *Pm8*, as well as genes that enhance yield and improve adaptability (Villareal et al. 1998). Out of the seven rye chromosomes, 1R is the only linkage group that is collinear over its entire length with another single homoeologous chromosome 1, as collinearity with wheat for other rye chromosomes has been disturbed by a series of translocations (Devos et al. 1993, Martis et al. 2013). This may explain why only the translocation that involved 1R was successfully used in commercial cultivars, although numerous translocations also carrying desirable genes have been generated from wheat-rye hybridization.

The 6VS·6AL wheat/*Haynaldia villosa* translocation is another widely used translocation that is carried by at least 10 commercial wheat cultivars in China (Chen et al. 1995, Cao et al. 2011). This translocation carries the powdery mildew resistance gene *Pm21,* which is derived from the chromosome 6VS of *H. villosa* and produces durable and broad spectrum resistance to wheat powdery mildew. Moreover, 6VS·6AL is also a translocation between homoeologous chromosomes, and its introduction into the newly developed cultivars has no obvious adverse effect on other agronomic traits, which is indicative of a genetically compensating translocation. Even when *Ph1* was not functional, 6D and 6V did not recombine; instead, the translocations between 6A and 6V chromosomes were retained (Qi et al. 2011). The preferential 6A-6V recombination also indicates that these chromosomes may have greater genetic affinity and hence may result in a better genetic compensation effect.

Cryptic Introgressions Used in Commercial Cultivars

Although most previous research focused on translocations with relatively big alien fragments that are easier to detect by cytogenetic-based approaches, introgression size can be highly variable. There are some cases of cryptic introgressions, wherein an introgression cannot be detected using cytogenetic-based approaches. The most successful utilization of a cryptic introgression for wheat cultivar development is exemplified by the Chinese cultivar 'Xiaoyan 6' (Wang 2011), which is a derivative of hybrids between common wheat and *Thinopyrum ponticum* (syn. *Agropyron elongatum, Lophopyrum ponticum, Elytrigia pontica;* $2n = 70$, $S^tS^tS^tS^tE^eE^eE^bE^bE^xE^x$). Xiaoyan 6 was grown on more than 10 million hectares in China from 1980-1995 and has been used as a core parent for wheat breeding in China over the past 20 years, producing more than 50 derivatives (Li et al. 2009). These derivatives have been grown on more than 20 million hectares and increased the total wheat grain production by 7.5 billion kg. The substantial contribution of Xiaoyan 6 to wheat production and breeding may be attributable to its valuable characteristics, which primarily include wide spectrum and durable resistance to stripe rust; tolerance to high temperatures, strong light, and hot-dry wind; and superior grain quality suitable for making traditional Chinese wheat foods (Li et al. 2009).

Recently, the commercial wheat cultivar Pubing 143 and some elite wheat lines with excellent agronomic characters, such as Pubing 2011, have been developed from distant hybridization with *Agropyron cristatum* (L.) Gaertn ($2n = 28$, PPPP) in China. Although some of the lines have no obvious cytogenetic markers for the P genome, they exhibit high grain number per spike because of the high floret number per spikelet, which is a typical character of its alien parent *A. cristatum* accession Z559 (Li et al. 1995) carried on chromosome 6P (Han et al. 2014).

With the development of molecular markers, cryptic introgressions can now be identified. An excellent example of such a cryptic translocation includes the transfer of *Lr57* and *Yr40* to wheat from *Ae. geniculata* Roth (Kuraparthy et al. 2007). In this study, leaf rust and stripe rust resistant introgression lines were developed by induced homoeologous chromosome pairing between wheat 5D and $5M^g$ of *Ae. geniculata* (U^gM^g). Genomic *in situ* hybridization (GISH) characterization of rust-resistant progenies revealed different types of introgressions, cytologically visible and invisible (cryptic alien introgressions), although these introgression types showed similar resistance. Molecular mapping revealed that cryptic introgressions that conferred resistance to leaf rust and stripe rust comprised less than 5% of the 5DS arm, and this region was designated T5DL·5DS-$5M^g$S, carrying the resistance genes *Lr57* and *Yr40* derived from *Ae. geniculata.* Cryptic introgressions were also revealed by molecular markers in hybridization of wheat with some other species, such as *Ae. sharonensis* (Millet et al. 2014), *Dasypyrum villosum* (Caceres et al. 2012), and *Thinopyrum intermedium* (Dong et al. 2004). These studies demonstrate the limited resolution of cytogenetic-based approaches for determining the presence of introgressions and underscore the need for more sensitive assays.

Considering linkage drag and gene dosage balance, especially when alien chromosomes are rearranged relative to homoeologous chromosomes of wheat, small introgressions may be better in wheat breeding. With the increase of genome and transcriptome information (e.g., Jia et al. 2013, Ling et al. 2013, IWGSC 2014, Li e al. 2014a), an increasing number of molecular markers will be developed. Utilizing these molecular markers for introgression identification will accelerate screening of a large number of interspecific hybrid progeny to identify very small introgressions. More small introgressions are expected to be exploited in wheat breeding in the future.

Alien Introgression Manipulation by the Ph *Gene System*

The *Ph* gene system, which controls diploid-like meiotic behavior by preventing homoeologous chromosome pairing, is important for allopolyploid wheat success because it ensures full fertility, disomic inheritance, and karyotypic stability (Sears 1976, Sánchez-Moran et al. 2001). This system includes the major pairing gene *Ph1* on chromosome 5B, the intermediate pairing gene *Ph2* on 3D, and several minor loci (Sears 1976). The *Ph1* locus is related to a cluster of genes similar to *Cdk2* (Yousafzai et al. 2010) and it has a downstream effect on the synapsis gene *TaAsy1,* reducing its expression level (Boden et al. 2009).

The increased Cdk-type activities can phenocopy the effect of deleting *Ph1* (Greer et al. 2012). A recent study showed that *C-Ph1* is a candidate gene for *Ph1,* because silencing of *C-Ph1* results in a phenotype characteristic of the

Ph1 mutation, including homoeologous chromosome pairing, multivalent formation, and disrupted chromosome alignment on the metaphase I plate (Bhullar et al. 2014). *Ph2* is involved in the progression of synapsis (Sutton et al. 2003, Prieto et al. 2004).

Ph1 and *Ph2* also prevent pairing between wheat and alien chromosomes in wheat-alien hybrids and thus restrict the production of wheat-alien translocations (Sears 1976, Martinez-Perez and Moore 2008). However, homoeologous pairing can occur when *Ph1* or *Ph2* do not work or increase Cdk-type activity (Greer et al. 2012). These manipulations of *Ph* genes can relieve the restrictions of homoeologous chromosome pairing and thus improve the efficiency of alien translocation development.

Compared with the randomness of spontaneous break-fusion (Lukaszewski and Gustafson 1983) and other approaches for inducing translocations, such as irradiation and the utilization of gametocidal genes (Endo 1990), the manipulation of the *Ph* system is advantageous because it can significantly increase the frequency of homoeologous chromosome pairing, thereby producing genetically compensating translocations (Qi et al. 2007). Because it is by far the most effective gene for translocations, *Ph1* manipulation has been extensively used for alien genetic introgression. Nevertheless, the efficiency of genetic manipulations of *Ph1* loci for alien gene transfer has not been as good as expected (Miller et al. 1998). The chromosome recombination frequencies of wheat-alien species seem to depend on the particular homoeologous group of concern, the overall genetic affinity of the alien chromosomes with respect to those of wheat, and the genetic distance between the target gene and centromere (Miller et al. 1998, Qi et al. 2007). A recent study indicated that in wheat-rye hybrids where homologues are absent, *Ph1* affects neither the level of synapsis nor the number of sites of mismatch repair protein MLH1. Thus, in the case of wheat-wild relative hybrids, *Ph1* may affect whether MLH1 sites are able to progress to crossover (Martin et al. 2014).

Exploiting new *Ph* genes that have different genetic mechanisms from *Ph1* is an alternative approach to further enhance the efficiency of alien gene transfers. *Ph2* is involved in the progression of synapsis (Prieto et al. 2004). Simultaneous mutations of *Ph1* and *Ph2* reinforce the *ph1b* effect and result in higher pairing frequencies than those achievable with *ph1b* alone in wheat-*Ae. variabilis* ($2n = 28$, UUS^LS^L) hybrids, although the increased pairing causes fertility reduction in hybrids (Ceoloni and Donini 1993).

phKL from Chinese common wheat landrace Kaixian-luohanmai (KL) can induce homoeologous pairing in hybrids with rye or *Ae. variabilis* at a level between those of hybrids that have Chinese Spring *ph1b* (CS*ph1b*) and CS*ph2b*/CS*ph2a* (Liu et al. 2014). There are additional phenotypic differences between *phKL* and *ph1b*: (1) *phKL* can promote homoeologous chromosome pairing in hybrids of KL with *Psathyrostachy shuashanica* ($2n = 14$, NsNs), but there is no promoting action for *ph1b* in hybrids of CS*ph1b* with *Psa. huashanica*, which is probably caused by a suppressor in *Psa. huashanica* (Kang et al. 2008); (2)

although the frequency of wheat-wheat associations was higher in CS*ph1b* × rye than in KL × rye, wheat-rye and rye-rye frequencies were higher in KL × rye than in CS*ph1b* × rye (Hao et al. 2011). These differences could result from different mechanisms of homoeologous chromosome pairing control between *phKL* and *ph1b*. Moreover, *phKL* seems to be a better choice than *ph1b* when transferring rye and *Psa. huashanica* genes.

Interspecific Manipulation of Ploidy for Wheat Improvement

Uniparent Genome Elimination Post-Pollination

Uniparental genomes may be eliminated during embryo development after fertilization of the egg by the sperm of another species in some hybrids of wheat with alien species, such as Triticeae species (*H. vulgare* and *H. bulbosum*) and more distantly related species (*Zea mays, Pennisetum glaucum, Sorghum bicolor, Coix lacryma-jobi, Imperata cylindrica*) (see review by Liu et al. 2014). This process can result in haploid wheat, which offers a shortcut for doubled haploid (DH) breeding, because a homozygous DH can be immediately generated from a haploid by chromosome doubling. Wheat-maize (*Z. mays*) hybridization is an important method for producing haploid wheat.

Uniparental chromosome elimination in hybrids results from the unequal interactions of centromeres from two parents with mitotic spindles. Centromeres are epigenetically specified by incorporation of the essential kinetochore protein CENH3 (Henikoff and Dalal 2005). The loss of CENH3 causes the failure of centromere formation and thus chromosomes egregation. Saneietal. (2011) observed the development of *H. vulgare* × *H. bulbosum* embryos and proposed a model of how the mitosis-dependent process of uniparental chromosome elimination affected the hybrid embryos. After fertilization, both parental CENH3 genes were transcriptionally active. CENH3 was then loaded into the centromeres of *H. vulgare* but not that of *H. bulbosum*. This led to *H. bulbosum* chromosome lagging and subsequent formation of micronuclei. Micronucleated *H. bulbosum* chromatin degraded and ahaploid *H. vulgare* embryo developed.

The artificial manipulation of the CENH3 protein is a new strategy for improving crop breeding efficiency. A study on intraspecific hybrids using *cenh3*-null mutants in *Arabidopsis thaliana* revealed that the loss of CENH3 leads to uniparental chromosome elimination (Ravi and Chan 2010). This study indicated that haploid plants can be easily generated through seeds by manipulating CENH3. The frequency of genome elimination by this CENH3-mediated method is higher than any previously reported interspecific hybridization approach; hence, this method can enhance the efficiency of DH breeding. Moreover, paternal chromosomes can shift into the maternal cytoplasm by crossing a female mutant with altered CENH3 proteins with a wild-type male. This method can be used to develop cytoplasmic male sterility, which has been widely used in the production of hybrid seeds (Chan 2010).

Genome Doubling by F_1 Hybrid Plants

Interspecific hybridization brings divergent genomes from different species together into amphihaploid (analogous to haploid) F_1 hybrids. Such hybrids are usually sterile because of the presence of only one copy of each homologous chromosome which remains unpaired during meiosis. Genome doubling (polyploidization) restores genome dosage and chromosome number to parent additivity in resulting allopolyploids and thus stabilizes the reproductive cycle by conferring bivalent chromosome pairing and fertility. Formation of unreduced gametes is a dominant mechanism for spontaneous genome doubling in interspecific hybrids of many Triticeae species (see review by Liu et al. 2014) or in haploid wheat plants (Jauhar 2007). Unreduced gametes result from meiotic restitution, in which meiotic cell division is converted into 'mitotic-like meiosis', or a meiosis resembling mitosis in that there is only one equational division (Zhang et al. 2007) which generates dyads instead of the normal tetrads at the end of meiosis. Because asynapsis is a key feature of unreduced gamete formation (Wang et al. 2010, Ressurreição et al. 2012), this phenomenon is also referred to as 'univalent-dependent meiotic non-reduction' (De Storme and Geelen 2013, Hao et al. 2014).

The production of unreduced gametes is genetically controlled, and some interspecific hybrid combinations can result in a high frequency of polyploidization. For example, different studies consistently showed that the *T. turgidum* durum line 'Langdon' produces a high frequency of unreduced gametes in its hybrids with *Ae. tauschii* (Xu and Joppa 2000, Matsuoka and Nasuda 2004, Zhang et al. 2010, Cai et al. 2010). A major gene, *QTug.sau-3B,* was recently mapped on chromosome 3B in *T. turgidum* × *Ae. tauschii* hybrids (Hao et al. 2014). This gene resulted in prolonged cell division during meiosis I and thus promoted the formation of unreduced gametes through the pathway of first division restitution. Based on the similarity in the phenotypic effects and syntenic relationships, it was suggested that *TAM* (*tardy asynchronous meiosis*) may be a candidate gene for *QTug.sau-3B.* In *Arabidopsis, TAM* encodes CYCA1; 2, and *TAM* mutants can abolish the second meiotic division and systematically produce unreduced gametes (d'Erfurth et al. 2010). Although the relationship between *QTug.sau-3B* and *TAM* requires further study, high frequencies of unreduced gametes may be related to the reduced *TAM* expression in wheat (Hao et al. 2014).

In crop genetics and breeding, unreduced gametes can help increase large-scale production of amphidiploids, which are widely used as the bridge of alien gene introgression into crops, and production of doubled haploids (DH). DH in wheat can quickly fix genetic recombination and thus enhance breeding efficiency. Genome doubling is always a bottleneck when using large-scale DH production for breeding purposes. Currently, it should be possible to transfer the major gene *QTug.sau-3B* into a desirable genotype for DH production. This gene from *T. turgidum* is also functional in derived hexaploid lines because meiotic restitution occurs in hybrids of synthetic hexaploid wheat with other species, such as rye (Zeng et al. 2014) and *Ae. variabilis* (Yang et al. 2010). In addition, some common wheat landraces exhibited spontaneous but low-frequency genome doubling in haploids (Jauhar 2007). If the major gene for haploid doubling, *QTug.sau-3B*, is combined with the manipulated CENH3 for haploid production, the efficiency of DH breeding will be further improved.

Single Genome Elimination During Hybrid Selfing

In addition to serving as a bridge for gene introgression by backcrossing, amphidiploids have special applications in crop genetics and breeding. The first man-made crop, triticale, is an amphidiploid between wheat and rye. However, octaploid triticale is known to be unstable and have low fertility with respect to meiotic and mitotic processes, resulting in chromosome elimination (Lukaszewski and Gustafson 1987). Such chromosome elimination may result in hexaploid triticale that contains chromosomes from A, B, and R genomes, with D-genome chromosomes being preferentially eliminated (Dou et al. 2006, Hao et al. 2013). Hao et al. (2013) developed a method for rapid production of hexaploid triticale via hybridization of hexaploid wheat with rye which involved two main components: (1) hybridization between hexaploid wheat with meiotic restitution gene(s) and rye for the spontaneous production of octaploid triticale (Zeng et al. 2014); and (2) selection for good fertility during the F3 and subsequent generations. Selection for high fertility that depends on cytologically stable conditions may help retain lines with eliminated chromosomes. In addition to complete hexaploid triticale with 28 intact A/B and 14 intact R chromosomes, this method can develop hexaploid triticales with other chromosome constitutions, including monosomic, substitution, and translocation lines.

The preferential elimination of the D genome has been reported in other hybrids, such as the amphiploid (genome $AABBDDU^kU^kS^kS^k$) between hexaploid wheat and *Ae. kotschyi* (Tiwari et al. 2010), and trigeneric hybrids between hexaploid wheat, rye, and *Psathyrostachys huashanica* (genome AABBDRNs) (Xie et al. 2012). Chromosome elimination during hybrid selfing only involves one genome, which differs from uniparental chromosome elimination post-pollination. The mechanism responsible for the preferential elimination of the whole D genome is unknown, although some related cytological behaviors are observed in octaploid triticale, such as the budding-like chromatin elimination from pollen mother cells (Kalinka et al. 2010), unequal chromosome division in somatic cells (Tang et al. 2012), and centromere loss of chromosome fragments (Hao et al. 2013).

The R genome may also be preferentially eliminated in some hybrids of hexaploid wheat × rye and result in the recovery of hexaploid wheat with the genome AABBDD. We called this process 'hybridization-based genome extraction' (HBGE). Hexaploid wheat may spontaneously appear in the self-pollinated progenies of octaploid triticale (Gupta and Priyadarshan 1982). In octaploid triticales, because many lines with univalents tend to revert back to hexaploid wheat, it was speculated that the univalents predominantly belong to rye chromosomes that are eliminated. In a recent study, hexaploid wheat was produced in the early generations of common wheat-rye hybrids (Yuan et al. 2014). A possible cytological mechanism for the quick loss of the rye R genome is if there is an equational + reductional pathway for gamete formation (Silkova et al. 2013). If equational division of wheat chromosomes and reductional division of rye chromosomes occurred, unreduced gametes for wheat and reduced gametes for rye will probably be produced from F1 hybrids. We herein refer to these wheat gametes as uniparental unreduced gametes to differentiate this phenomenon from biparental unreduced gametes. The unions of uniparental unreduced gametes are expected to result in partial F2 amphidiploids that contain all or most of the wheat chromosomes and monosomic rye chromosomes that are eliminated during selfing.

Although the recovery of hexaploid wheat by alien genome elimination prevents gene introgression, HBGE may be a useful tool to induce *de novo* genetic variation that broadens genetic diversity for wheat improvement and can be induced by the 'genome shock' that results from interspecific hybridization (McClintock 1984). As reviewed by Ma and Gustafson (2008), wheat hybrids with rye have higher rates of genomic change than wheat hybrids with some other related species. Yuan et al. (2014) obtained extracted bread wheat genomes with altered high-molecular-weight glutenin subunit alleles through wheat-rye hybridization. Moreover, new wheat lines produced via HBGE can be derived from the same F1 hybrid plant. These lines are desirable for studying the evolutionary biology of interspecific hybridization, because an F1 plant originates from the union of a female and male gamete; therefore, the variations in its derivatives can be attributed to distant hybridization.

Conclusions

Triticeae contains a large number of species with diverse genome constitutions, and interspecific hybridization of these species exhibits a wide variety of biological mechanisms that can affect the outcome of interspecific hybridization and thus contribute to species evolution. In practical application, the biological mechanisms of interspecific hybridization provide basic principles for genetic manipulation of wheat improvement. With the application of updated molecular approaches, artificial allopolyploids and interspecific hybrids will continue to be exploited in theoretical investigations of biological mechanisms related to interspecific hybridization. New knowledge obtained will in turn boost wheat breeding through interspecific hybridization. In the near future, re-synthesized hexaploid wheat will play an important role in high-yield breeding of common wheat through homologous recombination. The utilization of high-throughput molecular markers in identification of alien introgressions will accelerate screening a large number of progeny produced by interspecific hybridization to find and trace tiny introgressions. In future, more small introgressions are expected to be used as operational modules in wheat breeding.

Acknowledgements

We thank Prof. Chi Yen and Junliang Yang for the comments on generic treatment. This work was supported by in part by the NSFC (Nos. 31271723, 91331117, 31271716), 863 HiTech Program (Nos. 2011AA100103, 2011AA100104, 2012AA10A308), Technology Innovation of Chinese Academy of Sciences (XDA08030106), and CAAS Knowledge Innovation Program.

References

Akhunov, E.D., S. Sehgal, H. Liang, S. Wang, A.R. Akhunova, G. Kaur, W. Li, K.L. Forrest, D. See, H. Simková, Y. Ma, M.J. Hayden, M. Luo, J.D. Faris, J. Dolezel and B.S. Gill. 2013. Comparative analysis of syntenic genes in grass genomes reveals accelerated rates of gene structure and coding sequence evolution in polyploid wheat. Plant Physiol. 161: 252-265.

Badaeva E.D., O.S. Dedkova, G. Gay, V.A. Pukhalskyi, A.V. Zelenin, S. Bernard and M. Bernard. 2007. Chromosomal rearrangements in wheat: their types and distribution. Genome 50: 907-926.

Barkworth, M.E., S.W.L. Jacobs and H.Q. Zhang. 2009. *Connorochloa*: a new genus in Triticeae. Breed Sci. 59: 685-686.

Belyayev, A. and O. Raskina. 2013. Chromosome evolution in marginal populations of *Aegilops speltoides*: causes and consequences. Ann. Bot. (Lond.) 111: 531-538.

Bento, M., J.P. Gustafson, W. Viegas and M. Silva. 2011. Size matters in Triticeae polyploids: larger genomes have higher remodeling. Genome 54: 175-183.

Bhullar, R., R. Nagarajan, H. Bennypaul, G.K. Sidhu, G. Sidhu, S. Rustgi, D. Wettstein and K.S. Gill. 2014. Silencing of a metaphase I-specific gene results in a phenotype similar to that of the pairing homeologous I (*Ph1*) gene mutations. Proc. Natl. Acad. Sci. USA 39: 14187-14192.

Birchler, J.A. and R.A. Veitia. 2012. Gene balance hypothesis: connecting issues of dosage sensitivity across biological disciplines. Proc. Natl. Acad. Sci. USA 109: 14746-14753.

Boden, S.A., P. Langridge, G. Spangenberg and J.A. Able. 2009. *TaASY1* promotes homologous chromosome interactions and is affected by deletion of *Ph1*. Plant J. 57: 487-497.

Bothmer R. von and B. Salomon 1994. Triticeae: a tribe for food, feed and fun. pp. 1-12. *In*: R.R.C. Wang, K.B. Jensen and C. Jaussi (eds.). Proc. 2nd Intern. Triticeae Symp. Utah State University Publication Design and Production, Logan.

Caceres, M.E., F. Pupilli, M. Ceccarelli, P. Vaccino, V. Sarri, C. De Pace and P.G. Cionini. 2012. Cryptic introgression of *Dasypyrum villosum* parental DNA in wheat lines derived from intergeneric hybridization. Cytogenet. Genome Res. 136: 75-81.

Cai, X., S.S. Xu and X.W. Zhu. 2010. Mechanism of haploidy-dependent unreductional meiotic cell division in polyploidy wheat. Chromosoma 119: 275-285.

Caldwell, K.S., J. Dvorak, E.S. Lagudah, E. Akhunov, M.C. Luo, P. Wolters and W. Powell. 2004. Sequence polymorphism in polyploid wheat and their d-genome diploid ancestor. Genetics 167: 941-947.

Cao, A., L. Xing, X. Wang, X. Yang, W. Wang, Y. Sun, C. Qian, J. Ni, Y. Chen, D. Liu, X. Wang and P. Chen. 2011. Serine/threonine kinase gene Stpk-V, a key member of powdery mildew resistance gene *Pm21*, confers powdery mildew resistance in wheat. Proc. Natl. Acad. Sci. USA 108: 7727-7732.

Ceoloni, C. and P. Donini. 1993. Combining mutations for two homoeologous pairing suppressor genes *Ph1* and *Ph2* in common wheat and in hybrids with alien Triticeae. Genome 36: 377-386.

Chagué, V., J. Just, I. Mestiri, S. Balzergue, A.M. Tanguy, C. Huneau, V. Huteau, H. Belcram, O. Coriton, J. Jahier and B. Chalhoub. 2010. Genome-wide gene expression changes in genetically stable synthetic and natural wheat allohexaploids. New Phytol. 187: 1181-1194.

Chan, S.W.L. 2010. Chromosome engineering: power tools for plant genetics. Trends Biotechnol. 28: 650-610.

Chen, P.D., L.L. Qi, B. Zhou, S.Z. Zhang and D.J. Liu. 1995. Development and molecular cytogenetic analysis of wheat-*Haynaldia villosa* 6VS/6AL translocation lines specifying resistance to powdery mildew. Theor. Appl. Genet. 91: 1125-1128.

Chen, W.J., X. Fan, B. Zhang, B.L. Liu, Z.H. Yan, L.Q. Zhang, Y.L. Zheng, H.G. Zhang and D.C. Liu. 2012. Novel and ancient HMW glutenin genes from *Aegilops tauschii* and their phylogenetic positions. Genetic Resour. Crop Evol. 59: 1649-1657.

David, J.L., E. Benavente, C. Brès-Patry, J.C. Dusautoir and M. Echaide. 2004. Are neopolyploids a likely route for a transgene walk to the wild The *Aegilops ovate* × *Triticum turgidum* durum case. Biol. J. Linn. Soc. 82: 503-510.

d'Erfurth, I., L. Cromer, S. Jolivet, C. Girard, C. Horlow, M. Simon, E. Jenczewski and R. Mercier. 2010. The CYCLIN-A CYCA1; 2/TAM is required for the meiosis I to meiosis II transition and cooperates with OSD1 for the prophase to first meiotic division transition. PLoS Genet. 6: e1000989.

De Storme, N. and D. Geelen 2013. Sexual polyploidization in plants-cytological mechanisms and molecular regulation. New Phytol. 198: 670-684.

De Storme, N. and A. Mason. 2014. Plant speciation through chromosome instability and ploidy change: cellular mechanisms, molecular factors and evolutionary relevance. Curr. Plant Biol. 1: 10-33.

Devos, K.M., M.D. Atkinson, C.N. Chinoy, H.A. Francis, R.L. Harcourt, R.M.D. Koebner, C.J. Liu, P. Masoj, D.X. Xie and M.D. Gale. 1993. Chromosomal rearrangements in the rye genome relative to that of wheat. Theor. Appl. Genet. 85: 673-680.

Dewey, D.R. 1984. The genome system of classification as a guide to intergeneric hybridization with the perennial Triticeae. pp. 209-279. *In*: J.P. Gustafson (ed.) Gene Manipulation in Plant Improvement. Plenum Publishing, New York.

Dong, Y., X. Bu, Y. Luan, M. He and B. Liu. 2004. Molecular characterization of a cryptic wheat-*Thinopyrum intermedium* translocation line: evidence for genomic instability in nascent allopolyploid and aneuploid lines. Genet. Mol. Biol. 27: 237-241.

Dou, Q.W., H. Tanaka, N. Nakata and H. Tsujimoto. 2006. Molecular cytogenetic analyses of hexaploid lines spontaneously appearing in octoploid Triticale. Theor. Appl. Genet. 114: 41-47.

Dou, Q.W., Z.G. Chen, Y.A. Liu and H. Tsujimoto. 2009. High frequency of karyotype variation revealed by sequential FISH and GISH in plateau perennial grass forage *Elymus nutans*. Breed. Sci. 59: 651-656.

Dubcovsky, J. and J. Dvorak. 2007. Genome plasticity a key factor in the success of polyploid wheat under domestication. Science 316: 1862-1866.

Dvorak, J., E.D. Akhunov, A.R. Akhunov, K.R. Deal and M.C. Luo. 2006. Molecular characterization of a diagnostic DNA marker for domesticated tetraploid wheat provides evidence for gene flow from wild tetraploid wheat to hexaploid wheat. Mol. Biol. Evol. 23: 1386-1396.

Dvorak, J. and H.B. Zhang. 1990. Variation in repeated nucleotide sequences sheds light on the phylogeny of the wheat B and G genomes. Proc. Natl. Acad. Sci. USA 87: 9640-9644.

Eckardt, N.A. 2014. Genome dominance and interaction at the gene expression level in allohexaploid wheat. Plant Cell 26: 1834-1834.

Endo, T.R. 1990.Gametocidal chromosomes and their induction of chromosome mutations in wheat. Jpn. J. Genet. 65: 135-152.

Escobar, J.S., C. Scornavacca, A. Cenci, C. Guilhaumon, S. Santoni, E.J. Douzery, V. Ranwez, S. Glémin and J. David. 2011. Multigenic phylogeny and analysis of tree incongruences in Triticeae (Poaceae). BMC Evol. Biol. 11: 181.

Feldman, M., A. Levy, B. Chalhoub and K. Kashkush. 2012. Genomic plasticity in polyploid wheat. pp. 109-136. *In*: P.S. Soltis and D.E. Soltis (eds.). Polyploidy and Genome Evolution. Springer-Verlag, Berlin, Heidelberg.

Grant, V. 1981. Plant speciation. Columbia University Press, New York.

Greer, E., A.C. Martín, A. Pendle, I. Colas, A.M.E. Jones, G. Moore and P. Shaw. 2012. The *Ph1* locus suppresses Cdk2-type activity during premeiosis and meiosis in wheat. Plant Cell 24: 152-162.

Gupta, P.K. and P.M. Priyadarshan. 1982. Triticale, present status and future prospects. Adv. Genet. 21: 255-345.

Han, H.M., L. Bai, J.J. Su, J.P. Zhang, L.Q. Song, A.N. Gao, X.M. Yang, X.Q. Li, W.H. Liu and L.H. Li. 2014. Genetic rearrangements of six wheat-*Agropyron cristatum* 6P addition lines revealed by molecular markers. PLoS ONE 9: e91066.

Hao, M., J. Luo, M. Yang, L. Zhang, Z. Yan, Z. Yuan, Y. Zheng, H. Zhang and D. Liu. 2011. Comparison of homoeologous chromosome pairing between hybrids of wheat genotypes Chinese Spring *ph1b* and Kaixian-luohanmai with rye. Genome 54: 959-964.

Hao, M., J. Luo, L. Zhang, Z. Yuan, Y. Yang, M. Wu, W. Chen, Y. Zheng, H. Zhang and D. Liu. 2013. Production of hexaploid triticale by a synthetic hexaploid wheat-rye hybrid method. Euphytica 193: 347-357.

Hao, M., J. Luo, D. Zeng, L. Zhang, S. Ning, Z. Yuan, Z. Yan, H. Zhang, Y. Zheng, C. Feuillet, F. Choulet, Y. Yen, L. Zhang and D. Liu. 2014. *Qtug.sau-3B* is a major quantitative trait locus for wheat hexaploidization. G3: Genes, Genomes, Genetics 4: 1943-1953.

Hegde, S.G. and J.G. Waines. 2004. Hybridization and introgression between bread wheat and wild and weedy relatives in North America. Crop Sci. 44: 1145-1155.

Henikoff, S. and Y. Dalal. 2005. Centromeric chromatin: what makes it unique? Curr. Opin. Genet. Dev. 15: 177-184.

Huang, S., A. Sirikhachornkit, X. Su, J.D. Faris, B.S. Gill, R. Haselkorn and P. Gornicki. 2002. Genes encoding plastid acetyl-CoA carboxylase and 3-phosphoglycerate kinase of the *Triticum/Aegilops* complex and the evolutionary history of polyploid wheat. Proc. Natl. Acad. Sci. USA 99: 8133-8138.

IWGSC, International Wheat Genome Sequencing Consortium. 2014. A chromosome-based draft sequence of the hexaploid bread wheat (*Triticum aestivum*) genome. Science 345: 1251788.

Jauhar, P.P. 2007. Meiotic restitution in wheat polyhaploid (amphihaploids): a potent evolutionary force. J. Hered. 98: 188-193.

Jia, J., S. Zhao, X. Kong, Y. Li, G. Zhao, W. He, R. Appels, M. Pfeifer, Y. Tao, X. Zhang, R. Jing, C. Zhang, Y. Ma, L. Gao, C. Gao, M. Spannagl, K.F.X. Mayer, D. Li, S. Pan, F. Zheng, Q. Hu, X. Xia, J. Li, Q. Liang, J. Chen, T. Wicker, C. Gou, H. Kuang, G. He, Y. Luo, B. Keller, Q. Xia, P. Lu, J. Wang, H. Zou, R. Zhang, J. Xu, J. Gao, C. Middleton, Z. Quan, G. Liu, J. Wang, International Wheat Genome Sequencing Consortium, H. Yang, X. Liu, Z. He, L. Mao and J. Wang. 2013. *Aegilops tauschii* draft genome sequence reveals a gene repertoire for wheat adaptation. Nature 496: 91-95.

Kalinka, A., M. Achrem and S.M. Rogalska. 2010. Cytomixis-like chromosomes/chromatin elimination from pollen mother cells (PMCs) in wheat-rye allopolyploids. Nucleus 53: 69-83.

Kang, H.Y., H.Q. Zhang, Y. Wang, Y. Jiang, H.J. Yuan and Y.H. Zhou. 2008. Comparative analysis of the homoeologous pairing effects of *phKL* gene in common wheat × *Psathyrostachy huashanica* Keng ex Kuo. Cereal Res. Comm. 36: 429-440.

Kawshara, T. and S. Taketa. 2000. Fixation of translocation 2A-4B infers the monophyletic of Ethiopian tetraploid wheat. Theor. Appl. Genet. 101: 705-710.

Kellogg, E.A., R. Appels and R.J. Mason-Gamer. 1996. When gene trees tell different stories: The diploid genera of Triticeae. Syst. Bot. 21: 312–347.

Kihara, H. 1930. Genomanalyse bei *Triticum* und *Aegilops*. Cytologia 1: 263-284.

Kihara, H. 1944. Discovery of the DD-analyser, one of the ancestors of *Triticum vulgare*. Agric. Hortic. (Tokyo) 19: 889–890.

Kilian, B., K. Mammen, E. Millet, R. Sharma, A. Gamer, F. Salamini, K. Hammer and H. Ozkan. 2011. pp. 1-76. *Aegilops*. *In*: C. Kole (ed). Wild crop relatives: genomic and breeding resources. Cereals. Springer-Verlag, Berlin, Heidelberg.

Kimber, G. and Y. Yen. 1988. Analysis of pivotal-differential evolutionary patterns. Proc. Natl. Acad. Sci. USA 85: 9106-9108.

Kuraparthy V., P. Chhuneja, H.S. Dhaliwal, S. Kaur, R.L. Bowden and B.S. Gill. 2007. Characterization and mapping of cryptic alien introgression from *Aegilops geniculata* with new leaf rust and stripe rust resistance genes *Lr57* and *Yr40* in wheat. Theor. Appl. Genet. 114: 1379-1389.

Li, A., D. Liu, J. Wu, X. Zhao, M. Hao, S. Geng, J. Yan, X. Jiang, L. Zhang, J. Wu, L. Yin, R. Zhang, L. Wu, Y. Zheng and L. Mao. 2014a. mRNA and small RNA transcriptomes reveal insights into dynamic homoeolog regulation of allopolyploid heterosis in nascent hexaploid wheat. Plant Cell 26: 1878-1900.

Li, A.L., S.F. Geng, L.Q. Zhang, D.C. Liu and L. Mao. 2015. Making the bread: insights from newly synthesized allohexaploid wheat. Mol. Plant. 8: 847-859.

Li, L.H., Y.S. Dong, R.H. Zhou, X.Q. Li and P. Li. 1995. Cytogenetics and self-fertility of hybrids between *Triticum aestivum* L. and *Agropyron cristatum* (L.) Gaertn. Chinese. J. Genet. 22: 105-112.

Li, J., H.S. Wan and W.Y. Yang. 2014b. Synthetic hexaploid wheat enhances variation and adaptive evolution of bread wheat in breeding processes. J. Syst. Evol. 52: 735-742.

Li, Z., B. Li and Y.P. Tong. 2009. The contribution of distant hybridization with decaploid *Agropyron elongatum* to wheat improvement in China. J. Genet Genomics 35: 451-56.

Ling, H.Q. et al. 2013. Draft genome of the wheat A-genome progenitor *Triticum urartu*. Nature 496: 87-90.

Liu, C.J., M.D. Atkinson, C.N. Chinoy, K.M. Devos and M.D. Gale. 1992. Nonhomoeologous translocations between group 4, 5 and 7 chromosomes within wheat and rye. Theor. Appl. Genet. 83: 305-312.

Liu, D., H. Zhang, L. Zhang, Z. Yuan, M. Hao and Y. Zheng. 2014. Distant hybridization: a tool for interspecific manipulation of chromosomes. pp. 25-42. *In*: A. Pratap and J. Kumar (eds). Alien Gene Transfer in Crop Plants, Vol. 1. Springer, New York.

Löve, A. 1984. Conspectus of the Triticeae. Feddes Repertorium 95: 425-521.

Lukaszewski, A.J. and J.P. Gustafson. 1983. Translocations and modifications of chromosomes in *Triticale* x wheat hybrids. Theor. Appl. Genet. 64: 239-248.

Lukaszewski, A.J. and J.P. Gustafson. 1987. Cytogenetics of triticale. pp. 41-93. *In*: J. Janick (ed). Plant Breeding Reviews, Vol. 5. AVI Publishing, New York.

Luo, J., M. Hao, L. Zhang, J. Chen, L. Zhang, Z. Yuan, Z. Yan, Y. Zheng, H. Zhang, Y. Yen and D. Liu. 2012. Microsatellite mutation rate during allohexaploidization of newly resynthesized wheat. Int. J. Mol. Sci. 13: 12533-12543.

Luo, M.C., K.R. Deala, E.D. Akhunova, A.R. Akhunovaa, O.D. Andersonb, J.A. Andersonc, N. Blaked, M.T. Clegge, D. Coleman-Derrb, E.J. Conleyc, C.C. Crossmanb, J. Dubcovskya, B.S. Gillf, Y.Q. Gub, J. Hadamf, H.Y. Heod, N. Huob, G. Lazob, Y. Maa, D.E. Matthewsg, P.E. McGuirea, P.L. Morrelle, C.O. Qualseta, J. Renfrob, D. Tabanaoc, L.E. Talbertd, C. Tiana, D.M. Tolenoe, M.L. Warburtonh, F.M. Youb, W. Zhanga and J. Dvoraka. 2009. Genome comparisons reveal a dominant mechanism of chromosome number reduction in grasses and accelerated genome evolution in Triticeae. Proc. Natl. Acad. Sci. USA 106: 15780-15785.

Ma, J., J. Stiller, Y. Wei, Y.L. Zheng, K.M. Devos, J. Doleel and C. Liu. 2014. Extensive pericentric rearrangements in the bread wheat (*Triticum aestivum* L.) genotype 'Chinese Spring' revealed from chromosome shotgun sequence data. Genome Biol. Evol. 6: 3039-3048.

Ma, X.F. and J.P. Gustafson. 2008. Allopolyploidization-accommodated genomic sequence changes in Triticale. Ann. Bot. 101: 825-832.

Mahelka, V., D. Kopecky and L. Patová. 2011. On the genome constitution and evolution of intermediate wheatgrass (*Thinopyrum intermedium*: Poaceae, Triticeae). BMC Evol. Biol. 11: 127.

Marcussen, T., S.R. Sandve, L. Heier, M. Spannagl, M. Pfeifer, IWGSC, K.S. Jakobsen, B.B.H. Wulff, B. Steuernagel, K.F.X. Mayer and O.A Olsen. 2014. Ancient hybridizations among the ancestral genomes of bread wheat. Science 345: 1250092.

Martin A.C., P. Shaw, D. Phillips, S. Reader and G. Moore. 2014. Licensing MLH1 sites for crossover during meiosis. Nature Commun. 5: 4580.

Martinez-Perez, E. and G. Moore 2008. To check or not to check? The application of meiotic studies to plant breeding. Curr. Opin. Plant Biol. 11: 222-227.

Martis, M.M., R. Zhou, G. Haseneyer, T. Schmutzer, J. Vrána, M. Kubaláková, S. Knig, K.G. Kugler, U. Scholz, B. Hackauf, V. Korzun, C.C. Schn, J. Doleel, E. Bauer, K.F.X. Mayer and N. Stein. 2013. Reticulate evolution of the rye genome. Plant Cell 25: 3685-3698.

Mason-Gamer, R.J. 2004. Reticulate evolution, introgression, and intertribal gene capture in an allohexaploid grass. Syst. Biol. 53: 25-37.

Mason-Gamer, R.J. 2008. Allohexaploidy, introgression, and the complex phylogenetic history of *Elymus repens* (Poaceae). Mol. Phylogenet. Evol. 47: 598-611.

Matsuoka, Y. and S. Nasuda. 2004. Durum wheat as a candidate for the unknown female progenitor of bread wheat: an empirical study with a highly fertile F1 hybrid with *Aegilops tauschii* Coss. Theor. Appl. Genet. 109: 1710-1717.

McClintock, B. 1984. The significance of responses of the genome to challenge. Science 226: 792-801.

McFadden, E.S. and E.R. Sears. 1944. The artificial synthesis of *Triticum spleta*. Rec. Genet. Soc. Am. 13: 26-27.

Mestiri, I., V. Chagué, A.M. Tanguy, C. Huneau, V. Huteau, H. Belcram, O. Coriton, B. Chalhoub and J. Jahier. 2010. Newly synthesized wheat allohexaploids display progenitor-dependent meiotic stability and aneuploidy but structural genomic additivity. New Phytol. 186: 86-101.

Miller, T.E., S.M. Reader, P.J. Shaw and G. Moore 1998. Towards an understanding of the biological action of the *Ph1* locus in wheat. pp. 17-19. *In*: A.E. Slinkard (ed). Proc. 9th Intern. Wheat Genet. Symp. Vol. 1. University Extension Press, Saskatoon.

Millet, E., J. Manisterski, P. Ben-Yehuda, A. Distelfeld, J. Deek, A. Wan, X. Chen and B.J. Steffenson. 2014. Introgression of leaf rust and stripe rust resistance from Sharon goatgrass (*Aegilops sharonensis* Eig) into bread wheat (*Triticum aestivum* L.). Genome 57: 309-316.

Mujeeb-Kazi, A., V. Rosas and S. Roldan. 1996. Conservation of the genetic variation of *Triticum tauschii* (Coss.) Schmalh. (*Aegilops squarrosa* auct. non L.) in synthetic hexaploid wheats (*T. turgidum* L. s.lat. × *T. tauschii*; $2n = 6x = 42$, AABBDD) and its potential utilization for wheat improvement. Genet. Resour. Crop Evol. 43: 129-134.

Murat, F., C. Pont and J. Salse. 2014. Paleogenomics in Triticeae for translational research. Curr. Plant Biol. 1: 34-39.

Nakamura, I., B. Rai, H. Takahashi, K. Kato, Y. I. Sato and T. Komatsuda. 2009. *Aegilops* section *Sitopsis* species contains the introgressive *PolA1* gene with a closer relationship to that of *Hordeum* than *Triticum-Aegilops* species. Breed. Sci. 59: 602-610.

Ogbonnaya, F.C., O. Abdalla, A. Mujeeb-Kazi, A.G. Kazi, S.S. Xu, N. Gosman, E.S. Lagudah, D. Bonnett, M.E. Sorrells and H. Tsujimoto. 2013. Synthetic hexaploids: Harnessing species of the primary gene pool for wheat improvement. Plant Breed. Rev. 37: 35-122.

Petersen, G., O. Seberg, M. Yde and K. Berthelsen. 2006. Phylogenetic relationships of *Triticum* and *Aegilops* and evidence for the origin of the A, B, and D genomes of common wheat (*Triticum aestivum*). Mol. Phylogenet. Evol. 39: 70-82.

Pont, C., F. Murat, S. Guizard, R. Flores, S. Foucrier, Y. Bidet, U.M. Quraishi, M. Alaux, J. Doleel, T. Fahima, H. Budak, B. Keller, S. Salvi, M. Maccaferri, D. Steinbach, C. Feuillet, H. Quesneville and J. Salse. 2013. Wheat syntenome unveils new evidences of contrasted evolutionary plasticity between paleo- and neodupli-cated subgenomes. Plant J. 76: 1030-1044.

Prieto, P., P. Shaw and G. Moore 2004. Homologue recognition during meiosis is associated with a change in chromatin structure. Nat. Cell Biol. 6: 906-908.

Qi, L., B. Friebe and B.S. Gill. 2006. Complex genome rearrangements reveal evolutionary dynamics of pericentromeric regions in the Triticeae. Genome 49: 1628-1639.

Qi, L.L., B. Friebe, P. Zhang and B.S. Gill. 2007. Homoeologous recombination, chromosome engineering and crop improvement. Chromosome Res. 15: 3-19.

Qi, L.L., M.O. Pumphrey, B. Friebe, P. Zhang, Q. Chen, R.L. Bowden, M.N. Rouse, Y. Jin and B.S. Gill. 2011. A novel Robertsonian translocation event leads to transfer of a stem rust resistance gene (*Sr52*) effective against race Ug99 from *Dasypyrum villosum* into bread wheat. Theor. Appl. Genet. 123: 159-167.

Ravi, M. and S.W.L. Chan. 2010. Haploid plants produced by centromere-mediated genome elimination. Nature 464: 615-618.

Ressurreição, F., A. Barão, W. Viegas and D. Margarida. 2012. Haploid independent unreductional meiosis in hexaploid wheat. pp. 321-330. *In*: A. Swan (ed). Meiosis-Molecular Mechanisms and Cytogenetic Diversity. InTech Press, Dublin.

Ribeiro-Carvalho, C., H. Guedes-Pinto, G. Igrejas, P. Stephenson, T. Schwarzacher and J.S. Heslop-Harrison. 2004. High levels of genetic diversity throughout the range of the Portuguese wheat landrace 'Barbela'. Ann. Bot. 94: 699-705.

Sánchez-Moran, E., E. Benavente and J. Orellana. 2001. Analysis of karyotypic stability of homoeologous pairing (*ph*) mutants in allopolyploid wheat. Chromosoma 110: 371-377.

Sanei, M., R. Pickering, K. Kumke, S. Nasuda and A. Houben. 2011. Loss of centromeric histone H3 (CENH3) from centromeres precedes uniparental chromosome elimination in interspecific barley hybrids. Proc. Natl. Acad. Sci. 108: 13373-13374.

Scoles, G., Q. Wang, J. Xiang, A. Gao, X. Yang, W. Liu, X. Li and L. Li. 2010. Analysis of chromosomal structural polymorphisms in the St, P, and Y genomes of Triticeae (Poaceae). Genome 53: 241-249.

Sears, E.R. 1976. Genetic control of chromosome pairing in wheat. Annu. Rev. Genet. 10: 31-51.

Silkova, O.G., I.G. Adonina, E.A. Krivosheina, A.I. Shchapova and V.K. Shumny. 2013. Chromosome pairing in meiosis of partially fertile wheat/rye hybrids. Plant Reprod. 26: 33-41.

Sun, G.L. 2014. Molecular phylogeny revealed complex evolutionary process in *Elymus* species. J Syst. Evol. 52: 706-711.

Sutton, T., R. Whitford, U. Baumann, C. M. Dong, J.A. Able and P. Langridge. 2003. The *Ph2* pairing homoeologous locus of wheat (*Triticum aestivum*): identification of candidate meiotic genes using a comparative genetics approach. Plant J. 36: 443-456.

Tang, Y.L., C.S. Li, X.L. Wu, C. Wu, W.Y. Yang, G. Huang and X.L. Ma. 2014. Accumulation of dry matter, canopy structure and photosynthesis of synthetic hexaploid wheat-derived high yielding varieties grown in Sichuan basin, China. Scientia Agricultura Sinica 47: 844-855. (in Chinese)

Tang, Z., S. Fu, B. Yang, H. Zhang and Z. Ren. 2012. Unequal chromosome division and inter-genomic translocation occurred in somatic cells of wheat-rye allopolyploid. J. Plant Res. 125: 283-290.

Tang, Z.X., Z.J. Yang, S.L. Fu, M.Y. Yang, G.R. Li, H.Q. Zhang, F.Q. Tan and Z. Ren. 2011. A new long terminal repeat (LTR) sequence allows the identify J genome from Js and St genomes of *Thinopyrum intermedium*. J. Appl. Genet. 52: 31-33.

Tiwari, V.K., N. Rawat, K. Neelam, S. Kumar, G.S. Randhawa and H.S. Dhaliwal. 2010. Random chromosome elimination in synthetic *Triticum-Aegilops* amphiploids leads to development of a stable partial amphiploid with high grain micro- and macronutrient content and powdery mildew resistance. Genome 53: 1053-1065.

Tomas, P.A., G.E. González, G.E. Schrauf and L. Poggio. 2012. Chromosomal characterization in native populations of *Elymus scabrifolius* from Argentina through classical and molecular cytogenetics (FISH–GISH). Genome 55: 591-598.

Villareal, R.L., O. Banuelos, A. Mujeeb-Kazi and S. Rajaram. 1998. Agronomic performance of chromosome 1B and T1BL.1RS, near-isolines in the spring bread wheat Seri M82. Euphytica 103: 195-202.

Wang, C.J., L.Q. Zhang, S.F. Dai, Y.L. Zheng, H.G. Zhang and D.C. Liu. 2010. Formation of unreduced gametes is impeded by homologous chromosome pairing in tetraploid *Triticum turgidum* x *Aegilops tauschii* hybrids. Euphytica 175: 323-329.

Wang, Q., H. Liu, A. Gao, X. Yang, W. Liu, X. Li and L. Li. 2012. Intergenomic rearrangements after polyploidization of *Kengyilia thoroldiana* (Poaceae: Triticeae) affected by environmental factors. PLoS One 7: e31033.

Wang, R.R.C. 2011. *Agropyron* and *Psathyrostachys*. pp. 77-108. *In*: Kole C (ed). Wild Crop Relatives: Genomic and Breeding Resources. Cereals. Springer-Verlag, Berlin Heidelberg.

Wang, R.R.C., R. von Bothmer, J. Dvorák, G. Fedak, I. Linde-Laursen and M. Muramatsu. 1994. Genome symbols in the Triticeae (Poaceae). pp. 29-34. *In*: R.R.C. Wang, K.B. Jensen and C. Jaussi (eds.). Proc. 2nd Intern. Triticeae Symp. Utah State University Publication Design and Production, Logan.

Wang, R.R. and B. Lu. 2014. Biosystematics and evolutionary relationships of perennial Triticeae species revealed by genomic analyses. J. Syst. Evol. 52: 697-705.

Weissmann S., M. Feldman and J. Gressel. 2005. Sequence evidence for sporadic intergeneric DNA introgression from wheat into a wild *Aegilops* species. Mol. Biol. Evol. 22: 2055-2062.

Xie, Q., H. Kang, S. Tao, D.L. Sparkes, X. Fan, Z. Cui, L. Xu, J. Huang, X. Fan, L. Sha, H. Zhang and Y.H. Zhou. 2012. Wheat lines derived from trigeneric hybrids of wheat-rye-*Psathyrostachys huashanica*, the potential resources for grain weight improvement. Aust. J. Crop. Sci. 6: 1550-1557.

Xu, S.J. and L.R. Joppa. 2000. First-division restitution in hybrids of Langdon durum disomic substitution lines with rye and *Aegilops squarrosa*. Plant Breed. 119: 233-241.

Yang, W., D. Liu, J. Li, L. Zhang, H. Wei, X. Hu, Y. Zheng, Z. He and Y. Zou 2009. Synthetic hexaploid wheat and its utilization for wheat genetic improvement in China. J. Genet. Genomics 36: 539-546.

Yang, Y., L. Zhang, Z. Yan, Y. Zheng and D. Liu. 2011. The cytological instability of neoallopolyploids suggesting a potent way for DNA introgression: The case of synthetic hexaploid wheat × *Aegilops peregrine*. African J. Agric. Res. 6: 1692-1697.

Yang, Y.W., L.Q. Zhang, Y. Yen, Y.L. Zheng and D.C. Liu. 2010. Cytological evidence on meiotic restitution in pentaploid F1 hybrids between synthetic hexaploid wheat and *Aegilops variabilis*. Caryologia 63: 354-358.

Yang, Z.J., C. Liu, J. Feng, G.R. Li, J.P. Zhou, K.J. Deng and Z.L. Ren. 2006. Studies on genome relationship and species-specific PCR marker for *Dasypyrum breviaristatum* in Triticeae. Hereditas 143: 47-54.

Yen, Y., P.S. Baenziger and R. Morris. 1996. Genomic constitution of bread wheat: current status. pp. 359-373. *In*: P.P. Jauhar (ed). Methods of Genome Analysis in Plants. CRC Press, Boca Raton.

Yen, C. and J.L. Yang. 2009. Historical review and prospect of taxonomy of tribe Triticeae Dumortier (Poaceae). Breed Sci. 59: 513-518.

Yen, C. and J.L. Yang. 2013a. Biosystematics of Triticeae. Vol. I. *Triticum-Aegilops* complex (Version II). Chinese Agricultural Press, Beijing. (in Chinese)

Yen, C. and J.L. Yang. 2013b. Biosystematics of Triticeae. Vol. II. Genera: *Secale, Tritiosecale, Pseudosecale, Eremopyrum, Henrardia, Taeniantherum, Heteranthelium, Crithopsis*, and *Hordeum* (Version II). Chinese Agricultural Press, Beijing. (in Chinese)

Yen, C. and J.L. Yang 2013c. Biosystematics of Triticeae. Vol. IV. Genera: *Stenostachys, Psathyrostachys, Leymus, Pseudoroegneria,* and *Roegeneria.* Chinese Agricultural Press, Beijing. (in Chinese)

Yen, C. and J.L. Yang 2013d. Biosystematics of Triticeae. Vol. V. Genera: *Campeiostachys, Elymus,Pascopyrum, Lophopyrum, Trichopyrum, Hordelymus, Festucopsis, Peridictyon,* and *Psammopyrum.* Chinese Agricultural Press, Beijing. (in Chinese)

Yen, C., J.L. Yang and B.R. Baum. 2013e. Biosystematics of Triticeae. Vol. III. Genera: *Kengyilia, Douglasdeweya, Agropyron, Australopyrum,* and *Anthosachne* (Version II). Chinese Agricultural Press, Beijing. (in Chinese)

Yen, C., J.L. Yang and Y. Yen. 2005. Hitoshi Kihara, Askell Löve, and the modern genetic concept of the genera in the tribeTriticeae (Poaceae). Acta Phytotaxonomic Sinica 43: 82-93.

Yousafzai, F.K., N. Al-Kaff and G. Moore. 2010. Structural and functional relationship between the *Ph1* locus protein 5B2 in wheat and CDK2 in mammals. Funct. Integr. Genomics 10: 157-166.

Yuan, Z., M. Liu, Y. Ouyang, X. Zeng, M. Hao, L. Zhang, S. Ning, Z. Yan and D. Liu. 2014. The detection of a de novo allele of the *Glu-1Dx* gene in wheat-rye hybrid offspring. Theor. Appl. Genet. 127: 2173-2182.

Zaharieva, M. and P. Monneveux. 2006. Spontaneous hybridization between bread wheat (*Triticum aestivum* L.) and its wild relatives in Europe. Crop Sci. 46: 512-527.

Zeng, D.Y., M. Hao, J.T. Luo, L.Q. Zhang, Z.W. Yuan, S.Z. Ning, Y.L. Zheng and D.C. Liu. 2014. Amphitelic orientation of centromeres at metaphase I is an important feature for univalent-dependent meiotic nonreduction, J. Genet. 93: 531-534.

Zhang, H.K., Y. Bian, X.W. Gou, Y.Z. Dong, S. Rustgi, B.J. Zhang, C.M. Xu, N. Li, B. Qi, F.P. Han, D. von Wettstein and B. Liu. 2013. Intrinsic karyotype stability and gene copy number variations may have laid the foundation for tetraploid wheat formation. Proc. Natl. Acad. Sci. USA. 110: 19466-19471.

Zhang, L.Q., Y. Yen, Y.L. Zheng and D.C. Liu. 2007. Meiotic restriction in emmer wheat is controlled by one or more nuclear genes that continue to function in derived lines. Sex. Plant Reprod. 20: 159-166.

Zhang, L.Q., Z.H. Yan, S.F. Dai, Q.J. Chen, Z.W. Yuan, Y.L. Zheng and D.C. Liu. 2008. The crossability of *Triticum turgidum* with *Aegilops tauschii*. Cereal Res. Commun. 37: 417-427.

Zhang, L.Q., D.C. Liu, Y.L. Zheng, Z.H. Yan, S.F. Dai, Y.F. Li, Q. Jiang, Y.Q. Ye and Y. Yen. 2010. Frequent occurrence of unreduced gametes in *Triticum turgidum-Aegilops tauschii* hybrids. Euphytica 172: 285-294.

Zhao, N., L. Xu, B. Zhu, M. Li, H. Zhang, B. Qi, C. Xu, F. Han and B. Liu. 2011. Chromosomal and genome-wide molecular changes associated with initial stages of allohexaploidization in wheat can be transit and incidental. Genome 54: 692-699.

Chapter 3

Potato Breeding through Ploidy Manipulations

Domenico Carputo[1,*] and Riccardo Aversano[2]

ABSTRACT

The cultivated potato (*Solanum tuberosum* L.) is a clonally propagated autotetraploid ($2n = 4x = 48$) that represents the third most important crop in the world. Breeding efforts are complicated by some typical genetic features including, among the others, tetrasomic inheritance, chromatid segregation, high levels of heterozygosity and multiple allelism. However, unique reproductive characteristics of cultivated and wild potatoes may facilitate breeders' work: the production of 2n gametes; the presence of an endosperm dosage system that predicts success of interspecific crosses; the possibility to easily extract haploids. After illustrating the genetics and reproductive characteristics of this important autotetraploid and the significance of its wild relatives, our chapter describes main breeding targets and reviews results obtained in manipulating whole chromosome sets, from the monoploid to the hexaploid level. Finally, the perspectives of molecular breeding are discussed.

Introduction

The cultivated potato (*Solanum tuberosum* L.) belongs to the highly diversified genus *Solanum* and represents the third most important food crop in the world after rice and wheat. It originated in the Andes of South America, where it has been cultivated for thousands of years. The Spanish introduced it to Europe in the second part of the XVI century, and from the Old Continent it then spread all over the world. The potato is a very versatile

[1] University of Naples Federico II, Department of Agricultural Sciences, Via Università 100, 80055 Portici, Italy, E-mail: carputo@unina.it
[2] University of Naples Federico II, Department of Agricultural Sciences, Via Università 100, 80055 Portici, Italy, E-mail: raversan@unina.it
* Corresponding author

crop, and as such is cultivated worldwide for either food, feed or industrial materials in more than 19 million ha (2013 data, www.faostat.fao.org), with New Zealand and the United States of America being the leaders in terms of tuber yield (more than 50 t/ha). Significantly, in the last 30 years, potato production in developing countries has risen to such a level that now about a third of global production takes places in these countries. Among cultivated plants, the potato is one of the most important contributors in terms of starch calories, and several antioxidants. On a dry weight basis, its protein content is very similar to that of cereals. In addition, it is also a very good source of fiber, vitamin C and elements such as Cu, K, P and Fe (Bradeen and Haynes 2011).

The cultivated potato is a clonally propagated autotetraploid ($2n = 4x = 48$). It can reproduce also sexually as an outbreeder with crossbreeding, therefore open pollination results in a mixture of seeds deriving from both cross- and self- pollinations. While diploid potatoes are generally self-incompatible, *S. tuberosum* and other polyploid potatoes are self-compatible, due to a phenomenon referred to as "competition interaction" because certain interactions that either weaken or suppress the incompatibility reaction take place in pollen grains carrying different S alleles (Frankel and Galun 1977). Being an autotetraploid, the cultivated potato possesses four copies of homologous chromosomes. Therefore, at a given locus, two homozygous (aaaa = nulliplex, AAAA = quadruplex) and three heterozygous (Aaaa = simplex, AAaa = duplex, AAAa = triplex) genotypes are possible based on the number of dominant alleles. Since there could be as many as four alleles per locus, three non-additive genetic effects are possible, i.e. first order (between two different alleles, e.g. A1A1A2A2), second order (between three different alleles, e.g. A1A1A2A3) and third order (between four different alleles, e.g. A1A2A3A4). In the tetrallelic A1A2A3A4 genotype this corresponds to 11 heteroallelic interactions. Multiple allelism in potato has been demonstrated for genes affecting either metabolic pathways or phenotypic traits. Van der Wall et al. (2001) found at least eight different alleles coding for granule-bound starch synthase I, responsible for amylose biosynthesis. Through SNP genotyping, Achenbach et al. (2009) demonstrated the existence of multiple alleles either at a single locus or at various linked loci controlling resistance to *Globodera pallida*. Recently, Obidiegwu et al. (2015) found multiple resistance alleles with different pathotype specificities at a major locus conferring resistance to *Synchytrium endobioticum*.

The characteristics above described, typical of any autotetraploid, make potato genetics quite complicated when compared to that of a diploid plant. There are also additional noteworthy constraints. In fact, depending on gene-centromere distance, alternative segregation patterns in the gametes are possible. Besides normal chromosome segregation, the possibility exists that portions of two sister chromatids merge in the same gamete. This occurs when recombination takes place between a given locus and the centromere. Due to double reduction, a triplex AAAa individual can produce aa gametes, with a frequency proportional to the distance between the locus and the centromere. In addition, as any clonal crop, the cultivated potato is highly heterozygous. Consequently, segregation of traits is expected following hybridization. Homozygous genotypes are difficult to obtain upon selfing due to inbreeding depression. This largely depends on the retention of deleterious and dysfunctional alleles. Finally, the cultivated potato has a narrow genetic base compared with that of its centre of origin and of wild species. This is mainly due to the limited number of adapted genotypes that were used to develop the *S. tuberosum* form of the cultivated potato (Bradshaw and Ramsay 2005). Additional causes of genetic erosion may be attributed to the well known epidemics of late blight that caused the great Irish famine and to the decreased fertility over time of clonally propagated varieties.

The highly heterozygous genome of the cultivated potato as well as its polyploid nature make *S. tuberosum* recalcitrant to current sequencing technologies and bioinformatics programs. However, recently the Potato Genome Sequencing Consortium (2011) sequenced roughly 10 percent of the heterozygous diploid clone RH89-039-16 (referred as to RH) using bacterial artificial chromosome (BAC) clones that were selected from the AFLP fingerprint physical map of this genotype. Even with a diploid genome, the degree of heterozygosity between the two homologous chromosomes was a challenge in constructing a complete minimal tiling path. Therefore, the PGSC solved this problem by sequencing an homozygous doubled monoploid (DM 1-3 516 R44, referred as to DM) of *S. tuberosum* Group Phureja ($2n = 2x = 24$), in which there were only two copies of each chromosome and, more importantly, each copy was identical. Using a combination of Sanger, Illumina, and Roche 454 sequence reads, a 727 Mb assembly was built. The dynamic nature of the potato genome, and more specifically of the haplotypes, was supported by an in-depth analysis of 334 kb of the three haplotypes that revealed an extensive amount of diversity, as shown by copy-number variants, presence/absence variants, and mutations resulting in dysfunctional alleles. Regions with distorted segregations were also identified as blocks on several chromosomes, depending on the populations. Notably, the whole-genome sequence (WGS) assembly also provided evidence that the genome has undergone extensive genome duplications through evolution and that the expansion of particular gene families has contributed to evolution of this crop. In particular, the distance-transversion rate at fourfold degenerate sites (4DTv) method was used to analyze paralogous gene pairs between syntenic blocks. Two peaks (4DTv ~0.36 and ~1.0) supported two WGD events, which were both already inferred in *Vitis vinifera, Arabidopsis thaliana* and *Populus trichocarpa*. The ancient WGD corresponds to the palaeo-hexaploidization (γ) event that clearly occurred in the common ancestor of the rosids and asterids (Tang et al. 2008), albeit it probability preceded the mono–dicotyledon split, about 185±55 million years ago. The recent duplication could be placed at ~67 million years ago, consistent with the WGD that occurred near the Cretaceous – Tertiary boundary ~65 million years ago (Busch et al. 2011).

Potato Relatives and Significant Reproductive Characteristics

The potato is unrivalled within cultivated crops in terms of related germplasm. Potatoes are grouped in 19 taxonomic series within section Petota, with about 200 species distributed from the southern part of United States up to southern Chile. All together, potatoes form a polyploid series with a basic chromosome number (x) of 12. Series Tuberosa is the most numerous, including all the cultivated potatoes as well as the wild species closely related to them. Wild species occur as diploids, triploids, tetraploids, pentaploids, and hexaploids, with about 70% being diploid. By contrast, the cultivated species extend to the pentaploid level only. Based on morphological and molecular analyses, four species of cultivated potatoes are accepted in the latest taxonomic treatment (Huamán and Spooner 2002; Spooner et al. 2007), namely (1) *S. tuberosum*, with two cultivar-groups: (a) the Andigenum Group of upland Andean genotypes containing diploids, triploids and tetraploids, and (b) the Chilotanum Group of lowland tetraploid Chilean landraces, from which our modern cultivars arise (Ames and Spooner 2008); (2) *S. ajanhuiri* (diploid), which originated from hybridization between diploid cultivars of the *S. tuberosum* Andigenum Group and the diploid wild species *S. boliviense* (= *S. megistacrolobum*); (3) *S. juzepczukii* (triploid) formed by hybridization between diploid cultivars of *S. tuberosum* Andigenum Group and the tetraploid wild species *S. acaule*, and (4) *S. curtilobum* (pentaploid), likely formed by hybridization between tetraploid forms of *S. tuberosum* Andigenum Group and *S. juzepczukii*. The allopolyploid origins of *S. juzepczukii*, and *S. curtilobum* and the diploid hybrid origin for *S. ajanhuiri* was ascertained by Rodriguez et al. (2010) using DNA sequence data of the *waxy* gene. The existence of many relatives is very attractive from the breeding standpoint in that one can look for traits that are present only in low frequencies or that the commercial varieties lack. This is very important considering all the productivity, resistance and quality traits that a new potato variety must possess to be commercially successful. A few examples of potato species carrying resistances to environmental stresses are reported in Table 1.

The polyploid cultivated *S. tuberosum* and its relatives show three essential biological features that make it possible to change ploidy levels quite easily: the production of gametes with unreduced chromosome number; the presence of an endosperm dosage system that regulates success of interploidy/interspecific crosses; and the possibility to easily obtain haploid plants. The following part of this section provides details on these important features. Several genotypes of cultivated and wild potatoes produce 2*n* gametes. These had a fundamental role in the origin and population biology of polyploid potatoes, in that crosses involving both n and 2*n* gametes linked all ploidy levels of *Solanum* species, providing the opportunity for gene flow all the way from diploid to hexaploid species. 2*n* gametes are the result of meiotic defects affecting various nuclear and cytoplasmic aspects of sporogenesis (chromosome pairing, centromere

Table 1. Example of potato species with sources of resistance to environmental stresses. Resistance data from the Germplasm Resources Information Network (GRIN) database.

Species	Origin	Ploidy/EBN	Reported resistances
S. commersonii	ARG, BRA, URU	2*x*/1EBN	Frost; Early blight; PVX; PVM; Flea beetle
S. demissum	GUA, MEX	6*x*/4EBN	Blackleg; Bacterial wilt; Late blight; Wart; Heat; Frost
S. agrimonifolium	GUA, HON, MEX	4*x*/2EBN	Ringrot; Leaf hopper
Solanum albicans	ECU, PER	6*x*/4EBN	Potato aphid; *Meloidogyne chitwoodi*; *Helminthosporium solani*
Solanum berthaultii	ARG, BOL	2*x*/2EBN	Ring rot; Verticillium wilt; Colorado potato beeatle; Leaf Hopper; Blackroot nematode
Solanum boliviense	ARG, BOL, PER	2*x*/2EBN	Late blight; Early blight; PVM; PVS; Wart; Flea beetle; Leaf Hopper; Frost
Solanum bulbocastanum	GUA, HON, MEX	2*x*/1EBN	Late blight; Blackleg; Bacterial wilt; PLRV; Ring rot; Verticillium wilt; Flea beetle; Golden potato cyst nematode
Solanum chacoense	ARG, BOL, BRA, PAR, PER, URU	2*x*/2EBN	PLRV; Verticillium wilt; Colorado potato beeatle; PVA; PVF; Horsenettle cyst nematode;
Solanum colombianum	COL, ECU, PER,VEN	4*x*/2EBN	Frost; Ring rot; Verticillium wilt
Solanum hjertingii	MEX	4*x*/2EBN	Late blight; PVY; Potato aphid; Leaf hopper; *Meloidogyne hapla*
Solanum kurtianum	ARG	2*x*/2EBN	Leaf Hopper; PLRV; Ring rot; *Rhizoctonia solani*; Verticillium wilt
Solanum microdontum	ARG, BOL	2*x*/2EBN	Late blight; Verticillium wilt; Heat

division, spindle formation or cytokinesis). For a review on meiotic variants leading to 2*n* gamete formation in potato see Carputo et al. (2000). 2*n* pollen production in potatoes is generally due to a parallel orientation of spindles at anaphase II (Fig. 1). Among mutations giving rise to 2*n* egg production, omission of second meiotic division is the most common (Fig. 1). The result of this mutation is a dyad with two 2*n* megaspores, one of which becomes the functional spore. Other mutations leading to 2*n* egg formation may affect megasporogenesis. They may cause failure of cytokinesis after the second division, an irregular anaphase II and delayed meiotic divisions, and an irregular spindle axis formation at metaphase I.

Recent advances in the field of molecular biology allowed the discovery in *A. thaliana* of several genes causing meiotic defects. The first gene (*AtPS1*) implicated in the parallel orientation of spindles and 2*n* pollen production in *A. thaliana* was isolated and characterized at the molecular level by d'Erfurth et al. (2008). This gene encodes a protein of about 1500 amino

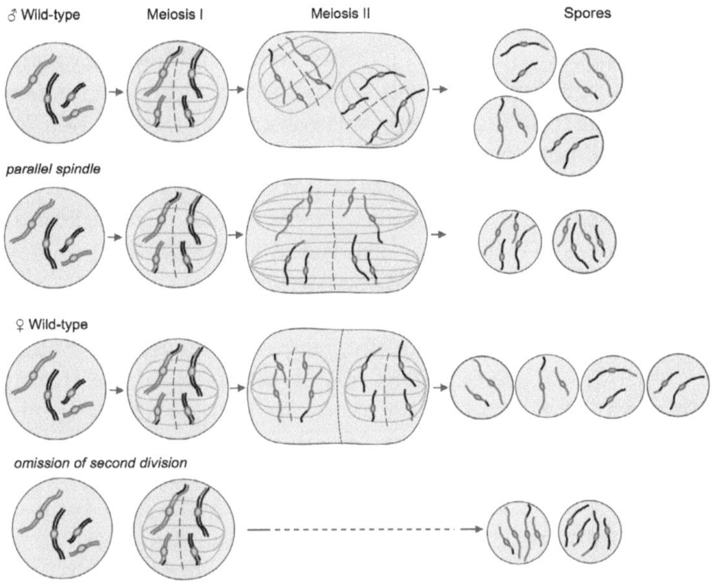

Figure 1. Schematic representation of chromosome behavior at meiosis in the wild type and in *parallel spindle* and *omission of second division* mutants producing 2*n* gametes. For simplification, only two pairs of chromosomes are represented. In the wild type micro- (♂) and macro-sporogenesis (♀), homologous chromosomes pair, recombine and then segregate to opposite spindle poles. Sister chromatids are separated at meiosis II, and haploid spores are formed. In *parallel spindle* mutant, recombined chromosomes that were separated during the first division are re-gathered into a single metaphase plate due to the parallel orientation of metaphase II spindles. During the *omission of second division* mutant meiosis, the first division occurs normally and the homologous chromosomes are separated, but lack of the second round of cell division enables sister chromatids to remain together in the same daughter cell, thus forming a dyad consisting of two 2*n* megaspores.

acids conserved throughout the plant kingdom, and it is a good candidate for the gene behind the *ps* locus in potato. Depending on the meiotic stage affected by the mutation, 2*n* gametes produced are genetically equivalent to two basic categories of nuclear restitution, first division restitution (FDR) and second division restitution (SDR) mechanisms (Peloquin et al. 1999). It should be pointed out that the genetic mode of 2*n* gamete formation and the cytological event are two separate entities. Thus, for example, the mutation parallel spindle occurs at the second meiotic division but it is equivalent to a FDR mechanism, whereas omission of the second meiotic division (occurring during first meiotic division) is equivalent to SDR. Progenies deriving from FDR and SDR 2*n* gametes are expected to be genetically different. For heterozygous loci in the parents, it has been estimated that with FDR mechanisms, all loci from the centromere to the first crossover will be heterozygous in the gamete, and 50% of loci between the first and the second crossover will be heterozygous in the gamete. If crossing over does not occur, then the intact genotype of the diploid parent will be incorporated into each $2n$ gamete. SDR gametes do not maintain the original genic combinations of the diploid parents. In fact, all heterozygous loci from the centromere to the first crossover will be homozygous in the gametes, and all loci between the first and the second crossover which are heterozygous in the parent, will be heterozygous in the gametes. FDR gametes are expected to strongly resemble each other and the parental clone they derive from. By contrast, SDR is expected to produce a heterogeneous population of highly homozygous gametes. It has been estimated that FDR $2n$ gametes transmit a high fraction of parental heterozygosity (roughly 75-80%), whereas the proportion of heterozygosity retained by the SDR mode of $2n$ gametogenesis is much lower (35-50%). Therefore, various levels of heterozygosity, variability and fitness are expected in polyploids obtained through sexual polyploidization. The genetic control of $2n$ gamete production has been generally attributed to the action of single recessive genes (Bretagnolle and Thompson 1995; Peloquin et al. 1999). These genes exhibit incomplete penetrance and variable expressivity, therefore their phenotypic expression may be significantly modified. This allows one plant to produce both n and $2n$ gametes, so that there can be continuous gene flow within and between ploidy levels.

Another important reproductive feature is the presence of a strong post-zygotic barrier to hybridization that causes endosperm breakdown. Normal endosperm development occurs when there is a balance of qualitative genetic factors (Endosperm Balance Number, EBN) between parents (Johnston et al. 1980). According to the model developed, each species has its own EBN, which is not necessarily a direct reflection of its ploidy. The EBN must be in a 2:1 maternal to paternal ratio in the hybrid endosperm for its normal development. Thus, successful interspecific hybridization occurs only when parents produce gametes with the same EBN (Fig. 2). Exceptions to the 2:1 ratio are possible, and inter-EBN crosses may be sometimes successful. This is probably the result of non-heritable random events such as multiple fertilizations of the central cell, mitotic abnormalities in the gametophyte, endomitosis of the polar nuclei in the endosperm and increase in the number of polar nuclei (Hanneman 1999). The EBN has been experimentally assigned to each *Solanum* species based on its behavior in crosses with EBN standards, and assuming the 2:1 ratio as a prerequisite for normal endosperm development. Cultivated *S. tuberosum* has been assigned 4EBN. Diploid species have been assigned either 1EBN (e.g. *S. bulbocastaanum, S. cardyophillum*) or 2EBN (e.g. all diploid cultivated species and most wild species, like *S. multidissectum, S. chacoense, S. verrucosum*). Tetraploid species have been assigned either 2EBN or 4EBN. All the hexaploid species, such as *S. oplocense*, are 4EBN. The essential role of EBN as an internal reproductive barrier to interspecific hybridization is that it creates effective barriers between sympatric species, leaving intact their genotypic integrity. For this reason, for example, $2x$ (1EBN) *S. commersonii* cannot be crossed with

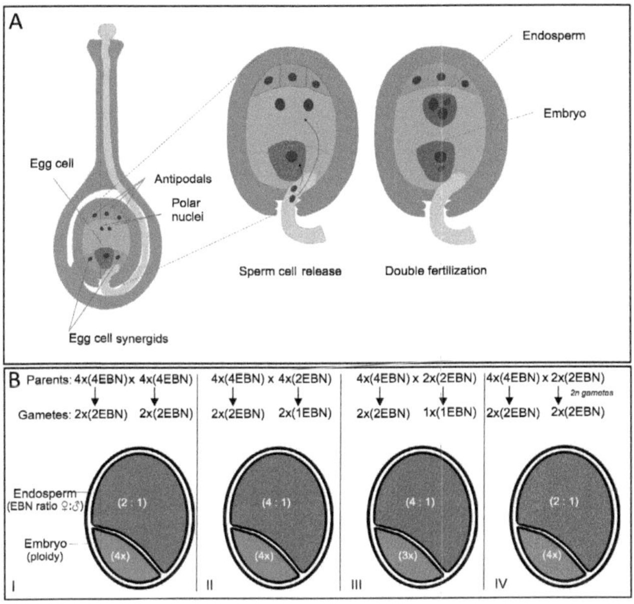

Figure 2. Double fertilization and EBN hypothesis. A. Structure of the male (germinating pollen grain) and female (embryo sac) gametophytes. The release of sperm cells and the double fertilization events are represented. B. Ploidy of embryo and endosperm, and female to male EBN ratio in the hybrid endosperm after intra(inter) ploidy and intra(inter) EBN crosses between *Solanum* species differing in ploidy and EBN (I-IV). The EBN and ploidy of both central cells and eggs of the female gametophyte, and EBN and ploidy of sperm nuclei are shown. Only crosses with a 2:1 maternal to paternal EBN ratio are successful (I and IV).

S. chacoense, which is also diploid but with 2EBN. On the other hand, *S. commersonii* hybridizes with diploid (1EBN) species. The EBN incompatibility barrier played an important role in the speciation of polyploid from diploid potatoes, complementing the role of 2*n* gametes. Indeed, it represents a powerful screen for 2*n* gametes during sexual polyploidization events (Carputo et al. 2003). From breeding perspectives, the EBN model is very attractive, and indeed makes it possible to logically predict the success or failure of interspecific crosses and allows hypotheses about offspring ploidy and EBN. This helps potato breeders to design crossing strategies for the exploitation of incongruent species.

The third essential biological feature of the potato that we want to emphasize here is that maternal haploids can be easily extracted following crosses with pollinator clones of *S. tuberosum* Group Phureja. Haploids, here defined as sporophytes with the gametophytic chromosome number, represent living gametes of a given genotype, and their production makes it possible to scale genomes down by half in a single step. As will be discussed below, their use in potato breeding is particularly important for germplasm enhancement. It is believed that maternal haploids originate from unfertilized egg cells of the female parent. Both sperm nuclei of the pollinator fuse with the polar nuclei of the central cell of the female gametophyte to form the primary endosperm nucleus. This stimulates the parthenogenetic division of the egg cell, which forms a haploid embryo. Refinements in the technique of maternal haploid extraction, such as the incorporation of markers into *S. tuberosum* Group Phureja and the selection of superior pollinators, make it possible to routinely produce these important genetic resources. Potato haploids can also be produced through anther culture. In this case anthers containing microspores at the end of the uninucleate stage are excised from flower buds and cultured *in vitro* under appropriate conditions. Responsive microspores undergo cell division and develop haploid embryos. It should be pointed out that there are two additional reproductive features of the potato to underline in this context: pollen-pistil incompatibility and nuclear-cytoplasmic male sterility. Most diploid potato species are self-incompatible because they possess a single polymorphic *S* locus determining gametophytic incompatibility. A second reproductive characteristic that guarantees species integrity in tuber-bearing *Solanum* species is nuclear-cytoplasmic male sterility, controlled by mitochondrial DNA. This isolating barrier is connected with specific interactions between nuclear genes of one species and cytoplasmic genes of another species. As a result of this interaction, the progeny is male sterile. The existence of nuclear-cytoplasmic male sterility has been reported as a result of either interactions between *S. tuberosum* cytoplasm and nuclear genetic factors of wild species (e.g. *S. infundibuliforme, S. raphanifolium, S. commersonii*) or species cytoplasm and nuclear genes of *S. tuberosum*. There is an important aspect to consider that is strongly related to nuclear-cytoplasmic male sterility and its role in sexual isolation: bumblebees, the pollination vector, do not visit male sterile plants, and these are consequently isolated.

Breeding Objectives

Breeders consider several traits during field and laboratory evaluation and selection of new potato clones. Among them, quality traits challenge potato breeders worldwide. Quality parameters can refer to both external characteristics, such as skin color, tuber size and shape and eye depth, and internal characteristics, comprising nutritional properties, cooking/after cooking properties, dry matter content, protein content and amylose/amylopectine ratio in starch. Breeding objectives related to quality for processed potatoes are normally different from those for fresh use. The former are well defined and are mainly related to dry matter content, level of reducing sugars during cold storage, starch type and, to some extent, tuber shape and size. The latter fall more in the category of external traits, even though eating quality traits such as after-cooking blackening are also important.

Concerns about environmental protection, agricultural sustainability and food safety also make breeding for resistance traits a major objective. In terms of biotic stress resistance, of primary importance is *Phytophthora infestans*, the causal agent of late blight. Race-specific resistance has been introgressed into the cultivated gene pool from hexaploid *S. demissum*. However, new *P. infestans* races have overcome this resistance. In recent years, additional resistances to various pathogens and pests (e.g. nematodes, *Colletotrichum coccodes, Leptinotarsa decemlineata, Streptomyces scabies, Rizochtonia solani, Verticillium spp., Clavibacter michiganensis* and *Erwinia chrysantemii*) have also become important breeding targets. Equally important is the resistance to abiotic stresses. In general, potato varieties show limited resistance to these kinds of environmental constraints. Among them, water and temperature stresses are deemed very important due to the trend towards increased areas of potato cultivation in warm climates (Levy and Veilleux 2007). These stresses negatively affect several parameters, including vegetative growth, photosynthesis and tuberization. Resistance to salinity is another breeding objective, given that the potato is classified as moderately sensitive to salt. Noteworthy are also the resistance to low temperatures and the capacity to cold harden. Since the potato enjoys cool climates, frost events before harvest or after planting may seriously hamper tuber production. This stress may be a limiting factor also in the Mediterranean area, where the potato is planted in winter, starting from November.

Development of Ploidy Series

Potato can enjoy an array of procedures that make it possible to scale genomic multiples up and down, thus producing ploidy series useful for breeding purposes and genetic studies. Particularly attractive is the possibility to produce synthetic polyploids, both allo- and autopolyploids. Since polyploid formation causes a series of structural and functional genetic changes, as well as epigenetic remodeling, polyploids offer the possibility to exploit several advantages (reviewed in Comai 2005). Among the direct advantages are heterosis, gene redundancy and loss of self-incompatibility. An additional advantage is particularly useful for potato breeders, given the differences in somatic chromosome number within *Solanum*. It is the possibility to overcome sexual barriers due to either ploidy or EBN differences between parental species. There are several methods to manipulate ploidy levels in potato. 2*n* gametes have considerable importance for breeding, and the potato is one of the few crops where 2*n* gametes have been extensively studied and exploited. Besides sexual polyploidization through 2*n* gametes, increase in chromosome sets in potato can also be achieved somatically following three possible strategies. The first relies upon the use of antimitotic agents (e.g. colchicine, trifluralin and oryzaline), successfully used not only in potato but also in other species like *Citrus reticulata, Lolium perenne* and *Buddleja globosa* (Dhooghe et al. 2011). In this type of polyploidization chromosome restitution takes place during mitosis and all the chromosomes of a cell are included in one daughter nucleus, giving rise to a cell with doubled chromosome number. Through this method, we have recently produced tetraploids of two noteworthy 1EBN diploid species, *S. bulbocastanum* and *S. commersonii* (Aversano et al. 2013). Developed materials proved to be useful not only for breeding, but also for investigating the phenotypic and molecular effects of early polyploidization underlying several species- and genotype-specific effects. Somatic polyploids can also be produced through shoot regeneration from *in vitro* cultured explants. In this case polyploid synthesis is achieved that exploits both the somaclonal variation induced by *in vitro* conditions and the natural endoduplication events typical of plant cells. Finally, polyploids can be generated through protoplast fusion. In this case, protoplasts are extracted from two different parental genotypes (either belonging to the same species or to two different species). Since potato is a model crop for tissue culture applications, somatic hybridization *via* protoplast fusion provides a powerful tool to produce genotypes integrating parental nuclear and cytoplasmic genomes. It is a multi-step process involving protoplast isolation and fusion, culture and regeneration of the post-fusion mixture, and identification of somatic hybrids among the regenerated shoots. As a result of the interspecific protoplast fusion, many important traits exhibit wide variation both in the somatic hybrids and in their progenies, and therefore further breeding efforts are necessary. A recent review by Rokka (2015) outlines some basic and practical aspects of somatic hybridization and its advantages.

We will now discuss results obtained in scaling genomic multiples up and down, and relate these manipulations to germplasm enhancement. Mainly, breeding methods based on sexual hybridization will be considered. These manipulations range from the monoploid to the hexaploid level.

The Monoploid Level ($2n = x = 12$)

Monoploid production can represent an interesting approach to develop homozygous genotypes and to obtain representatives of gametes at the total plant level that have no lethal/deleterious alleles. Monoploids can be obtained parthenogenetically through pollinations with *S. tuberosum* Group Phureja, as first reported by Breukelen et al. (1975) in *S. tuberosum*. Alternatively, tissue culture of anthers/ovules can be employed. Monoploids and their derivatives can be used for both basic and applied research. A recent, well-known example on the use of monoploids is that related to the aforementioned release of the potato genome sequence, where the authors used a doubled monoploid of *S. tuberosum* Group Phureja to sequence and assemble the genome. Another noteworthy example is the use of a monoploid of *S. tuberosum* Group Phureja to generate a polyploid homozygous series to study the effects of autopolyploidization at the phenotypic and transcriptomic level (Stupar et al. 2007). To improve efficiency in breeding,

Wenzel et al. (1979) proposed a scheme involving two cycles of haploid extraction (from $4x$ to $2x$, and from $2x$ to $1x$) followed by chromosome doubling of the monoploids. In this way homozygous diploids are produced and then intercrossed to yield heterozygous diploids. The resynthesis of highly heterozygous tetraploid genotypes can then be easily accomplished. One main bottleneck in the use of potato monoploids in breeding programs is that they are sterile. Thus, strategies alternative to sexual hybridization must be followed for their utilization. One of the first examples related to the use of monoploids for breeding is given by their production to select amylose–free mutants (*amf*) (Jacobsen et al. 1991). Exploitation of monoploids for breeding purposes was also attempted by Lightbourn and Veilleux (2007), who proposed the so-called monoploid sieve (i.e. the possibility to eliminate lethal/unfavorable alleles by exposing them at the monoploid level). Based on second-cycle doubled monoploids crossed with heterozygous diploids, the authors produced inter-monoploid somatic hybrids with complex pedigrees (including *S. tuberosum* Group Phureja, *S. tuberosum* Group Tuberosum, *S. chacoense* and *S. stenotomum*).

***The Diploid Level* (2*n* = 2*x* = 24)**

This ploidy level is typical of most wild potato species, such as *S. chacoense, S. multidissectum, S. commersonii, S. pinnatisectum* and *S. tarijense.* It also characterizes the somatic chromosome complement of haploids from *S. tuberosum* and other tetraploid species. Following the pioneering work of Hougas and Peloquin (1958), several studies have been undertaken on the production, characterization and use of haploids from *S. tuberosum* as well as other tetraploid species and hybrids. Peloquin et al. (1996) provided evidence that in maternally derived parthenogenetic haploids of *S. tuberosum,* the pollinator influenced haploid frequency via its effect on the endosperm. Camadro et al. (1992) used haploids of *S. acaule* to investigate chromosome pairing of this allotetraploid species. Haploid populations derived from heterozygous tetraploids of *S. tuberosum* can be used in segregation analysis to study complex polygenic traits, and as such represent unique material for simpler genetic studies. With this in mind, Kotch et al. (1992) analyzed the genetics of tuber yield-related traits, determining both additive and non-additive components. Similarly, Hutten et al. (1995) calculated the segregation ratios for resistance to *G. rostochiensis* of haploids derived from several varieties and breeding lines. Potato haploids may also represent sources of variability by expressing alleles that are hidden in the polyploid parent, as suggested by Ercolano et al. (2004). For a review on the use of haploids see Rokka (2009).

For practical breeding, *S. tuberosum* haploids are widely used in interspecific sexual hybridization programs. First of all, they efficiently capture the genetic diversity of $2x$ (2EBN) *Solanum* species through $2x \times 2x$ crosses with these species (Fig. 3). This is felt to be very important given that wild species often do not tuberize under long day conditions, whereas most haploid-species hybrids do (Kittipadukal et al. 2012). In other words, *S. tuberosum* haploids have become a useful tool for breeders to indirectly evaluate tuber characteristics of wild species. Development of haploid-species hybrids also allows breeding at the diploid level before returning to the tetraploid condition, with the large advantages of disomic rather than tetrasomic inheritance patterns, and requirement of a smaller population size under selection. *S. chacoense, S. gourlay* and *S. spegazzini* were crossed to haploids of Argentinean varieties to select hybrids carrying superior tuber-related traits (Santini et al. 2000). Similarly, Ortega and Carrasco (2005) used potato haploids to put resistance to PVY and dry matter content of *S. berthaultii, S. gourlay, S. tarijense* and *S. vernei* into a usable form. Haploid-species hybrids can be used in sexual polyploidization crossing schemes to re-establish the chromosome number of the cultivated potato and gradually reduce the wild genome content (see section dedicated to the tetraploid level). The diploid level has also been recently proposed as a revolutionary potato breeding strategy (Lindhout et al. 2011). This is based upon the replacement of the out-breeding and clonal propagation of the tetraploid potato into an F1 hybrid system with true seed. The strategy will be possible through both the development of superior diploids and the exploitation of a gene (*Sli*) which overcomes gametophytic self-incompatibility, thus allowing the production of inbred parents. This would provide opportunities for producing selfed progenies, thus eliminating undesirable hidden alleles.

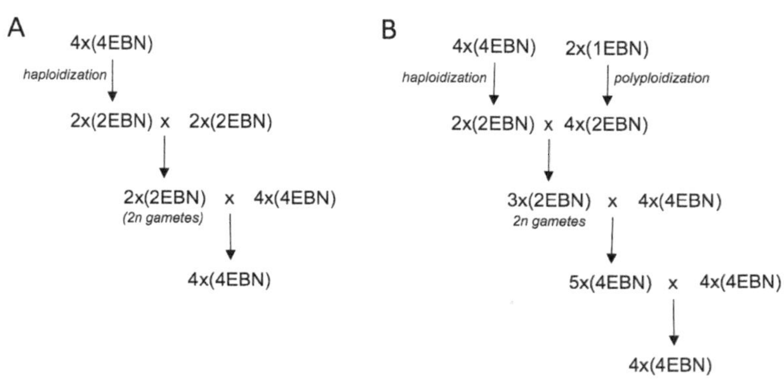

Figure 3. Breeding strategies of ploidy manipulation. A) The analytic breeding scheme involves first the production of *S. tuberosum* haploids (2*x*(2EBN)) prior to crossing with compatible 2*x*(2EBN) species. Once the resulting diploid hybrids that produce 2*n* gametes at acceptable frequencies are selected for traits of interest, the return to the tetraploid level of the cultivated potato is achieved through sexual polyploidization. B) In the "bridge ploidy approach" a tetraploid clone of a 2*x*(1EBN) species is produced and crossed to a S. tuberosum haploid (2*x*(2EBN)). The resulting 3*x*(2EBN) F_1 hybrids are used as a genetic "bridge" and backcrossed to 4*x*(4EBN) to obtain 5*x* bridge hybrids. These are functional tetraploids and thus can be used as both female and male parents in further backcrosses with 4*x*(4EBN) *S. tuberosum* to produce tetraploid improved clones.

The Triploid Level **(2*n* = 3*x* = 36)**

Although often regarded as dead ends, triploids did play a key role in the evolutionary pathways of several angiosperms (Ramsey and Schemske 1998). In potato, triploids are much more common than previously thought. Six species are known only as triploids, and several diploid species also have triploid cytotypes (Hijmans et al. 2007). The triploid block may not be complete in potato, and crosses between tetraploids and diploids can produce triploid progeny (Jackson et al. 1978). In addition, in 2*x*(2EBN) – 2*x*(1EBN) crosses the required 2:1 ratio in the endosperm favors 2*n* gametes of the parent with lower EBN, resulting in the formation of a viable endosperm associated with a triploid embryo. It is likely that triploids are produced regularly in nature and that they continue to hybridize with *Solanum* relatives. They probably represent a link between 2*x* and 4*x* species, guaranteeing gene flow from tetraploids to diploids when they produce haploid gametes following 3*x* × 2*x* crosses. Triploids were often generated as intermediate hybrids (i.e. ploidy bridges) to overcome EBN incongruities. 4*x*(2EBN) species such as *S. acaule, S. fendleri, S. hiertingii, S. papita, S. polytrichon, S. stoloniferum* have been crossed to 2*x*(2EBN) genotypes to produce triploids (Adiwilaga and Brown 1991). Since 2*n* gamete formation is very common in potato, triploids with this feature could be selected and used in subsequent breeding efforts. Triploids were also induced through 4*x*(2EBN) × 2*x*(2EBN) crosses involving somatically doubled 2*x*(1EBN) species. *S. commersonii* is an incongruent species with many resistance traits, including an exceptionally high resistance to low temperatures and capacity to cold harden. Following somatic tetraploidization of *S. commersonii*, 4*x* – 2*x* crosses yielded 2*n* egg producing triploids that allowed the production of subsequent fertile progenies (Fig. 2B) (Carputo et al. 1997). A similar approach has been reported by Sanetomo et al. (2014) to circumvent hybridization barriers between the potato and 1*x*(1EBN) *S. pinnatisectum*, which carries resistance to late blight. Breeding efforts exist also for triploid production from 2*x*(1EBN) – 2*x*(2EBN) crosses, where the species with lower EBN produced 2*n* gametes. For example, Dinu et al. (2005) used this scheme to get triploids between *S. verrucosum* and 2*x*(1EBN) *S. pinnatisectum*.

The Tetraploid Level **(2*n* = 4*x* = 48)**

Besides the cultivated potato *S. tuberosum*, tetraploids include some of its wild relatives like *S. fendleri, S. acaule, S. oxycarpum, S. stoloniferum*. The origin of tetraploid *S. tuberosum* has been explained by hybridization between diploid wild species followed by polyploidization involving 2*n* gametes. The two hypotheses related to the origin of cultivated potato have been recently reviewed by Spooner et al. (2014). According to the multiple origin hypothesis, potato landraces have two main separate origins, involving different species in different geographic areas. By contrast, the restricted origin hypothesis

Figure 4. Field evaluation of tuber traits produced by advanced clones.

suggests that potato domestication took place between Colombia and Bolivia, with a subsequent expansion into Colombia/Venezuela and Chile.

Conventional potato breeding at the tetraploid level is based on intermating of well-adapted tetraploids followed by selection of superior clones among large populations (including thousands of seedlings). Generally, at the beginning a single tuber is collected from each plant. Then, repeated cycles of selection are carried out in successive years, with a decreased number of clones grown and an increased number of plants evaluated per clone (Fig. 4).

It should be pointed out that according to this scheme, it can take 10 years of evaluation and selection before a clone is released as new variety. The biological and reproductive characteristics of the potato allow several additional methods to synthesize new tetraploids. First of all, somatic doubling of diploid genotypes is possible using anti-mitotic agents. This type of polyploidization does not give rise to new allelic interactions, limiting the transmission of all the parental heterozygosity. Therefore, if a heterozygous diploid A1A2 is doubled, a balanced diallelic tetraploid is formed which has only one first order interaction, like its progenitor, and high levels of inbreeding. Tetraploidization can be achieved also by sexual polyploidization through 2*n* gametes, with different genetic consequences related to 2*n* gamete formation. This pathway definitely results in greater variability, fitness, and heterozygosity than somatic doubling. In addition, the availability of so many traits in the wild germplasm, the widespread occurrence of 2*n* gametes, and the ease with which diploid 2*n* gamete-producers hybridize with 4*x* potato strongly favor sexual polyploidization as a valuable tool for germplasm transfer from the diploid to the tetraploid level. This breeding approach has two main components: haploid-species hybrids, that are produced as previously described in the section on the diploid ploidy level; and 2*n* gametes, that are the means to transmit diversity (Fig. 3). Superior diploid-wild species hybrids that form 2*n* gametes are crossed to cultivated potato in 4*x*-2*x* crosses (unilateral sexual polyploidization, USP) or intercrossed in 2*x*-2*x* crosses (bilateral sexual polyploidization, BSP) to generate tetraploid offspring. So USP is based on crosses where only one parent produces 2*n* gametes (either 2*n* pollen or 2*n* eggs), whereas the other already has a 48-chromosome complement, whereas BSP involves crosses where the female parent forms 2*n* eggs and the male parent forms 2*n* pollen. The greatest difficulty with this latter method is that both parents must produce 2*n* gametes. The exploitation of 2*n* gametes provides breeders the unique possibility of accumulating non-additive genetic effects at the tetraploid level in the offspring, whereas the asexual reproduction system of the potato allows the immediate fixation of heterosis. The extent of allelic interaction transmission from haploid-species hybrids to the tetraploid progeny depends mainly on the mode of 2*n* gamete production (FDR vs. SDR), whereas the creation of new interactions depends mainly on the allelic diversity of parents involved. The potential of these breeding schemes in potato has been demonstrated (mainly by USP) for traits such as tuber yield (Buso et al. 1999; Alberino et al. 2004), stress tolerance (Watanabe et al. 1992; Sterrett et al. 2003; Frost et al. 2006) and processing quality (Hutten et al. 1996; Hayes and Thill 2002; Jansky et al. 2011). Watanabe et al. (1992) proposed a breeding strategy to produce tetraploids from *S. acaule*, a 4*x*(2EBN) species. It involves sexual crosses 4*x*(2EBN) – 4*x*(4EBN) followed by a second pollination with *S. tuberosum* Group Phureja and embryo rescue. Tetraploids may be also produced from somatic hybridization between *S. tuberosum* haploids and diploid species. For example, Przetakiewicz et al. (2007) produced somatic hybrids from a wide range of parental combinations involving diploid lines containing various species (e.g. *S. chacoense, S. vernei, S. stoloniferum*) in a *S. tuberosum* pedigree. Iovene et al. (2007) electrofused *S. bulbocastanum* with *S. tuberosum* haploids to produce a population of tetraploid interspecific hybrids.

The Pentaploid (2n = 5x = 60) *and Hexaploid* (2n = 6x = 72) *Levels*

The most direct method to produce pentaploids is by 4*x*(4EBN) – 6*x*(4EBN) crosses. To introgress resistance to rot-knot nematodes of hexaploid species *S. hougasii*, Janssen et al. (1997) produced resistant pentaploids that were successfully backcrossed as females to *S. tuberosum*. Pentaploids were also generated following 4*x*(4EBN) – 3*x*(2EBN) crosses where the triploid parent produces 2*n* gametes (Brown and Adiwilaga 1990); Adiwilaga and Brown 1991; Carputo 2003). 2*n* gametes serve to both equalize EBN between parents and to produce functional, balanced gametes. In most cases, pentaploids were produced through the function of 2*n* pollen (4*x* × 3*x* crosses). This guarantees the production of euploids when the meiotic mutation responsible for 2*n* pollen formation is parallel spindles. In fact, after the chromosomes of the triploid distribute randomly into the telophase I nuclei, the parallel spindle mechanism in the second division ensures symmetric incorporation of 36 chromosomes in each pair of 2*n* microspores. Pentaploid *S. curtilobum* was likely produced from the union of 2*n* gametes from triploid *S. juzepczukii* and normal gametes from tetraploid *S. tuberosum* Group Andigena. This hybrid was recreated artificially using this scheme (Hawkes 1990). *S. curtilobum* is fertile and able to hybridize with *S. tuberosum*. Pentaploid landraces, therefore, are not genetically isolated from the rest of the cultivated gene pool. Consequently, some level of genetic recombination is likely to occur in hybrids between landraces.

The hexaploid level represents the natural ploidy level of a few wild species. Among them, *S. demissum* is one of the most famous due to its high resistance to *P. infestans*. Hexaploids can be easily obtained through the union of 2*n* gametes from triploid genotypes. New hexaploids can be efficiently produced also by 4*x*(2EBN) – 4*x*(4EBN) crosses if the parent with lower EBN produces 2*n* gametes. The F_1 hexaploid offspring produced through these matings can be used as a genetic bridge and crossed to *S. tuberosum*. Direct crossing between virus resistant 4*x*(2EBN) *S. acaule* and 4*x*(4EBN) *S. tuberosum* was possible through the production of 2*n* eggs (Camadro and Espinillo 1990). Similarly, hexaploids were yielded from 4*x*(2EBN) *S. fendleri* × 4EBN *S. tuberosum* (Janssen et al. 1997). Hexaploids were also produced by somatically doubling the chromosome number of triploids. (Culley et al. 2002), for example, performed crosses between *S. hjertingii* (4*x*2EBN) and a 2*x*(2EBN) hybrid. Since triploid offspring were sterile, they selected hexaploids occurring from *in vitro* spontaneous somatic chromosome doubling. Finally, the hexaploid level may be potentially produced by 3*x*-6*x* crosses in which there is a selective screen for 2*n* gametes from the 3*x* parent.

The Molecular Breeding Perspective

The complex genetics associated with the polyploid nature of the cultivated potato has limited the understanding of the inheritance of genetic traits in potato compared to diploid crops. The advent of molecular markers facilitated a quantum leap in elucidating the molecular basis of disease resistance and quality traits. To date, many markers have been mapped in potato, leading to significant progress especially for late blight, virus and nematode resistances, chip color, dry matter, cold-induced sweetening, and processing quality (reviewed in Ramakrishnan et al. 2015). However, although mapping studies (at the diploid and tetraploid levels) have been conducted since the late 1980s, molecular-assisted selection (MAS) has not alleviated the difficulty of working with a polyploid crop where the full classification of multiple alleles at a given locus is still challenging. Indeed, relatively few markers have successfully been used in MAS programs in tetraploid potato breeding. Molecular markers in potato breeding have been also used to choose the most suitable genotype to be backcrossed among those combining the useful gene from the wild species with the lowest wild genome content. This strategy, known as negative marker-assisted selection (Barone 2004), was successfully applied in potato to the aforementioned breeding scheme aimed at transferring useful traits from *S. commersonii* to the cultivated gene pool (Carputo et al. 1997).

The potential for markers to analyze whole genomes, to improve estimation of genetic variance, recover elite backgrounds, and select superior varieties has not been fully realized in potato. This is due to several factors, among them the availability of markers developed in experimental, diploid or interspecific populations rather than in adapted tetraploid germplasm (Ortega and Lopez-Vizcon 2012). The availability of new resources should help to facilitate the adoption of marker-assisted selection. In this regard, the sequencing of *S. tuberosum* Group Phureja and its wild potato species *S. commersonii* (Aversano et al. 2015) genomes give an unprecedented opportunity to develop molecular markers, holding promise for providing allelic variants of genes contributing to important quantitative traits in potato. In particular, the Infinium High-Confidence SNP (69K) array and its subset, the SolCAP 8303 Infinium SNP array, were developed about the same time as the first complete genome sequence. The SNP array combines data from cDNA sequencing of one diploid and six tetraploid cultivars (Atlantic, Premier Russet, Snowden, Bintje, Kennebec, and Shepody). The SNPs are a subset of almost 69,000 high confidence SNPs that were mapped to the PGSC v2.1.11 DM Pseudomolecules (Hamilton et al. 2011; Felcher et al. 2012; Douches et al. 2014). This SNP array facilitated the mapping of novel SNPs/quantitative trait loci (QTLs) for late blight resistance (Lindqvist-Kreuze et al. 2014), glycoalkaloid biosynthetic pathways (Manrique-Carpintero et al. 2013), tuber shape and eye depth (Prashar et al. 2014), and sugar content (Schreiber et al. 2014). Using the Infinium 8303 Potato Array, (Hirsch et al. 2013) surveyed a wide range of cultivars, including historical and newly released cultivars, advanced breeding lines, genetic stocks, and several wild species. Population structure analysis allowed the separation of cultivated potato lines from wild species and genetic stocks and provided a clear separation between potato market classes. In addition, several of these loci were associated with genes related to carbohydrate metabolism and tuber development.

References

Achenbach, U., J. Paulo, E. Ilarionova, J. Lübeck, J. Strahwald, E. Tacke, H.-R. Hofferbert and C. Gebhardt. 2009. Using SNP markers to dissect linkage disequilibrium at a major quantitative trait locus for resistance to the potato cyst nematode *Globodera pallida* on potato chromosome V. Theor. Appl. Genet. 118: 619-629.

Adiwilaga, K.D. and C.R. Brown. 1991. Use of 2*n* pollen-producing triploid hybrids to introduce tetraploid Mexican wild species germplasm to cultivated tetraploid potato gene pool. Theor. Appl. Genet. 81: 645-652.

Alberino, S., D. Carputo, G. Caruso, M.R. Ercolano and L. Frusciante. 2004. Field performance of families and clones obtained through unilateral sexual polyploidization in potato (*Solanum tuberosum*). Adv. Hort. Sci. 18: 47-52.

Ames, M. and D.M. Spooner. 2008. DNA from herbarium specimens settles a controversy about origins of the European potato. Am. J. Bot. 95: 252-257.

Aversano, R., F. Contaldi, M.R. Ercolano, V. Grosso, M. Iorizzo, F. Tatino, L. Xumerle, A. Dal Molin, C. Avanzato, A. Ferrarini, M. Delledonne, W. Sanseverino, R. Aiese Cigliano, S. Capella-Gutierrez, T. Gabaldón, L. Frusciante, J.M. Bradeen and D. Carputo. 2015. The *Solanum commersonii* genome sequence provides insights into adaptation to stress conditions and genome evolution of wild potato relatives. Plant Cell 27(4): 954-968.

Aversano, R., I. Caruso, G. Aronne, V. De Micco, N. Scognamiglio and D. Carputo. 2013. Stochastic changes affect *Solanum* wild species following autopolyploidization. J. Exp. Bot. 64(2): 625-635.

Barone, A. 2004. Molecular marker-assisted selection for potato breeding. Am. J. Potato Res. 81: 111-117.

Bradeen, J.M. and K.G. Haynes. 2011. Introduction to potato. pp. 1-19. *In*: J.M. Bradeen and K. Chittaranjan (eds.). Genetics, genomics and breeding of potato, Science Publishers, Enfield, NH.

Bradshaw, J.E. and G. Ramsay. 2005. Utilization of the Commonwealth potato collection in potato breeding. Euphytica 146: 9-19.

Bretagnolle, F. and J.D. Thompson. 1995. Gametes with somatic chromosome number: mechanisms of their formation and role in the evolution of autopolyploid plants. New Phytol. 129: 1-22.

Brown, C.R. and K. Adiwilaga. 1990. Introgression of *Solanum acaule* germplasm from the endosperm balance number 2 gene pool into the cultivated endosperm balance number 4 potato gene pool via triplandroids. Genome 33: 273-278.

Busch, B.L., G. Schmitz, S. Rossmann, F. Piron, J. Ding, A. Bendahmane and K. Theres. 2011. Shoot branching and leaf dissection in tomato are regulated by homologous gene modules. Plant Cell 23: 3595-3609.

Buso, J.A., F.J.B. Reifscneider, L.S. Boiteux and S.J. Peloquin. 1999. Effects of 2*n*-pollen formation by first meiotic division restitution with and without crossover on eight quantitative traits in $4x - 2x$ potato progenies. Theor. Appl. Genet. 98: 1311-1319.

Camadro, E.L. and J.C. Espinillo. 1990. Germplasm transfer from the wild tetraploid species *Solanum acaule* Bitt. to the cultivated potato, *S. tuberosum* L. using 2*n* eggs. Am. Potato J. 67: 737-749.

Camadro, E.L., R.W. Masuelli and M.C. Cortes. 1992. Haploids of the wild potato *Solanum acaule* spp. *acaule*: generation, meiotic behavior, and electrophoretic pattern for the aspartate aminotransferase system. Genome 35: 431-435.

Carputo, D., B. Basile, T. Cardi and L. Frusciante. 2000. *Erwinia* resistance in backcross progenies of *Solanum tuberosum* x *S. tarijense* and *S. tuberosum* (+) *S. commersonii* hybrids. Potato Res. 43: 135-142.

Carputo, D. 2003. Cytological and breeding behavior of pentaploids derived from $3x \times 4x$ crosses in potato. Theor. Appl. Genet. 106: 883-888.

Carputo, D., Frusciante, L., Peloquin, S.J. 2003. The role of $2n$ gametes and endosperm balance number in the origin and evolution of polyploids in the tuber-bearing Solanums. Genetics 163: 287-294.

Carputo, D., A. Barone, T. Cardi, A. Sebastiano, L. Frusciante and S.J. Peloquin. 1997. Endosperm balance number manipulation for direct *in vivo* germplasm introgression to potato from a sexually isolated relative (*Solanum commersonii* Dun.). Proc. Natl. Acad. Sci. USA. 94: 12013-12017.

Comai, L. 2005. The advantages and disadvantages of being polyploid. Nat. Rev. Genet. 6: 836-846.

Culley, D.E., B.B. Dean and C.R. Brown. 2002. Introgression of the low browning trait from the wild Mexican species *Solanum hjertingii* into cultivated potato (*S. tuberosum* L.). Euphytica 125: 293-303.

d'Erfurth, I., S. Jolivet, N. Froger, O. Catrice, M. Novatchkova, M. Simon, E. Jenczewski and R. Mercier. 2008. Mutations in AtPS1 (*Arabidopsis thaliana Parallel Spindle* 1) lead to the production of diploid pollen grains. PLoS Genet. 4(11): 1-9.

Dhooghe, E., K. van Laere, T. Eeckhaut, L. Leus and J. van Huylenbroeck. 2011. Mitotic chromosome doubling of plant tissues *in vitro*. Plant Cell Tiss. Org. 104: 359-737.

Dinu, I.I., R.J. Hayes, R.G. Kynast, R.L. Phillips and C.A. Thill. 2005. Novel inter-series hybrids in *Solanum*, section Petota. Theor. Appl. Genet. 110: 403-415.

Douches, D., C.N. Hirsch, N.C. Manrique-Carpintero, A.N. Massa, J. Coombs, M. Hardigan, D. Bisognin, W. De Jong and C.R. Buell. 2014. The contribution of the *Solanaceae* coordinated agricultural project to potato breeding. Potato Res. 57: 215-224.

Ercolano, M.R., D. Carputo, J. Li, L. Monti, A. Barone and L. Frusciante. 2004. Assessment of genetic variability of haploids extracted from tetraploid ($2n = 4x = 48$) *Solanum tuberosum*. Genome 47: 633-638.

Felcher, K.J., J.J. Coombs, A.N. Massa, C.N. Hansey, J.P. Hamilton, R.E. Veilleux, C.R. Buell and D.S. Douches. 2012. Integration of two diploid potato linkage maps with the potato genome sequence. PLoS ONE 7: e36347.

Frankel, R. and E. Galun. 1977. Pollination mechanisms, reproduction, and plant breeding. Springer, New York.

Frost, K.E., S.H. Jansky and D.I. Rouse. 2006. Transmission of Verticillium wilt resistance to tetraploid potato via unilateral sexual polyploidization. Euphytica 149: 281-287.

Hamilton, J.P., C.N. Hansey, B.R. Whitty, K. Stoffel, A.N. Massa, A. van Deynze, W.S. De Jong, D.S. Douches and C.R. Buell. 2011. Single nucleotide polymorphism discovery in elite North American potato germplasm. BMC Genomics 12: 302.

Hanneman, Jr R.E. 1999. The reproductive biology of the potato and its implications for breeding. Potato Res. 42: 283-312.

Hawkes, J.G. 1990. The Potato - evolution, biodiversity and genetic resources. Belhaven Press, London.

Hayes, R.J. and C.A. Thill. 2002. Introgression of cold (4 C) chipping from $2x$ (2 Endosperm Balance Number) potato species into $4x$ (4EBN) cultivated potato using sexual polyploidization. Am. J. Potato Res. 6: 421-431.

Hijmans, R.J., T. Gavrilenko, S. Stephenson, J. Bamberg, A. Salas and D.M. Spooner. 2007. Geographical and environmental range expansion through polyploidy in wild potatoes (*Solanum* section Petota). Glob. Ecol. Biogeogr. 16: 485-495.

Hirsch, C.N., C.D. Hirsch, K. Felcher, J. Coombs, D. Zarka, A. van Deynze, W. De Jong, R.E. Veilleux, S. Jansky, P. Bethke, D.S. Douches and C.R. Buell. 2013. Retrospective view of North American potato (*Solanum tuberosum* L.) breeding in the 20th and 21st centuries. G3 3: 1003-1013.

Hougas, R.W. and S.J. Peloquin. 1958. The potential of potato haploids in breeding and genetic research. Am. Potato J. 35: 701-707.

Huamán, Z. and D.M. Spooner. 2002. Reclassification of landrace populations of cultivated potatoes (*Solanum* sect. Petota). Am. J. Bot. 89: 947-965.

Hutten, R.C.B., M.G.M. Schippers, J. Eising, P.M. van Til, J.G.Th. Hermsen and E. Jacobsen. 1996. Analysis of parental effects on mean vine maturity and chip colour of $4x.2x$ potato progenies. Euphytica 88: 175-179.

Hutten, R.C.B., W.J.J. Soppe, J.G.T. Hermsen and E. Jacobsen. 1995. Evaluation of dihaploid populations from potato varieties and breeding lines. Potato Res. 38: 77-86.

Iovene, M., S. Savarese, T. Cardi, L. Frusciante, N. Scotti, P.W. Simon and D. Carputo. 2007. Nuclear and cytoplasmic genome composition of *Solanum bulbocastanum* (+) *S. tuberosum* somatic hybrids. Genome 50: 443-450.

Jackson, M.T., P.R. Rowe and J.G. Hawkes. 1978. Crossability relationships of Andean potato varieties of three ploidy levels. Euphytica 27: 541-551.

Jacobsen, E., M.S. Ramanna, D.J. Huigen and Z. Sawor. 1991. Introduction of an amylose-free (*amf*) mutant into breeding of cultivated potato, *Solanum tuberosum* L. Euphytica 53: 247-253.

Jansky, S.H., A. Hamernik and P.C. Bethke. 2011. Germplasm Release: tetraploid clones with resistance to cold-induced sweetening. Am. J. Potato Res. 88: 218-225.

Janssen, G.J.W., A. van Norel, B. Verkerk-Bakker, R. Janssen and J. Hoogendoorn. 1997. Introgression of resistance to root-knot nematodes from wild Central American *Solanum* species into *S. tuberosum* spp. *tuberosum*. Theor. Appl. Genet. 95: 490-496.

Johnston, S.A., T.P.M. den Nijs, S.J. Peloquin and R.E.Jr. Hanneman. 1980. The significance of genic balance to endosperm development in interspecific crosses. Theor. Appl. Genet. 57: 5-9.

Kittipadukal, P., P.C. Bethke and S.H. Jansky. 2012. The effect of photoperiod on tuberisation in cultivated 3 wild potato species hybrids. Potato Res. 55: 27-40.

Kotch, G.P., R. Ortiz and S.J. Peloquin. 1992. Genetic analysis by use of potato haploid populations. Genome 35: 103-108.

Levy, D. and R.E. Veilleux. 2007. Adaptation of potato to high temperatures and salinity - a review. Am. J. Potato Res. 84: 487-506.

Lightbourn, G.J. and R.E. Veilleux. 2007. Production and evaluation of somatic hybrids derived from monoploid potato. Am. J. Potato Res. 8: 425-435.

Lindhout, P., D. Meijer, T. Schotte, R.C.B. Hutten, R.G.F. Visser and H.J. van Eck. 2011. Towards F1 hybrid seed potato breeding. Potato Res. 54: 301-312.

Lindqvist-Kreuze, H., M. Gastelo, W. Perez, G.A. Forbes, D. de Koeyer and M. Bonierbale. 2014. Phenotypic stability and genome-wide association study of late blight resistance in potato genotypes adapted to the tropical highlands. Phytopathology 104: 624-633.

Manrique-Carpintero, N.C., J.G. Tokuhisa, I. Ginzberg, J.A. Holliday and R.E. Veilleux. 2013. Sequence diversity in coding regions of candidate genes in the glycoalkaloid biosynthetic pathway of wild potato species. G3 (Bethesda) 3: 1467-1479.

Obidiegwu, J., R. Sanetomo, K. Flath, E. Tacke, H.-R. Hofferbert, A. Hofmann, B. Walkemeier and C. Gebhardt. 2015. Genomic architecture of potato resistance to *Synchytrium endobioticum* disentangled using SSR markers and the 8.3k SolCAP SNP genotyping array. BMC Genet. 16: 1-16.

Ortega, F. and A. Carrasco. 2005. Germplasm enhancement with wild tuber-bearing species: introgression of PVY resistance and high dry matter content from *Solanum berthaultii, S. gourlayi, S. tarijense* and *S. vernei*. Potato Res. 48: 97-104.

Ortega, F. and C. Lopez-Vizcon. 2012. Application of molecular marker-assisted selection (MAS) for disease resistance in a practical potato breeding program. Potato Res 55: 1-13.

Peloquin, S.J., A.C. Gabert and R. Ortiz. 1996. Nature of "pollinator" effect in potato (*Solanum tuberosum* L.) haploid production. Ann. Bot-London 77: 539-542.

Peloquin, S.J., L. Boiteux and D. Carputo. 1999. Meiotic mutants of the potato: valuable variants. Genetics 153: 1493-1499.

Potato Genome Sequencing Consortium. 2011. Genome sequence and analysis of the tuber crop potato. Nature 475: 189-195.

Prashar, A., C. Hornyik, V. Young, K. McLean, S.K. Sharma, M.F.B. Dale and G.J. Bryan. 2014. Construction of a dense SNP map of a highly heterozygous diploid potato population and QTL analysis of tuber shape and eye depth. Theor. Appl. Genet. 127: 2159-2171.

Przetakiewicz, J., A. Nadolska-Orczyk, D. Kuć and W. Orczyk. 2007. Tetraploid somatic cybrids of potato (*Solanum tuberosum* L.) obtained from diploid breeding lines. Cell Mol. Biol. Lett. 12: 253-267.

Ramakrishnan, A.P., C.E. Ritland, R.H. Blas Sevillano A. and Riseman. 2015. Review of potato molecular markers to enhance trait selection. Am. J. Potato Res. 92: 455-472.

Ramsey, J. and D.W. Schemske. 1998. Pathways, mechanisms, and rates of polyploid formation in flowering plants. Annu. Rev. Ecol. Sys. 29: 467-501.

Rodriguez, F., M. Ghislain, A.M. Clausen, S.H. Jansky and D.M. Spooner. 2010. Hybrid origins of cultivated potatoes. Theor. Appl. Genet. 121: 1187-1198.

Rokka, V.-M. 2015. Protoplast technology in genome manipulation of potato through somatic cell fusion. pp. 217-235. *In*: X.-Q. Li, D.J. Donnelly and T.G. Jensen (eds.) Somatic genome manipulation. Springer, New York, USA.

Rokka, V.-M. 2009. Potato haploids and breeding. pp 199-208. *In:* A. Touraev, B.P. Forster, S.M. Jain (eds.). Advances in haploid production in higher plants. Springer, New York,.

Sanetomo, R., S. Akino, N. Suzuki and K. Hosaka. 2014. Breakdown of a hybridization barrier between *Solanum pinnatisectum* Dunal and potato using the *S* locus inhibitor gene (*Sli*). Euphytica 1: 119-132.

Santini, M., E.L. Camadro, O.N. Marcellán and L.E. Erazzú. 2000. Agronomic characterization of diploid hybrid families derived from crosses between haploids of the common potato and three wild Argentinean tuber-bearing species. Am. J. Potato Res. 77: 211-218.

Schreiber, L., A.C. Nader-Nieto, E.M. Schönhals, B. Walkemeier and C. Gebhardt. 2014. SNPs in genes functional in starch-sugar interconversion associate with natural variation of tuber starch and sugar content of potato (*Solanum tuberosum* L.). G3 (Bethesda) 4(10): 1797-1811.

Spooner, D.M., M. Ghislain, R. Simon, S.H. Jansky and T. Gavrilenko. 2014. Systematics, diversity, genetics, and evolution of wild and cultivated potatoes. Bot. Rev. 80: 283-383.

Spooner, D.M., J. Núñez, G. Trujillo, M.D.R. Herrera, F. Guzman and M. Ghislain. 2007. Extensive simple sequence repeat genotyping of potato landraces supports a major reevaluation of their gene pool structure and classification. Proc. Natl. Acad. Sci. USA. 104: 19398-19403.

Sterrett, S.B., M.R. Henninger, G.C. Yencho, W. Lu, B.T. Vinyard and K.G. Haynes. 2003. Stability of internal heat necrosis and specific gravity in tetraploid · diploid potatoes. Crop Sci. 43: 790-796.

Stupar, R.M., P.B. Bhaskar, B.S. Yandell, W.A. Rensink, A.L. Hart, S. Ouyang, R.E. Veilleux, J.S. Busse, R.J. Erhardt, C.R. Buell and J. Jiang. 2007. Phenotypic and transcriptomic changes associated with potato autopolyploidization. Genetics 176: 2055-2067.

Tang, H., X. Wang, J.E. Bowers, R. Ming, M. Alam and A.H. Paterson. 2008. Unraveling ancient hexaploidy through multiply-aligned angiosperm gene maps. Genome 18(12): 1944-1954.

van Breukelen, E.W.M., M.S. Ramanna and Th. J.G. Hermsen. 1975. Monohaploids ($2n = x = 12$) from autotetraploid *Solanum tuberosum* ($2n = 4x = 48$) through two successive cycles of female parthenogenesis. Euphytica 24: 567-574.

van der Wall, M.H.B.J., E. Jacobsen and R.G.F. Visser. 2001. Multiple allelism as a control mechanism in metabolic pathways: GBSSI allelic composition affects the activity of granule-bound starch synthase I and starch composition in potato. Mol. Genet. Genom. 265: 1011-1021.

Watanabe, K, H.M. El-Nashaar and M. Iwanaga. 1992. Transmission of bacterial wilt resistance by First Division Restitution (FDR) $2n$ pollen via $4x \times 2x$ crosses in potatoes. Euphytica 60: 21-26.

Wenzel, G., O. Schieder, T. Przewozny, S.K. Sopory and G. Melchers. 1979. Comparison of single cell culture-derived *Solanum tuberosum* L. plants and model for their application in breeding programs. Theor. Appl. Genet. 55: 49-55.

Chapter 4

Polyploid Induction Techniques and Breeding Strategies in Poplar

Xiang-Yang Kang[1]

ABSTRACT

Polyploid breeding is one of the most effective approaches for improvement of poplars, as allotriploid poplar varieties exhibit good growth and wood quality properties. Several methods have been developed in poplars in order to produce allopolyploids efficiently, including hybridization with spontaneous $2n$ gametes, crossing with $2n$ pollen or eggs induced by colchicine or high temperature treatment, and chromosome doubling of zygotes or somatic cells. In the practice of polyploid poplar breeding, knowledge of reproductive biology characteristics is the key to improving chromosome doubling efficiency. Furthermore, utilization of both heterosis and vigor due to increased ploidy via sexual polyploidization is very important in poplar polyploid breeding. However, due to the normal segregation of trait phenotypes during sexual polyploidization of poplar, strong selection based on a large polyploid population is necessary for polyploid varieties.

Introduction

Poplar is a general term for the genus *Populus* L. (Salicaceae), which comprises more than 100 dioecious and deciduous tree species. Poplar is easy to hybridize. In nature, although the majority of poplar species are diploid ($2n = 2x = 38$), triploid plants were discovered in *Populustremula* (Nilsson-Ehle 1936), *P. alba* (Van Dillewijn 1939), *P. balsamifera* (Gurreiro 1944), *P. tremuloides* (Van Buijtenen et al. 1957) and *P. tomentosa* (Zhu et al. 1998). Poplar is one of the most widely distributed and most adaptable trees in the world, possessing the characteristics of fast growth and easy asexual reproduction. It is also an important species for building timber stands, shelter forests, and

[1] P.O. BOX 118, Beijing Forestry University, No. 35, Tsinghua East Road, Haidian District, Beijing, 100083, P.R. China, Email: kangxy@bjfu.edu.cn

for urban and rural greening (Xu 1988). *Populus* is also of interest for polyploid breeding. Plants with lower chromosome numbers, where vegetative mass is harvested and seeds are produced by cross-pollination are generally more suitable for breeding by chromosome doubling (Lewis 1980). Additionally, two more characteristics contributing to the success of polyploid breeding were proposed by Dewey (1980): perennial growth habit and vegetative reproduction, which reduces the dependence on seed production of polyploid plants. Obviously, *Populus* ($2n = 2x = 38$) has the above five characteristics.

Poplar polyploid breeding was initiated by the discovery of a giant form of triploid *Populus tremula* ($2n = 3x = 57$) (Nilsson-Ehle 1936), which became a research hotspot in the 1960s and 1970s. However, subsequently only a few allotriploids were able to be produced through hybridization between tetraploid and diploid species of *P. tremula* and *P. tremuloides* in America and Germany (Einspahr and Wyckoff 1975, Weisgerber et al. 1980, Einspahr 1984). Since then, due to the rise of genetic engineering, as well as the low utilization coefficient of triploid poplar resulting from the characteristics of low basic wood density (easily broken by wind), studies on polyploidy breeding in poplar became scarce. However, in the 1990s, a number of auto triploid varieties of *P. tomentosa* with fast growth and high wood quality were successfully produced by pollinating with spontaneously produced unreduced pollen ($2n$, with the somatic chromosome number) in China (Zhu et al. 1995). The outstanding performance of triploid *P. tomentosa* in growth and wood properties then prompted re-recognition of the role of polyploid breeding in genetic improvement of poplar.

In recent years, researches on induction of poplar polyploids have increased, and the efficiency of polyploid induction has significantly improved as a result (Zhang and Li 1992, Kang et al. 1999, 2000a, b, 2004, Li et al. 2000, 2001, Huang et al. 2002, Li et al. 2006, Li 2007, Li et al. 2008, Kang and Wang 2010, Wang et al. 2010, 2012a, b, 2013, Cai and Kang 2011, Huang et al. 2015). A series of famous poplar interspecific hybrids which were widely cultivated in China (and worldwide) have also been proven to be triploids (Chen et al. 2004, Zhang et al. 2004, 2005). The induction and utilization of poplar polyploids has already demonstrated great potential. At this point, what kind of breeding strategies should be carried out to improve the efficiency and effectiveness of polyploid breeding in poplar? In other words, how can poplar polyploids be efficiently induced, and how can poplar polyploids with relatively high genetic gains be obtained? Based on many years of experience in polyploid breeding in poplar, and combined withrelated research in poplar and other plants, polyploid induction techniques and breeding strategies in poplar are discussed.

The Advantages and Applications of Allotriploids in the Genetic Improvement of Poplar

Extensive interspecific hybridization accompanied by chromosome doubling always produces all triploids in *Populus*. Chromosome doubling in plants often initiates morphological and physiological changes, and can result in advantages of improved growth and quality traits. In terms of growth, poplar polyploids grow faster than diploids, especially in early growth stages. For example, the height, diameter at breast height (DBH), and volume of spontaneous triploid *P. tremula* were 10%, 11%, and 36% higher respectively than diploids of the same age under the same environmental conditions (Nilsson-Ehle 1936); the growth of triploid hybrids between *P. tremula* and *P. tremuloides* was about as twice as that of diploid hybrids (Einspahr 1984); triploid hybrid "Astria" was 22% taller and 25% larger in diameter than that of diploid poplar (Weisgerber et al. 1980) and the growth in volume of triploids of *P. tomentosa* was as 2-3 times that of ordinary diploids (Zhu et al. 1995). Some poplar varieties which have been widely deployed in China were also detected to be triploids, such as *P.* × *canadensis* 'I-214' from Italy, *P.* × *canadensis* 'Sacrau 79' from Germany, *P.* × *euramericana* 'Zhonglin-46', *P.* × *euramericana* 'Wuhei-1', *P.* × *liaohenica, P.* × *langfangensis*-3 and *P. alba* × *P. berolinensis* (Chen et al. 2004, Zhang et al. 2004, 2005). All of these identified triploids were superior clones which were selected from filial generations on the basis of growth index and presumably originated from spontaneous 2*n* gametes.

Due to the typically larger size of cells in the polyploid poplar, the fiber length is increased and the number of cells per volume and cell surface area are decreased, leading to relatively decreased lignin and pentos an content with obvious advantages in terms of wood fiber length andchemical composition. Therefore, not only the growth index, but also the wood texture of poplar is optimized by polyploidy. Fibre length and proportion in triploid hybrids between *P. tremula* and *P. tremuloides* were shown to be 18% and 20% higher than diploids of the same age respectively (Einspahr and Wyckoff 1975, Einspahr 1984), which indicates that those triploid hybrids can provide excellent raw material for pulp. Average fiber length of 5-yr-old triploid *P. tomentosa* was 52.4% longer than that of diploids, the lignin content was only 16.71%, which was 17.9% lower than that of diploids, and the α-cellulose content of triploid *P. tomentosa* was 53.21%, which was 5.8 % higher than that of diploids with the same age (Yao and Pu 1998, Fang et al. 2001).

Comparative trials have recently been completed among triploid hybrid clones of white poplar (Table 1). Triploid hybrid clones of B301 – B333 originated from pollination with spontaneous 2*n* pollen of *P. tomentosa*, and triploid clones of BLXZ-1 and BLXZ-2 were derived from pollination with 2*n* pollen of *P. alba* × *P. glandulosa* induced by colchicine solution. The stem volume, fiber length, and holocellulose content of these triploid clones were 79.6%, 20.7% and 1.3% higher than that of the diploid control (clone of 1319), respectively. However, the lignin content was 21.7% lower than that of the diploid control. Compared with diploids, triploid clones have a slightly lower wood basic density. With combined vigor and heterosis, triploid poplar could comprehensively realize multiple improvement objectives, such as growth

Table 1. Comparison of growth and wood properties in triploid poplar hybrid clones.

Cross combination	Clones	Volume of timber (m^3/hm^2)	Wood basic density (g/cm^3)	Fiber length (mm)	Holo-cellulose content (%)	Lignin content (%)
(*P. tomentosa* × *P. bolleana*) × (*P. alba* × *P. glandulosa*)	BLXZ-1	103.35^{b}	0.3555^{a}	0.854^{a}	85.90^{g}	17.42^{a}
	BLXZ-2	121.50^{a}	0.3393^{b}	0.820^{abc}	85.46^{fg}	17.72^{a}
(*P. tomentosa* × *P. bolleana*) × *P. tomentosa*	B301	85.80^{c}	0.3041^{de}	0.795^{d}	83.13^{bc}	19.31^{bc}
	B302	88.35^{c}	0.3282^{bc}	0.799^{d}	84.44^{ef}	18.05^{a}
	B303	80.40^{c}	0.3190^{cd}	0.822^{abc}	83.58b^{cde}	19.14^{bc}
	B304	78.90^{c}	0.3235^{bc}	0.845^{ab}	84.42^{def}	18.99^{b}
	B305	89.25^{c}	0.3283^{bc}	0.803^{bcd}	83.90^{cde}	19.00^{b}
	B306	81.75^{c}	0.3235^{bc}	0.844^{ab}	84.53^{ef}	19.04^{b}
	B307	84.45^{c}	0.3156^{cd}	0.820^{abc}	82.98^{abc}	20.09^{c}
	B312	77.70^{c}	0.2991^{e}	0.800^{cd}	82.06^{a}	21.09^{d}
(*P. alba* × *P. glandulosa*) × *P. tomentosa*	B330	43.65^{d}	0.3126^{cde}	0.770^{d}	83.32^{bcd}	19.33^{bc}
	B331	84.00^{c}	0.3384^{b}	0.841^{abc}	83.25^{bc}	19.63^{bc}
	B333	76.95^{c}	0.3372^{b}	0.822^{abc}	83.09^{abc}	20.04^{c}
Mean value of triploids		84.31	0.3249	0.818	83.85	19.14
Diploid control (*P. tomentosa*)	1319	46.95^{d}	0.3616^{a}	0.678^{e}	82.75^{ab}	24.44^{e}

Note: Different lower-case letters in the same column indicate significant differences for the same trait between these clones at a significance level of 0.05 (LSD test).

and wood properties, in only one round of breeding. The production of triploid poplar varieties characterized with fast growth, longer fibers, lower lignin and higher fiber content would have significant value in the efficient development of fiber timber, energy and carbon sink forests.

The Mainways and Methods for Polyploid Poplar Production

Polyploid breeding of plants embraces the breeding of spontaneous and artificial polyploids. In general, spontaneous poplar polyploids have experienced long-term natural selection owing to their vigor in adaptation and then have possessed with strong genetic stability and adaptability. However, the number of spontaneous poplar polyploids is limited. Identifying these polyploids from the huge natural population is also difficult, ultimately limiting the utilization of spontaneous polyploids. Thus, as well as cross-pollination between plants with different ploidy levels, artificial chemical and physical induction of chromosome doubling in reproductive and somatic cells is a practical way to obtain poplar polyploids. Specifically, poplar polyploidy scan be obtained by hybridization with spontaneous or artificially-induced 2*n* pollen, or by chromosome doubling of the megaspore, embryo sac, zygote or somatic cells.

Triploid Poplar Breeding by Hybridizing with Spontaneous 2n Gametes

Hybridizing with spontaneous 2*n* gametes is historically the most economic and fastest route for producing triploids. In terms of *Populus*, Seitz (1954) first discovered that bisexual flowers of *P. canescens* (female inflorescences with some anthers) could produce some large-sized pollen without meiosis. Self-pollination with these large-size pollen resulted in about 1% triploid production. Manzos (1960) obtained two triploid hybrids by screening for heavier pollen grains from *P. balsamifera* and pollinating female flowers with these pollen grains. Mohrdiek (1976) also produced some triploids, of which the ground diameter, DBH and height were 7-8% higher than that of diploids, by pollinating *P. canescens, P. alba,* and *P. tremula* with 2*n* pollen of *P. canescens*. Zhu et al. (1995) obtained 26 allotriploids with superior growth and wood quality by pollinating *P. tomentosa* × *P. bolleana* and *P. alba* × *P. glandulosa* with spontaneous 2*n* pollen of *P. tomentosa*. However, the utilization of spontaneous 2*n* pollen is difficult, due to its low rate of production and poor competitive ability with n pollen for fertilization.

Unlike 2*n* pollen, 2*n* female gametes cannot be easily detected by morphologic observation. Generally, 2*n* female gametes are detected based on the ploidy levels of hybrid progeny. Few researchers have reported poplar polyploids obtained by hybridization involving spontaneous 2*n* female gametes. However, it was reported that a tetraploid of *P.* × *euramericana* was produced by hybridizing with spontaneous 2*n* pollen, and using simple sequence repeat (SSR) analysis, it was proved that the female parent had produced spontaneous 2n female gametes via the mechanism of first division restitution (FDR) (Xi et al. 2012). Since 2*n* female gametescan overcome the problem of poor competitive ability in fertilization, 2*n* female gametes could be highly valuable for research and utilization, if female plants producing 2*n* gametes could be detected. Some spontaneous or artificial poplar varieties which have been detected to be triploids (Chen et al. 2004, Zhang et al. 2004, 2005) might have originated from spontaneous 2*n* female gametes. Unfortunately, the female parents for most of the triploid varieties have been unable to be identified.

The pollen larger than 37 μm is considered as 2*n* pollen (Mashkina et al. 1989a; Zhang & Li 1992). A relatively small number of researches have been conducted on the mechanisms of 2*n* gamete formation in poplar. From years of observation of pollen of 18 *P. tomentosa* clones from different locations, most of the male plants were found to produce 2*n* pollen (Table 2, 3; Fig. 1). Furthermore, the formation of 2*n* pollen, which is thought to be controlled mainly by genetic factors with subsidiary environmental effects, varied between different clones and plants with different ages. Based on cytological observation, $2n$ pollen of *P. tomentosa* was speculated to derive from two pathways: parallel spindle formation and abnormal cytokinesis resulting in the disordered formation of the cell plate (Kang 1996, 2002). Zhang and Kang (2010) found that the direction of the spindles in metaphase II was parallel, leading to dyad and triad formation at the tetrad stage of meiosis. Subsequently, using indirect immunofluorescence observation, it was speculated that $2n$ pollen formation was mainly caused by premature cytokinesis at the second meiotic division or by the fusion of poles of the spindles in some cells induced by abnormal orientation of microtubules (Zhang and Kang 2010). Additionally, occurrence of spontaneous $2n$ pollen was also reported in sections Tacamahaca and Aigeiros of *Populus*, including *P. pseudo-simonii*, *P. simonii*, *P. simonii* × *P. nigra* 'Tongliao', *P. simonii* × *P. euphratica*, *P. deltoids* and some clones of *P. deltoids* × *P. nigra* (Wang and Kang 2009, Wang et al. 2015, Tian et al. 2015). Based on the analysis of microtubular cytoskeletons during microsporogenesis, it was speculated that abnormal cytokinesis resultingfrom lack of phragmoplasts in telophase II was the major cytological mechanism for the formation of $2n$ pollen (Fig. 2) (Wang and Kang 2009).

Table 2. Production of 2n pollen grains in the Chinese white poplar clones tested.

Provenance	No. of clones	No. of clones produced $2n$ pollen	Frequency of $2n$ pollen	
			Average (%)	Range (%)
Anhui	5	5	1.8	0.7~6.0
Beijing	4	4	7.0	3.0~10.8
Gansu	5	5	3.5	2.6~15.5
Hebei	62	62	2.3	0.0~19.6
Henan	47	44	3.1	0.0~21.9
Jiangsu	5	5	7.5	3.7~11.2
Ningxia	1	1	—	—
Shanxi	60	60	2.1	0.6~15.6
Shaanxi	35	32	2.9	0.0~17.3
Sum	224	218		

Table 3. Analysis of the incidence and frequency of 2n pollen grains in Chinese white poplar within and among indigenous populations.

Source of variation	d.f.	SS	MS	F
Provenance	8	6110.18	763.77	2.31*
Clone/Provenance	223	73613.17	330.18	5.94**
Error	3128	174079.50	55.63	
Sum	3359	253760.90		

*,**Significant at $P = 0.05$ and 0.01 levels, respectively.

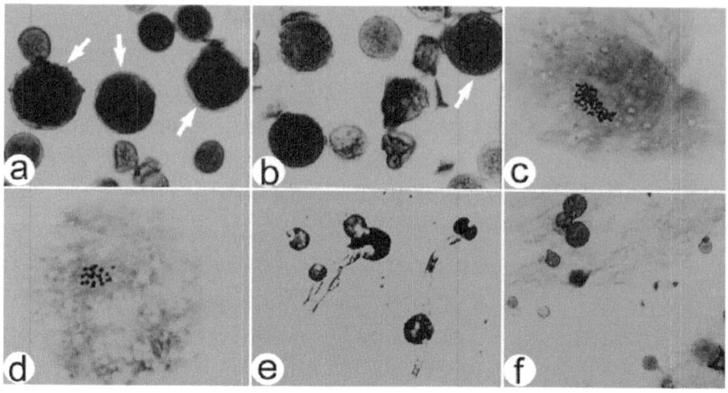

Figure 1. Morphology of spontaneous 2*n* pollen of *Populustomentosa*.
a, b. Spontaneous 2*n* pollen of *P. tomentosa* (arrows indicate 2*n* pollen); c. Chromosome number of 2*n* pollen ($n = 2x = 38$); d. Chromosome number of normal pollen ($n = x = 19$); e. 2*n* pollen germination on the culture medium; f. 2*n* pollen germination on the stigma of *P. tomentosa* × *P. bolleana*.

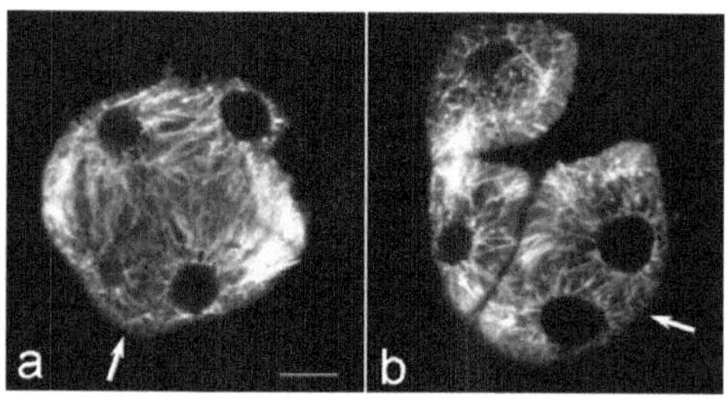

Figure 2. Lack of Phragmoplasts and Abnormal Cytokinesis During Meiosis of Microspore Mother Cells of *P. simonii* and *P. simonii* × *P. nigra* 'Tongliao'.
a. Atypical telophase II, showing the secondary phragmoplast formed between one pair of non-sister nuclei, and the lack of development between the other pair (arrows); b. Microtubular arrays in a triad, the arrow indicates the lack of a cell plate. Bar = 5 μm.

Triploid Poplar Breeding by Crossing with Artificial 2n Pollen

The fastest method for artificial polyploid breeding in trees is the artificial induction of chromosome doubling in pollen. Using colchicine solution, artificial 2*n* pollen was firstly obtained in *P. tremula* and *P. tremuloides*, and triploid hybrids were produced by hybridizing with the 2*n* pollen (Johnsson and Eklundh 1940). Subsequently, a certain ratio of 2*n* pollen was obtained by colchicine solution treatment in *P. deltoides*, *P. balsamifera*, *P. alba*, *P. tomentosa* × *P. bolleana*, *P. alba* × *P. glandulosa*, *P. ussuriensis*, and *P. pseudo-simonii* in succession (Gulyaeva and Sviridova 1979, Zhang and Li 1992, Kang 1996, Kang et al. 1999, Huang et al. 2002, Li et al. 2006, Wang 2009). The efficiencies of colchicine, propyzamide, oryzalin and trifluralin solution for producing 2*n* pollen were compared by Huang et al. (2002), and it was proved that colchicine solution was the most efficient (although relatively expensive) reagent. Alternatively, as a substitute, the rate of 2*n* pollen induction was 84.4% when using propyzamide solution with a concentration of 200 μmol/L and treating four times (Huang et al. 2002).

In the artificial induction of pollenchromosome doubling, good knowledge of the cytogenetic features of the studied species can help us get twice the result with half the effort. It was indicated that the pachytene stage of meiosis was the most efficient period for treatment of pollen mother cells chromosome doubling insection *Populus*, increasing the rate of $2n$ pollen to 88% and hence significantly improving the efficiency of $2n$ pollen production (Figs. 3 and 4) (Kang 1996, Kang et al. 1999). Reaching a ratio of 100% artificial $2n$ pollen is likely impossible, due to mixing with a certain amount of normal pollen. Unfortunately, the rate of triploids obtained by hybridizing with artificial $2n$ pollen is often less than 0.1%, indicating that competitiveness of participation in fertilization after pollination is different between $2n$ pollen and normal haploid pollen. The poor competitive ability of fertilization of $2n$ pollen was found to be mainly caused by the relatively slow germination of $2n$ pollen (Kang and Zhu 1997). To solve this problem, various methods were applied to improve the competitive ability of the $2n$ pollen. Taking advantage of the difference insensitivity to ^{60}Co-γ ray radiation between $2n$ and normal pollen proved to be an efficient way (Table 4), and 16 triploid hybrids including the new triploid varieties BLXZ-1 and BLXZ-2 were obtained by using this method in section *Populus* (Kang 1996, Kang et al. 2000a).

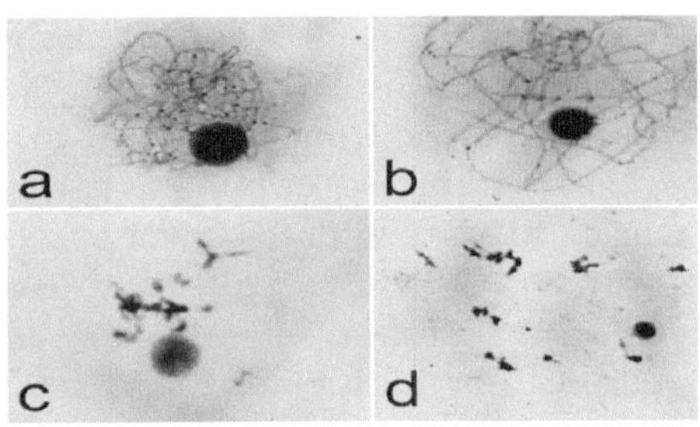

Figure 3. Prophase I of Microspore Mother Cells in *P. tomentosa* × *P. bolleana*.
a. leptotene; b. pachynema; c. diplonema; d. diakinesis.

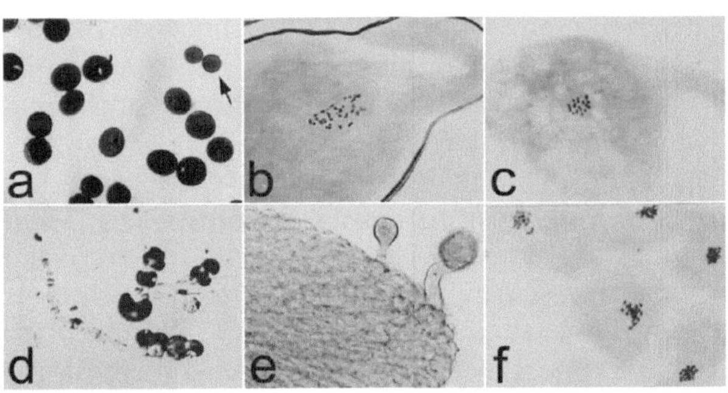

Figure 4. Pollen Chromosome Doubling by Colchicine Solution in *P. tomentosa* × *P. bolleana*.
a. The induced $2n$ pollen (the arrow indicates the normal pollen); b. Chromosome number of $2n$ pollen ($n = 2x = 38$); c. Chromosome number of normal pollen ($n = x = 19$); d–e. Induced $2n$ pollen germination on the culture medium and the stigma; f. Agglomerated chromosomes after treatment with colchicine solution.

Table 4. Percentage (%) of diploid and haploid pollen in which germination and sperm movement into the pollen tube occurred after different ^{60}Co-γ ray radiation doses.

Radiation dose (rad)	$p = 72.1\%$		$p = 85.1\%$		$p = 38.7\%$	
	2*n* pollen	*n* pollen	2*n* pollen	*n* pollen	2*n* pollen	n pollen
0	10.7	26.4	32.4	65.6	7.4	15.8
420	12.4	30.0	29.2	75.0	11.4	31.2
840	17.9	39.5	28.6	59.2	12.5	28.1
1,260	21.9	31.6	—	—	—	—
1,470	12.1	17.2	36.5	58.3	15.5	19.8
2,100	16.9	25.0	40.2	64.9	—	—
2,520	10.0	15.2	34.6	55.1	9.3	14.9
3,360	8.2	13.3	28.1	47.2	—	—
4,200	—	—	20.0	36.4	7.8	12.7
5,040	7.8	16.7	25.7	28.6	—	—

Compared with chemical treatment, physical treatment has the advantages of simplicity, low cost, and the possibility to induce a large amount of material at once. The rate of 2*n* pollen induction was increased to 94.4% by treating microsporocytes at prophase I in *P. deltoides*, *P. balsamifera*, *P. alba*, and *P. alba* × *P. tremula* with 38-40°C for 1.5-2 h (Mashkina et al. 1989a). Kang et al. (2000b) proved that the period from diakines is to metaphase I was the optimal treatment period for inducing 2*n* pollen via high temperatures and the rate of 2*n* pollen was up to 87.6% with the lower numbers of pollen grains. Results for *P. pseudo-simonii* and *P. simonii* × *P. nigra* 'Tongliao' were similar (Wang 2009). However, from current results, yield of 2*n* pollen is shown to be very difficult to guarantee by high temperature treatments, and further study of related technology is necessary.

In practice however, the effect of crossing with 2*n* pollenon induction of triploid poplars was limited, as the rate of triploids obtained by hybridizing with artificial 2*n* pollen was low because of poor competition of 2*n* pollen (Kang et al. 2000a). Hence, it is necessary to further explore methods of eliminating competition from normal pollen or enhancing the competitiveness of 2*n* pollen in future studies.

Triploid Poplar Breeding by Megaspore Chromosome Doubling

Unreduced megaspores (2*n* megaspores) can also be induced by physical and chemical treatment during meiosis of megasporocytes. In this case, there may be a 100% possibility of triploid hybrid formation by pollinating 2*n* female gametes with normal pollen. It was reported that triploids were obtained by treating female buds (cultured in water for 1-5 d) with colchicine solution and high temperature and then pollinating these female gametes with normal pollen. Based on the developmental stages of the female buds, it was inferred that megasporechromosome doubling was derived from the inhibition of metaphase I or metaphase II (Li et al. 2000, 2001). Compared with pollen chromosome doubling, the effective meiotic stages for inducing megaspore chromosome doubling were more difficult to determine, as instant determination of the meiosis stage in megasporocytes is not possible: unlike using squashed anthers as experimental material and an aceto-carmine solution as the stain to instantly determine the meiosis stage of microsporocytes, it takes several days to investigate the meiotic process in megasporocytes by observing paraffin sections. Ultimately, the induction of 2*n* megaspores and its application in breeding has so far been limited, as research on megaspore chromosome doubling has been conducted only through batch treatment and chromosome ploidy screening (Li et al. 2000, 2001), entailing a heavy workload, low yield of triploids and poor repeatability.

To solve the problem of timely determination of the meiotic process occurring in megasporocytes, and to subsequently avoid blindness in application of chromosome doubling treatments, is key to the breakthrough of megaspore chromosome doubling as atriploid breeding technology. A close association between developmental processes in female and male flower buds of *P. alba* × *P. glandulosa* has been identified (Li et al. 2005, Li 2007). When the meiotic stage of the microsporocytes developed into tetrads, the meiotic stage of the megasporocytes was just beginning leptotene; and when the meiotic stage of the microsporocytes developed into pollen with a single nuclear adjacent to cell wall, the meiotic stage of the megasporocytes was just beginning pachytene (Table 5). Subsequently it was shown that the problem of instant determination of the meiosis process in megasporocytes could be solved by referring to the developmental process of the male gametes. As a result, 16.7% triploids were obtained by treating female flower buds with 5% colchicine solution when the meiotic stage of the microsporocytes developed into tetrads (Li 2007, Li et al. 2008). This method prompted increased practicality of triploid poplar breeding by megaspore chromosome doubling. However, some poplar varieties lack a male reference. Thus, another method for instant determination of the meiotic stage in megasporocytes was supplied by cytological analysis combined with morphological characteristics of female flower buds, allowing 159 triploids to be obtained by chemical and physical treatments (Wang, 2009). It was also proved that the period from pachytene to diplotene in megasporocytes was the optimal treating period for megaspore chromosome doubling using colchicine solution (the rate of triploids was 13.04%), while the period from pachytene to diplotene was the optimal treating period for megaspore chromosome doubling using high temperature treatment (the rate of triploids was up to 60%).

Based on either the morphological characteristics of female flower buds or the developmental process of male gametes, the treatment of megaspore chromosome doubling was still not as well timed as that of pollen chromosome doubling in which instant determination of the meiotic stage is possible.

Table 5. Association between microsporocyte and megasporocyte development in *P. alba* × *P. glandulosa*.

The meiosis stages of microsporocyte and the male gamete development			The meiosis stages of megasporocyte	
The meiosis stages	**Color of anthers**	**Unfolded catkin/ bud ratio**	**The meiosis stages**	**Unfolded catkin/bud ratio**
Tetrad	yellow-green	1/3	Leptotene	1/5
Tetrad	yellow-green	1/2	Leptotene, late leptotene	1/4
Tetrad, mid-uninucleate microspore	yellow	1/2	Late leptotene	1/2
Pollen with a single nucleus adjacent to cell wall	reddish	3/4	Late leptotene, pachytene	1/2
Pollen with a single nuclear adjacent to cell wall	red	>3/4	Diplotene, pachytene	1/2
Binucleate pollen	deep red	>3/4	Late leptotene, pachytene, diakinesis, metaphase I	>1/2

However, as a result chromosome doubling treatments on megasporocytes could be focused to a shorter time period to reduce workload and to improve the efficiency of triploid production.

Triploid Poplar Breeding by Embryo Sac Chromosome Doubling

The embryo sac development of *Populus* is the "*Polygonum*" type, which is the most common type of embryo sac development. *Polygonum* type plants comprise 70% of studied plants (Maheshwari 1950), and have two stages: megasporogenes is and female gametogenesis (Drews et al. 1998, Hu 2005). The formation of the mature embryo sac is initiated from a functional megaspore via three rounds of mitotic division that each offer the possibility to induce 2*n* eggs.

Using *P. alba* × *P. glandulosa* as a female parent and *P. tomentosa* as a male parent and immersing female catkins in colchicine solution after pollination induced up to 57.1% triploids, with best results obtained by treating female catkins 24-36 h after pollination (Kang et al. 2004). The optimal treatment period for inducing triploids after pollination is the embryo sac development stage, according to a study of *P. alba* × *P. glandulosa* embryology (Li 2007). As the two sperm cells are formed 24 h before pollination, it can be concluded that the obtained triploids derived from colchicine solution treatment after pollinationoriginated from 2*n* female gametes derived from embryo sac chromosome doubling (Table 6).

Table 6. Analysis of embryo sac development stages after pollination in *P. alba* × *P. glandulosa*.

Hours after pollination (h)	Uninucleate embryo sacs (%)	Binucleate embryo sacs (%)	Tetranucleate embryo sacs (%)	Octanucleate embryo sacs (%)	Mature embryo sacs (%)	Embryo and free endosperm nuclei (%)
12	65.2	34.8				
24	40.9	54.5	4.5			
36	33.3	44.4	5.6	11.1	5.6	
48		18.2	27.3	18.2	36.3	
60			16.7	33.3	50.0	
72					33.3	76.7
96						100
120						100

The process of embryo sac development in *P. pseudo-simonii* × *P. nigra* 'Zheyin #3' was estimated by timing from its optimal pollination period, and 68 triploid hybrids of section Tacamahaca of *Populus* were obtained by embryo sac chromosome doubling (Wang 2009 Wang et al. 2010, 2012b). It was also shown that the tetranucleate embryo sac stage was the optimal treating stage for embryo sac chromosome doubling in *P. pseudo-simonii* × *P. nigra* 'Zheyin #3'. The rate of triploid production by embryo sac chromosome doubling via colchicine solution in these hybrids was up to 66.7%, with an optimal treatment period of 54-66 h after pollination, while embryo sac chromosome doubling via high temperature treatment resulted in 40% triploid production, with an optimal treatment period of 66-72 h after pollination. Moreover, for *P. adenopoda* and *P. tomentosa,* embryo sac chromosome doubling via high temperature treatment produced up to 80% triploids (Lu et al. 2013, Kang et al. 2015).

Obviously, the timing of treatment after pollination is also key to the success of embryo chromosome doubling. Currently, the ideal and realistic choice is to estimate the treatment period by timing from the optimal pollination period. However, this approach is deeply influenced by environmental effects. It is necessary to find a more effective method for determining the developmental stages of embryo sacs at the technical level.

Tetraploid Poplar Breeding By Zygote or Somatic Cell Chromosome Doubling

Somatic cell chromosome doubling was one of the earliest available methods for production of polyploid plants. Previous studies on induction of tetraploid plants mainly focused on treating seeds, apical buds, zygotes, immature embryos and callus with physical and chemical approaches (Mattila 1961, Pesina 1963, Einspahr 1965, Mashkina et al. 1989b, Wang 2009, Ewald et al.

2009, Cai and Kang 2011, Wang 2014). However, it is very difficult to induce chromosome doubling in all cells of an organism due to a synchronicity of cell division during the induction process, resulting in production of mixoploids or chimera with tissue of different ploidy levels. To avoid production of chimeras, tetraploid plants were produced by treating leaves with colchicine during regeneration of adventitious shoots from leaf explants of diploid *P. pseudo-simonii*. The highest efficiency of tetraploid induction (14.6%) was achieved by treating leaf explants that were pre-cultured for 6 d. All 36 tetraploids obtained were stable after being maintained *in vitro* for five cycles of subculture (Cai and Kang 2011). Similarly, chromosome doubling was induced by treating the leaves with colchicine during adventitious shoot regeneration from leaf explants of allotriploids in section Leuce of *Populus*, and the highest efficiency of hexaploid production was 3.57% (Wang 2014).

It is also noteworthy that chromosome doubling can be directly induced from zygotes. It was reported that tetraploids were produced by colchicine treatment during the first zygote division in *P. alba* (Mashkina et al. 1989b). However, because an indeterminate period of dormancy occurs in different plant species before entry into mitosis after the formation of the zygote in angiosperms (Hu 2005), the determination of the effective treatment period for zygotic chromosome doubling is still unresolved. Related studies were hence usually conducted through batch processing day by day with heavy workloads, low production rates and poor repeatability. Fast and accurate determination of the first zygotic division period is the key to successful induction of tetraploids. A method with which to instantly identify the zygote development stage using cotton-like fibres in the ovaries has been proposed (Wang et al. 2013). It was shown that when this cotton enclosed the base and middle of the ovules but did not cover all of the ovules, the majority of zygotes had finished the dormancy period and started the first mitotic division (Fig. 5). Based on chromosome doubling treatments during this period, six allotetraploids were obtained by treating with colchicine solution in (*P. pseudo-simonii* × *P. nigra* 'Zheyin #3') × *P.* × *beijingensis*, and 25 allotetraploids were obtained by treating with high temperature (Wang et al. 2013). Additionally, the relationship between zygote division stages and developmental changes of the cotton-like fiber in ovaries was also analyzed in *P. adenopoda*, and 32 tetraploids were obtained by inhibiting the first zygote division with high temperature exposure. The highest efficiency of tetraploid induction was 14.12%, showing that high temperature exposure during the first zygote division is an ideal method for tetraploid induction in *Populus* (Lu et al. 2014).

In summary, although there are many studies on tetraploid induction, the material for most of this somatic chromosome doubling research was stem tips or seeds, often producing mixoploids or chimeras due to asynchronism of cell divisions. To solve this problem, chromosome doubling of single cells could be undertaken by chemical and physical treatment of the zygote with the first division or adventitious shoots regenerated from leaf explants.

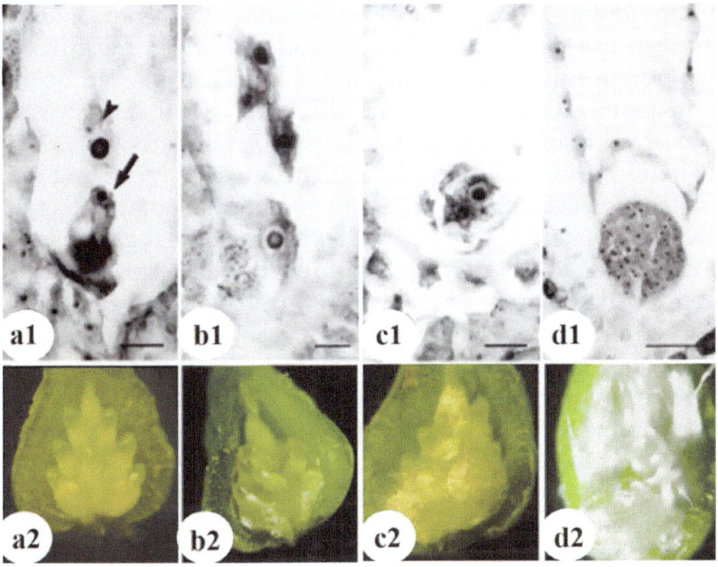

Figure 5. Development of zygote and cotton in ovary of (*Populus pseudo-simonii* × *P. nigra 'Zheyin #3'*) × *P.* × *beijingensis*

a1. Double fertilization, arrow showing the fusion between one sperm and the egg, arrow head showing the fusion between one sperm and the central cell; b1. Resting zygote cell; c1. Two-celled embryo; d1. Spherical embryo. Bar = 10 μm. a2. Initiation of cotton around the funiculus in ovary, but able to be seen by the naked eye; b2. The cotton covers the funiculus lightly and can be seen by the naked eye; c2. The cotton begins to enclose the ovules, but does not fill the ovary; d2. The cotton encloses the ovules completely and gradually fills the ovary.

Important Aspects of Chromosome Doubling and Polyploidy Breeding in Poplar

Knowledge of Reproductive Biology Characteristics is the Key to Improving Chromosome Doubling Efficiency in Poplar

Chromosome doubling is a very complex process, and so far, the mechanism of disturbance of normal cell division by physical and chemical treatment is not fully clear. At the cellular level, chromosome doubling has been mainly attributed to the effect of applying physical or chemical treatment to the formation of the spindle and the cell plate. When the spindle assembly function of tubulin is inhibited, the formation and contraction mechanism of the spindle is affected, or when the polarity of cell is disturbed, cells in metaphase would be prevented from entering anaphase, and consequently will form nuclei with double the chromosome number. In addition, although chromosomes may be properly separated to different poles, the number of chromosomes in the cell may also be multiplied due to the disordered division of the cytoplasm when the assembly functionality of the cell plates is inhibited (Jordan and Wilson 1999, Kang et al. 2000b). Therefore, both for chromosome doubling in gametes and for somatic chromosome doubling, the best induction efficiency can only be obtained by a timely treatment.

The spindle formation and contraction mechanisms can be affected by chemical reagents such as colchicine solution, which interferes with cells in metaphase and consequently forms nuclei with doubled chromosome number. Therefore, in terms of inducing chromosome doubling of female or male gametes by treating with colchicine solution, some time is needed for the colchicine to penetrate and affect the flower buds, regardless of whether injection or immersion is used. Identifying the best treatment period can guarantee that the metaphase stage of cell division is affected by chemical reagents. The period from late leptotene to pachytene of megasporegenesis and microsporegenesis is the optimal treatment period for colchicine solution (Kang et al. 1999, Li 2007, Wang 2009, Wang et al. 2012a). If the treatment period is early, the impact of the chemical reagents will be reduced by cell division auto-regulation, necessitating increased treatment duration that causes damage to flower buds and results in a waste of manpower and material. If the treatment period is late, the metaphase of cell division will be completed, and the chemical reagents cannot fulfil their roles (Kang et al. 1999). Hence, determination of the optimal treatment period is very important for improving the efficiency of chromosome doubling.

Chromosome doubling induced by physical treatments such as high temperature are advantageous in that they comprise simple, low-cost operations that can process a large number of materials at once. At the cellular level, the formation of the spindle and the cell plate are affected after applying physical treatments such as high temperature. The period from diplotene to metaphase I of megasporogenesis and microsporogenesis was found to be the optimal high temperature treatment period for megaspore and pollen chromosome doubling (Kang et al. 2000a, Wang et al. 2012a). The optimal physical treatment period for high temperature was later than that of colchicine, because heat is transmitted more directly and quickly than chemical reagents. During spindle formation close to metaphase I, the spindle structure can be changed and its function disordered by high temperature treatment. As a consequence, $2n$ gametes can be obtained by chromosome doubling without chromosomes moving to the two poles.

The Utilization of Heterozygosity in Poplar Polyploid Breeding Should Attract More Attention

The performance of polyploids seems to rely strongly on heterozygosity (Bingham 1980). Besides the genomic dosage effect caused by addition of a whole set of chromosomes, heterosis can also be intensified by the union of heterozygous gametes to produce allopolyploids. The performance of poplar polyploids is closely related to their heterozygosity. Autotetraploid poplar has been shown to have poor breeding performance in practice (Mattila 1961, Pesina 1963). All spontaneous or artificial poplar triploids with outstanding performance were the product of interspecific hybridization, such as *P.* × *canadensis* 'I-214', triploid hybrids "Astria", triploid *P. tomentosa*, *P.* × *euramericana* 'Zhonglin-46' and *P. alba* × *P. berolinensis* (Weisgerber et al. 1980, Zhu et al. 1995, Chen et al. 2004, Zhang et al. 2004, 2005). However, the height and diameter growth in triploids derived from gametic chromosome doubling were higher than those of diploids from the interspecific hybrid combination, and there were differences in growth during the seedling stage among triploids from different interspecific hybrid combinations (Kang 2002, Li 2007, Wang 2009, Wang et al. 2012b). Parental selection is very important in poplar polyploid breeding (Kang and Wang 2010).

The heterozygosity of polyploid offspring not only depends on the selection of both parents, but is also closely related to the selection of chromosome doubling pathways. In terms of chromosome doubling in plants, types of 2*n* gametes can be divided into FDR (first-division restitution), SDR (second-division restitution) and PMR (post-meiotic restitution) types, according to stages of nuclear restitution. The heterozygosity of different types of 2*n* gametes varies dramatically due to their different formation mechanisms. FDR 2*n* gametes are more than twice as effective as SDR 2*n* gametes in the transmission of heterozygosity (Hermsen 1984, Peloquin et al. 2008). Using RFLP markers, it was shown that the heterozygosity transmitted through FDR and SDR 2*n* gametes in potato was 71.4% and 31.8%, respectively (Barone et al. 1995). In *Populus*, it was reported that parental heterozygosity transmitted by SDR and PMR 2*n* eggs were 39.58% and 35.90%, respectively. In addition, the maternal heterozygosity transmitted by FDR 2*n* eggs (74.80%) was significantly higher than that of both SDR and PMR 2*n* eggs ($P<0.01$) (Dong et al. 2015).

Generally, due to their increased capability in transmitting parental heterozygosity, FDR 2*n* gametes have higher utilization value in breeding (Bingham 1980, Peloquin et al. 2008). However, in addition to the process of 2*n* gamete formation, the level of heterozygosity transmitted via 2*n* gametes depends on whether homologous recombination occurs during meiosis. Homologous recombination reduces the heterozygosity transmitted by FDR 2*n* gametes and increases the heterozygosity transmitted by SDR and PMR 2*n* gametes (Dong et al. 2015). Although parental heterozygosity transmitted by SDR and PMR 2*n* eggs is reduced compared to FDR 2*n* eggs, loci with the greatest frequency of homologous recombination tend to be located in regions with high numbers of protein-coding genes (Dong et al. 2014), and hence specific target genes related to traits of interest can be heterozygous in SDR and PMR 2*n* eggs due to homologous recombination. This indicates that the breeding value of SDR and PMR 2*n* gametes is not necessarily lower than that of FDR 2*n* gametes in individual selection.

On the basis of poplar breeding in practice, growth traits of triploid hybrids resulting from FDR 2*n* gametes were found to be significantly better than those of triploid hybrids from SDR and PMR 2*n* gametes. Some special genotypes of triploid hybrids from SDR and PMR 2*n* gametes also have outstanding performance (Kang and Wang 2010). It should be suggested that for poplar with asexual reproduction, all three types of 2*n* gametes should have a certain value. What is worthy of attention is that there have been nearly a hundred years of accumulated crossbreeding in poplar, and almost all of the important species of poplar have been crossbred. Based on experience, crossbreeding appears beneficial for improving the effectiveness of poplar polyploid breeding and in selecting parent combinations with prominent heterosis.

A Strategy of "Large Population, Strong Selection" Will Guarantee Success in Poplar Polyploid Breeding

Current poplar breeding research is still in the conventional breeding stages of seed selection, introduction and hybridization, which provide limited opportunities to further improve poplar growth and wood properties. By combining ploidy vigor and heterosis, polyploidy breeding could play a crucial role in poplar genetic improvement.

Triploid breeding can produce new varieties with superior growth and wood quality in poplar, as demonstrated by previous research (Weisgerber et al. 1980, Zhu et al. 1995, Chen et al. 2004, Zhang et al. 2004, 2005). However, solving the technical problems of chromosome doubling is not enough for polyploid breeding: further selection of triploid hybrids is also necessary. The reason is that growth traits of poplar triploid hybrids from sexual polyploidization within the same cross show normal segregation for trait phenotypes due to the effect of heterozygosity, and hence there is a big difference in growth between different triploid individuals. In addition, although growth traits of some triploids are excellent, triploids may still have other problems that need to be addressed, such as low wood basic density and the potential to be easily broken by wind. Allotriploids of *P. tremula* × *P. tremuloides* from America have been eliminated from most breeding programs because they are easily broken by wind after planting.

A recently completed comparative trial showed differences in growth and wood properties between different cross combinations and between different triploid individuals within the same cross combination (Table 1). In terms of different cross combinations, the performance of triploid clones of BLXZ-1 and BLXZ-2 which derived from pollinating *P. tomentosa* × *P. bolleana* with 2*n* pollen of *P. alba* × *P. glandulosa* induced by colchicine solution were 120% and 159% respectively higher than the diploid control in growth of timber volume. Additionally, compared with a diploid control, triploid clones of BLXZ-1 and BLXZ-2 had a high holocellulose content, low lignin and similar wood basic density. At present, triploid clones of BLXZ-1 and BLXZ-2 have passed the national examination and approval of improved varieties of forest trees, subsequent to increased popularization and application in China. On the other side, as an example of different triploid individuals within the same cross combination, triploid hybrids of (*P. tomentosa* × *P.*

bolleana) × *P. tomentosa* originating from pollination with spontaneous 2*n* pollen show differences in growth volume, fiber morphology and chemical composition between different triploid individuals, especially for wood basic density, which for triploid hybrid clones of B312 was less than 0.3 g/cm^3.

Therefore, after determining the optimal cross combinations between parent lines and the chromosome doubling pathway, the number of triploid hybrids should be increased as much as possible through technological optimization of chromosome doubling and maintenance of hybrid seedlings in the poplar breeding process. Only on the basis of "large population, strong selection", is there an increased possibility of breeding additional new and excellent triploid varieties of poplar. As a prediction, and as a result of the worldwide shortage of forest resources and intensification of environmental problems, polyploid breeding will play a more important role in future breeding of new varieties with excellent performance in wood property, growth volume and stress resistance.

References

Barone, A., C. Gebhardt and L. Frusciante. 1995. Heterozygosity in 2*n* gametes of potato evaluated by RFLP markers. Theor. Appl. Genet. 91: 98-104.

Bingham, E.T. 1980. Maximizing heterozygosity in autopolyploids. pp. 471-489. *In*: W.H. Lewis (ed.). Polyploidy: Biological Relevance. Plenum Press, New York.

Cai, X. and X.Y. Kang. 2011. *In vitro* tetraploid induction from leaf explants of *Populus pseudo-simonii* Kitag. Plant Cell Rep. 30: 1771-1778.

Chen C.B., L.W. Qi, S.G. Zhang, S.Y. Han, X.L. Li, W.Q. Song and R.Y. Chen. 2004. The karyotype analysis of triploid poplar. Journal of Wuhan Botanical Research 22: 565-567. (in Chinese)

Dewey, D.R. 1980. Some applications and misapplications of induced polyploidy in plant breeding. pp. 445-470. *In*: W.H. Lewis (ed.). Polyploidy: Biological Relevance. Plenum Press, New York.

Dong, C.-B., Y.-J. Suo, J. Wang and X.-Y. Kang. 2015. Analysis of transmission of heterozygosity by 2*n* gametes in *Populus* (Salicaceae). Tree Genet. Genomes 11: 799.

Dong, C.-B., J.-F. Mao, Y.-J. Suo, L. Shi, J. Wang, P.-D. Zhang and X.-Y. Kang. 2014. A strategy for characterization of persistent heteroduplex DNA in higher plants. Plant J. 80: 282-291.

Drews, G.N., D. Lee and C.A. Christensen. 1998. Genetic analysis of female gametophyte development and function. Plant Cell 10: 5-17.

Einspahr, D.W. 1965. Colchicine treatment of newly formed embryos of quaking Aspen. For. Sci.11: 456-459.

Einspahr, D.W. and G.W. Wyckoff. 1975. Aspen hybrids promise future source of Lake States fiber. Pulp and paper 49: 118-119.

Einspahr, D.W. 1984. Production and utilization of triploid hybrid aspen. Iowa State J. Res. 58: 401-409.

Ewald, D., K. Ulrich, G. Naujoks and M.B. Schröder. 2009. Induction of tetraploid poplar and black locust plants using colchicine: chloroplast number as an early marker for selecting polyploids *in vitro*. Plant Cell Tiss. Organ Cult. 99: 353-357.

Fang, G.G., Y.J. Deng and P. Li. 2001. Evaluation of pulping properties of triploid *Populustomentosa*. Forestry science and technology management 87-90.

Gulyaeva, E.M. and A.D. Sviridova. 1979. Method of producing diploid pollen in forest trees. U.S.S.R. Patent # 664, 617.

Gurreiro, M.G. 1944. The silvicultural improvement of *Populus*. Publ. Serv. Flor. Aquic., Portugal, 11(1/2): 53-117.

Hermsen, J.G. 1984. Mechanisms and genetic implications of 2*n*-gamete formation. Iowa State J. Res. 58: 421-434.
Hu, S.Y. 2005. Reproductive Biology of Angiosperm. Higher Education Press, Beijing. (in Chinese)
Huang, Q.J., Z.Y. Zhang and X.Y. Kang. 2002. 2*n* pollen of *P. tomentosa* × *P. bolleana* induced by four antimicrotubule agents. Journal of Beijing Forestry University 24: 12-15. (in Chinese)
Huang, Z., C. Xu, Y. Li, P. Wang, Y. Li and X. Kang. 2015. Induction of somatic embryogenesis by anther-derived callus culture and plantlet ploidy determination in poplar (*Populus* × *beijingensis*). Plant Cell Tiss. Organ Cult. 120: 949-959.
Johnsson, H. and C. Eklundh. 1940. Colchicine treatment as a method in breeding hardwood species. Svensk Papp Tidn. 43: 373-377.
Jordan, M.A. and L. Wilson. 1999. The use and action of drugs in analyzing mitosis. Method Cell Biol. 61: 267-295.
Kang, N., F.Y. Bai, P.D. Zhang, Y.W. Luo and X.Y. Kang. 2015. Induction of chromosome doubling of embryo sac in *Populus tomentosa* by high temperature exposure to produce hybrid triploids. Journal of Beijing Forestry University 37: 79-86. (in Chinese)
Kang, X.Y. 1996. Cytogenesis and triploid breeding of Chinese white poplar. Ph.D. Thesis, Beijing Forestry University, Beijing. (in Chinese)
Kang, X.Y. and Z.T. Zhu. 1997. A study on the 2*n* pollen vitality and germinant characteristics of white poplars. Acta Botanica Yunnanica 19: 395-401. (in Chinese)
Kang, X.Y., Z.T. Zhu and H.B. Lin. 1999. Study on the effective treating period for pollen chromosome doubling of *Populus tomentosa* × *P. bolleana*. Scientia Silvae Sinicae 35: 21-24. (in Chinese)
Kang, X.Y., Z.T. Zhu and Z.Y. Zhang. 2000a. Suitable period of high temperature treatment for 2*n* pollen of *Populus tomentosa* × *P. bolleana*. Journal of Beijing Forestry University 22: 1-4. (in Chinese)
Kang, X.Y., Z.T. Zhu and H.B. Lin. 2000b. Radio sensitivity of different ploidy pollen in poplar and its application. Acta Genetica Sinica 27: 78-82. (in Chinese)
Kang, X.Y. 2002. Mechanism of 2*n* pollen occurring in Chinese white poplar. Journal of Beijing Forestry University 24: 67-70. (in Chinese)
Kang, X.Y., P.D. Zhang, P. Gao and F. Zhao. 2004. Discovery of a new way of poplar triploids induced with colchicine after pollination. Journal of Beijing Forestry University 26: 1-4. (in Chinese)
Kang, X.Y. and J. Wang. 2010. Studies on Techniques of Polyploid Induction in *Populus* spp. Science Press, Beijing. (in Chinese)
Lewis, W.H. 1980. Polyploidy: Biological Relevance. Plenum Press, New York.
Li, K.L., J. Xiao, G.F. Liu and Z.X. Li. 2006. Optimization of inducing 2*n* pollen grain of *Populusus suriensis* Kom by colchicines. Journal of Nuclear Agricultural Sciences 20: 282-286. (in Chinese)
Li, Y.H., J. Ma and X.Y. Kang. 2005. Stages of MMC meiosis and its timely discrimination of white poplars. Journal of Beijing Forestry University 27: 70-74. (in Chinese)
Li, Y.H. 2007. Chromosome doubling of female gametes in white poplars. Ph.D. Thesis, Beijing Forestry University, Beijing. (in Chinese)
Li, Y.H., X.Y. Kang, S.D. Wang, Z.H. Zhang and H.W. Chen. 2008. Triploid induction in *Populus alba* × *P. glandulosa* by chromosome doubling of female gametes. Silvae Genet. 57: 37-40.
Li, Y., Z.T. Zhu, Y.T. Tian, Z.Y. Zhang and X.Y. Kang. 2000. Obtaining triploids by high and low temperature treating female flower buds of white poplar. Journal of Beijing Forestry University 22: 7-12. (in Chinese)
Li, Y., Z.T. Zhu, Y.T. Tian, Z.Y. Zhang and X.Y. Kang. 2001. Studies on obtaining triploids by colchicine treating female flower buds of white poplar. Scientia Silvae Sinicae 37: 68-74. (in Chinese)
Lu, M., P.D. Zhang and X.Y. Kang. 2013. Induction of 2*n* female gametes in *Populus adenopoda* Maxim by high temperature exposure during female gametophyte development. Breeding Sci. 63: 96-103.

Lu, M., P.D. Zhang, J. Wang, X.Y. Kang, J.Y. Wu, X.J. Wang and Y. Chen. 2014. Induction of tetraploidy using high temperature exposure during the first zygote division in *Populus adenopoda* Maxim. Plant Growth Regul. 72: 279-287.

Maheshwari, P. 1950. An Introduction to the Embryology of Angiosperms. McGraw-Hill, New York.

Manzos, A.M. 1960. Fast-growing form of *Populus balsamifera* obtained by pollinating female flowers with fractionated pollen of the same species. Dokl. Akad. Nauk. SSSR. 130: 433-435.

Mashkina, O.S., L.M. Burdaeva and M.M. Belozerova. 1989a. Method of obtaining diploid pollen of woody species. Lesovedenie 1: 19-25.

Mashkina, O.S., L.M. Burdaeva and L.N. V'yunova. 1989b. Experimental mutagenesis and polyploidy in breeding forest trees. Lesnaya Genetika 136-137.

Mattila, R.E. 1961. On the production of the tetraploid hybrid Aspen by colchicine treatment. Hereditas 47: 631-640.

Mohrdiek, O. 1976. Progeny studies in poplars of the sections Aigeiros, Tacamahaca and Leuce, with recommendations for further breeding work. Thesis, Georg August Universität, Göttingen, German Federal Republic.

Nilsson-Ehle, H. 1936. Note regarding the gigas form of *Populus tremula* found in nature. Hereditas 21: 372-382.

Peloquin, S.J., L.S. Boiteux, P.W. Simon and S.H. Jansky. 2008. A chromosome-specific estimate of transmission of heterozygosity by 2*n* gametes in potato. J. Hered. 99: 177-181.

Pesina, K. 1963. Experimental induction of polyploidy in poplars. Preslia 35: 101-109.

Seitz, F.W. 1954. The occurrence of triploids after self-pollination of anomalous and rogynous flowers of a grey poplar. Z. Forstgenet. 3: 1-6.

Tian, J., J.H. Wang, L. Dong, F. Dai and J. Wang. 2015. Pollen variation as a response to hybridisation in *Populus* L. section *Aigeiros* Duby. Euphytica. Doi: 10.1007/s10681-015-1507-z.

Van Buijtenen, J.P., P.N. Joranson and D.W. Einspahr. 1957. Naturally occurring triploid quaking aspen in the United States. Proc. Soc. Amer. For.: 62-64.

Van Dillewijn, C. 1939. Cytology and breeding of *Populus*. Ned. Boschb. Tijdschr. 12: 470-481.

Wang, J. 2009. Techniques of polyploidy induction in *Populus* spp. (Section Tacamahaca). Ph.D. Thesis, Beijing Forestry University, Beijing. (in Chinese)

Wang, J. and X.Y. Kang. 2009. Distribution of microtubular cytoskeletons and organelle nucleoids during microsporogenesis in a 2*n* pollen producer of hybrid *Populus*. Silvae Genet. 58(5/6): 220-226.

Wang, J., X.Y. Kang, D.L. Li, H.W. Chen and P.D. Zhang. 2010. Induction of diploid eggs with colchicine during embryo sac development in *Populus*. Silvae Genet. 59: 40-48.

Wang, J., D.L. Li and X.Y. Kang. 2012a. Induction of unreduced megaspores with high temperature during megasporogenesis in *Populus*. Ann. Forest Sci. 69: 59-67.

Wang, J., X.Y. Kang and D.L. Li. 2012b. High temperature-induced triploid production during embryo sac development in *Populus*. Silvae Genet. 61: 85-93.

Wang, J., L. Shi, S. Song, J. Tian, X.Y. Kang. 2013. Tetraploid production through zygotic chromosome doubling in *Populus*. Silva Fenn. 47: 932.

Wang, J., H.L. You, J. Tian, Y.F. Wang, M.H. Liu and W.L. Duan. 2015. Abnormal meiotic chromosome behavior and gametic variation induced by intersectional hybridization in *Populus* L. Tree Genet. Genom. 11(3): 61.

Wang, P.Q. 2014. Studies on hexaploid induction of poplar. M.S. Thesis, Beijing Forestry University, Beijing. (in Chinese)

Weisgerber, H., H.M. Rau, E.J. Gartner, G. Baumeister, H. Kohnert and L. Karner. 1980. 25 years of forest tree breeding in Hessen. Allg. Forestz. 26: 665-712.

Xi, X.J., D. Li, W.T. Xu, L.Q. Guo, J.F. Zhang and B.L. Li. 2012. 2*n* egg formation in *Populus* × *euramericana* (Dode) Guinier. Tree Genet. Genomes 8: 1237-1245.

Xu, W.Y. 1988. Poplar. Heilongjiang People's Publishing House, Harbin. (in Chinese)

Yao, C.L. and J.W. Pu. 1998. Timber characteristics and pulp properties of the triploid of *Populus tomentosa*. Journal of Beijing Forestry University 20: 18-21. (in Chinese)

Zhang, S., L. Qi, C. Chen, X. Li, W. Song, R. Chen and S. Han. 2004. A report of triploid *Populus* of the section Aigeiros. Silvae Genet. 53: 69-75.

Zhang, S., C.B. Chen, S.Y. Han, X.L. Li, J.Z. Ren, Y.Q. Zhou, W.Q. Song, R.Y. Chen and L.W. Qi. 2005. Chromosome numbers of some *Populus* taxa from China. Acta Phytotaxonomica Sinca 43: 539-544. (in Chinese)

Zhang, Z.H. and X.Y. Kang. 2010. Cytological characteristics of numerically unreduced pollen production in *Populus tomentosa* Carr. Euphytica 173: 151-159.

Zhang, Z.Y. and F.L. Li. 1992. Studies on chromosome doubling and triploid breeding of white polar (I) – The techniques of the pollen chromosome doubling. Journal of Beijing Forestry University 14: 52-58. (in Chinese)

Zhu, Z.T., H.B. Lin and X.Y. Kang. 1995. Studies on allotriploid breeding of *Populus tomentosa* B301 clones. Scientia Silvae Sinicae 31: 499-505. (in Chinese)

Zhu, Z.T., X.Y. Kang and Z.Y. Zhang. 1998. Studies on selection of natural triploid of *Populus tomentosa*. Scientia Silvae Sinicae 34: 22-31.

Chapter 5

Musa Interspecific Hybridization and Polyploidy for Breeding Banana and Plantain (Musaceae)

Rodomiro Ortiz[1]

ABSTRACT

Banana and plantain (*Musa* spp.) are perennial, giant herbs grown mostly in small plots and orchards. The main centers of *Musa* diversity are in the tropics: from South Asia to Polynesia. All edible cultivars derived from the diploid species *M. acuminata* and *M. balbisiana*, which contribute the A and B genomes, respectively. Advances in DNA-aided analysis of diversity are providing new insights into *Musa* diversity, filling gaps and unraveling relationships among species and cultivars, which are often complex owing to interspecific hybridization, heterozygosity, and polyploidy. Seed set in triploid bananas and plantains – whose fruit develop through parthenocarpy– is very low due to high levels of sterility, thus making crossbreeding of both cultigens difficult, although some cultivars produce seed after crossing with diploid bananas. The genetic enhancement of banana and plantain is aimed at triploid cultivars whose multiplication is by vegetative means. The challenge is to breed using triploid cultigens – which are mostly sterile – as source populations, then release genetic variation by interploidy $3x \times 2x$ crossing involving interspecific hybridization to obtain tetraploid hybrids, and thereafter using interploid $4x \times 2x$ crossing with the aim of producing a non-seed bearing triploid cultivar. Interspecific hybridization, ploidy manipulations via $2n$ gametes, embryo culture, rapid *in vitro* multiplication, field-testing and selection led to development and identification of suitable plantain bred-germplasm for the West and Central Africa lowlands, and of beer or cooking bananas for the East African highlands. Genetic knowledge gives essential information to design crossing blocks targeting specific trait improvement according to combining ability, and may allow breeding gains to be predicted based on heritability and genetic correlations among traits.

[1] Swedish University of Agricultural Sciences, Department of Plant Breeding, Sundsvagen 14 Box 101, SE 23053 Alnarp, Sweden, Email: rodomiro.ortiz@slu.se

DNA markers, the genetic maps based on them, and the recent sequencing of the *Musa* genome offer means for gaining further insights into banana and plantain genetics, and to identify genes that may accelerate their betterment.

Introduction

The world produced 106.7 million t of banana with 21 t ha^{-1} average yield in 2013, while the plantain harvest was 37.9 million t, resulting from a yield of 6.9 t ha^{-1} (FAO 2015). The banana average annual growth rates (%) between 1961 and 2013 were 1.71 for area, 3.01 production and 1.28 for yield; whereas they were 1.54, 1,84 and 0.30 for cooking banana and plantain, respectively, in same period (FAO 2015). Most banana and plantain harvests are from relatively small plots and kitchen or backyard gardens.

Bananas and plantains are native to the tropics of Asia and Oceania but are found today throughout the tropics and subtropics. They are grown as perennial crops and their harvests are all year round in the tropics. According to FAOSTATS (FAO 2015), the main banana producers by country are India, China, Philippines, Brazil, and Ecuador; and for cooking banana and plantain are Uganda, Cameroon, Ghana, Colombia and Rwanda. The top banana exporters are Ecuador, Costa Rica, Philippines, Colombia and Guatemala; and the USA, Germany, Benelux (Belgium/Netherlands/ Luxembourg) and Japan are among the main importers (FAO 2015). Together they rank among the 10 most important food staples – particularly in the developing world – and the dessert type is among the most popular fruit elsewhere. Plantain and banana are very important in the diets of people living in the humid-lowland tropics worldwide, and in mid-altitude agro-ecologies of the Great Lakes of Africa. The average world consumption is 15 kg per head of population: 13 kg for the developed world and 21 kg for the developing world.

Musa Diversity and Banana/Plantain Cultigen Pools

The main centers of *Musa* diversity are in the tropics; from South Asia to Polynesia. The two main diploid ($2n = 2x = 22$ chromosomes) *Musa* species are *M. acuminata* and *M. balbisiana*, which contribute the A and B genomes, respectively. There are AA, BB, AB, AAA, AAB, AAAA, AAAB and AABB banana, AAA and ABB cooking banana, AAA beer banana and AAB plantain cultigens. Most triploid cultivars are almost completely sterile and develop fruit by parthenocarpy; i.e., without seeds due to lack of fertilization. Mutations reduce fertility in populations with predominantly clonal propagation, thus leading to sexual dysfunction and loss of sex (Barrett 2015). Although the crop originated in Southeast Asia, the West African lowlands and East African highlands are secondary centers of diversity for the triploid ($2n = 3x = 33$) plantain and East African highland banana cultigens, respectively. This may have resulted from accumulation of somatic mutations and thereafter through human selection during the long growth history of banana in Africa. Somatic mutations can influence fitness in clonal plants and may explain genetic diversity in sterile clonal populations (Barrett 2015).

Important qualitative morphological descriptors for distinguishing *Musa* species and cultivars are persistence of male bud and hermaphrodite flowers; pigmentation of the pseudostem foliage, petiole and male flower; pseudostem blotching and waxiness; and leaf orientation. Significant variation can be noticed for quantitative morphological descriptors such as pseudostem girth, height of tallest sucker, number of fruits and fruit sizes, which show a high heritability (>0.8), high repeatability (>2) and low coefficient of variation (9-15%) with the exception of the height of the tallest sucker (Ortiz 1997a). Principal component analysis of quantitative traits allows establishment of a more objective taxonomic relationship between cultivar groups and subgroups in triploid *Musa* germplasm (Osuji et al. 1997a). Fruit traits, number of hermaphrodite flowers, total number of leaves, plant girth at 50 cm, and days to flowering and harvest were the major discriminating traits among the polyploid cultigens. There were easily defined cultigen clusters for AAB plantains, AAA dessert bananas and ABB cooking bananas. The AAB starchy bananas were separated into two subgroups, with one group close to the AAB plantains and the other group close to the ABB cooking bananas. Quantitative descriptors were also used to develop a phenotypic diversity index (Ortiz et al. 1998a), which was further used for germplasm clustering and identification of duplicates. The phenotypic distance index, which was calculated as the average difference between each pair of accessions for all quantitative descriptors, was significantly different between *Musa* cultigen clusters. The between-cluster variance was larger than the within-cluster variance, and the ratio of these variances suggests little gene flow among triploid cultigen clusters via pollen, thus explaining the high population differentiation in this vegetatively propagated crop with very low male fertility, and confirming that the variation within each cultigen cluster arose mainly through mutations.

The specific genotype of the *Musa* accessions determines most of the quantitative trait variation, which is often significantly influenced by both the environment and the genotype-by-environment interaction (Ortiz and Vuylsteke, 1998a). The most productive cultivars are the Cavendish dessert bananas (AAA) and the giant French plantains (> 30 t ha^{-1} $year^{-1}$). Short-cycling cultivars show early flowering, while tall cultivars have wide plant girth and many leaves, as determined by phenotypic correlations. The number of fruits per bunch depends on the number of hands (nodal fruit clusters). Bunch weight is the most important potential yield component but growth cycle length (or days to harvest) is also important in AAA dessert bananas.

Index descriptors – based on the combination of two traits – are useful for indirect selection, particularly for characters difficult to score or showing a high coefficient of variation. For example, the index of pseudostem condensation (i.e., the ratio between plant height and total number of leaves, or the average of the distance between the insertion in the pseudostem of two leaf petioles) facilitates selection of short cultivars. Dwarf cultivars have low values for this trait (< 9 cm leaf^{-1}).

There are five clone sets among East African highland bananas; namely, Nfuuka, Musakala, Nakabulu, Nakitembe and Mbidde (Karamura 1999). Bunch type plus orientation, fruit size plus shape, male bud tip and floral bracts in the male inflorescence rachis are used to cluster cultivars in each of the first four clone sets, while the cultivars of the last clone set have a bitter and stringent pulp that shows sticky brown excretions. Plantain groups are defined by their inflorescence type and fruit number plus size as French, French Horn, False Horn, and Horn plantains (Swennen et al. 1995). Each group subdivides further by pseudostem height and leaf number into giant, medium and small cultivars. Giant and small plantains have over 38 and below 32 leaves, respectively. Path analysis was used to establish relationships between growth characteristics and yield potential, and to attempt defining ideotypes for plantain breeding (Ortiz and Langie 1997). There are few shared pathways determining yield potential among plantain landraces, suggesting that plantains possess different genes controlling similar pathways or different traits contributing to yield potential. An ideotype depends on each plantain group and production system, and therefore both should be taken into account when selecting hybrid germplasm.

Sources of Variation

The genus *Musa* includes sections *Eumusa, Rhodochlamys, Callimusa, Australimusa* and *Ingentimusa*. *Callimusa* and *Australimusa* show the same chromosome number ($2n = 20$), while *Eumusa* and *Rhodochlamys* have a basic chromosome number of 11 ($2n = 22$). *Callimusa* and *Rhodochlamys* are mostly ornamentals because they lack fruit parthenocarpy, *Autralimusa* includes abaca or Manila hemp (*M. textilis*) and edible Fe'i banana cultivars while *M. acuminata* and *M. balbisiana* belong to *Eumusa*.

Advances in DNA-aided analysis of diversity are providing new insights into *Musa* diversity, filling gaps and unraveling relationships among species and cultivars, which are often complex owing to interspecific hybridization, heterozygosity, and polyploidy (Ortiz 2011). Genome identification is possible by genomic *in situ* hybridisation. Four putative genomes (A, B, S for *M. schizocarpa* and T for *Australimusa*) are recognized in the genus, which has a relatively small genome size ranging from 550 to 612 Mbp (Pillay et al. 2004). Nonetheless, B genome microsatellite (SSR) markers are able to identify homoeologous loci in AA and AAA genomes, and A genome-derived SSR markers frequently detect homoeologous loci in B genome accessions (Buhariwalla et al. 2005). Various DNA markers have been broadly used for determining diversity in cultigens and their crop wild relatives. For example, the sequence-related amplified polymorphism (SRAP) technique and amplified fragment length polymorphisms (AFLP) both revealed variation between sections and species, but the relationships within Eumusa species and subspecies varied according to the DNA marker system used (Youssef et al. 2011). SRAP markers exhibited approximately three-fold more specific and unique bands than AFLP markers. SRAP markers are able to discriminate amongst *M. acuminata, M. balbisiana* and *M. schizocarpa* within the Eumusa section, and between plantain and cooking banana cultigens. Comparative analysis of phenotypic and genotypic diversity among plantain landraces demonstrated, however, that plantain landrace groupings based on morphotype do not match with overall genetic divergence, which was low in plantains (Crouch et al. 2000). This finding confirms that plantains evolved through somatic mutations from a relatively small number of introductions to Africa.

Diploid genetic resources remain an important source of useful variation for *Musa* germplasm enhancement, particularly for broadening the very narrow genetic base of banana and plantain (Ortiz and Swennen 2014), which likely results from 20 to 25 meiosis events. Crop wild relatives and diploid cultivars are source populations for selection of promising parents in subsequent cycles of recurrent selection in crossbreeding schemes. Population improvement through phenotypic recurrent selection appears to be based on the elimination of deleterious recessive alleles (Ortiz and Vuylsteke 1994a). For example, the International Institute of Tropical Agriculture (IITA, Nigeria) released diploid banana hybrids with good combining ability and resistance to black Sigatoka and burrowing nematodes after crossing diploid accessions belonging to different subspecies of *M. acuminata* to diversify and broaden host plant resistance to pathogens and pests (Tenkouano et al. 2003). Likewise, IITA made available plantain-derived diploid hybrids with black Sigatoka resistance (Vuylsteke and Ortiz 1995a). These hybrids ensued from crossing triploid plantain landraces with a wild diploid banana, and as a result of the union of haploid gametes ($n = x$) from both parents. Their breeding value was determined by crossing with primary tetraploids ($2n = 4x = 44$). Some of the resulting offspring showed more plantain-like traits and may be important for improving fruit quality.

Interspecific Hybridization and Ploidy Manipulations

Seed set in triploid bananas and plantains is very low due to high levels of sterility, thus making crossbreeding of both cultigens difficult, although some cultivars produce seed after crossing with diploid bananas (Ortiz and Vuylsteke 1995a). *Musa* breeding often requires 1000 seeds produced from more than 1000 hand pollinations of 200 plants (0.12 ha), to obtain one selected tetraploid plantain-derived hybrid per year (Vuylsteke et al. 1997). The highest seed set rates are noted in ABB cooking bananas followed by some French plantains (Vuylsteke et al. 1993), and in a few East African highland bananas belonging to the Nfuuka clone set (Ssesuliba et al. 2006a). Pistil abnormalities make most East African highland bananas sterile.

Most plantain-derived hybrids after an interploidy $3x \times 2x$ crossing are diploids, and a few are tetraploids, which suggests production of both n (= x) and $2n$ (= $3x$) eggs in plantain cultivars. Segregation of SSR loci in $2n$ gametes of the triploid cultigen demonstrates the occurrence of recombination during the formation of its $2n$ megaspores (Crouch et al. 1998). A moderate population size along with segregation and recombination can be used to facilitate trait introgression.

The plantain cultivar and climatic factors affect seed set. Stigma development stage and banana hand position also influence seed production in East African highland bananas (Ssesuliba et al. 2006). Pollinating middle hands containing highly receptive stage III stigma leads to greater seed set. Style length seems to be another important factor affecting seed set (Ssesuliba et al. 2005), as pollen grains traveling short distances may have more chance of fertilizing ovules.

Tetraploid plantain-derived seed and further *in vitro* seed germination occur more frequently after hand pollinations made at high temperatures, high solar radiation and low relative humidity, which appear to favor $2n$ egg production relative to haploid eggs. Low embryo rescue rates were noted in hybrid seeds derived from East African highland bananas (Ssesuliba et al. 2006a), which may relate to endosperm breakdown causing high levels of embryo abortion. Tetraploid plantain-derived hybrids are usually used for crossing with diploid accessions or cultivars to obtain triploid seed. Natural open pollination of the same tetraploid hybrids can also generate viable seed (Ortiz and Crouch 1998). Hence, plantain-derived tetraploid and diploid hybrids can be included in breeding programs based on their combining ability in an isolated poly-crossbreeding nursery where their natural pollinators are abundant.

Hybrid seed production depends on crossing fertile female cultigens with accessions producing viable pollen. Diploid hybrids and landraces have more pollen than triploid and tetraploid cultigens (Dumpe and Ortiz 1996), which accounts for the use of the former as male parents because the quantity of pollen influences pollen germination. Nonetheless, triploid cultivars producing $2n$ pollen provide a means for further breeding through ploidy manipulation. The amount of pollen produced and its viability are also highly correlated, which suggests that closely associated genetic factors control both traits. The node along the rachis from which the pollen is taken seems to affect pollen stainability – a proxy of its fertility – in some diploid and triploid bananas (Ssesuliba et al. 2008), suggesting node position should be considered when collecting pollen for further use in crossing.

The environment influences pollen quality and quantity. Seasonal variation of pollen stainability occurs in diploid banana cultigens and plantain-derived diploid hybrids (Ortiz et al. 1998b). Solar radiation positively affects pollen stainability, which is not significantly correlated with seed set. Solar radiation, temperature, total pan-evaporation rainfall and minimum relative humidity are significantly associated with seasonal variation in $2n$ pollen production. Although one dominant gene controls $2n$ pollen production in *Musa* (Ortiz 1997b), knowing the best time of the year and identifying the best male-fertile $2n$ pollen-producing diploid accession will facilitate polyploid synthesis via sexual polyploidization, which explains the formation of *Musa* polyploid cultigens and provides means for their genetic enhancement. The occurrence of $2n$ pollen in diploid bananas suggests that unilateral sexual polyploidization ($2n \times n$) could be involved in the origin of triploid cultigens (Ortiz 1997b). Further introgression of desired alleles from diploid species to polyploid cultigens can be easily achieved via unilateral or bilateral ($2n \times 2n$) sexual polyploidization.

From Hybrid Seedlings to On-Station Field Testing

The genetic enhancement of banana and plantain is aimed at triploid cultivars whose multiplication is by vegetative means. The challenge is to breed using triploid cultigens – which are mostly sterile – as source populations, then release genetic variation by interploidy $3x \times 2x$ crossing involving interspecific hybridization to obtain tetraploid hybrids, and thereafter using interploid $4x \times 2x$ crossing with the aim of producing a non-seed bearing triploid cultivar (Ortiz et al. 1995a). *Musa* breeding is also limited at the practical level by time, plant material and land-management resources. It takes at least 2 years to complete a seed-to-seed crop cycle. Moreover, there are very few viable seedlings after interploid $3x \times 2x$ crossing. Furthermore, each seedling-derived plant grows in 6 m^{-2} in the field for evaluation.

Target Traits

After almost a century of crossbreeding, bred-hybrids that meet the banana export trade quality demands are lacking (Tenkouano et al. 2010b). The hybrids should show host plant resistance to various pathogens and pests: bacteria, fungi, insects, nematodes and viruses. *Mycosphaerella* species producing Sigatoka leaf spots, *Fusarium oxysporum* f. sp. *cubense* causing Panama disease or wilt, and *Xanthomonas campestris* or bacterial wilt, *Banana bunchy top virus* and *Banana streak virus* (BSV) are among the main pathogens; while the main pests are those resulting from damages by banana weevil *Cosmopolites sordidus*, the burrowing nematode *Radopholus similis*, lesion nematode, *Pratylenchus goodeyi* and banana spiral nematode *Helicotylenchus multicinctus*. The hybrids should also be high yielding per unit area and time (t ha^{-1} $year^{-1}$), photosynthetically efficient, early maturing, and short cycling between consecutive harvests, as well as show short stature and strong roots to survive damaging winds.

Crossbreeding

Banana and plantain breeding should attempt to mimic the evolutionary development of *Musa* cultigens (Ortiz 1997c). Following this approach, female fertile triploid cultivars are mated with diploid wild species, landraces or breeding stocks with desired traits. The primary hybrids ensuing from such a crossing scheme are mostly tetraploids or diploids. Those showing the desired attributes are selected and used in crossing blocks to produce high-yielding, secondary triploid hybrids with host plant resistance to pathogens and pests, appropriate phenology plus plant architecture, and adequate fruit quality that meets end-user demands. Recurrent selection is also practiced at the diploid level to obtain breeding stocks.

The previously known intractability of plantain to genetic enhancement was challenged after identification of various seed-fertile cultivars, which led to crossbreeding and production of several tetraploid hybrids in a short period in the 1990s (Vuylsteke et al. 1993a). These tetraploid hybrids were selected on the basis of their host plant resistance to black Sigatoka, high bunch weight, large parthenocarpic fruits and improved rationing. The wild accession "Calcutta 4" from *M. acuminata* spp. *burmannicoides* was the diploid male parent of most of these selected hybrids, which shows that the inferior bunch characteristics of this wild banana were generally not transmitted to its tetraploid hybrid offspring. The plantain cultivars differ in their combining ability for host plant resistance to black sigatoka, growth, bunch weight and fruit traits. Likewise tetraploid siblings are always significantly different for various traits, thus confirming the occurrence of segregation and recombination during the modified megasporogenesis leading to the formation of $2n$ eggs in the plantain parents.

Interspecific hybridization, ploidy manipulations via $2n$ gametes, embryo culture, rapid *in vitro* multiplication, field-testing and selection led to development and identification of suitable plantain bred-germplasm for the West and Central Africa lowlands. Bunches are harvested at full maturity after hand pollination and ripened with acetylene in a room for 4 days. Thereafter, ripe fruits are peeled and their seeds extracted by squashing. Seed embryos are excised aseptically after 2 days and germinated *in vitro*. Seedlings are subsequently planted in early evaluation trials after acclimatizing them in the greenhouse. Chromosome counts on root tips using a squashing technique (Osuji et al. 1997b), stomatal density and size (Vandehout et al. 1995) or flow cytometry (Ortiz and Swennen 2014) are used for determining ploidy in hybrid offspring. Ploidy may affect both plant height and fruit traits. Tetraploid and diploid hybrids are the most promising for further breeding of secondary triploid hybrids, while aneuploids provide means for further characterization of the *Musa* genome and physical mapping.

Field Plot Techniques

Field-testing requires appropriate plot techniques (i.e., number of plants per experimental plot and plot shape) and enough replications to ensure both accuracy and precision. The number of replications required depends on management practices. For example, single rows of 5 and 20 plants suffice to detect significant true mean differences for bunch weight under alley cropping or under monoculture, respectively (Ortiz 1995). Replication number may also vary according to management practices, population size for testing and plot size. For example, a true bunch weight mean difference of 15% may be found to be significant when using a randomized complete block design with two replications of 10 plants $plot^{-1}$ in trials under alley cropping with multispecies hedgerows, which maintains diversity and is highly efficient for testing bred-*Musa* germplasm. The optimum plot size depends also on the trait being tested, the germplasm and the plant growth cycle (Nokoe and Ortiz 1998). For example, about 4 plants $plot^{-1}$ may suffice to assess host plant resistance to black Sigatoka in plantain and its hybrids (Ortiz and Vuylsteke 1994b), while 13 plants $plot^{-1}$ and 15 plants $plot^{-1}$ appear to be adequate to determine host plant response to black Sigatoka in dessert bananas (Nokoe and Ortiz 1998) and in East African highland bananas (Okoro et al. 1997), respectively.

Successful allele frequency change in a segregating population being bred for host plant resistance is a function of the precision in identifying and isolating offspring bearing the gene(s) under selection. The magnitude of 'escapes' affects host plant resistance breeding and depends on the screening protocol; i.e., the lower the selection intensity, the lower the response to selection, as established by population genetics. The host response to a pathogen or pest should be therefore defined by a sound approach such as using reference standard accessions or cultivars, and measuring whether the segregating offspring are significantly different from the standards. For example, hybrids showing a youngest leaf spotted (YLS) below 8 (as the surrounding susceptible check) are rated as black Sigatoka susceptible, while those with YLS of 8 to 10 are regarded as less susceptible and above 10 as partially resistant to black Sigatoka (Ortiz and Vuylsteke 1994b) when testing under natural infection, which is the most common and preferred method to assess incidence and severity of black Sigatoka (Craenen and Ortiz 2003). Similarly, banana weevil resistance assessed by measuring damage and infestation levels (Ortiz et al. 1995b) and host plant resistance to burrowing nematode based on the inoculation of individual roots (Dochez et al. 2009) are determined using scale thresholds for susceptibility and resistance according to reference standard accessions and cultivars of known host response.

Bred-hybrids should meet market standards to ensure their adoption. Hence, testing of fruit quality is mandatory before engaging in large-scale multi-environment trials (Ferris et al. 1999). Fruits traits of promising hybrids along with cultivars and landraces are evaluated at harvest for both the plant and ratoon crops. Fruit number, size and weight, pulp percentage, unripe and ripe texture, dry matter content, peel thickness, ripening time as measured by changes in peel color, daily weight loss rate, and changes in soluble solids are among the traits evaluated to further select hybrids with high fruit quality and great potential for market acceptance.

On-Station Trials

The whole breeding cycle (from crossing to cultivar release) takes about 15 years. Vision, patience, ingenuity and core commitment foster success in banana and plantain long-term breeding (Ortiz 2001), which is still a technically difficult endeavor. Screening the germplasm for fertility, manipulating ploidy and the intensive application of tissue culture techniques overcome some hindrances related to *Musa* genetic enhancement. Field-testing starts with early evaluation non-replicated trials (EET) that include different numbers of plants per genotype, and are evaluated for plant height (including dwarfism), host plant resistance, fruit parthenocarpy and bunch weight in at least two production cycles: plant crop and first ratoon (Ortiz and Vuylsteke 1995b). Less than 1% of EET seedlings are selected for the next evaluation steps. The next step is station preliminary yield trials (PYT) that evaluate selections from EET for host plant resistance, rationing, bunch weight, fruit size and quality in replicated plots (2 or 3) of 4 to 5 plants during both plant crop and first ratoon. Cropping system affects PYT selection efficiency (Tenkouano et al. 2010a). Selection based on bunch weight of 10 kg or above under alley cropping led to identifying more promising bred-hybrids than selection under mono-cropping, but selection under the latter produces genotypes suitable for cultivation under the former method.

Multi-Environment Trials, Cultivar Development and Delivery

The selected PYT clones showing desired traits including proper fruit quality are thereafter included in multi-environment trials (3-5 locations for at least two production cycles) in the target population of environments to assess their adaptability and stability for edible yield (Ortiz and Vuylsteke, 1995b). Multi-environment testing also provides means for in-depth knowledge on host plant resistance and consumer acceptability of the fruit quality of newly bred hybrid germplasm (Ortiz et al. 1997). Best-bet PYT selections may also be included in on-farm participatory testing. Nationally coordinated trials (NCT) vary according to each country as per their requirements for official cultivar release and for the trait. For example, plant height and bunch weight are traits with high repeatability, while days to harvest and fruit filling time show low repeatability across locations (Tenkouano et al. 2012b). Plant height, due to its highest repeatability, requires the smallest plot size among all traits, irrespective of location.

Multi-Environment Trials

The genotype-by-environment interaction (GEI) and bred-germplasm stability are determined after analyzing multi-environment trials (Ortiz and Tenkouano 2011). The GEI includes the genotype-by-location and the genotype-by-crop cycle (or year or season) interactions and may affect genotype ranking across environments: a clonal phenotype corresponding to a specific genotype may vary from year to year in the same location, or from location to location within an agro-ecology in the same year. The understanding of GEI therefore allows, its management, efficient selection schemes and planning of multi-environment testing prior to cultivar release (Ortiz 2013). *Musa* breeding requires assessing newly bred germplasm through multi-environment testing to select promising clones in target agro-ecologies, while stability analysis of edible yield potential (kg ha^{-1} $year^{-1}$) facilitates identification of stable, high-yielding hybrids.

Site rationalization should be taken into account for multi-environment testing, which is more profitable than single site evaluation over several years in a *Musa* breeding station (Ortiz and de Cauwer 1998-1999). Nonetheless, bred-germplasm performance must be evaluated at least until the first ratoon cycle because bananas and plantains are perennial crops. Genetic correlation of the same trait between locations allows estimation of spatial repeatability that could be used to determine the number of sites required for multi-environment testing of bred-germplasm. Furthermore, correlated responses across environments can also assist in avoidance of highly similar selection sites, e.g. within an agro-ecology, as selection at one site may be generalized to the other location.

The additive main effects and multiplicative interaction (AMMI) model accounts for a significant percentage of the GEI in *Musa* multi-environment trials (de Cauwer and Ortiz 1998). AMMI demonstrated that tetraploid plantain-derived hybrids achieved high yield potential due to their short growth cycle, while the high yield potential of cooking bananas ensued from their fast sucker development. Likewise, AMMI along with stability analysis assists in identifying bred-germplasm with homeostatic and genotypic responses to environmental changes, plus adaptation to specific niches (Ortiz 1998). Stability analysis of bunch weight and edible yield potential based on the phenotypic coefficient of variation allows selection of high and stable-yielding hybrids (de Cauwer et al. 1995), whereas high-yielding hybrids with specific adaptation are selected in the environments where the respective stress(es) occur(s).

Bred-Germplasm Releases

Germplasm registration through journal articles should be actively pursued to place in the public domain polyploid bred-hybrids showing host plant resistance, high edible yield and other interesting traits, and of diploid breeding stocks with good combining ability, and to provide landmarks of banana and plantain breeding advances. For example, improved tropical plantain tetraploid hybrids or TMPx (Vuylsteke et al. 1993b), tetraploid plantain hybrid cultivars PITA 9 (Vuylsteke et al. 1995) and PITA 14 (Ortiz and Vuylsteke 1998b), tetraploid starchy banana hybrid cultivar BITA 3 (Ortiz and Vuylsteke 1998c), secondary triploid hybrids (Ortiz et al. 1998b), plantain-derived diploid hybrids (Vuylsteke and Ortiz 1995) and diploid banana breeding stocks (Tenkouano et al. 2003) were made available by IITA by their registration in the journal *HortScience*.

TMPx cultivars combine black sigatoka resistance, regulated suckering and high bunch weight, and show a high resemblance to their female triploid plantain parent (Vuylsteke et al. 1993). They are both male- and female-fertile and can be used in crossing blocks to produce secondary triploids. PITA 9 derives from the French somaclonal variant of the False Horn plantain Agbagba and the wild banana Calcutta 4 (Vuylsteke et al. 1995). It shows high edible yield in the West African humid forest and good fruit quality. PITA 9 makes accessible the sterile False Horn plantain group to crossbreeding. The short-cycling hybrid black sigatoka resistant plantain hybrid PITA 14 does not show symptoms to both BSV and *Cucumber Mosaic Virus* after several years of cultivation in a site where both are widespread. Its short cycling and rapid ratooning owing to regulated suckering behavior results in high yield potential, which led PITA 14 to be the most promising tetraploid plantain hybrid cultivar in Nigeria. The starchy banana BITA 3 is a hybrid derived from crossing the AAB banana Laknau, which resembles plantains, and the diploid banana Tjau Lagada, which has a long bunch with many hands (Ortiz and Vuylsteke 1998c). This high yielding tetraploid hybrid cultivar shows low partial resistance to black sigatoka, tolerance to BSV, and large fruit that are very similar in length to that of Agbagba – among the most preferred plantain cultivars in Nigeria, where big fruit is a key quality trait. TM3x are secondary triploid hybrid bred-germplasm showing black sigatoka resistance and significantly out-yielding their tetraploid hybrid parents and triploid landrace grandparents (Ortiz et al. 1998b). The highest yielding male-sterile TM3x PITA 16 has a pendulous dense bunch with curved angular bottlenecked fruits, few persistent neutral flowers, imbricated male bud and regulated suckering behavior. Its pedigree reflects the importance of germplasm exchange in plantain and banana breeding because it involves as triploid grandparents an African plantain and a wild Asian banana, and as a male parent a diploid stock bred in Central America whose ancestry includes wild bananas and cultivars from Asia and the Pacific. Crossing diploid accessions belonging to *M. acuminata* subspecies broadens the genetic base and brings together important traits into a single breeding stock.

In the mid-1990s, the National Agricultural Research Organization (NARO) of Uganda and IITA began crossbreeding of cooking and juicy East African highland bananas. They announced about two years ago the availability of secondary triploid hybrids – known as NARITA – with host plant resistance to pests and high edible yield (Tushemereirwe et al. 2014). Most NARITA bananas had an average bunch weight of 17.8 kg at trials in Central Uganda, while the bunch weight of the widely grown local landrace was 11 kg (Tushemereirwe et al. 2015). The edible yield of many NARITA bananas was significantly above their founder East African highland grandparent (i.e., heterosis ranging from 10 to 300%), which shows the significant breeding gains.

Delivering Bred-Germplasm

Hybrid cultivars should be included in national propagule multiplication schemes only after official multi-locational and advance trials along with on-farm testing show their value for cultivation and use, and that they are distinct (Ortiz 1997d). *Musa* propagule production starts after indexing source stocks to main pathogens and pests. The indexed stocks from the tissue culture laboratory are pre-basic propagules, while the sucker-derived propagules, which are also indexed for all known pathogens using recommended protocols in the field nursery or small orchard, are basic propagules. This dual system minimizes the risk of losing the true-to-type stock due to somaclonal variation after micropropagation. The banana and plantain breeding programs provide a detailed description of bred-germplasm to monitor their growth during the multiplication of the basic propagules. Diagnostic DNA markers can be also useful for bred-germplasm characterization (Ortiz et al. 1998b). Indexed sucker-derived propagules or foundation propagules are the planting materials in multiplication plots grown in isolated clean fields. These foundation propagules are also the source of planting materials for plots of indexed and clean sucker-derived propagules or registered propagules in decentralized multiplication centers. Registered propagules are grown in fields with easy access to farmers, and they are the source of planting materials for certified propagation plots in public or farmers' fields. Propagation plots from registered propagules are used with the only aim of producing registered stocks which inspectors verify for true-to-type and sanitary status for sucker propagation before distributing them to other farmers. These suckers are regarded as certified stocks after a satisfactory inspection.

Quantitative Genetics, Genomics and *Musa* Breeding

Genetic knowledge gives essential information to design crossing blocks targeting specific trait improvement according to combining ability, and may allow breeding gains to be predicted based on heritability and genetic correlations among traits (Tenkouano et al. 2010). Pedigree and DNA marker-facilitated analysis also provide means for assessing inbreeding status and relatedness of breeding stocks, which allows maximum recombinative heterosis to be pursued through a ploidy manipulation crossbreeding approach.

The coefficients of relationship between secondary triploid offspring and diploid parents are invariably greater than the coefficients of relationship between offspring and tetraploid parents (Tenkouano et al. 2012a). Hence, the percentage of parental performance that is predictably transmitted to the offspring is larger in diploids than in tetraploids irrespective of the ploidy level of the offspring. This is not surprising, because high ploidy states allow for greater effects of non-additive interactions on gene expression relative to additive effects. Heritability estimates based on parent–offspring regression may show a bias, when intra- and inter-generation ploidy polymorphisms ensue from an unusual megasporogenesis. Nonetheless, heritability estimates for bunch weight indicate that parental selection requires progeny testing.

Additive genetic variation affects bunch weight, fruit filling time, fruit length, plant height, and leaf number, while non-additive variation accounts significantly for suckering behavior and fruit circumference in plantain-derived secondary triploid hybrids (Tenkouano et al. 2012c). Maternal general combining ability (GCA) effects explain most of the additive genetic variation for plant height and leaf number, thus suggesting that selection for both traits should be done on the primary tetraploid hybrids. Paternal GCA effects are the main cause of genetic variation for fruit filling time, bunch weight, and fruit length, which suggests that selection for these traits should be done on diploid breeding stocks. Dosage and epistatic effects are also very important for time to fruit filling and fruit number in secondary triploid hybrids (Tenkouano et al. 2012a). Little recombinant heterotic arises if GCA is significantly above specific combining ability (Tenkouano et al. 1998a), and the $4x \times 2x$ crossbreeding scheme should aim to accumulate favorable alleles in both $4x$ and $2x$ breeding populations through reciprocal recurrent selection with the goal of using outstanding stocks from each in further crossing blocks.

DNA Marker-Aided Breeding

DNA markers, the genetic maps based on them, and the recent sequencing of the *Musa* genome offer means for gaining further insights into banana and plantain genetics, and to identify genes that may accelerate their betterment (Ortiz and Swennen 2014). For example, most DNA markers detect a high level of polymorphism between parental genotypes and within their hybrid offspring, but there is a poor correlation between estimates of genetic similarity based on different DNA marker types (Crouch et al. 1999a). Comparative DNA marker-based analysis of full-sib hybrids with their parental genotypes show that tetraploid hybrids are generally more closely related to their triploid parents than their diploid full-sibs (Crouch et al. 1999b). Such an analysis may also allow identification of promising tetraploid and diploid hybrids for further use in subsequent breeding of secondary triploid hybrids. However, the edible yield and other traits with complex inheritance in secondary triploid hybrids do not correlate significantly with genetic similarity indices based on either pedigree or molecular data (Tenkouano et al. 1998b), which suggests a lack of adequate genetic models for populations with intergenerational genome size polymorphism. Pedigree-based estimates of parent-offspring relationships are significantly different than those ensuing from DNA marker data (Tenkouano et al. 1999a). DNA marker-based contribution of triploid maternal accessions to their diploid-derived offspring was larger than expected from available

Musa meiosis models. Progeny prediction of bunch weight seems to be best when based on genealogical distance and equal parental contribution, while predicted fruit size appears to be most accurate when DNA marker data are used under the assumption of an unequal parental contribution (Tenkouano et al. 1999b). Although DNA markers provide a more accurate description of genetic relatedness, pedigree-based analysis are still useful for the selection of prospective parental combinations in *Musa* breeding. Nonetheless, developing high-density DNA markers for early selection of priority traits may increase genetic gains per unit of time, thus accelerating cultivar releases.

References

Barrett, SCH. 2015. Influences of clonality on plant sexual reproduction. Proc. Natl. Acad. Sci. (USA) 112: 8859-8866.

Buhariwalla, H.K. R.L. Jarret, B. Jayashree, J.H. Crouch and R. Ortiz. 2005. Isolation and characterization of microsatellite markers from *Musa balbisiana*. Mol. Ecol. Notes 5: 327-330.

Craenen K. and R. Ortiz. 2003. Genetic improvement for a sustainable management of resistance. pp. 181-198. In L. Jacome, P. Lepoivre, D. Marin, R. Ortiz, R. Romero & J.V. Escalant (eds.) *Mycosphaerella Leaf Spot Diseases of Bananas: Present Status and Outlook*. International Network for the Improvement of Banana and Plantain, Montpellier, France.

Crouch, H.K., J.H. Crouch, S. Madsen, D.R. Vuylsteke and R. Ortiz. 2000. Comparative analysis of phenotypic and genotypic diversity among plantain landraces (*Musa* spp., AAB group). Theor. Appl. Genet. 101: 1056-1065.

Crouch, H.K., J.H. Crouch, R.L. Jarret, P.B. Cregan and R. Ortiz. 1998. Segregation at microsatellite loci in haploid and diploid gametes of *Musa*. Crop Sci. 48: 211-217.

Crouch, J.H., H.K. Crouch, H. Constandt, A. Van Gysel, P. Breyne, M. Van Montagu, R.L. Jarret and R. Ortiz. 1999a. Comparison of PCR-based molecular marker analyses of Musa breeding populations. Mol. Breed. 5: 233-244.

Crouch, J.H., H.K. Crouch, A. Tenkouano and R. Ortiz. 1999b. VNTR-based diversity analysis of 2*x* and 4*x* full-sib *Musa* hybrids. Electronic J. Biotechnol. 2: 130-139.

de Cauwer, I. and R. Ortiz. 1998. Analysis of the genotype environment interaction in *Musa* trials. Expl. Agric. 34: 177-188.

de Cauwer, I., R. Ortiz and D. Vuylsteke. 1995. Genotype-by-environment interaction and phenotypic stability of Musa germplasm in West and Central Africa. African Crop Sci. J. 3: 425-432.

Dochez, C., A. Tenkuano, R. Ortiz, J. Whyte and D. De Waele. 2009. Host plant resistance to *Radopholus similis* in a diploid banana hybrid population. Nematology 11: 329-335.

Dumpe, B.P. and R. Ortiz. 1996. Apparent male fertility in *Musa* germplasm. HortScience 31: 1019-1022.

FAO. 2015. FAOSTAT. Food and Agriculture Organization of the United Nations, Rome, Italy. http: //faostat3.fao.org/browse/Q/*/E

Ferris, R.S.B., R. Ortiz and D. Vuylsteke. 1999. Fruit quality evaluation of plantains, plantain hybrids, and cooking bananas. Postharvest Biol. Tech. 15: 73-81.

Karamura, D.A. 1999. Numerical taxonomic studies of the East African highland banana (*Musa* AAA-East Africa) in Uganda. International Network for the Improvement of Banana and Plantain, Montpellier, France.

Nokoe, S. and R. Ortiz. 1998. Optimum plot size for banana trials. HortScience 33: 130-132.

Okoro, J.U., R. Ortiz and D. Vuylsteke. 1998. Field plot techiques for black Sigatoka evaluation in East African highland bananas. Tropicultura 15: 186-189.

Ortiz, R. 1995. Plot techniques for assessment of bunch weight in banana trials under two systems of crop management. Agron. J. 87: 63-69.

Ortiz, R. 1997a. Morphological variation in *Musa* germplasm. Genet. Resour. Crop Evol. 44: 393-404.
Ortiz, R. 1997b. Occurrence and inheritance of 2*n* pollen in Musa. Ann. Bot. 79: 449-453.
Ortiz, R. 1997c. Secondary polyploids, heterosis and evolutionary crop breeding for further improvement of the plantain and banana genome. Theor. Appl. Genet. 94: 1113-1120.
Ortiz, R. 1997d. A delivery system of improved banana and plantain propagules. InfoMusa 6 (2): 14-15.
Ortiz, R. 1998. AMMI and stability analyses of bunch mass in multilocational testing of *Musa* germplasm in sub-Saharan Africa. J. Amer. Soc. Hort. Sci. 123: 623-627.
Ortiz, R. 2001. Dirk R. Vuylsteke: *Musa* scientist and humanitarian. Plant Breed. Rev. 21: 1-25.
Ortiz, R. 2013. Conventional banana and plantain breeding. Acta Horticulturae 986: 177-194.
Ortiz, R. and J.H. Crouch. 1998. The efficiency of natural and artificial pollinators in plantain (*Musa* spp. AAB group) hybridization and seed production. Ann. Bot. 80: 693 -695.
Ortiz, R. and I. de Cauwer. 1998-1999. Genotype-by-environment interaction and testing environments for plantain and banana (*Musa* spp. L.) breeding in West Africa. Tropilcultura 16-17: 97-102.
Ortiz, R., R.S.B. Ferris and D.R. Vuylsteke. 1995a. Banana and plantain breeding. pp. 110-146. *In*: S. Gowen (ed.) Bananas and Plantains. Chapman & Hall, London, United Kingdom.
Ortiz, R. and H. Langie. 1997. Path analysis and ideotypes for plantain breeding. Agron. J. 89: 988-994.
Ortiz, R., S. Madsen and D. Vuylsteke. 1998a. Classification of African plantain landraces and banana cultivars using a phenotypic distance index of quantitative descriptors. Theor. Appl. Genet. 96: 904-911.
Ortiz, R. and R. Swennen. 2014. From crossbreeding to biotechnology-facilitated improvement of banana and plantain. Biotechnol. Adv. 32: 158-169.
Ortiz R. and A. Tenkouano. 2011. Genotype by environment interaction and *Musa* improvement. pp. 235-247. *In* M. Pillay and A. Tenkouano (eds.) Banana Breeding: Constraints and Progress. CRC Press, Boca Raton, Florida.
Ortiz, R., F. Ulburghs and J.U. Okoro. 1998. Seasonal variation of apparent male fertility and 2*n* pollen production in plantain and banana. HortScience 22: 146-148.
Ortiz, R. and D. Vuylsteke. 1994a. Inheritance of albinism in banana and plantain (*Musa* spp.) and its significance in breeding. HortScience 29: 903-905.
Ortiz, R. and D. Vuylsteke. 1994b. Inheritance of black sigatoka disease resistance in plantain-banana (*Musa* spp.) hybrids. Theor. Appl. Genet. 89: 146-152.
Ortiz, R. and D. Vuylsteke. 1995a. Factors influencing seed set in triploid *Musa* spp. L. and production of euploid hybrids. Ann. Bot. 75: 151-155.
Ortiz, R. and D. Vuylsteke. 1995b. Recommended experimental designs for selection of plantain hybrids. InfoMusa 4 (1): 11-12.
Ortiz, R. and D. Vuylsteke. 1998a. Quantitative variation and phenotypic correlations in banana and plantain. Scientia Horticulturae 72: 239-253.
Ortiz, R. and D.Vuylsteke. 1998b. 'PITA-14': a black sigatoka-resistant tetraploid plantain hybrid with virus tolerance. HortScience 33: 360-361.
Ortiz, R. and D. Vuylsteke, 1998c. 'BITA-3': a starchy banana with partial resistance to black Sigatoka and tolerance to streak virus. HortScience 33: 358-359.
Ortiz, R., D. Vuyksteke, H. Crouch and J. Crouch. 1998b. TM3x: Triploid black sigatoka-resistant *Musa* hybrid germplasm. HortScience 33: 362-365.
Ortiz, R., D. Vuylsteke, B. Dumpe and R.S.B. Ferris. 1995b. Banana weevil resistance and corm hardness in *Musa* germplasm. Euphytica 86: 95-102.
Ortiz, R., D. Vuylsteke, R.S.B. Ferris, J.U. Okoro, A. N' Guessan, O. B. Hemeng, D.K. Yeboah, K. Afreh-Nuamah, E.K.S. Ahiekpor, E. Foure, B.A. Adelaja, M. Ayodele, O.B. Arene, F.E.O. Ikiediugwu, A.N. Agbor, A.N. Nwogu, E. Okoro, G. Kayode, I.K. Ipinmoye, S. Akele and A. Lawrence. 1997. Developing new plantain varieties for Africa. Plant Varieties Seeds 10: 39-57.
Osuji, J.O., B.E. Okoli, D. Vuylsteke and R. Ortiz. 1997a. Multivariate pattern of quantitative trait variation in triploid banana and plantain cultivars. Scientia Horticulturae 71: 197-202.

Osuji, J.O., D. Vuylsteke and R. Ortiz. 1997b. Ploidy variation in hybrids from interploidy $3x \times 2x$ crosses. Tropicultura 15: 37-39.

Pillay, M., A. Tenkouano, G. Ude & R. Ortiz. 2004. Molecular characterization of genomes in *Musa* and its applications. pp. 271-286. *In*: S.M. Jain and R. Swennen (eds.) Banana Improvement: Cellular, Molecular Biology and Induced Mutations. Science Publishers, Inc, Enfield, New Hampshire.

Ssesuliba, R., M. Magambo, D. Talangera, D. Makumbi, A. Tenkouano, P. Rubaihayo, and M. Pillay. 2006b. Biological factors affecting seed production in East African highland bananas. J. Crop Improvement 16: 67-79.

Ssesuliba, R., P. Rubaihayo, A. Tenkouano, D. Makumbi, D. Talangera, and M. Magambo. 2005. Genetic diversity among East African highland bananas for female fertility. African Crop Sci. J. 13: 13-26.

Ssesuliba, R., D. Talengera, D. Makumbi, P. Namanya, A. Tenkouano, W. Tushemereirwe and M. Pillay. 2006a. Reproductive efficiency and breeding potential of East African highland (*Musa* AAA-EA) bananas. Field Crops Res. 95: 250-255.

Ssesuliba, R.N., A. Tenkouano and M. Pillay. 2008. Male fertility and occurrence of $2n$ gametes in East African highland bananas (*Musa* spp.). Euphytica 164: 153-162.

Swennen, R., D. Vuylsteke and R. Ortiz. 1995. Phenotypic diversity and patterns of variation in West and Central African plantain (*Musa* spp. AAB Group Musaceae). Econ. Bot. 43: 320-327.

Tenkouano, A., J.H. Crouch and R. Ortiz. 2012a. Additive relationships and parent-offspring regression in *Musa* germplasm with intergeneration genome size polymorphism. *Scientia Horticulturae* 136: 69-74.

Tenkouano, A., J.H. Crouch, H.K. Crouch and R. Ortiz. 1998b. Genetic diversity, hybrid performance and combining ability for yield in *Musa* germplasm. Euphytica 102: 281-288.

Tenkouano A., J.H. Crouch, H.K. Crouch, D. Vuylsteke and R. Ortiz. 1999a. Comparison of DNA marker and pedigree-based methods of genetic analysis of plantain and banana (*Musa* spp.) clones. I. estimation of genetic relationships. Theor. Appl. Genet. 98: 62-68.

Tenkouano A., J.H. Crouch, H.K. Crouch, D. Vuylsteke and R. Ortiz. 1999b. Comparison of DNA marker and pedigree-based methods of genetic analysis of plantain and banana (*Musa* spp.) clones. II. Predicting hybrid performance. Theor. Appl. Genet. 98: 69-75.

Tenkouano, A., R. Ortiz and S. Nokoe. 2012b. Repeatability and optimum trial configuration for field-testing of banana and plantain. Scientia Horticulturae 140: 39-44.

Tenkouano, A., R. Ortiz and D. Vuylsteke. 1998a. Combining ability for yield and plant phenology in plantain-derived populations. Euphytica 104: 151-158.

Tenkouano, A., R. Ortiz and D. Vuylsteke. 2012c. Estimating genetic effects in maternal and paternal half-sibs from tetraploid-diploid crosses in *Musa* spp. Euphytica 185: 195-201.

Tenkouano, A., H.O. Oselebe and R. Ortiz. 2010a. Selection efficiency in *Musa* L. under different cropping systems. Aus. J. Crop Sci. 4: 74-80.

Tenkouano, A., M. Pillay, K. Tomekpe & R. Ortiz. 2010b. Breeding techniques. In Pillay, M. & A. Tenkouano (eds.) *Banana Breeding: Constraints and Progress*. CRC Press, Boca Raton, Florida. pp. 181-200

Tenkouano, A., D. Vuylsteke, J. Okoro, D. Makumbi, R. Swennen and R. Ortiz. 2003. Diploid banana hybrids TMB2x5105-1 and TMB2x9128-3 with good combining ability, resistance to black Sigatoka and nematodes. HortScience 38: 468-472.

Tushemereirwe W., M. Batte, M. Nyine, R. Tumuhimbise, A. Barekye, S. Tendo, D. Talengera, J. Kubiriba, J. Lorenzen, R. Swennen and B. Uwimana. 2015. Performance of NARITA Hybrids in the Preliminary Yield Trial for Three Cycles in Uganda. National Agricultural Research Organization – International Institute of Tropical Agriculture, Kampala, Uganda.

Tushemereirwe W., M. Batte, M. Nyine, R. Tumuhimbise, A. Barekye, S. Tendo, J. Kubiriba and R. Swennen. 2014. Performance of NARITA Hybrids in the Preliminary Yield Trial in Uganda. National Agricultural Research Organization – International Institute of Tropical Agriculture, Kampala, Uganda.

Vuylsteke, D., R. Ortiz, R.S.B. Ferris and R. Swennen. 1995. 'PITA-9': a black sigatoka resistant hybrid from the 'False Horn' plantain gene pool. HortScience 30: 395-397.
Vuylsteke, D.R. and R. Ortiz. 1995. Plantain-derived diploid hybrids (TMP2x) with black Sigatoka resistance. HortScience 30: 147-149.
Vuylsteke, D.R., R. Ortiz, R.S.B. Ferris and J.H. Crouch. 1997. Plantain improvement. Plant Breed. Rev. 14: 268-320.
Vuylsteke, D.R., R.L. Swennen and R. Ortiz. 1993a. Development and performance of black Sigatoka-resistant tetraploid hybrids of plantain (*Musa* spp., AAB group). Euphytica 65: 33-42.
Vuylsteke, D.R., R.L. Swennen and R. Ortiz. 1993b. Registration of 14 improved tropical *Musa* plantain hybrids with black sigatoka resistance. HortScience 28: 957-959.
Youssef, M., A.C. James, R. Rivera-Madrid, R. Ortiz and R.M. Escobedo-Gracia Medrano. 2011. *Musa* genetic diversity revealed by SRAP and AFLP. Mol. Biotechnol, 47: 189-199.

Chapter 6

Strawberry (Plants in the Genus *Fragaria*)

Tomohiro Yanagi[1,*] and Yuji Noguchi[2]

ABSTRACT

Cultivated strawberries are allo-octoploids of interspecific hybrid origin. In this chaper, we focus on five main topics: 1) an overview of species and interspecific hybrids in the genus *Fragaria*; 2) historical use of interspecific and intergeneric hybrids in strawberry breeding; 3) practical methods for producing interspecific hybrids; 4) practical methods to evaluate hybridity and chromosome doubling; and 5) examples from our recent breeding programs using wild strawberries. We describe breeding approaches using the decaploid interspecific hybrid cultivar 'Tokun' in Japan, as well as breeding using *Fragaria chiloensis* 'CHI-24-1' which exhibits a unique photoperiodic reaction to flower initiation. The present article is helpful to understand the botanical aspects of wild strawberries, useful techniques for strawberry breeding, and recent progress in strawberry breeding using interspecific hybrids.

Introduction

As an introduction to this chapter, a story detailing the development of the modern cultivated strawberry (*Fragaria* × *ananassa* Duchesne ex Rozier) will be told. The first protagonist of this story is Amédée-François Frézier (1682-1773), who was a French military engineer and mathematician. He was appointed by a French minister to secretly perform a reconnaissance mission, and entered into Peru and Chile in 1712 under the guise of a merchant. This area was ruled by Spain at that time. He conducted intelligence operations

[1] Faculty of Agriculture, Kagawa University, Ikenobe 2393, Miki-cho, Kita-gun, Kagawa, 761-0795, Japan, E-mail: yanagi@ag.kagawa-u.ac.jp

[2] Breeding and Genome Research Division, NARO Institute of Vegetable and Tea Science (NIVTS) in Japan, 360 Kusawa, Ano, Tsu, Mie, 514-2392 Japan, E-mail: ynogu@affrc.go.jp

* Corresponding author

for two years and returned with various achievements. For example, he corrected maps of the coastal area of Peru and Chile and collected information about the life and customs of the local people (Frézier 1717). Furthermore, he noticed the presence of strawberry fruits called 'Frutilla' by the Spanish people at Conception in Chile. The size, color, and taste of the fruit were completely different from those of that time in France. The 'Frutilla' was a strawberry cultivar grown by the local growers in Chile. He mentioned in his book that the fruits were usually as big as a whole walnut, and sometimes as large as a small egg. He subsequently returned with five 'Frutilla' plants of *F. chiloensis* (although he recorded the plant as *F. chiliensis* in his book) to France in 1714 (Darrow 1966, Wilhelm and Sagan 1974). After returning to France, he attempted to produce the fruits of the 'Frutilla' but did not appear to have had much success. The yield of the'Frutilla' was very low because it had only female flowers. Although the exact date is unclear, plants of *F. virginiana* (scarlet strawberry) were used for pollination in 'Frutilla' production. Thereafter, the hybrid strawberry plants, characterized by perfect flowers and the production of good fruits, were found in various places. Antoine Nicolas Duchesne (1747-1827), who was a French botanist, demonstrated that the new strawberries were interspecific hybrids between *F. chiloensis* and *F. virginiana*, judging from the morphological characteristics (Duchesne 1766). The new strawberries were then classified as *F.* × *ananassa.* As can be understood from this story, modern cultivated strawberries are of interspecific hybrid origin. After more than 250 years, the production of the cultivated strawberry has currently expanded around the world. According to the FAO statistics in 2013, more than 70 countries produce strawberry fruits totaling more than 7.7 million tons per year. The cultivated strawberry has become one of the most important crops for growers and consumers, globally.

Basic Information for Species and Interspecific Hybrids in the Genus *Fragaria*

Morphology of the Plants in the Genus Fragaria

Almost all plants in the genus *Fragaria*, generally termed "strawberry," are herbaceous perennials and have common morphological characteristics. The stem is of rosette morphology with a short and cylindrical shape and has leaves, roots, and inflorescences (Fig. 1). The stem has main and some lateral short branches, and is generally called the "crown". When strawberry plants are grown under long day-length and high temperature conditions, many long and slender stems, which are called "stolons" in botanical terms, are produced from the short stems. At the terminal portion of the stolon, a vegetatively propagated plant can form that has leaves and roots, and can produce secondary stolons. In general, the stolon is called a "runner" and the vegetatively propagated plant that is produced is called a "daughter plant".

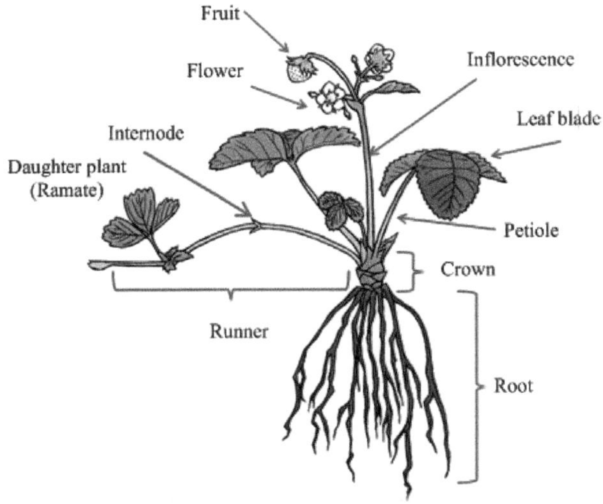

Figure 1. General appearance of plants in the genus *Fragaria*.

Some strawberry cultivars can produce over 200 daughter plants in a summer season. When the daughter plants are separated from the mother plants, the daughter plants can be grown as individuals. Strawberry growers generally utilize the vegetatively propagated clones for strawberry production, except under specific circumstances.

Strawberry plants have leaves composed of a petiole and three leaflets arranged pinnately. Strawberry plants produce leaves at an interval of seven to ten days by 2/5 phyllotaxy. Strawberry plants have a fibrous root system, which is likely monocotyledonous. Strawberry plant inflorescences are compound dichasia. A flower has a dome-shaped receptacle with a large number of pistils at the center, and some stamens (approximately 20-30) and five white petals at the periphery. As an exception, *F. iinumae* has seven petals in one flower. The fruit of the strawberry plant is composed of a portion of enlarged receptacle and has many achenes. Every achene has one seed. The fruit color of strawberry plants is red. However, some of the plants in *F. vesca*, *F. nilgirrensis*, *F. chiloensis*, and *F.* × *ananassa* produce white fruit.

Species in the Genus Fragaria

To illustrate the species in the genus *Fragaria*, all species listed in some books and review papers published after 1966 are summarized in Table 1. From the list, it is apparent that the number of the species has been under continual review.

A total of 14 diploid species ($2n = 2x = 14$ chromosomes) are reported in the literature. Among them, eight species, namely *F. daltoniana*, *F. iinumae*, *F. mandshurica*, *F. nilgerrensis*, *F. nipponica*, *F. nubicola*, *F. vesca*, and *F. viridis* are listed in every example in the literature, and appear to have identical

Table 1. The taxonomic species in the genus *Fragaria* as listed in different studies.

Ploidy level	Species name	Darrow	Staudt	Han-cock	Folta & Davis	Hu-mmer & Han-cock	Staudt	Hu-mmer et al.
		1966	1989	1999	2006	2009	2009	2011
2x	*F. bucharica*	–	–	–	O	O	O	O
	F. chinensis	–	–	–	–	–	O	O
	F. daltoniana	O	O	O	O	O	O	O
	F. gracilis	–	–	O	O	O	–	–
	F. hayatae	–	–	–	–	–	O	–
	F. iinumae	–	O	O	O	O	O	O
	F. mandshurica	–	O	O	O	O	O	O
	F. nilgerrensis	O	O	O	O	O	O	O
	F. nipponica	-	O	O	O	O	O	O
	F. nubicola	O	O	O	O	O	O	O
	F. pentaphylla	-	-	O	O	O	O	O
	F.vesca	O	O	O	O	O	O	O
	F. viridis	O	O	O	O	O	O	O
	F. yezoensis	–	O	O	O	O	–	–
4x	*F. corymbosa*	–	O	O	O	O	O	O
	F. gracilis	–	–	–	–	O	O	O
	F. moupinensis	–	O	O	O	O	O	O
	F. orientalis	O	O	O	O	O	O	O
	F. tibetica	–	–	–	O	O	O	O
6x	*F. moschata*	O	O	O	O	O	O	O
8x	*F. × ananassa*	O	O	O	O	O	O	O
	F. chiloensis	O	O	O	O	O	O	O
	F. iturupensis	–	O	O	O	O	O	–
	F. ovalis	O	–	–	–	–	–	–
	F. virginiana	O	O	O	O	O	O	O
10x	*F. cascadensis*	–	–	–	–	–	–	O
	F. iturupensis	–	–	–	–	–	–	O

* 'O' denotes listed and '–' denotes not listed in each study.

morphological characteristics between listings. These can be regarded as independent species. Originally, *F. bucharica* was recorded by Losinskaja (1926) as a species independent of *F. nubicola*. Subsequently, Staudt (2006) also suggested a difference between *F. nubicola* and *F. bucharica*, and classified *F. bucharica* as a distinct species from *F. nubicola*. In addition, *F. chinensis* was first recorded by Losina-Losinskaja (1926). Staudt (2009) classified *F. chinensis* as a distinct species. However, in the Flora of China (Li et al. 2003), *F. chinensis* is recognized as a synonym of *F. vesca*. In actual fact, few botanical studies have been performed for *F. chinensis*. It is therefore necessary to re-examine the classification of *F. chinensis*. *Fragaria gracilis*, which was first recorded by Losina-Losinskaja (1926), is classified as diploid by Hancock (1999) but tetraploid by Staudt (2009). Dr. Lei, who is affiliated with Shenyang Agricultural University in China, indicated that there are diploid and tetraploid strains of *F. gracilis* (personal communication). It is therefore also necessary to re-examine the classification of *F. gracilis*. *Fragaria hayatae* was first recorded by Makino (1912). Subsequently, *F. hayatae* was treated as a distinct species in Staudt (2009). However, *F. hayatae* was classified as a synonym of *F. nilgerrensis* in the Flora of China, although two Japanese botanists indicated that *F. hayatae* is a distinct species as a proviso in the same book. Naruhashiet al. (1999) also suggested that *F. hayatae* is a different species from *F. nilgerrensis*, judging from the karyotype and flower color. *Fragaria yezoensis* was first recorded in Hara (1944). However, Naruhashi and Iwata (1988) later classified *F. yezoensis* as a synonym of *F. nipponica* by the comparisons of morphological characteristics.

Five species are listed among the tetraploids of the genus *Fragaria* (Table 1). Among them, *F. corymbosa* is recognized as a distinct species in Losina-Losinskaja (1926), Studut (1989), Lei et al. (2006), and Lei et al. (2009). However, in Li et al. (2003), *F. corymbosa* is classified as a synonym of *F. orientalis*. Five octoploid species have to date been reported. However, *F. ovalis* has been classified as *F. virginiana*. The *F. iturupensis* has been classified as a decaploid (Hummer et al. 2009, Hummer et al. 2011). Furthermore, as indicated in many studies, *F. virginiana* and *F. chiloensis* are classified as different species; however, crossability between these two species is nearly 100%. For this reason, these species might be united under one species from the point of view of sexual reproduction. Hammer and Pistrick (2003) proposed a new species named *F.* × *rosea*, which is an intergeneric hybrid between *F.* × *ananassa* and *Potentilla palustris*.

Interspecific Hybrids in Nature

In plants of the genus *Fragaria*, natural interspecific hybrids have been reported in six studies (Table 2). In Finland, an example of a tetraploid (4*x*) strawberry plant was isolated and identified as an interspecific hybrid between *F. vesca* and *F. viridis* by differences in morphological characteristics and by molecular biological analysis (Ahokas 1999 and 2002). In addition, diploids (2*x*) and triploids (4*x*) of *F.* × *bifera* have been found in nature (Staudt et al. 2003). *Fragaria* × *bifera* was confirmed as a hybrid between *F. vesca* and *F. viridis* by morphological and molecular biological analysis. Pentaploid (5*x*) strawberry hybrids were identified in the Inner Mongolia region of China (Lei et al. 2005), and allopentaploid (5*x*) (Nosrati et al. 2011a) and alloheptaploid (7*x*) (Nosrati et al. 2013) strawberries were found in Germany. In North America, pentaploid (5*x*), hexaploid (6*x*), and nanoploid (9*x*) interspecific hybrids between *F. vesca* and *F. chiloensis* were identified (Bringhurst and Khan 1963, Bringhurst and Senanayake 1966). These hybrids were classified as *F.* × *bringhurstii* Spec. Nov. (Staudt 1999a). However, few studies have listed *F.* × *bringhurstii* as a species in the genus *Fragaria*. In addition, natural hybrids between *F. virginiana* and *F. chiloensis* have been identified in Canada (Staudt 1999a). Decaploids (10*x*) of interspecific hybrids between *F. virginiana* and *F. vesca* were classified under a new species named *F. cascadensis* (Hummer 2012).

Table 2. Natural interspecific hybrid plants in the genus *Fragaria*.

Ploidy level	Scientific name	Distribution	Literature
4*x*	–	Finland	Ahokas 1999
2*x*, 3*x*	*F.* × *bifera*	France, Germany, Finland	Ahokas 2002
			Staudt et al. 2003
4*x*	–	Inner Mongolia, China	Lei et al. 2005
5*x*	–	Bavaria, Germany	Nosrati et al. 2011a
5*x*, 6*x*, 9*x*	*F.* × *bringhurstii**	California, USA	Bringhurst and Khan 1963
			Bringhurst and Senanayake 1966
7*x*	–	Bavaria, Germany	Nosrati et al. 2013
8*x*	*F. ananassa* ssp. *cuneifolia*	Vancouver, Canada	Staudt 1999a
10*x*	*F. cascadensis*	Oregon, USA	Hummer 2012

*The plants were classified as *F.* × *bringhurstii* (Staudt 1999a).

Relatedness Between Species

Crossability Among Species

Relatedness between species can be evaluated by crossability between species. If species are closely related, healthy hybrids can be obtained with high frequency. The crossability among species in the genus *Fragaria* has been poorly described (Table 3). *Fragaria chiloensis, F. virginiana,* and *F.* × *ananassa* 8*x* species are characterized by complete crossability between them, as mentioned before. In addition, these species can produce a small number of vigorous hybrids with *F. vesca* and *F. moschata*. Of the 2*x* species,

Table 3. Interspecific crossibility in the genus *Fragaria*. Source: Dowrick and Williams (1959), Fadeeva (1966), Hancock et al. (1991), Bors and Sullivan (2005a); Bors and Sullivan (2005b), Hancock and Luby (1993), Noguchi et al. (1997), Nosrati et al. (2010), Rho et al. (2012), Sargent et al. (2004), Staudt et al. (2009), and Yanagi et al. (2010).

Seed parent		Pollen parent													
		VE	VI	NU	NIP	NIL	II	DA	PE	OR	MO	VI	CH	AN	IT
2*x*	*F. vesca* (VE)	1*	1, 3	1	3	1, 3	4	1	1	3	3	2, 3	2, 3	2, 3	–
	F.viridis (VI)	1, 4	1	2	1	4	–**	–	–	3	2, 3	–	–	3	–
	F.nubicola (NU)	4	1	1	1	4	–	–	–	–	4	–	–	–	–
	F. nipponica (NIP)	4	1	1	1	4	–	–	–	3	–	–	–	–	–
	F.nilgerrensis (NIL)	4	4	4	3	1	–	–	–	–	–	–	–	3	–
	F. iinumae (II)	–	–	–	–	–	–	–	–	–	–	–	–	–	–
	F. daltoniana (DA)	–	–	–	–	–	–	–	–	–	–	–	–	–	–
	F. pentaphylla (PE)	–	–	–	–	–	–	–	–	–	–	–	–	–	–
4*x*	*F.orientalis* (OR)	2, 3	2, 3	–	–	–	–	–	–	1	2, 3	–	–	–	–
6*x*	*F.moschata* (MO)	3	2, 3	2	–	–	–	–	–	2, 3	1	3	3	–	–
8*x*	*F. virginiana* (VI)	2, 3	3	–	–	–	–	–	–	–	2, 3	1	1	–	–
	F. chiloensis (CH)	2, 3	3	–	–	–	–	–	–	–	2, 3	1	1	–	–
	F. x ananassa (AN)	2, 3	3	–	–	3	3	–	–	–	2, 3	1	1	–	3
10*x*	*F. iturupensis* (IT)	–	–	–	–	–	–	–	–	–	–	–	–	3	–

*1: Full seed set, viable F1 plants; 2: Partial seed set, viable F1 plants; 3: No viable F^1 plants or sterile F1; 4: No seed set.**No data

F. vesca is crossable with various 2*x*, 4*x*, 6*x*, and 8*x* species (Yarnell 1931, Evans and Jones 1967, Hancock and Luby 1993, Bors and Sullivan 2005b). *Fragaria viridis, F. nubicola,* and *F. nipponica* can be crossed to each other, but crossability of these species with other 2*x* species and polyploids remains unclear. It is necessary to clarify the crossability among species in the genus *Fragaria* using some different strains in each species.

Genome Analysis Using Chromosome Pairing in the Pollen Mother Cell

Genome analysis by chromosome pairing is one method that can be used to clarify the relatedness between species. The genome composition of allopolyploid plants can be identified by the observation of chromosome pairing in the pollen mother cells of interspecific hybrids between the allopolyploid and the diploid species that appears to be a genome donor to the allopolyploid (Lilienfeld 1951). In the case of the octoploid (8*x*) strawberries, many interspecific hybrids were produced to clarify the genome composition (Table 4). Yarnell (1931), Dogadkina (1941), Fedorova (1946), Ellis (1962), and Senanayake and Bringhurst (1967) developed various pentaploid (5*x*) hybrids between *F. vesca* (2*x*) and *F.* × *ananassa* (8*x*) or *F. chiloensis* (8*x*). They then determined that *F. vesca* was one of the genome donors of the 8*x* species, as the pentaploids (5*x*) had 14 bivalent and seven univalent chromosomes in the pollen mother cells. In this case, it can be considered that the seven chromosomes from *F. vesca* could produce seven bivalent pairs with the seven chromosomes from the octoploids (8*x*). In addition, the remaining seven bivalent chromosome pairs were produced by the different subgenomes in the octoploids (8*x*). Senanayake and Bringhurst (1967) subsequently proposed that: 1) the genome composition of octoploids (8*x*) was AAA′A′BBBB; 2) the A genome was *F. vesca;* and 3) the remaining A′A′BBBB genomes were unknown. In addition, Fedorova (1946) reported that the octoploids (8*x*) had two sets of chromosomes from *F. orientalis* (4*x*) and one set from each of *F. nipponica* (2*x*) and *F. vesca* (2*x*). Furthermore, Byrne and Jelenkovic (1976) determined that: 1) *F.* × *ananassa* had 28 bivalent chromosome pairs in the pollen mother cell and 2) the B genome was from *F. nubicola,* judging from the results of chromosome pairing by the pentaploid (5*x*) hybrids between *F. nubicola* and *F.* × *ananassa.* Bringhurst (1990) subsequently proposed a new genome composition of AAA′A′BBB′B′, as 28 bivalent pairs with no tetravalents were observed in the pollen mother cells of the octoploids (8*x*) as indicated by Byrne and Jelenkovic (1976). Bringhurst (1990) suggested that diploidization had occurred in the evolution of the octoploid (8*x*) species in the genus *Fragaria.* From these results, *F. vesca* seems to have higher relatedness to octoploid (8*x*) strawberries than the other diploids.

Molecular Biological Analysis

After the 1990s, many studies attempted to elucidate the genome composition and phylogenetic relationships between species in the genus *Fragaria* using

Table 4. Genome analysis using pollen mother cells in the species of the genus *Fragaria*.

Literature	Seed parent	Chromosome number	Pollen parent	Chromosome number	Chromosome number of hybridsv	Chromosome pairing
Ichijima 1926	*F. vesca*	14	*F. virginiana*	56	35	7 II + 21 I
Yarnell 1931	*F. vesca*	14	*F. virginiana*	56	35	17 II + 1 I
	F. vesca	14	*F. virginiana*	56	35	14 II + 1 III + 1 IV
Dogadkina 1941	*F. vesca*	14	*F.* × *ananassa*	56	35	14 II + 7 I
Fedorova 1946	*F. vesca*	14	*F.* × *ananassa*	56	35	14 II + 7 I
Ellis 1962	*F. vesca*	14	*F. virginiana*	56	35	10.5 (II or III) + 14 I
	F. vesca	14	*F.* × *ananassa*	56	35	14 (II or III) + 7 I
Senanayake and Bringhurst 1967	*F. chiloensis*	56	*F. vesca*	14	35	10 II + 15 I
	F. vesca	14	*F. virginiana*	56	35	14 II + 7 I
	F. chiloensis	56	*F. viridis*	14	35	7 II + 21 I
Byrne and Jelenkovic 1976	*F.* × *ananassa*	56	*F.* × *ananassa*	56	56	28 II
	F. × *ananassa*	56	*F. nubicola*	14	35	11 II + 9 I + 1 IV

molecular biological methods. Various DNA analysis methods indicated that the octoploid species were allopolyploids (Lerceteau-Köhler et al. 2003, Rousseau-Gueutin et al. 2009, Kunihisa et al. 2011, Sargent et al. 2012, Honnjo et al. 2013, Isobe et al. 2013). Harrison et al. (1997) and Potter et al. (2000) suggested that *F. iinumae* (2*x*) is theoriginal progenitor species of the genus *Fragaria*, judging from the results of molecular biological analysis using cytoplasmic and nuclear DNA. Mahoney et al. (2010), using mitochondrial DNA analysis, reported that *F. iinumae* (2*x*) is a genome donor of the octoploids (8*x*). In addition, Njuguna et al. (2013), using chloroplast DNA analysis, indicated that *F. vesca sub. bracteata* (2*x*) is a cytoplasmic genome donor of the octoploids (8*x*). Tennessen et al. (2014) demonstrated that the four genomes of the octoploids were composed of one from *F. vesca* (2*x*), one from *F. iinumae* (2*x*), and the remaining two from *F. iinumae*-related species. In the case of the other species, *F. vesca* and *F. nubicola* were phylogenetically close to *F. orientalis, F. moschata, F. chiloensis,* and *F. virginiana* (Potter et al. 2000). Lin and Davis (2000) indicated that *F. viridis* and *F. vesca* are genome donors to *F. moschata* but not *F. nubicola* using chloroplast DNA analysis. Hirakawa et al. (2014) mentioned that *F. vesca* and *F. nubicola* are more genetically close to *F.* × *ananassa* than *F. iinumae* and *F. nipponica* based on a phylogenetic tree obtained by DNA analysis. From these studies, it can be considered that *F. vesca* and *F. iinumae* might be genome donors to the octoploid species in the genus *Fragaria*.

Analysis of Morphological Characteristics

The relatedness among species in the genus *Fragaria* can be evaluated by comparison of morphological characteristics. Staudt (2009) indicated that the species of the genus *Fragaria* were divided into three groups by chromosome number, branching of stolons, pollen morphology, sex expression, and crossability. Group 1 comprised 2*x F.* × *bifera, F. bucharica, F. mandshurica, F. nilgerrensis, F. vesca,* and *F. viridis*, 4*x F. orientalis*, 6*x F. moschat*, and 8*x F. chiloensis, F. virginiana,* and *F. iturupensis*. Group 2 comprised 2*x F. chinensis, F. nipponica, F. nubicola,* and *F. pentaphylla,* and 4*x F. corymbosa, F. gracilis F. moupinensis,* and *F. tibetica*. The remaining 2*x* species *F. daltoniana, F. hayatae,* and *F. iinumae* could not be grouped.

As mentioned in the previous section, there are different opinions regarding the taxonomic classification of *F. chinensis, F. corymbosa,* and *F. iturupensis*. In addition, Nathewet et al. (2009b and 2010) conducted karyotype analysis to illustrate the phylogenetic relationships between the species. Sargen et al. (2004) also divided diploids in the genus *Fragaria* into three groups on the basis of differences in morphological characteristics and crossability.

Table 5. Artificial interspecific hybrids used in the literature.

Literature	Parental species ♀	Parental species ♂	Ploidy level of inter-specific hybrids
Ahmadi and Bringhurst (1992)	*F. chiloensis*	*F. vesca*	5*x*
	F. virginiana	*F. viridis*	5*x*
	6*x* (*F. chiloensis* × *F. vesca*)	*F.* × *ananassa*	10*x*
	12*x* (16*x* (*F.* × *ananassa*) × *F.* × *ananassa*)	*F.* × *ananassa*	10*x*
	16*x F.* × *ananassa*	4*x F. vesca*	10*x*
	12*x F. moschata*	*F.* × *ananassa*	10*x*
	12*x F. moschata*	*F. chiloensis*	10*x*
	12*x F. moschata*	*F. virginiana*	10*x*
	9*x* (12*x F. moschata* × 6*x F. moschata*)	10*x* ?	10*x*
Barritt and Shanks (1980)	*F.* × *ananassa*	*F. chiloensis*	8*x*
Coman and Popescu (2008)	*F.* × *vescana*	*F. moschata*	
	F. × *ananassa* × *F. vesca*	*F. moschata*	
Evans (1982a)	Synthetic 8*x* (*F. moschata, F. nubicola*)		8*x*
Evans (1982b)	Synthetic 8*x* (*F. vesca, F. virisdis* and *F. moupinensis*)		8*x*
Evans (1982c)	Synthetic 8*x* (*F.* × *ananssa, F. moschata, F. nubicola, F. vesca, F. virisdis* and *F. moupinensis*)		8*x*
Harbut et al. (2009 and 2012)	Synthetic 8*x* (*F. moschata, F. nubicola, F. vesca*)		8*x*
	Synthetic 8*x* (*F. moschata, F. nubicola*)		8*x*
	Synthetic 8*x* (*F. orientalis, F. moschata, F. nubicola*)		8*x*
	Synthetic 8*x* (*F. nubicola, F. vesca, F. pentaphylla*)		8*x*
	Synthetic 8*x* (*F. orientalis, F. nilgirrensis, F. vesca*)		8*x*
	Synthetic 8*x* (*F. moschata, F. nubicola,F. virisdis*)		8*x*
	F. × *ananassa*	Synthetic 8*x* (*F. orientalis, F. nilgirrensis, F. vesca*)	8*x*
	F. × *ananassa*	Synthetic 8*x* (*F. moschata, F. nubicola, F. virisdis*)	8*x*
	F. × *ananassa*	Synthetic 8*x* (*F. moschata, F. nubicola, F. vesca*)	8*x*
	F. × *ananassa*	*F.* × *ananassa* × Synthetic 8*x* (*F. moschata, F. nubicola, F. virisdis*)	8*x*

cont.

Hancock et al. (2002)	*F. virginiana*	*F.* × *ananassa*	8*x*
Hancock et al. (2010)	*F. virginiana*	*F. chiloensis*	8*x*
Ichijima (1926)	*F. vesca*	*F. virginiana*	5*x*
Kishaba et al. (1972)	*F.* × *ananassa*	*F. chiloensis*	8*x*
	F. × *ananassa*	*F. virginiana*	8*x*
Lei et al. (2009)	*F.* × *ananassa*	Unknown 5*x*	12*x*
Luby et al. (2008)	*F. virginiana*	*F. chiloensis*	8*x*
Longley (1926)	*F. vesca*	*F. chiloensis*	5*x*
Milenkovic et al. (2014)	*F.* × *ananassa*	*F. chiloensis*	8*x*
Mochizuki et al. (2002)	*F. nilgirrensis*	*F.* × *ananassa*	10*x*
	F. vesca	*F.* × *ananassa*	10*x*
Noguchi et al. (2009)	*F.* × *ananassa*	*F. nilgerrensis*	10*x*
Noguchi and Yamada (2014)	*F.* × *ananassa*	*F. nilgerrensis*	10*x*
Sasaki and Takeuchi (2008).	6*x* (*F. vesca* × *F.* × *ananassa*)	*F. vesca*	5*x*
Schulze et al. (2013)	*F. vesca*	*F.* × *ananassa*	5*x*
Senanayake and Bringhurst (1967)	*F. vesca*	*F. virginiana*	5*x*
	F. chiloensis	*F. viridis*	5*x*
	F. chiloensis	*F. vesca*	5*x*
	F. vesca	*F. virginiana*	6*x*
	F. vesca 4*x*	*F. chiloensis*	6*x*
	F. chiloensis	*F. vesca* 4*x*	6*x*
Stegmeir et al. (2010)	*F. virginiana*	*F. chiloensis*	8*x*

Table 6. Cross-pollination to produce artificial interspecific hybrids in the genus *Fragaria*.

	Parent lines		Reference
	♀	♂	
2*x* × 2*x*	*F. mandschurica*	*F. viridis*	Lei et al. (2002).
	F. nilgerrensis	*F. vesca*	Evans and Jones (1967), Yarnell (1931), Rho et al. (2012)
	F. nilgerrensis	*F. viridis*	Evans and Jones (1967)
	F. nilgerrensis	*F. nubicola*	Evans and Jones (1967)
	F. nubicola	*F. vesca*	Evans and Jones (1967)
	F. nubicola	*F. nillgirrensis*	Evans and Jones (1967)
	F. nubicola	*F. viridis*	Evans and Jones (1967)
	F. nubicola	2*x* (*F. vesca* × *F. viridis*)	Evans and Jones (1967)
	F. nubicola	2*x* (*F. vesca* × *F. nubicola*)	Evans and Jones (1967)
	F. vesca	*F. nilgerrensis*	Bors and Sullivan (2005a), Evans and Jones (1967), Sargent et al. (2004), Rho et al. (2012)
	F. vesca	*F. nubicola*	Bors and Sullivan (2005a), Evans and Jones (1967), Nosrati et al. (2011b), Sargent et al. (2004).
	F. vesca	*F. pentaphylla*	Bors and Sullivan (2005a)
	F. vesca	*F. viridis*	Bors and Sullivan (2005a), Evans and Jones (1967), Nosrati et al. (2011b), Sargent et al. (2004).
	F. vesca	2*x* (*F. vesca* × *F. viridis*)	Evans and Jones (1967)
	F. vesca	2*x* (*F. vesca* × *F. viridis*)	Evans and Jones (1967)
	F. vesca	*F. daltoniana*	Nosrati et al. (2011b), Sargent et al. (2004).
	F. vesca	*F. nipponica*	Nosrati et al. (2011b), Sargent et al. (2004).
	F. vesca	*F. iinumae*	Sargent et al. (2004).
	F. vesca	*F. pentaphylla*	Sargent et al. (2004).
	F. vesca	2*x* (*F. vesca* × *F. nubicola*)	Sargent et al. (2004).
	F. vesca	2*x* (*F. vesca* × *F. viridis*)	Sargent et al. (2004).
	F. viridis	*F. vesca*	Evans and Jones (1967)
	F. viridis	*F. nillgirrensis*	Evans and Jones (1967)

cont.

	F. viridis	*F. nubicola*	Evans and Jones (1967)
	F. viridis	2*x* (*F. vesca* × *F. viridis*)	Evans and Jones (1967)
	F. viridis	2*x* (*F. vesca* × *F. nubicola*)	Evans and Jones (1967)
	F. viridis	*F. mandschurica*	Lei et al. (2002).
	2*x* (*F. vesca* × *F. viridis*)	*F. vesca*	Evans and Jones (1967)
	2*x* (*F. vesca* × *F. viridis*)	*F. nillgirrensis*	Evans and Jones (1967)
	2*x* (*F. vesca* × *F. viridis*)	*F. viridis*	Evans and Jones (1967)
	2*x* (*F. vesca* × *F. viridis*)	*F. nubicola*	Evans and Jones (1967)
	2*x* (*F. vesca* × *F. viridis*)	2*x* (*F. vesca* × *F. nubicola*)	Evans and Jones (1967)
	2*x* (*F. vesca* × *F. nubicola*)	*F. vesca*	Evans and Jones (1967)
	2*x* (*F. vesca* × *F. nubicola*)	*F. nillgirrensis*	Evans and Jones (1967)
	2*x* (*F. vesca* × *F. nubicola*)	*F. viridis*	Evans and Jones (1967)
	2*x* (*F. vesca* × *F. nubicola*)	*F. nubicola*	Evans and Jones (1967)
	2*x* (*F. vesca* × *F. nubicola*)	2*x* (*F. vesca* × *F. nubicola*)	Sargent et al. (2004).
2*x* × 4*x*	*F. mandschurica*	*F. orientalis*	Lei et al. (2004)
2*x* × 5*x*	*F. viridis*	5*x*	Lei et al. (2004)
2*x* × 6*x*	*F. nubicola*	*F. moschata*	Bors and Sullivan (2005b), Evans (1974)
	F. viridis	*F. moschata*	Bors and Sullivan (2005b), Evans (1974)
	F. vesca	*F. moschata*	Evans (1974), Fedorova (1934), Mangelsdorf and East (1927)
	F. nilgerrensis	*F. moschata*	Evans (1974), Yarnell (1931)
	2*x* (*F. vesca* × *F. viridis*)	*F. moschata*	Evans (1974)
2*x* × 8*x*	*F. nilgirrensis*	*F.* × *ananassa*	Evans (1974), Rho et al. (2012)
	F. nilgerrensis	*F. chiloensis*	Evans (1974)
	F. nilgerrensis	*F. virginiana*	Evans (1974)
	F. nilgerrensis	8*x* (*F. moupiensis* × *F. nubicola*)	Evans (1974)
	F. nubicola	*F.* × *ananassa*	Evans (1974)
	F. nubicola	*F. chiloensis*	Evans (1974)
	F. nubicola	*F. virginiana*	Evans (1974)
	F. nubicola	8*x* (*F. moupiensis* × *F. nubicola*)	Evans (1974)
	F. viridis	*F.* × *ananassa*	Evans (1974)

cont.

	F. viridis	*F. chiloensis*	Evans (1974)
	F. viridis	*F. virginiana*	Evans (1974)
	F. viridis	8x (*F. moupiensis* × *F. nubicola*)	Evans (1974)
	F. vesca	*F.* × *ananassa*	Evans (1974), Morishita et al. (1996), Rho et al. (2012), Richardson (1914), Schulze et al. (2011)
	F. vesca	*F. chiloensis*	East (1930), Evans (1974), Mangelsdorf and East (1927), Richardson (1914), Yarnell (1931), Yanagi et al. (2009.), Yanagi et al. (2010)
	F. vesca	*F. virginiana*	East (1930), East (1934), Mangelsdorf and East (1927), Evans (1974), Yarnell (1931)
	F. vesca	8x (*F. moupiensis* × *F. nubicola*)	Evans (1974)
	2x (*F. vesca* × *F. viridis*)	*F.* × *ananassa*	Evans (1974)
	2x (*F. vesca* × *F. viridis*)	*F. chiloensis*	Evans (1974)
	2x (*F. vesca* × *F. viridis*)	*F. virginiana*	Evans (1974)
	2x (*F. vesca* × *F. viridis*)	8x (*F. moupiensis* × *F. nubicola*)	Evans (1974)
4x	*F. orientalis*	*F. mandschurica*	Lei et al. (2004)
	F. orientalis	5x	Lei et al. (2004)
	F. vesca 4x	*F.* × *ananassa*	Morishita et al. (1996)
	4x (*F. vesca* × *F. viridis*)	*F. moschata*	Evans (1974)
	4x (*F.nilgerrensis* × *F. viridis*)	*F. moschata*	Evans (1974)
	4x (*F. moschata* × *F. nubicola*)	*F. moschata*	Evans (1974)
	4x (*F. vesca* × *F. viridis*)	*F. chiloensis*	Evans (1974)
	4x (*F. vesca* × *F. viridis*)	*F. virginiana*	Evans (1974)
	4x (*F. vesca* × *F. viridis*)	*F.* × *ananassa*	Evans (1974)
	4x (*F. vesca* × *F. viridis*)	8x (*F. moupiensis* × *F. nubicola*)	Evans (1974)
	4x (*F. moschata* × *F. nubicola*)	*F. chiloensis*	Evans (1974)

cont.

	4x (*F. moschata* × *F. nubicola*)	*F. virginiana*	Evans (1974)
	4x (*F. moschata* × *F. nubicola*)	*F.* × *ananassa*	Evans (1974)
	4x (*F. moschata* × *F. nubicola*)	8x (*F. moupiensis* × *F. nubicola*)	Evans (1974)
	F. viridis 4x	*F.* × *ananassa*	Trajkovski (1997)
	F. vesca 4x	*F.* × *ananassa*	Trajkovski (1997)
5x	5x (*F. chiloensis* × *F. vesca*)	*F. chiloensis*	Bringhurst and Gill (1970)
	5x (*F. chiloensis* × *F. vesca*)	*F. chiloensis*	Bringhurst and Gill (1970)
	6x (*F. chiloensis* × *F. vesca*)	*F. chiloensis*	Bringhurst and Gill (1970)
	9x (*F. chiloensis* × *F. vesca*)	*F. chiloensis*	Bringhurst and Gill (1970)
	5x (*F. vesca* × *F. virginaia*)	5x (*F. vesca* × *F. virginaia*)	East (1934)
	5x (*F. vesca* × *F.* × *ananassa*)	*F. vesca* 4x	Ellis (1962)
	5x (*F. vesca* × *F.* × *ananassa*)	*F.* × *ananassa*	Ellis (1962)
	5x (*F. vesca* × *F.* × *ananassa*)	F. vesca 8x	Ellis (1962)
	5x (*F. vesca* × *F.* × *ananassa*)	*F.* × *ananassa* 16x	Ellis (1962)
	5x (*F. vesca* 4x × *F.* × *ananassa*)	*F.* × *ananassa*	Ellis (1962)
	5x	*F. viridis*	Lei et al. (2004)
	5x	*F. orientalis*	Lei et al. (2004)
	5x	*F.* × *ananassa*	Lei et al. (2004)
6x	*F. moschata*	*F. nilgirrensis*	Yarnell (1931)
	F. moschata	*F. nipponica*	Lilienfeld (1933)
	F. moschata	*F. nubicola*	Bors and Sullivan (2005b)
	F. moschata	*F. viridis*	Bors and Sullivan (2005b)
	F. moschata	*F. vesca*	Mangelsdorf and East (1927), Yarnell (1931)
	F. moschata	*F. virginiana*	Waldo and Darrow (1928)
	6x (*F.* × *ananassa* × *F. vesca* 4x)	*F.* × *ananassa*	Scott (1951)

cont.

	6*x* (*F.* × *ananassa*× *F. vesca* 4*x*)	*F.* × *ananassa*	Scott (1951)
	6*x* (*F.* × *ananassa* × *F. vesca* 4*x*)	6*x* (*F.* × *ananassa* × *F. vesca* 4*x*)	Scott (1951)
	6*x* (*F.* × *ananassa* × *F. vesca* 4*x*)	*F.* × *ananassa*	Trajkovski (1993)
	6*x* (*F. vesca* 4*x* × *F.* × *ananassa*)	*F.* × *ananassa*	Morishita et al. (1996)
	6*x* (*F.* × *ananassa* × *F. vesca* 4*x*)	*F.* × *ananassa*	Bauer (1993)
	6*x*	*F.* × *ananassa*	Lei et al. (2004)
7*x*	7*x* (*F.* × *ananassa* × *F. moschata*)	7*x* (*F.* × *ananassa* × *F. moschata*)	Fedorova (1934)
	7*x* (*F.* × *ananassa* × *F. moschata*)	*F.* × *ananassa*	Trajkovski (1993)
8*x*	*F.* × *ananassa*	*F. nilgerrensis*	Noguchi et al. (1997), Noguchi et al. (2002), Rho et al. (2012)
	F. × *ananassa*	*F. iinumae*	Noguchi et al. (1997)
	F. × *ananassa*	*F. vesca*	Li et al. (2000), Rho et al. (2012), Marta et al. (2004)
	F. × *ananassa*	*F. vesca* 4*x*	Scott (1951)
	F. × *ananassa*	5*x*	Lei et al. (2004)
	F. × *ananassa*	6*x* (*F.* × *ananassa* × *F. vesca* 4*x*)	Bauer (1993)
	F. × *ananassa*	6*x*	Lei et al. (2004)
	F. × *ananassa*	*F. moschata*	Fedorova (1934)
	F. × *ananassa*	6*x* (*F.* × *ananassa* × *F. vesca* 4*x*)	Scott (1951)
	F. × *ananassa*	*F. chiloensis*	Yanagi et al. (2005)
	F. × *ananassa*	*F. virginiana*	Powers (1944 and 1945)
	F. × *ananassa*	*F. iturupensis*	Staudt et al. (2009)
	F. chiloensis	*F. vesca*	Yarnell (1931), Mangelsdorf and East (1927)
	F. chiloensis	*F. nilgirrensis*	Yarnell (1931)
	F. chiloensis	*F. virginiana*	Mangelsdorf and East (1927)
	F. virginiana	*F. vesca*	Yarnell (1931)

cont.

	F. virginiana	*F. moschata*	Mangelsdorf and East (1927), Yarnell (1931)
	F. virginiana	*F. chiloensis*	Mangelsdorf and East (1927)
10*x*	*F. iturupensis*	*F.* × *ananassa*	Staudt et al. (2009)
	F. vescana	*F. vescana*	Trajkovski (1997)
	F. vescana	9*x* (*F. vescana* × *F.* × *ananassa*)	Trajkovski (1997)
	F. vescana	*F. moschata*	Trajkovski (1997), Coman and Popescu (2008)
	F. vescana	6*x* (*F. vesca* 4*x* × *F.* × *ananassa*)	Trajkovski (1997)
	F. × *ananassa*	7*x* (*F.* × *ananassa* × *F. moschata*)	Trajkovski (1997)
	10*x* (*F.* × *ananassa* × *F. nilgirrensis*)	*F.* × *ananassa*	Noguchi et al. (2002)
	10*x* (*F.* × *ananassa* × *F. nilgirrensis*)	10*x* (*F.* × *ananassa* × *F. nilgirrensis*)	Noguchi et al. (2011)

Breeding Using Interspecific and Intergeneric Hybrids in the Genus *Fragaria*

The artificial interspecific hybrids used in the cited studies are summarized in Table 5. Additionally, the cross-pollination combinations used to produce artificial interspecific hybrids are listed in Table 6.

Breeding of 8x F. × ananassa Using Wild 8x Octoploids

Early History of Breeding Using Wild Octoploids (8x)

A retrospective view of the history of strawberry production worldwide over the last 250 years reveals that at various time points opportunities arose for hybridization between cultivated and wild octoploids. Thomas A. Knight and Michael Keens selected various cultivars from hybridizations between *F. chiloensis* and *F. virginiana* in the UK during the early 19th century (Darrow 1966). Albert F. Etter also conducted strawberry breeding in North California in the USA during the early 20th century using many wild strains of *F. chiloensis* native to Cape Mendocino in the Californian coastal area (Wilhelm and Sagen 1974). His cultivars contributed to the increment in yield of strawberry production in California State. C.L. Powers and A.C. Hildreth, who worked in the USDA, also tried to use *F. virginiana* ssp. *glauca* to breed cold-tolerant cultivars (Darrow 1966, Hildreth and Powers 1941).

Breeding After the Mid-20th Century

R.S. Bringhurst and V. Voth of the University of California bred day-neutral cultivars using *F. virginiana* ssp. *glauca,* which was native to the Wasatch Mountains of Utah State in the USA (Bringhurst and Voth 1982 and 1984, Ahmadi et al. 1990). The day-neutral strawberry extended the harvesting period of strawberries in California State. Thereafter, to enlarge the gene pool of the cultivated strawberry, *F. chiloensis* and *F. virginiana* were used for interspecific hybridization between 8*x* species in order to reconstruct the cultivated strawberry (Hancock and Luby 1993, Staudt 1997, Hancock et al. 2002, Luby et al. 2008, Hancock et al. 2010, Stegmeir et al. 2010.). Hancock et al. (2001) determined that some clones of *F. chiloensis* had a resistance to red stele, leaf spot, and powdery mildew. Barritt and Shanks (1980) and Milenkovic et al. (2014) indicated that a strain of *F. chiloensis* exhibited resistance to aphids (*Chaetosiphon fragaefolii*), which was heritable in strawberry cultivars created by interspecific hybridization. Kishaba et al. (1972) indicated that *F. virginiana ssp. glauca* had a two-spotted spider mite resistance which was heritable in hybrids obtained by crossing with non-resistant cultivars.

Breeding of 8x and 10x Cultivars Using Non-8x Wild Species

Darrow (1966) suggested three methods to produce 8*x* and 10*x* cultivars derived from interspecific hybrids between 2*x F. vesca* and 8*x* cultivars. In case of the interspecific 10*x* cultivar, the 4*x F. vesca* was initially bred from the 2*x F. vesca* by colchicine treatment to induce chromosome doubling. Thereafter, the 4*x F. vesca* was hybridized with the 8*x* cultivar to obtain 6*x* hybrids, which were subsequently crossed with 8*x* cultivars. A few decaploids were obtained through hybridization between unreduced gametes from the 6*x* hybrids and reduced gametes from the 8*x* cultivar. Bringhurst and Gill (1970) identified the occurrence of unreduced gamete production in several cross combinations in the genus *Fragaria.*

The 10*x* cultivars of 'Spadeka' (Bauer and Bauer 1979), 'Annelie' (Trajkovski 1997), 'Sara' (Trajkovski 1997), 'Florika' (Bauer and Weber 1989) and 'Rebecka' (Trajkovski 2002) were bred using this method. Bauer and Weber (1989) classified the decaploids as a new species of *F.* × *vescana.* Another method of breeding the 10*x* cultivar has also been reported (Bringhurst and Voth 1982). In this method, the 2*x* species is initially hybridized with the 8*x* species to obtain 5*x* hybrids. Subsequently, the 5*x* hybrids are treated with colchicine to induce chromosome doubling and resulting 10*x* plants are selected. *Fragaria vesca* and the octoploid species were combined in an attempt to produce decaploids (Ahmadi and Bringhurst 1992, Morishita et al. 1996, Sasaki and Takeuchi 2008). Noguchi et al. (1997) also produced some decaploids between 2*x F. iinumae* and *F. nilgerrensis* and 8*x* cultivars. Evans (1977) advocated the use of synthetic octoploids because of the difficulty of breeding decaploids with 8*x* cultivars, and he indicated four methods of producing synthetic 8*x* plants. Evans (1982a) bred a strain named 'Guelph

SO1', which was selected from colchicine-treated 4*x* hybrids between 6*x* *F. moschata* and 2*x* *F. vesca*. He also created a synthetic octoploid named 'Guelph SO2' using 2*x* *F. vesca*, 2*x* *F. viridis* and 4*x* *F. moupinensis* (Evans 1982b). He initially created 2*x* interspecific hybrids between *F. vesca* and *F. viridis*, following which 4*x* interspecific hybrids were selected from the 2*x* hybrids treated with colchicine to induce chromosome doubling and then crossed with 4*x* *F. moupinensis*. The 8*x* hybrids between the three species were selected from 4*x* hybrids treated with colchicine to induce chromosome doubling.

Breeding Using Intergeneric Hybrids

Intergeneric hybrids can be produced by hybridization between plants in the genera of *Potentilla* and *Fragaria*. The combinations of the species derived from these genera to produce hybrids are listed in Table 7. In most of these research studies, plants of the genus *Potentilla* were used as the pollen parent. From the results of several hybridization experiments, hybrids were obtained at a higher rate when the hybridization was conducted with a higher ploidy seed parent than with a higher ploidy pollen parent. Two commercial cultivars of 'Frel Pink Panda™' and 'Serenata' were derived from hybridization between *P. palustris* and *F.* × *ananassa* (Mabberley, 2002). Faghir et al. (2014) made clear by DNA analysis that *P. fruticosa* was the closest plant to the plants in the genus *Fragaria*.

Practical Methods of Breeding Using Interspecific Hybrids

Methods for Cross-Pollination

Petals and stamens of flowers from the seed parent must be manually eliminated using hands or forceps just a few days before the flower opens. At this time, it is important not to break the anthers or spread the pollen in the flower. In addition, to minimize unexpected pollination by the pollen of the seed parent, it is necessary to wash the pistils by spraying water containing surfactant. The emasculated flower must then be covered with a paper bag to prevent unexpected pollination. Because the flower of the diploid species is too small to cover with a paper bag, cotton may be used instead of the paper bag, and gently wound around the flower. Each flower should be tagged with the cultivar name and emasculation date. Stamens are removed from the flower of a pollen parent just a few days before opening, and placed in a petri dish. Pollen from opened flowers should be avoided to prevent unexpected pollination. Pollen may be retrieved within a few days from the dish, which is maintained at room temperature. When the dish is manually shaken, small yellow particles are visible to the naked eye. The fertilizing ability of strawberry pollen is maintained for a several days even at room temperature. However, if the pollen is desiccated, placed in sealed bottle and refrigerated, it remains viable for over one month. Pollination can be effected

Table 7. Cross-pollination to produce artificial intergeneric hybrids with the genus *Fragaria*.

Reference	Parental species		Ploidy level of hybrids obtained
	♀	♂	
Asker (1970)	*F. vesca*	*Potentilla reptans*	No seedlings obtained
		P. ererta	No seedlings obtained
		P. fruticosa (2*x*)	No seedlings obtained
		P. palustris	Matromorph/maternal seedlings only
		P. anserina	Matromorph/maternal seedlings only
		P. rupestris	No seedlings obtained
		P. argentea	No seedlings obtained
		P . interinedia	No seedlings obtained
		P . rrantzii	No seedlings obtained
		P . tabernaemontani	No seedlings obtained
	F. moschata	*P. frutirosa* (2*x*)	Some hybrids
		P. erccta	No seedlings obtained
		P. palustris	No seedlings obtained
	F. × *ananassa*	*P. fruticosa* (2*x*)	Hybrids
		P. rupestris	No seedlings obtained
		P. erecta	Matromorph/maternal seedlings only
		P. anglica	Matromorph/maternal seedlings only
		P. fruticosa (4*x*)	Matromorph/maternal seedlings only
		P. davurica	Matromorph/maternal seedlings only
		P. palustris	No seedlings obtained
		P. anserina	Matromorph/maternal seedlings only
Asker (1971)	*F. moschata*	*P. fruticosa*	4*x*
		P. erecta	No seeds obtained
		P. palustris	No seeds obtained

cont.

Ellis (1962)	*F. × ananassa*	*P. fruticosa* (2*x*)	5*x*
		P. erecta	No seedlings obtained
		P. reptans	No seeds obtained
		P. sterilis	No seeds obtained
		P. palustris	7*x*
		P. anglica	No seedlings
	10*x* (*F. vesca* × *F.* × *ananassa*)	*P. fruticosa* (2*x*)	6*x*
		P. erecta	No seeds obtained
		P. reptans	No seeds obtained
		P. sterilis	No seeds obtained
		P. palustris	No seeds obtained
		P. anglica	No seedlings obtained
	F. vesca	*P. fruticosa* (2*x*)	No seedlings obtained
		P. palustris	No seeds obtained
	F. vesca 4x	*P. fruticosa* (2*x*)	3*x*
		P. palustris	No seeds obtained
Hughes and Janick (1974)	*F.* × *ananassa*	*P. anserina*	4*x* and 8*x*
		P. fruticosa	4*x*, 5*x* and 8*x*
Jelenkovic et al. (1984)	*F. × ananassa*	*P . fruticosa*	8*x*
	F. moschata	*P . fruticosa*	No seedlings obtained
	F. virginiana	*P . fruticosa*	No seedlings obtained
	F. × ananassa	*P. anserina*	8*x*
	F. virginiana	*P. anserina*	No seedlings obtained
Macfarlane and Jones (1985).	*F. moschatu*	*P. fruticosa* (2x)	4*x* and aneuploids
Mangelsdorf and East (1927)	*F. vesca*	*Duchesnea indica*	Very few hybrids
	F. vesca	*P. nepalensis*	Very few hybrids
Marta et al. (2004)	*F. × ananassa*	*Duchesnea indica*	Very few hybrids
	F. × ananassa	*F. vesca*	Very few hybrids
Sayegh and Hennerty (1993)	*F. × ananassa*	*P. fruticosa* (2*x*)	5*x*, 4*x* and aneuploid
Senanayake & Bringhurst (1967)	*F. chiloensis*	*P. glandulosa*	5*x*

by applying pollen with a soft paintbrush a few days after emasculation; the pollination date and pollen parent should be recorded on the tag. After pollination, the pollinated flower should be covered with a paper bag or cotton for at least two weeks, after which the mature fruit can be harvested to take achene (seeds). A small number of seeds (<100) may be extracted from a fruit using forceps. However, if many seeds are simultaneously collected from many fruits, a blender with a dulled blade may be used. Healthier seeds are obtained using a blender.

Methods of Chromosome Doubling

Pollination between $2x$ and $4x$ species or between $2x$ and $8x$ species for example will yield odd-ploidy plants with $2n = 3x$ and $2n = 5x$. These odd-ploidy plants cannot produce healthy progeny by sexual reproduction. In such cases, a technique for doubling the number of chromosomes is important. Dermen and Darrow (1938) and Hull (1960) indicated that a higher percentage of the seedling chromosome numbers were doubled when a germinating seed was dipped in a 0.2% colchicine solution for 8 to 24 hours. Evans (1977) and Sebastiampillai and Jones (1976) reported similar results using 2.0% colchicine solution for 24 to 48 hours at the shoot apex under conditions of high humidity. An apex of a runner linked to the parent plant can also be used for the colchicine treatment (Ahokas 1998). Furthermore, two-step *in vitro* culture using an apical meristem can be used for producing chromosome-doubled plants. Various chromosome-doubling treatments have been conducted — for example, a primary medium containing 0.05% colchicine for 48 hours (Niemirowicz-Szczytt and Ciupka 1986); 100 ppm colchicine for 50 days (Morishita, et al. 1996); and 100 ppm colchicine for 10 days (Sasaki and Takeuchi 2008). The apical meristem is then transferred into a normal medium. Zhang et al. (2009) used *in vitro* culture of leaf segments dipped in a 0.3% colchicine solution for four days. In this case, some aneuploids were produced.

Emergence of Apomixis

Mangelsdorf and East (1927) investigated the inheritance of petal color (pink and white) and fruit color (red and white) in *F. vesca*. They conducted pollination between two heterozygous genotypes in the seed parent and all recessive genotypes in the pollen parent, and found that the segregation of offspring was the same as that of offspring obtained by self-pollination of the heterozygous seed parent. They demonstrated that seedlings originated from a phenomenon of chromosome doubling and apomixis occurring in the early stages of embryo development. Although they mentioned that the same phenomenon occurred in pollination failure to produce matromorphs, they indicated that it often occurred when a diploid species was used as the seed parent. The same phenomena were later confirmed by several reports (East 1930 and1934, Aalders 1964, Yanagi et al. 2009). Jelenkovic et al. (1984)

found some seedlings with the appearance of maternal plants in progeny resulting from intergeneric hybrids between *Fragaria* and *Potentilla.* He considered that the plants were produced by pollination failure. Sukhareva et al. (2002) reported that apomicts could be obtained by pollination between 10*x F. vescana* and 2*x*, 4*x* and 6*x* species in the genera *Potentilla* and *Duchesnea.* However, Nosrati et al. (2010) observed no evidence of apomixis by DNA analysis of the interspecific seedlings. He attributed the pollination failure to the production of maternal plants stimulated by interspecific hybridization. However, Kojima and Nagato (1997) reported that offspring derived from *Allium*-type apomixis had the same phenotypic variation obtained by self-pollination. This phenomenon deserves closer inspection in the case of matromorphy in the genus *Fragaria.*

Methods to Evaluate Hybridity and the Effect of Chromosome Doubling Treatments

Chromosome Observations

The hybridity of a plant which is derived from the hybridization between two species of different ploidy levels can be elucidated by counting the number of chromosomes in the somatic cells. For example, if a hybrid is obtained from a cross between a diploid ($2n = 2x = 14$) and an octoploid ($2n = 8x = 56$), counting the chromosomes in a somatic cell to reveal that the plant is a pentaploid ($2n = 5x = 35$) will confirm that the plant is a hybrid. In addition, the most efficient method to determine the effect of chromosome doubling treatment is to count the number of chromosomes in a somatic cell. If a decaploid ($2n = 10x = 70$) is obtained by colchicine treatment of pentaploids ($2n = 5x = 35$), it can be regarded that the colchicine treatment is completed. However, chromosome counting in plants in the genus *Fragaria* is very difficult, as the chromosomes in the somatic cell are very small (< 2 μm), and it is difficult to individually separate chromosomes without any overlap on a glass slide. For the higher ploidy level such as the octoploid (8*x*) and decaploid (10*x*) species, the difficulty involved in counting the chromosome number increases. However, we developed an easier method to count the number of chromosomes in somatic cells of the higher ploidy level species (Nathewet et al. 2007, Nathewet et al. 2009a). This method is a modification of that reported by Iwatsubo and Naruhashi (1989 and 1991).

Slide Sample Preparation

The selection of healthy roots, which have an overall white surface and pale yellow color at the terminal portion, is required. Approximately 2 cm of the root tip is cut using forceps at approximately 5:00 pm, dipped into 2 mM 8-hydroxyquinoline solution by 6:00 pm at room temperature, and subsequently stored in a refrigerator at 4°C for 15 h. The application of the 2 mM 8-hydroxyquinoline solution treatment is important to prevent formation of spindle fibers and to increase the number of cells which contain metaphase chromosomes. The roots are subsequently treated with 1:3 acetic acid and ethanol mixture solution for 40 minutes at room temperature to allow fixation, and are then stored in a 70% ethanol solution at –20°C. In addition, the roots should not be washed just before fixation: washing results in a dramatic decrease in the number of cells which contain metaphase chromosomes, as washing causes a restart of the cell division. In addition, the fixation solution must be mixed just before the treatment. Below, two chromosome observation methods are described that use a light microscope and fluorescent microscope, respectively.

Observation by Light Microscope

The root tip is softened in 1 N HCl solution at room temperature for 2 h, transferred to the 1 N HCl solution at 60°C for 11.5 minutes, and rinsed for a few minutes in distilled water. The single root tip is subsequently placed on a glass slide using forceps. The water around the root must be removed by absorption using a filter paper. The root is cut and the terminal portion < 1 mm retained, and stained with a drop (approximately 10 μL) of the 1.5% lacto-propionic orcein for a few seconds. The root is then placed in the space made by a cover slip and the razor blade that is filled with the staining solution. The cover slip is then gently tapped using a wooden chopstick to disintegrate the root tip into invisible particles. The razor is then removed and the root is completely covered with the cover slip. The slide is then warmed for approximately 1 s on a spirit lamp, wrapped with a piece of a filter paper, and using a thumb, a pressure is applied to the cover slip. Chromosomes are observed using a light microscope at 40 × and 100 × magnifications. Chromosomes which are well separated at the metaphase stages are selected and photographed using a digital camera at the 100 × magnification. Examples images are shown in Fig. 2.

Observation by Fluorescence Microscope

The root tip is digested in a small Petri dish using an enzyme mixture at 37°C for 30 minutes. We use an enzyme mixture containing 4% cellulose Onozuka RS (Yakult Co. Ltd., Tokyo), 0.3% pectolyase Y-23 (Seishin Pharmaceutical Co. Ltd., Tokyo), 2.1% macerozyme R10, and 1 mM ethylene diamine tetra-acetic acid (EDTA). The root is then rinsed briefly in distilled water. A single root tip is then carefully transferred onto a glass slide using a Pasteur pipette, retaining water around the root. During this process, maximum concentration must be maintained, as the root tip is very soft and easy to destroy. The root tip is cut using a filter paper to retain <1 mm of the terminal portion, treated with a drop of 60% acetic acid for 3 minutes, then crushed with a pair of fine forceps into invisible particles. A cover slip is then placed

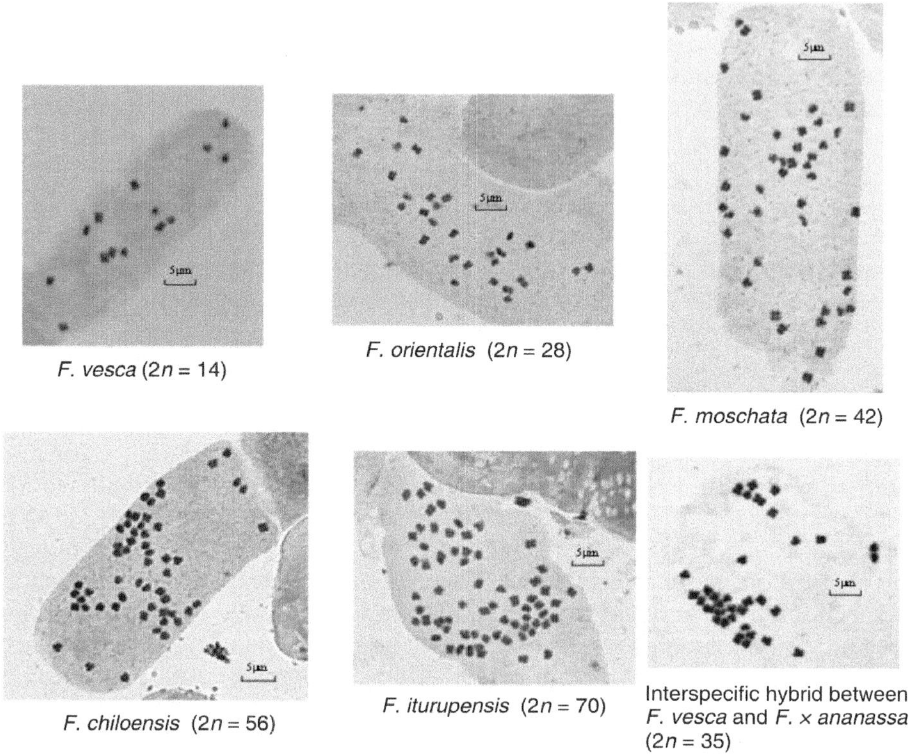

Figure 2. Examples of chromosome images in *Fragaria* root tip cells (magnification of 1 000 ×) observed by a light microscope (1.5% lacto-propionic orcein stain).

on the root tip, and warmed using an alcohol lamp for 1 s. The cover slip is then tapped gently with a wooden chopstick and pressure is applied using a thumb. The slide sample is subsequently frozen at −80°C for at least 5 minutes, and then the cover slip is removed using a razor blade. The slide sample is air-dried, soaked for 5 min in a series of alcohol solutions (75, 80, and 90%), air-dried once again, washed with 2 × saline sodium citrate buffer (SSC: 0.03 M sodium citrate and 0.3 M NaCl pH 7), and then stained with 2.0 mg/mL 4, 6-diamidino-2-phenylindole (DAPI) in the dark for 5 minutes at room temperature. The slide sample is subsequently dipped in the 2 × SSC for 5 minutes at room temperature, washed using distilled water, and dried in the dark. The slide sample is then covered using a coverslip with 30 μL DABCO [1% 1,4-diazobicyclo (2, 2, 2)-octane in 90% glycerol in a phosphate-buffered saline solution (PBS)]. Examples images are shown in Fig. 3. It appears that it is easier to obtain clear chromosome images using a fluorescent microscope than by using a light microscope.

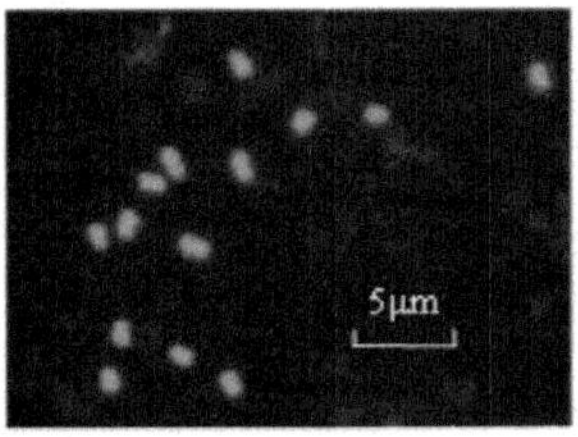

F. vesca (2*n* = 14)

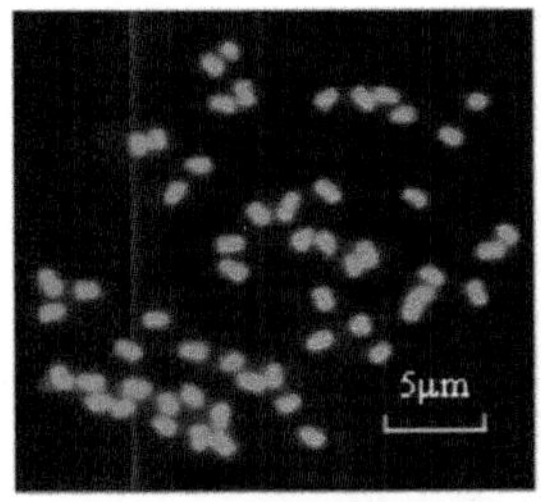

F. × *ananassa* (2*n* = 56)

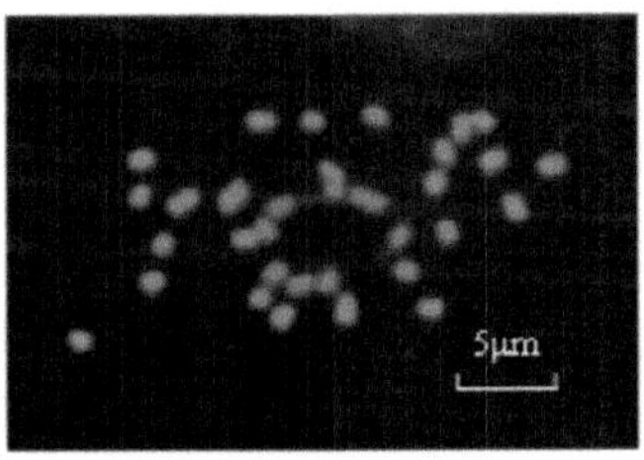

Interspecific hybrid between *F. vesca* and *F.* × *ananassa* (2*n* = 5*x* = 35)

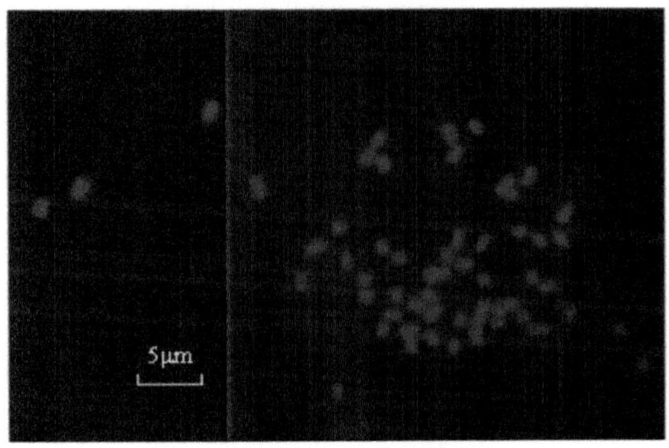

Interspecific hybrid between *F. vesca* and *F.* × *ananassa* (2*n* = 9*x* = 63)

Figure 3. Examples of chromosome images in *Fragaria* root tip cells (magnification of 1 000 ×) observed by a fluorescent microscope (DAPI stain).

Flow Cytometry Analysis

By using a flow-cytometer, relative DNA quantity in single nuclei can be evaluated by the strength of fluorescent light produced from the nucleus. If many nuclei extracted from cells in the same plant are measured, then a peak level of the relative DNA quantities can be obtained (Fig. 4). It can be considered that the relative DNA quantity will increase linearly by ploidy level. Flow-cytometry analysis is useful to determine the ploidy level of many seedlings at once. However, it is impossible to identify using flow cytometry an aneuploid plant that has for example one chromosome more or less than a complete set. Here, we describe the flow cytometry method using DAPI solution. In addition, some researchers have used propidium iodide (PI) staining for flow cytometry (Nehra et al. 1991, Akiyama et al. 2001, Brandizzi et al. 2001), and mithramycin staining (Nyman and Wallin, 1992) to measure the cell nuclear DNA content.

Samples are prepared from approximately 1 cm^2 of an unfolded fresh young leaf segment in a bud of a strawberry plant. The leaf is stored in

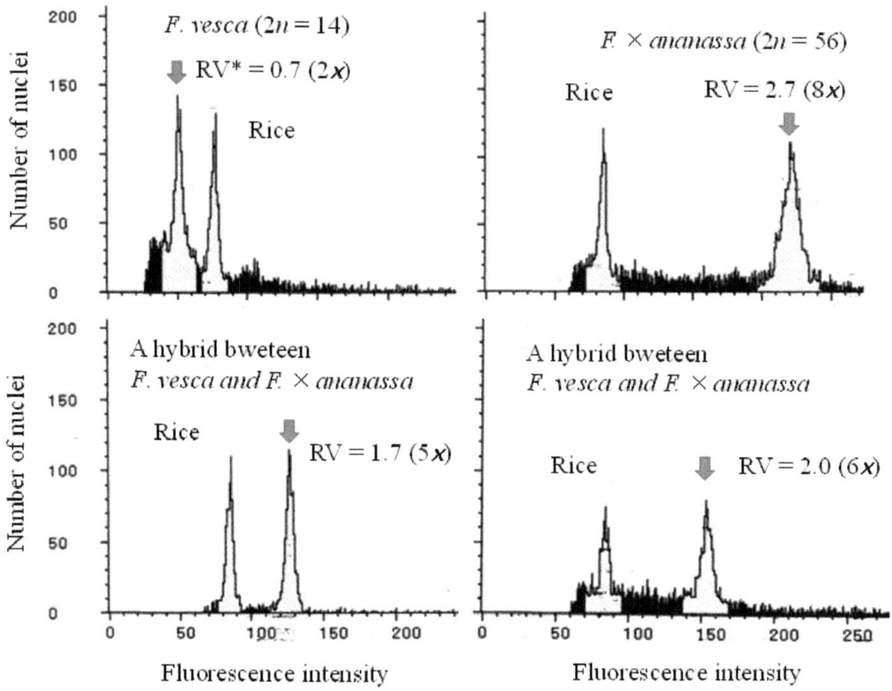

Figure 4. Examples of results obtained by flow cytometry analysis using plants in the genus *Fragaria*. RV* denotes the relative value calculated by the peak fluorescent intensity of strawberry plants divided by those of rice in each measurement.

crushed ice just before the measurement. In addition, we always use rice leaf as an internal standard. The leaf segments of strawberry and rice are chopped 30 times using a razor blade in a plastic Petri dish with a drop (approximately 0.2 mL) of commercial buffer solution (plant high-resolution DNA kit type P, Partec GmbH, Munster, Germany). A crude nuclear sample is maintained for 15 minutes at room temperature, and filtered through 20 mm nylon mesh. A 1 mL staining solution containing 10 mM Tris, 50 mM sodium citrate, 2 mM $MgCl_2$, 1% (w/v) PVP K-30, 0.1% (v/v) Triton X-100, and 2 mg DAPI is added to the filtered nuclear sample. The sample is stored for 2 minutes at room temperature. The sample is then analyzed using a flow cytometer (PAS, Partec GmbH). Examples are shown in Fig. 4.

DNA Analysis

The evidence obtained by DNA analysis is important to confirm the hybridity of a seedling produced by interspecific hybridization. In this section, the use of Cleaved Amplified Polymorphic Sequence (CAPS) markers for this purpose is explained. Using this method, the hybridity of a seedling can be confirmed by showing the existence of a specific DNA segment which is also present in the pollen parent.

CAPS DNA Analysis

Kunihisa et al. (2003, 2005, and 2009) developed special CAPS DNA markers for the plants in the genus *Fragaria*. If the appropriate CAPS maker is selected, the genetic relationships between parents and their progenies can be determined. Almost every marker is locus-specific to a single genome/homologous chromosome pair and is codominant, distinguishing between homozygotes and heterozygotes. A CAPS marker has a pair of forward and reverse primers and a restriction enzyme with which the primers can clip single allelic pairs. In a homozygote, A type DNA segments have specific base sequences that cannot be cut by the restriction enzyme, and are represented as a single electrophoretic band. In a homozygote, B-type DNA segments can be cut, and two smaller bands are produced. If the larger and the two smaller bands are identifiable, the DNA segments are regarded as the heterozygotetype. For example, if the electrophoretic banding patterns obtained using a CAPS marker indicate a homozygote A-type in the seed parent and a homozygote B-type in the pollen parent, the hybrid progenies must have the heterozygote type. Hybridity of the seedling produced by the interspecific hybridization can then be confirmed by the CAPS DNA analysis. The method which we utilize is explained in detail below.

DNA is extracted from 100 mg of young leaf tissue and purified using a plant mini kit (DNeasy; Qiagen Inc.). Certain appropriate CAPS markers are selected, as illustrated in Table 8. The PCR amplification is conducted in a 0.02 mL solution containing 1–10 ng of extracted genomic DNA, 200 mM each of dNTP, 10 mM Tris–HCl (pH 8.3), 50 mM KCl, 1.5 mM $MgCl_2$, 2.5 U Taq polymerase, and 1 nM of each primer. Amplification using a thermal cycler is conducted under the following program: 5 minutes at 94°C; 35 cycles of 30 s at 94°C, 30 s at 55°C, and 30 s at 72°C, as well as an extra extension for 5 minutes at 72°C. The amplified DNA solution is treated directly with 4 U of the appropriate endonuclease in a 0.01 mL volume. Polymorphism is detected by separating the whole volume of treated DNA on 1.5% agarose gel [1 × TBE (Tris borate EDTA)] containing ethidium bromide, and DNA bands are visualized under a ultra-violet trans-illuminator.

Table 8. Details of Cleaved Amplified Polymorphic Sequence (CAPS) markers used for hybridity analysis in *Fragaria* (Kunihisa et al. 2005).

Marker name	Primer name (Fw = forward; Rv = reverse)	Primer sequence	Restriction enzyme/ Reaction temperature
DFR-Hin6 I (DFR-HhaI)	DFR-Fw DFR-Rv	GAGACCCTGGTCCGTCG CCTCCGAACTGTCTTTGCTTTGAG	Hin6 I 37°C
APX-Mlu I	APX-Fw APX-Rv	GTGGTCACACCTTGGTGC AGTATAATATTTAAGCAGAATGCAGACTTC	Mlu I 37°C
CHI-Pvu II	CHI-Fw CHI-Rv	AGGAGTTGACAGAGTCGGTTG GACTTGTGAGTATGATAGTCTGCTG	Pvu II 37°C
F3H-NcoI(N)	F3H-Fw(N) F3H-Rv(N)	ACCATGGACATGTGAGTATACTTT ACTAAGGAACTCATACTCAACCA	Nco I 37°C
F3H-Eam1104 (EarI)(N)	F3H-Fw(N) F3H2-Rv(N)	ACCATGGACATGTGAGTATACTTT CCCAAATAATGTGTCAATACATATACGAT	Eam1104 I 37°C
F3H2-Hpa II(N)	F3H-Fw(N) F3H2-Rv(N)	ACCATGGACATGTGAGTATACTTT CCCAAATAATGTGTCAATACATATACGAT	*Hpa* II 37°C
F3H2-DdeI(N)	F3H-Fw(N) F3H2-Rv(N)	ACCATGGACATGTGAGTATACTTT CCCAAATAATGTGTCAATACATATACGAT	*Dde* I 37°C
F3H3-AccI(N)	F3H3-Fw F3H3-Rv(N)	TAATAGGGTCTAGGTGCGTGG ACCCAAATAATGTGTCAATACATATAAGAC	*Acc* I 37°C
CTI1-Hinf I	CTI1-Fw CTI1-Rv	TTCTAATGATCAACACCTACTTTCCC GTAGCCCACCCGCCTG	*Hinf* I 37°C
MSR-Alu I	MSR-Fw MSR-Rv	AGCACTTTCACCATAGGCATAATC CCTTGAGCATAAATGAACTGGCA	*Alu* I 37°C
PGPA-Acc I (N)	PGP-FwA PGP-RvA(N)	CCTCACCTTCCTCGAGCTC AAGTCTATCCGATCAAAGTTCATG	*Acc* I 37°C
PGPA-Rsa I (N)	PGP-FwA PGP-RvA(N)	CCTCACCTTCCTCGAGCTC AAGTCTATCCGATCAAAGTTCATG	*Rsa* I 37°C
PGPB-Rsa I	PGP-FwB PGP-RvB	ACCTCACCTTCCTTGAGCTT GACAAGTCTATCCGATCAAAGTTCATA	*Rsa* I 37°C
APX2-Dra I	APX2-FwA APX2-Rv	CAGAGGCCTCATCGCCG TCAGGTCCACCGGTGACC	*Dra* I 37°C
APX3-Dra I (N)	APX3-Fw(N) APX2-Rv	GGCCTCATCGCCGAG TCAGGTCCACCGGTGACC	*Dra* I 37°C
APX4-Taq I (N)	APX4-Fw(N) APX2-Rv	CTCCGATCCCTATCTTTTCTTT TCAGGTCCACCGGTGACC	*Taq* I 65°C

cont.

AUB-Hin6 I (Hha I) (N)	AUB-Fw(N) AUB-Rv(N)	GGGTGTTTGTGAATTRGTTTGC TACATACTGCCCCCCAGA	*Hin6* I 37°C
OLP-Dde I	OLP-Fw OLP-Rv	TGTGTCCAAAACCGATCAGTATTGC TCTTTCAGAGTGGTACGTACCCC	*Dde* I 37°C
CTI2-Mbo I (N)	CTI2-Fw(N) CTI2-Rv(N)	CAAAGCATGCATGATCGTAGTG CTCCGATTGCCTTACCCGC	*Mbo* I 37°C
CTI2-Bsh1236 I (N)	CTI2-Fw(N) CTI2-Rv(N)	CAAAGCATGCATGATCGTAGTG CTCCGATTGCCTTACCCGC	*Bsh1236* I 37°C
CYT-BsaB I (N)	CYT-Fw(N) CYT-Rv	CCAGCCATAATGTCTTAC CCGTACTTGAGCCTATCTGACTGG	*BsaB* I 60°C
tRNA-BseG I (FokI)	tRNA-Fw tRNA-Rv	CATTTCACAAACAGATCTGAGCGG TTATTTGAACTGGTGACACGAGGA	*BseG* I 55°C
PYDA-Hae III	PYD-FwA PYD-RvA	CTTTCAGGTAAGGAACATGATCAAG GTAAGAACTTAACAAAACCATAATCTCTCTA	*Hae* III 37°C
PYDA-Cfr13 I	PYD-FwA PYD-RvA	CTTTCAGGTAAGGAACATGATCAAG GTAAGAACTTAACAAAACCATAATCTCTCTA	*Cfr13* I 37°C
PYDB-Hae III (N)	PYD-FwB(N) PYD-RvB(N)	CAACTTTGAGTCTTTATGATGAATTGA ACCAAGTAGAAACTTACGTTAAGTTA	*Hae* III 37°C

An example of the results of CAPS DNA analysis is shown in Fig. 5. The illustrated CAPS DNA analysis was performed using the *Fragaria vesca* (2*x*) 'Baron solemacher', *F.* × *ananassa* (8*x*) 'Nyoho', and 23 interspecific hybrids (5*x*) between them. The CAPS marker PYDB-HaeIII was used. In this case, 'Baron solemacher' and 'Nyoho' had zero and two bands (the heterozygote type), respectively. The 23 plants then had one band. This result indicates that the 23 plants are interspecific hybrids between the two parents.

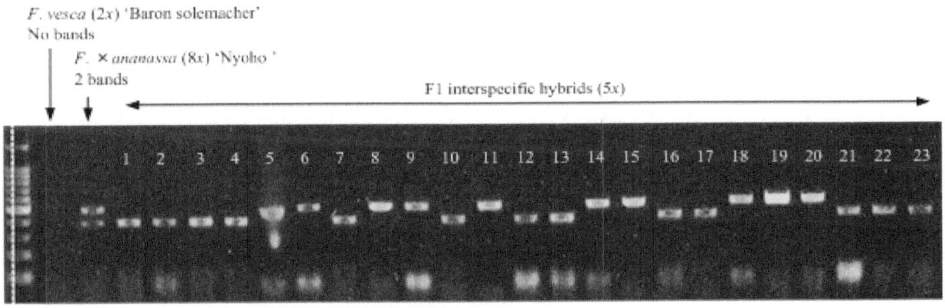

Figure 5. An example of CAPS DNA analysis results.

Cleaved Amplified Polymorphic Sequence (CAPS) DNA analysis was performed on seed parent *Fragaria vesca* (2*x*) 'Baron solemacher', pollen parent *F.* × *ananassa* (8*x*) 'Nyoho' and the 23 interspecific hybrids (5*x*) produced between them. The CAPS marker PYDB-HaeIII was used. In this case, 'Baron Solemacher' and 'Nyoho' had zero and two bands (the heterozygote type), respectively. The plants in the F_1 generation then had one band. This result indicated that the 23 plants are interspecific hybrids between the two parent lines.

Recent Progress of our Breeding Program Using the Decaploid Interspecific Hybrid (10x) *Cultivar 'Tokun' in Japan*

Fragaria nilgerrensis is native to south eastern Asia and has various unique characteristics with regard to plant shape (Ichijima 1930). The species is regarded as the most distantly related wild species to other species (Arulsekar and Bringhurst 1983, Harrison et al. 1997, Njuguna et al. 2013). For this reason, *F. nilgerrensis* was selected as breeding material because it was expected that the interspecific decaploid produced using *F. nilgerrensis* would have high-seed fertility. Darrow (1966) indicated that *F. nilgerrensis* has a low value for use in breeding because of its tasteless fruit. However, as mentioned in various studies (Staudt et al. 1975, Staudt 1999b, Weber and Ulrich 2009, Zhao et al. 2014), the mature fruit of *F. nilgerrensis* has a valuable sweetness and strong fragrance similar to that of banana and peach fruit. Subsequently, the introduction of the aroma to the strawberry cultivar was carried out.

An interspecific hybridization between a seed parent of *F.* × *ananassa* 'Toyonoka,' (which has a good aroma in Japanese strawberries) and a pollen parent of *F. nilgerrensis* was performed. Vigorous hybrids (5*x*) were then selected, and chromosome doubling treatments were performed using the B5 medium with 100 $mg{\cdot}L^{-1}$ colchicine according to the method reported by Morishita et al. (1996). Out of the amphi-decaploid (10*x*) hybrids, a strain named 'TN13-125' was selected (Noguchi et al. 2002). The 10*x* hybrid had a shape similar to that of the plant 'Toyonoka,' but it was hairy similar to the pollen parent *F. nilgerrensis*. The fruit weight and acidity, sugar, and vitamin C concentration of the cone fruit was similar to the cultivated strawberries, but the cone fruit was less firm and pale red in color with less luster.

The aromatic volatile constituents of 'TN13-125', 'Toyonoka,' and *F. nilgerrensis* were analyzed by the headspace method. The relative value of the peak area of ethyl acetate was lower than that of ethyl n-butyrate in 'Toyonoka.' In the case of *F. nilgerrensis* (which has a peach-like aroma) the relative value of ethyl acetate was higher than that of ethyl n-butyrate. The aromatic volatile constituent of 'TN13-125' (also witha peach-like aroma) was then similar to that of *F. nilgerrensis* (Noguchi et al. 2002). Although 'TN13-125' showed some disadvantages, such as lack of fruit firmness and less appealing appearance, it had an attractive and distinctive aroma and the sugar and acidity contents of the fruit were almost equal to those of the cultivated strawberries. 'TN13-125' obtained a Japanese patent registration (No. 13534), and the new cultivar was named 'Kurume IH1' in December 2005. This cultivar was the first globally produced decaploid (10*x*) interspecific hybrid using *F. nilgerrensis*. However, the cultivars were utilized as a gardening cultivar in households and not for commercial production in Japan because of low productivity and poor fruit appearance.

Table 9. Assessment of volatile compounds in 'Tokun' and related *Fragaria* cultivars.

Compound	Retention time (minutes)	Cultivar			
		Tokun	Toyonoka	Kurume IH	Karen-berry
	Concentration (ppb)				
ethyl butanoate	14.09	69	22	40	35
hexanal	16.68	243	150	277	161
cis-3-hexenal	20.33	4	15	9	21
ethyl hexanoate	25.62	525	26	22	15
trans-3-hexenol	32.32	3	3	2	3
cis-3-hexenol	33.30	16	13	18	32
cis-2-hexenol	34.33	167	155	211	236
acetic acid	36.73	495	572	349	182
cis-2-nonenal	38.71	4	8	7	7
trans-2-nonenal	39.98	8	14	2	7
linalool	40.51	65	510	33	168
octanol	41.03	60	9	153	7
propanoic acid	40.74	3	74	25	8
trans, cis-2, 6-nonadienal	42.10	3	13	5	5
2,5-dimethyl-4-methoxy-2H-furan-3-one	42.45	1193	15	8	328
2-methylpropanoic acid	41.82	154	186	269	254
butanoic acid	44.36	312	241	28	277
2-methylbutanoic acid	45.85	41	2294	952	1429
γ-hexalactone	46.76	86	51	41	15
δ-hexalactone	50.26	53	22	4	7
hexanoic acid	52.24	8824	8378	8245	2630
γ-octalactone	54.68	32	19	29	2
δ-octalactone	56.52	57	16	32	5
2,5-dimethyl-4-hydroxy-2H-furan-3-one	58.66	6713	5612	5259	368
nerolidol	58.74	78	857	22	83
γ-decalactone	62.26	76	81	37	69
δ-decalactone	63.89	131	19	257	4
γ-dodecalactone	69.20	114	187	279	269
δ-dodecalactone	70.76	154	6	35	6
vanillin	75.19	1	35	17	22

Figure 6. Fruits of *Fragaria* 10*x* interspecific hybrid 'Tokun'.

Figure 7. *Fragaria* polyploid interspecific hybrid breeding lines 'CHI-24-1' (A and B) and 'T-18-2' (C and D).

As a next step, decaploid (10*x*) F_1 plants that exhibited vigorous growth and larger fruits could be obtained from hybridization between the previously obtained decaploid (10*x*) interspecific hybrids (Noguchi et al. 2009). Subsequently, to breed a new decaploid (10*x*) hybrid with characteristics of good flavor and fruit appearance, breeding was initiated using a *Fragaria* × *ananassa*, 'Karen-berry,' which had the desired characteristics as a pollen parent. According to the same procedure used for the 'TN13-125' hybrid, a new 10*x* interspecific hybrid named 'K58N7-21' was selected from the progeny. 'K58N7-21' had a weaker fruit aroma, but an attractive shiny red fruit. Furthermore, an F_1 hybrid strain of 'DH0604-1-19' was selected from the progeny derived from a cross between the seed parent of 'K58N7-21' and the pollen parent of 'Kurume IH1.' The 'DH0604-1-19' hybrid was registered as a new cultivar named 'Tokun' (Patent No. 13534) during 2011 after a 2-year trial, as it had a strong aroma and good fruit quality for commercial production levels (Noguchi et al. 2011).

In general, 'Toyonoka' is well known to have a strong fragrance among the Japanese strawberries. Fukuhara et al. (2005) indicated that 2, 5-dimethyl-4-hydroxy-2H-furan-3-one (DMHF) contributed to the aroma of 'Toyonoka' because it has a sweet caramel-like aroma. In addition, Li et al. (2009) mentioned that the chemical constituents influencing the odor, particularly DMHF, hexanoic acid, γ-dodecalactone, and esters were detected at relatively high levels; and that this could explain why 'Toyonoka' strawberries have abundant aroma. In comparison with ordinary strawberry cultivars in Japan, 'Kurume IH1' has lesser linalool (orange odor), butanoic acid (cheesy odor), and 2-methylbutanoic acid (sweet odor), but more octanol (herbalodor) as well as octalactone (coconut odor), δ-decalactone, and dodecalactone groups (peach odor). In addition, 'Tokun' has a very attractive flavor because of lesser linalool and 2-methylbutanoic acid but more 2,5-dimethyl-4-hydroxy-2H-furan-3-one, δ-decalactone, and 2,5-dimethyl-4-methoxy-2H-furan-3-one contents as well as octalactone and dodecalactone groups (Table 9). 'Tokun' is used for the commercial production as well as for cultivating gardens in homes because of its unique and attractive flavor. The fruits (Fig. 6) are sold as a luxury market fruit, and are usually used for gifting.

The purpose of the breeding as presented here was to produce a new *F.* × *ananassa* cultivar (8x) using the wild strawberry strain *F. chiloensis* CHI-24-1 (Fig. 7 a and b) in which flowerinitiation occurs under 24 h day length treatment (continuous light) for 25 days.

Flowering initiation of the CHI-24-1 plants is restricted to autumn, similar to the ordinary short-day type of June-bearing strawberry cultivars which grow under natural conditions in Japan. However, Yanagi and Oda (1991) noticed that CHI-24-1 (the strain was named chiloensis 2 in their paper) produced inflorescences when grown under 24 h day length conditions (continuous lighting at night by incandescent lamps). In addition, the floral initiation of CHI-24-1 under the 24 h day length condition was induced in both the parent and the asexually propagated daughter plants linked by runners

(Yanagi et al. 2005). There are few reports regarding the same phenomenon in strawberries; however, Collins and Barker (1964) and Collins (1966) indicated that a strawberry cultivar named 'Sparkle' produced inflorescences from the daughter plants attached to the parent plant by runners, even though it was of a short-day type (Austin et al. 1961, Porlingis and Boynton 1961, Moore and Hough 1962). In addition, far red light (approximately 735 nm) was effective in inducing flower initiation under 24 h day length conditions (Yanagi et al. 2006). Furthermore, the characteristics of floral initiation under 24 h day length were inherited by more than 50% of the F1 hybrids between the 'CHI-24-1' and a short-day type Japanese strawberry cultivar named 'Nyoho' (Yanagi et al. 2005 and 2016). These reports indicated that the flower initiation gene(s) controlling flower initiation under 24 h day length exist in certain wild and cultivated strawberry plants. Nevertheless, few reports have described a similar phenomenon in other higher plants.

In Japan, various attempts have been made for over 20 years to increase strawberry production under the longer day length and higher temperature conditions during summer and early autumn. Some domestic ever-bearing strawberry cultivars are used for production. However, the number of daughter plants produced from ever-bearing strawberry cultivars during summer is generally lower than that produced from June-bearing cultivars. Therefore, for the purpose of improving summer and early autumn strawberry production, attempts are being made to breed a new cultivar (using CHI-24-1 plants as a genetic resource) in which floral initiation occurs under 24 h day length with employment of artificial lighting as well as high production of runners under natural day length (Fig. 8).

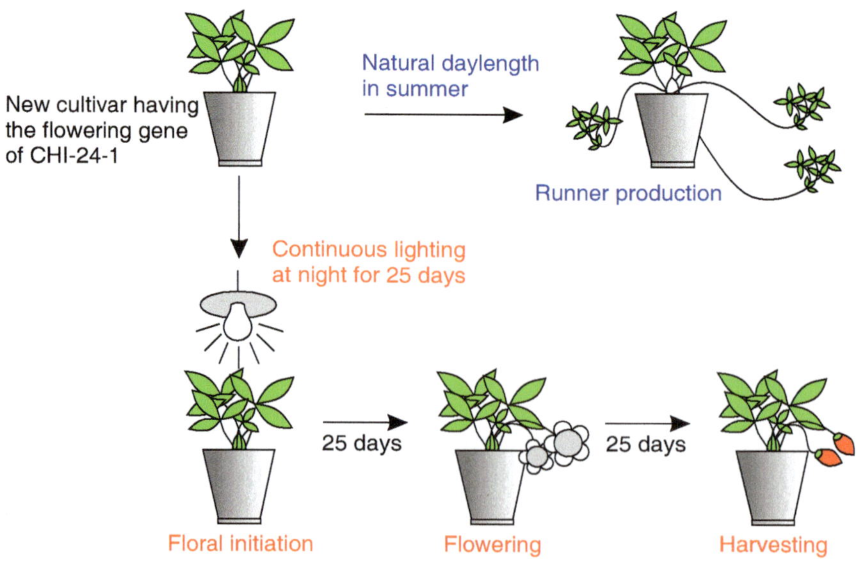

Figure 8. A novel strawberry cultivar containing flowering traits from line 'CHI-24-1'.

An interspecific hybridization between *F.* × *ananassa* 'Nyoho' as the seed parent and 'CHI-24-1' as the pollen parent was conducted (CHI-24-1 was a male plant). In addition, among the hybrids in which flower initiation occurred under 24 h day length, the three strains CN18, CN46, and CN73 were selected according to fruit quality. As the next step, pollination between seed parents 'Asuka Ruby,' 'Nyoho,' and 'Toyonoka' and the three strains was conducted. A strain named T-18-2 (Fig. 7 C and D) was then selected for due to inflorescence production under 24 h day length and fruit quality. The strain produced many runners under natural day length conditions during summer and initiated flower buds under continuous lighting for 25 days. The flowers bloom after 30 days and the fruits mature after 30 days. The characteristics of the flowering and fruiting were similar to expectations. The average fruit weight was slightly smaller; however, sweetness was improved compared with the ordinary strawberry cultivars. The breeding program has now made progress in producing the next generation of the T-18-2.

Conclusions

In 1899, the Viscount Hayato Fukuba, chairman of the Shinjuku Imperial Botanical Garden in Japan, released a new strawberry cultivar called 'Fukuba'. He used the seeds of a French cultivar 'General Chanzy' as the breeding material. Now, after 117 years, more than ten new strawberry cultivars are registered every year in Japan.

During the first half of the 20th century, various strawberry cultivars were introduced from foreign countries for the purpose of production and breeding materials. However, in the last half of the century, the production of new varieties was hampered because only the leading Japanese cultivars were used for breeding materials. The new strawberry cultivars have similar tendencies and minor characteristics with little genetic diversity.

Therefore, interspecific and intergeneric hybridization is important to enlarge the genetic diversity. The process of hybridization can produce strawberry varieties that are more resistant to disease, promote the ever-bearing habit, and produce high yields and high-quality fruits. To accomplish this, it is important to introduce various genes from wild diploids (2*x*), tetraploids (4*x*), hexaploids (6*x*), and octoploids (8*x*) and to clearly understand the genotypic variation of many wild strawberry plants.

References

Aalders, E.L. 1964. Production of maternal-type plants through crosses to apomictic species. Nature. 204: 101-102.

Ahmadi, H., R.S. Bringhurst and V. Voth. 1990. Modes of inheritance of photoperiodism in *Fragaria*. J. Amer. Soc. Hort. Sci. 115: 146-152.

Ahmadi, H. and R.S. Bringhurst. 1992. Breeding strawberries at the decaploid level. J. Am. Soc. Hort. Sci. 117: 856-862.

Ahokas, H. 1998. A method to induce polyploids from adult strawberry plants, *Fragaria* spp., avoiding tissue culture *in vitro*. Plant Breeding. 117: 500-502.

Ahokas, H. 1999. Spontaneous tetraploidy in strawberry (*Fragaria sp.*, Rosaceae). Nord. J. Bot. 19: 227-234.

Ahokas, H. 2002. Factor controlling the north-edge distribution of *Fragaria viridis* and its *F. vesca* hybrids in south Finland. Acta Hort. 567: 385-388.

Akiyama, Y., Y. Yamamoto, N. Ohmido, M. Ohshima and K. Fukui. 2001. Estimation of the nuclear DNA content of strawberries (*Fragaria* spp.) compared with *Arabidopsis thaliana* by using dual-step flow cytometry. Cytologia. 66: 431-436.

Arulsekar, S. and R.S. Bringhurst. 1983. Strawberry. pp. 391-400. *In*: S.D. Tanksley and T.J. Orton (eds.). Isozymes in Plant Genetics and Breeding, Part B. Elsevier Science Publishers, Amsterdam.

Asker, S. 1970. An intergeneric *Fragaria* × *Potentilla* hybrid. Hereditas. 64: 135-139.

Asker, S. 1971. Some viewpoints on *Fragaria* × *Potentilla* intergeneric hybridization. Hereditas. 90: 181-190.

Austin, M.E., V.G. Shutak and E.P. Christopher. 1961. Response of Sparklestrawberry to inductive cycle. Proc. Am. Soc. Hort. Sci. 77: 372-375.

Barritt, B.H. and C.H. Shanks. 1980. Breeding strawberries for resistance to the aphids *Chaetosiphon fragaefolii* and *C. thomasi*. Hort Science. 15: 287-288.

Bauer, R. and A. Bauer. 1979. Hybridzuechtung in der Gattung Fragaria. Spadeka, eine neue Sorte mit dem Aroma der Waldbeere. Erwerbsobstbau. 21: 151-161.

Bauer, A. 1993. Progress in breeding decaploid *Fragara* × *vescana*. Acta Hort. 348: 60-64.

Bauer, A. and H.E. Weber. 1989. *Ribes* × *nidigrolaria* R. &. A. Bauer und *Fragaria* × *vescana* R. & A. Bauer-Beschreibung zweier Hybridarten. Osnabrücker naturwissenschaftliche Mitteilungen 15: 49-58 (in German with English abstract).

Bors, R.H. and J.A. Sullivan. 2005a. interspecific hybridization of *Fragaria moschata* with two diploid species, *F. nubicola* and *F. viridis*. Euphytica. 143: 201-207.

Bors, R.H. and J.A. Sullivan. 2005b. Interspecific hybridization of *Fragaria vesca* subspecies with *F. nilgerrensis, F. nubicola, F. pentaphylla*and *F. viridis*. J. Amer. Soc. Hort. Sci.130: 418-423.

Brandizzi, F., C. Forni, A. Frattarelli and C. Damianot. 2001. Comparative analysis of DNA nuclear content by flow cytometry on strawberry plants propagated via runners and regenerated from meristem and callus cultures. Plant Biosystems. 135: 169-174.

Bringhurst, R.S. and D.A. Khan. 1963. Natural pentaploid *Fragaria chiloensis-F. vesca* hybrids in coastal California and their significance in polyploidy *Fragaria* evolution. Am. J. Bot. 50: 658-661.

Bringhurst, R.S. and Y.D.A. Senanayake. 1966. The evolutionary significance of natural *Fragaria chiloensis* × *F. vesca* hybrids resulting from unreduced gametes. Am. J. Bot. 53: 1000-1006.

Bringhurst, R.S. and T. Gill. 1970. Origin of *Fragaria* polyploids. II. Unreduced and doubled unreduced gametes. Am. J. Bot. 57: 969-976.

Bringhurst, R.S. and V. Voth. 1982. Hybridization in strawberries. California Agriculture 36: 25.

Bringhurst, R.S. and V. Voth. 1984. Breeding octoploid strawberries. Iowa State J. Res. 58: 371-381.

Bringhurst, R.S. 1990. Cytogenetics and evolution in American *Fragaria*. HortScience. 25: 879-881.

Byrne, D.and G. Jelenkovic. 1976. Cytological diploidization in the cultivated octoploid strawberry *F.* × *ananassa*. Can. J. Genet. Cytol. 18: 653-659.

Collins, W.B. 1966. Floral initiation in strawberry and some effects of red and far-red radiation as components of continuous light. Can. J. Bot. 44: 663-668.

Collins, W.B. and W.G. Barker. 1964. A flowering response of strawberry to continuous light. Can. J. Bot. 42: 1309-1311.

Coman, M. and A. Popescu. 2008. Interspecific hybridization for the introduction of fruit flavor from wild *Fragaria* into commercial strawberry. Acta Hort. 842: 511-514.

Darrow, G. 1966. The Strawberry: history, breeding and physiology. Holt, Rinehart and Winston, New York, Chicago, 447 p.

Dermen, H. and G.M. Darrow. 1938. Colchicine-induced tetraploid and 16-ploid strawberries. Proc. Amer. Soc. Hort. 36: 300-301.

Dogadkina, N.A., 1941. A contribution to the question of genome relations in some species of *Fragaria.* C.R. Acad. Sci. URSS. 30: 166-168.

Dowrick, G.J. and H. Williams. 1959. Species crosses in the genus *Fragaria.* John Innes. Horticultural Institution: 9-10.

Duchesne, A.N. 1766. Histoire naturelle des fraisiers. Didot le jeune, Paris, France.

East, E.M. 1930. The origin of the plants of maternal type which occur in connection with interspecific hybridizations. Proc. Natl. Acad. Sci. U.S.A. 16: 377-380.

East, E.M. 1934. A novel type of hybridity in *Fragaria.* Genetics. 19: 167-174.

Ellis, J.R. 1962. *Fragaria-Potentilla* hybridization and evolution in *Fragaria.* Proc. Linnean Soc. Lond. 173: 99-106.

Evans, W.D. 1974. Evidence of a crossability barrier in diploid × hexaploid and diploid × octoploid crosses in the genus *Fragaria.* Euphytica. 23: 95-100.

Evans, W.D. 1977. The use of synthetic octoploids in strawberry breeding. Euphytica 6: 497-503.

Evans, W.D. 1982a. Guelph SO1 synthetic octoploid strawberry breeding clone. HortScience. 17: 833-834.

Evans, W.D. 1982b. Guelph SO2 synthetic octoploid strawberry breeding clone. HortScience 17: 834.

Evans, W.D. 1982c. The production of multispecific octoploids from *Fragaria* species and the cultivated strawberry. Euphytica. 31: 901-907.

Evans, W.D. and J.K. Jones. 1967. Incompatibility in *Fragaria.* Can. J.Genet. Cytol. 9: 831-836.

Fadeeva, T.S. 1966. Problems of comparative plant genetics: Communication1. Principles of genome analysis (with reference to the genus *Fragaria*). Genetika. 2: 6-16.

Faghir, M.B., F. Attar, A. Farazmand and S.O. Kazempour. 2014. Phylogeny of the genus *Potentilla* (Rosaceae) in Iran based on nrDNA ITS and cpDNA trnL-F sequences with a focus on leaf and style characters' evolution. Turkish J. Bot. 38: 417-429.

Fedorova, N. 1934. Polyploid inter-specific hybrids in the genus *Fragaria.*Genetica. 16: 524-541.

Fedorova, N. 1946. Crossability and phylogenetic relations in the main European species of *Fragaria.* Comp. Rend. Acad. Sci. USSR. 53: 545-547.

Folta, K.M. and T.M. Davis. 2006. Strawberry genes and genomics. Crit. Rev. in Plant Sci. 25: 399-415.

Frézier, A.F. 1717. A voyage to the South-Sea, and along the Coasts of Chili and Peru with a Postscript by Dr. Edmund Halley, J. Boyer, London.

Fukuhara, K., X. Li, M. Okamura, K. Nakahara and Y. Hayata. 2005. Evaluation of odorants contributing to 'Toyonoka' strawberry aroma in extracts using an adsorptive column and aroma dilution analysis. J. Japan. Soc. Hort. Sci. 74: 300-305.

Hammer, K. and K. Pistrick. 2003. New versus old scientific names in strawberries (*Fragaria* L.). Genetic Resources and Crop Evolution. 50: 789-791.

Hancock, J.F., J.L. Maas, C.H. Shanks, P.J. Breen and J.J. Luby. 1991. Strawberries (*Fragaria*). Acta Hort. 290: 489-546.

Hancock, J.F. and J.J. Luby. 1993. Genetic resources at our doorstep: the wild strawberries. Bioscience. 43: 141-147.

Hancock, J.F. 1999. Strawberries. Crop production science in horticulture series, No. 11. CABI, Wallingford, UK.

Hancock, J.F., P.W. Callow, A. Dale, J.J. Luby, C.E. Finn, S.C. Hokanson and K.E. Hummer. 2001. From the Andes to the Rockies: Native strawberry collection and utilization. Hort Science. 36: 221-225

Hancock J.F., J.J. Luby, A. Dale, P.W. Callow, S. Serceand A. El-Shiek. 2002. Utilizing wild *Fragariavirginiana* in strawberry cultivar development: inheritance of photoperiod sensitivity, fruit size, gender, female fertility and disease resistance. Euphytica. 126: 177-184.

Hancock, J.F., C.E.Finn, J.J. Luby, A.Dale, P.W. Callow and S. Serçe. 2010. Reconstruction of the strawberry, *Fragaria × ananassa*, using genotypes of *F. virginiana* and *F. chiloensis*. Hort Science. 45: 1006-1013.

Hara, H. 1944. *Fragaria yezoensis* H. Hara, J. Jap. Bot. 20: 118.

Harbut, R.H., J.A. Sullivan, J.T.A. Proctor and H.J. Swartz. 2009. Early generation performance of *Fragaria* species hybrids in crosses with cultivated strawberry. Can. J. Plant Sci. 89: 1117-1126.

Harbut, R.H., J.A. Sullivan, J.T.A. Proctor and H.J. Swartz. 2012. Net carbon exchange rate of *Fragaria* species, synthetic octoploids, and derived germplasm. J. Amer. Soc. Hort.Sci. 137: 202-209.

Harrison, E.R., J.J. Luby and G.R.Furnier. 1997. Chloroplast DNA restriction fragment variation among strawberry (*Fragaria spp.*) taxa. J. Am. Soc. Hortic. Sci. 122: 63-68.

Hildreth, A.C. and L. Powers. 1941. The Rocky Mountain strawberry as a source of hardiness. Proc. Amer. Soc. Hort. Sci. 38: 410-412.

Hirakawa H,K. Shirasawa, S. Kosugi, K. Tashiro, S. Nakayama, M. Yamada, M. Kohara, A. Watanabe, Y. Kishida, T. Fujishiro, H. Tsuruoka, C. Minami, S. Sasamoto, M. Kato, K. Nanri, A. Komaki, T. Yanagi, G. Qin, F. Maeda, M. Ishikawa, S. Kuhara, S. Sato, S. Tabata and S.N. Isobe. 2014. Dissection of the octoploid strawberry genome by deep sequencing ofthe genomes of *Fragaria* species. DNA Res. 21: 169-181.

Honjo, M., S. Yui and M. Kunihisa. 2013. Observation of the disomic inheritance of four allelic pairs in the Octoploid. Hort Science. 48: 948-954.

Hughes, H.G. and J. Janick. 1974. Production of tetrahaploid in the cultivated strawberry. HortScience. 9: 442-444.

Hull, J. 1960. Development of colchicine-induced 16-ploid breeding lines in *Fragaria*. Proc. Amer. Soc. Hort. Sci. 75: 354-359.

Hummer, K.E. and J.F. Hancock. 2009. Strawberry genomics: botanical history, cultivation, traditional breeding, and new technologies, Chap. 11. In: Folta K.M., Gardiner S.E. (eds) Plant genetics and genomics of crops and models, vol 6: Genetics and genomics of Rosaceae. Springer, Germany, pp 413-435.

Hummer, K.E., P. Nathewetand T. Yanagi. 2009. Decaploidy in *Fragaria iturupensis*(Rosaceae). Amer. J. Bot. 96: 713-716.

Hummer, K.E., J.D. Postman, N. Bassiland P. Nathewet. 2011. Chromosome numbers and flow cytometry of strawberry wild relatives. Acta Hort. 948: 169-174.

Hummer, K.E., N.V. Bassil and W. Njuguna. 2011. *Fragaria*. In: C. Kole (ed.) Wild Crop Relatives: Genomics and Breeding Resources, Temperate Fruits. Berlin Germany: Springer 6: 17-44.

Hummer, K.E. 2012. A new species of *Fragaria* (Rosaceae) from Oregon. J. Bot. Res. Inst. Texas. 6: 9-15.

Ichijima, K. 1926. Cytological and genetic studies on *Fragaria*. Genetics. 11: 590-604.

Ichijima, K. 1930. Studies on the genetics of *Fragaria*. Z Indukt Abstamm Vererbung sl. 55: 300-347.

Isobe S.N., H. Hirakawa, S. Sato, F. Maeda, M. Ishikawa, T. Mori, Y. Yamamoto, K. Shirasawa, M. Kimura, M. Fukami, F. Hashizume, T. Tsuji, S. Sasamoto, M. Kato, K. Nanri, H. Tsuruoka, C. Minami, C. Takahashi, T. Wada, A. Ono, K. Kawashima, N. Nakazaki, Y. Kishida, M. Kohara, S. Nakayama, M. Yamada, T. Fujishiro, A. Watanabe and S. Tabata. 2013. Construction of an integrated high density simple sequence repeat linkage map in cultivated strawberry (*Fragaria × ananassa*) and its applicability. DNA Res. 20: 79-92.

Iwatsubo, Y. and N. Naruhashi. 1989. Karyotypes of three species *Fragaria* (Rosaceae). Cytologia. 54: 493-497.

Iwatsubo, Y. and N. Naruhashi. 1991. Karyotypes of *Fragaria nublicola* and *F. daltoliana* (Rosaceae). Cytologia. 56: 453-457.

Jelenkovic, G., M.L. Wilson and P. J. Harding. 1984. An evaluation of intergeneric hybridization of *Fragaria spp. × Potentilla spp.* as a means of haploid production. Euphytica. 33: 143-152.

Kishaba A.N., V. Voth, A.F. Howland, R.S. Bringhurst and H.H. Toba. 1972. Two spotted spider mite resistance in California strawberries. J. Econ. Entomol. 65: 117-119.

Kojima, A. and Y. Nagato. 1997. Discovery of highly apomictic and highly amphimictic dihaploids in Allium tuberosum. Sex. Plant Reprod. 10: 8-12.

Kunihisa, M., N. Fukino and S. Matsumoto. 2003. Development of cleavage amplified polymorphic sequence (CAPS) markers for identification of strawberry cultivars. Euphytica. 134: 209-215.

Kunihisa, M., N. Fukino and S. Matsumoto. 2005. CAPS markers improved by cluster-specific amplification for identification of octoploid strawberry (*Fragaria* × *ananassa* Duch.) cultivars, and their disomic inheritance. Theor. Appl.Genet. 110: 1410-1418.

Kunihisa, M., H. Ueda, N. Fukino and S. Matsumoto. 2009. DNA marker for identificationof strawberry (*Fragaria* × *ananassa* Duch.) cultivars based on probability theory. J. Jpn Soc. Horticult. Sci. 78: 211-217.

Kunihisa, M. 2011. Studies using DNA markers in *F.* × *ananassa*: genetic analysis, genome structure, and cultivar identification. J. Jpn. Soc. Hort. Sci. 80: 231-43.

Lei, J., H. Dai and M. Deng. 2004. Studies on the hexaploid interspecific hybrid and its backcross with cultivars in strawberry. Acta Hort. Sinica. 31: 496-498 (In Chinese with English abstract).

Lei J., H. Dai, M. Deng, L. Wu and W. Hu. 2002. Studies on the interspecific hybridization in the genus *Fragaria*. Acta Hort. Sinica. 29: 519-523 (In Chinese with English abstract).

Lei J., Y. Li, G. Du, H. Dai and M. Deng. 2005. A natural pentaploid strawberry genotype from the Changbai Mountains in northeast China. HortScience. 40: 1194-1195.

Lei J., H. Dai, C. Tan, M. Deng, M. Zhao and Y. Qian. 2006. Studies on the taxonomy of the strawberry (*Fragaria*) species distributed in China. Acta Hort. Sinca 33: 1-5.

Lei, J., H. Dai, X. Bi and M. Deng. 2009. Studieson the classification of wild strawberry species distributed in China and their utilization by interspecific hybridization. Acta Hort. 842: 415-418 (In Chinese with English abstract).

Lerceteau-Köhler E, G. Guérin, F. Laigretand B. Denoyes-Rothan. 2003. Characterization of mixed disomic and polysomic inheritance in the octoploid strawberry (*Fragaria* × *ananassa*) using AFLP mapping. Theor. Appl. Genet. 107: 619-628.

Li, C.L., H. Ikeda and H. Ohba. 2003. *Fragaria* Linnaeus. In: Z.Y. Wu, P.H. Raven, and D.Y. Hong, editors. Flora of China. Beijing: Science Press, St. Louis: Missouri Botanical Garden Press. 337-340.

Li, X.X., K. Fukuhara and Y. Hayata. 2009. Concentrations of character impact odorants in 'Toyonoka' strawberries quantified by standard addition method and PQ column extraction with GC-MS analysis. J. Japan. Soc. Hort. Sci. 78: 200-205.

Li, Y., Hou, X., Lin, L., Jing, S., Deng, M., 2000. Abnormal pollen germination and embryo abortion in the interspecific cross, *Fragaria* × *ananassa* × *F. vesca* as related to cross-incompatibility. J. Jpn. Soc. Hort. Sci. 69: 84-89.

Lilienfeld, F. 1933. Karyologische und genetische Studien an Fragaria L. Ein tetraploider fertiler Bastard zwischen F. nipponica ($n = 7$) und F. elatior ($n = 21$). Jpn. J. Bot. 6: 425-458 (in German).

Lilienfeld, F.A. 1951. H. Kihara: Genome-analysis in *Triticum* and *Aegilops*. X. Concluding review. Cytologia. 16: 101-123.

Lin, J. and T.M. Davis. 2000. S1 analysis of long PCR heteroduplexes: Detection of chloroplast indel polymorphisms in *Fragaria*. Theor. Appl.Genet. 101: 415-420.

Longley, A.E., 1926. Chromosomes and their significance in strawberry classification. J. Agric. Res. 15: 559-568.

Losina-Losinskaja, A.S. 1926. Review of the genus *Fragaria* L., Proc. Bot. Garden USSR (Известия Ботанического сада АН СССР). 25: 47-88 (in Russian).

Luby, J.J., J.F. Hancock, A. Dale and S. Serçe. 2008. Reconstructing *Fragaria* × *ananassa* utilizing wild *F. virginiana* and *F. chiloensis*: Inheritance of winter injury, photoperiod sensitivity, fruit size, female fertility and disease resistance in hybrid progenies. Euphytica. 163: 57-65.

Mabberley, D.J. 2002. *Potentilla* and *Fragaria* (Rosaceae) reunited. Telopea. 9: 793-801.

Macfarlane S.W.H. and J. K. Jones. 1985. Intergeneric crosses with *Fragaria* and *Potentilla*. I. Crosses between *Fragaria moschata* and *Potentilla fruticosa*. Euphytica. 34: 725-735.

Mahoney, L.L., M.L. Quimby, M.E. Shields and T.M. Davis. 2010. Mitochondrial DNA transmission, ancestry, and sequences in *Fragaria*. Acta Hort. 859: 301-308.

Makino, T.1912. Observations on the flora of Japan. Bot. Mag. (Tokyo). 26: 282-290.

Mangelsdorf, A.J. and E.M. East. 1927. Studies on the genetics of *Fragaria*. Genetics. 12: 307-339.

Marta, A.E., E.L. Camadro, J.C. Diaz-Ricci and A.P. Castagnaro. 2004. Breeding barriers between the cultivated strawberry, *Fragaria* × *ananassa*, and related wild germplasm. Euphytica. 136: 139-150.

Milenkovic, S., M. Pesakovic, D. Marcic and D. Milosevic. 2014. Strawberry resistance to the aphid *Chaetosiphon fragaefolii* Cockerell (Homoptera: Aphididae). Pesticidi I Fitomedicina. 29: 267-273.

Mochizuki, T., Y. Noguchi and K. Sone. 2002. Processing quality of decaploid strawberry lines derived from *Fragaria* × *ananassa* and diploid wild species. Acta Hort. 567: 239-242.

Moore, J.M. and L.F. Hough. 1962. Relationships between auxin levels, time offloral induction and vegetative growth of the strawberry. Proc. Am. Soc.Hort. Sci. 81: 255-264.

Morishita, M., O. Yamakawa and T. Mochizuki. 1996. Studies of interspecific hybrids of strawberry. Bull. Natl. Res. Inst. Veg. Ornament. Plants Tea Ser. A. 11: 69-95 (in Japanese with Englishabstract).

Naruhashi, N. and T. Iwata. 1988. Taxonomic reevaluation of *Fragaria nipponica* Makino and allied species. J. Phytogeogr. Taxon. 36: 59-64.

Naruhashi, N., Y. Iwatsubo and C.I. Peng. 1999. Cytology, flower morphology and distribution of *Fragaria hayatai* Makino (Rosaceae). J. Phytogeogr. Taxon. 47: 139-143.

Nathewet, P., T.Yanagi, K.Sone, S.Taketaand N. Okuda. 2007. Chromosome observation method at metaphase and pro-metaphase stages in diploid and octoploid strawberries. Sci. Hortic. 114, 133-137.

Nathewet, P., T. Yanagi, Y. Iwatsubo, K. Sone, T. Takamura and N. Okuda. 2009a. Improvement of staining method for observation of mitotic chromosomes inoctoploid strawberry plants. Sci. Hortic. 120: 431-435.

Nathewet P.,T. Yanagi, K.E. Hummer,Y. Iwatsubo and K. Sone. 2009b. Karyotype analysis in wild diploid, tetraploid and hexaploid strawberries, *Fragaria* (Rosaceae). Cytologia. 74: 355-364.

Nathewet P., K.E. Hummer, T. Yanagi, Y. Iwatsubo and K. Sone. 2010. Karyotype analysis in octoploid and decaploid wild strawberries in *Fragaria* (Rosaceae). Cytologia. 75: 277-288.

Nehra, N.S., K.K. Kartha and C. Stushnoff. 1991. Nuclear DNA content and isozyme variationsin relation to morphogenic potential of strawberry (*Fragaria* × *ananassa*) callus cultures. Can. J. Bot. 69: 239-244.

Niemirowicz-Szczytt, K.W.P. and B. Ciupka. 1986. *In vitro* colchicine treatment of interspecific sterile *Fragaria* l. hybrid meristems and screening of regenerated plants for fruiting ability. Genetica Polonica. 27: 315-324.

Njuguna, W, A. Liston, R. Cronn, T-L. Ashman and N. Bassil. 2013. Insights into phylogeny, sex function and age of *Fragaria* based on whole chloroplast genome sequencing. Mol. Phylogenet. Evol. 66: 17-29.

Noguchi, Y., T. Mochizuki and K. Sone. 1997. Interspecific hybrids originated from crossing Asian wild strawberries (*Fragaria nilgerrensis* and *F. iinumae*) to *F.* × *ananassa*. Hortscience 32: 438-439.

Noguchi, Y., T. Mochizuki and K. Sone. 2002. Breeding of a new aromatic strawberry by interspecific hybridization *Fragaria* × *ananassa* × *F. nilgerrensis*. J. Jpn. Soc. Hort. Sci. 71: 208-213.

Noguchi, Y., T. Muro and M. Morishita. 2009. The possibility of using decaploid interspecific hybrids (*Fragaria* ×*ananassa* × *F. nilgerrensis*) as a parent for a new strawberry. Acta Hort. 842: 447-450.

Noguchi, Y., M. Morishita, T. Muro, A. Kojima, Y. Sakata, T. Yamada and K. Sugiyama. 2011. 'Tokun': a new aromatic decaploid interspecific hybrid strawberry. Bulletin of the National Institute of Vegetable and Tea Science. 10: 59-67 (In Japanese with English abstract).

Noguchi, Y. and T. Yamada. 2014. Flower bud initiation of a decaploid strawberry 'TOKUN' by night chilling and short day treatment. Acta Hort. 1049: 907-910.

Nosrati, H., A.H. Priceand C.C. Wilcock. 2010. No evidence of apomixis in matroclinal progeny from experimental crosses in the genus *Fragaria* (strawberry) based on RAPDs. Euphytica. 171: 193-202.

Nosrati, H., A.H. Price, P. Gerstberger and C.C. Wilcock. 2011a. Identification of a natural allopentaploid hybrid *Fragaria* (Rosaceae), new to Europe. New J. Bot. 1: 88-92.

Nosrati, H., A.H. Price and C.C. Wilcock. 2011b. Relationship between genetic distances and postzygotic reproductive isolation in diploid *Fragaria* (Rosaceae). Bio. J. Linnean Soc. 104: 510-526.

Nosrati, H., A.H. Price, P. Gerstberger and C.C. Wilcock. 2013. The first report of an alloheptaploid from the genus *Fragaria* (Rosaceae). New J. Bot. 3: 205-209.

Nyman, M. and A. Wallin. 1992. Improved culture technique for strawberry (*Fragaria* × *ananassa* Duch.) protoplasts and the determination of DNA content inprotoplast derived plants. Plant Cell Tissue Organ Cult. 30: 127-133.

Potter, D., J.J. Luby and R.E. Harrison. 2000. Phylogenetic relationships among species of *Fragaria* (Rosaceae) inferred from non-coding nuclear and chloroplast DNA sequences. Syst. Bot. 25: 337-48.

Porlingis, I.C. and D. Boynton. 1961. Growth response of the strawberry plant, *Fragaria chiloensis var. ananassa*, to gibberellic acid and to environmental conditions. Proc. Am. Soc. Hort. Sci. 78: 261-269.

Powers, L. 1944. Meiotic studies of crosses between *Fragaria ovalis* and *F. ananassa*. J. Agr. Res. 69: 435-448.

Powers, L. 1945. Strawberry breeding studies involving cross between the cultivated varieties (× *Fragaria ananassa*) and the Native Rocky Mountain Strawberry (*F. ovalis*). Jour. Agr. Res. 70: 95-122.

Rho R., Y.J. Hwang, H.I. Lee, K.B. Lim and C.H. Lee. 2012. Interspecific hybridization of diploids and octoploids in strawberry. Sci. Hort. 134: 46-52.

Richardson, C.W., 1914. A preliminary note on the genetics of *Fragaria*. J. Genet. 3: 171-177.

Rousseau-Gueutina, M., A. Gaston, A. Aïnouche, M.L. Aïnouche, K. Olbricht, G. Staudt, L. Richard and B. Denoyes-Rothan. 2009. Tracking the evolutionary history of polyploidy in *Fragaria* L. (strawberry): new insights from phylogenetic analyses of low-copy nuclear genes. Mol. Phylogenet. Evol. 51: 515-530.

Sargent, D.J., A.M. Hadonou, M.J. Wilkinson, N.H. Battey, J.A.Hawkins and D.W. Simpson. 2004. Cross-species amplification and phylogenetic reconstruction using *Fragaria* microsatellite primers. Acta Hort. 649: 87-92.

Sargent, D.J., T. Passey, N. Šurbanovski, E. Lopez Girona, P. Kuchta, J. Davik, R. Harrison, A. Passey, A.B. Whitehouse and D.W. Simpson. 2012. A microsatellite linkage map for the cultivated strawberry (*Fragaria* × *ananassa*) suggests extensive regions of homozygosity in the genome that may have resulted from breeding and selection. Theor. Appl. Genet. 124: 1229-1240.

Sargent, D.J., M. Geibel, J. Hawkins, M.J. Wilkinson, N.H. Battey and D.W. Simpson. 2004. Quantitative and qualitative differences in morphological traits revealed between diploid *Fragaria* species. Ann. Bot. 94: 787-796.

Sasaki, M. and T. Takeuchi. 2008. Chromosome doubling of odd number polyploid strawberry [*Fragaria ananassa*] and characteristics of the resulting line. Bulletin of the Shizuoka Research Institute of Agriculture and Forestry (Japan) 1: 11-19 (In Japanese with English abstract).

Sayegh, A.J. and M.J. Hennerty. 1993. Intergeneric hybrids of *Fragaria* and *Potentilla*. Acta Hort. 348: 151-154.

Schulze, J., P. Stoll, A. Widmer and A. Erhardt. 2011. Searching for gene flow from cultivated to wild strawberries in Central Europe. Ann Bot. 107: 699-707.

Scott, D.H. 1951. Cytological studies on polyploids derived from tetraploid *Fragaria vesca* and cultivated strawberries.Genetics. 36: 311-331.

Sebastiampillai, A.R. and J.K. Jones. 1976. Improved techniques for the induction and isolation of polyploidsin the genus *Fragaria*. Euphytica. 25: 725-732.

Senanayake, Y.D.A. and R.S. Bringhurst. 1967. Origin of *Fragaria* polyploids. I. Cytological analysis. Amer. J. Bot. 54: 221-228.

Stegmeir, T.L., C.E. Finn, R.M. Warner and J.F. Hancock. 2010. Performance of an elite strawberry population derived from wild germplasm of *Fragaria chiloensis* and *F. virginiana*. Hort Science. 45: 1140-1145.

Staudt, G. 1989. The species of *Fragaria*, the taxonomy and geographical distribution. Acta Hort. 265: 23-33.

Staudt, G. 1997. Reconstitution of *Fragaria × ananassa*; the effect of *Fragaria virginiana* Cytoplasm. Acta Hort. 439: 55-62.
Staudt, G. 1999a. Systematics and geographic distribution of the American strawberry species. Taxonomic studies in the genus *Fragaria* (Rosaceae: Potentilleae). Univ. Calif. Publ. Bot. 81: 1-129.
Staudt, G. 1999b. Notes on Asiatic *Fragaria* species: *Fragaria nilgerrensis* Schltdl. *ex* J. Gay. Bot. Jahrb. 121: 297-310.
Staudt, G. 2006. Himalayan species of *Fragaria* (Rosaceae). Bot. Jahrb. Syst. 126: 483-508.
Staudt, G. 2009. Strawberry biogeography, genetics and systematics. Acta Hort. 842: 71-84.
Staudt, G., F. Drawert and R. Tressl. 1975. Gaschromatographic-mass spectrometric differentiation of aroma substances from strawberry species. II. *Fragaria nilgerrensis*. Z. Pflanzenzucht. 75: 36-42 (in German with English abstract).
Staudt, G., L.M. DiMeglio, T.M. Davis and P. Gerstberger. 2003. *Fragaria × bifera*: origin and taxonomy. Bot. Jahrb. Syst. 125: 53-72.
Staudt, G., S. Schneider, P. Scheewe, D. Ulrich and K. Olbricht. 2009. *Fragaria iturupensis*: a New Source for Strawberry Improvement? Acta Hort. 842: 479-482.
Sukhareva, N.B., S.O. Baturin and K. Trajkovski. 2002. Induction of apomixis in *Fragaria vescana*. Acta Hort. 567: 231-234.
Tennessen, J., R.Govindarajulu, T.-L.Ashman and A. Liston. 2014. Evolutionary origins and dynamics of octoploid strawberry subgenomes revealed by dense targeted capture linkage maps. Genome Biology and Evolution. 6: 3295-3313.
Trajkovski, K. 1993. Progress report in *Fragaria* species hybridization at Balsgard, Sweden. Acta Hort. 348: 131-136.
Trajkovski, K. 1997. Further work on species hybridization in *Fragaria* at Balsgard. Acta Hort. 439: 67-74.
Trajkovski, K. 2002. Rebecka, a day-neutral *Fragaria × vescana* variety from Balsgard. Acta Hort. 567: 177-178.
Waldo, G. and G. Darrow. 1928. Hybrids of the Hautbois strawberry. Fertile hybrids between *Fragaria moschata* and *Fragaria virginiana*. J. Hered. 19: 509-510.
Weber, M. and D. Ulrich. 2009. Characterization of the aroma pattern of the wild strawberry *Fragaria nilgerrensis* (Schltdl.) as a precondition for the use in strawberry breeding. Mitt. Julius Kühn-Inst. 419: 36-40 (in German with English abstract).
Wilhelm, S. and J.E. Sagen. 1974. A history of the strawberry from ancient gardens to modern markets. University of California Div. of Agr. Sci., Berkeley, CA.
Yanagi, T. and Y. Oda. 1991. Characteristics of inflorescence production and their classification in cultivated and wild octoploid strawberries grown under long-day photoperiod. J. Jpn. Soc. Hort. Sci. 59: 737-743 (In Japanese with English abstract).
Yanagi, T., N. Okuda and T. Takamura. 2005. Introgression of unique characteristics of floral initiation under 24 hour day-length of *Fragaria chiloensis* 'CHI-24-1' into *F × ananassa*. Euphytica. 144: 79-84.
Yanagi, T., T. Yachi, N. Okuda and K. Okamoto. 2006. Light quality of continuous illuminating at night to induce floral initiation of *Fragaria chiloensis* L. CHI-24-1. Sci. Hort. 109: 309-314.
Yanagi, T., M. Matsumura, P. Nathewet, N. Okuda and K. Sone. 2009. Morphological and polyploidy variation of strawberry plants obtained by cross pollination of *Fragaria vesca* and *F. × ananassa*. Acta Hort. 842: 427-430.
Yanagi, T., K.E. Hummer, T. Iwata, K. Sone, P. Nathewet and T. Takamura. 2010. Aneuploid strawberry ($2n = 8x + 2 = 58$) was developed from homozygous unreduced gamete (8x) produced by second division restitution in pollen. Sci. Hort. 125: 123-128.
Yanagi, T., N. Okuda and K. Okamoto. 2016. Effects of light quality and quantity on flower initiation of Fragaria chiloensis L. CHI-24-1 grown under 24 h day-length. Sci. Hort. 202:150-155.
Yarnell, S.H. 1931. Genetic and cytological studies on *Fragaria*. Genetics. 16: 422-454.
Zhang, J., G. Li, Y. Qiao, M. Wang and J. Li. 2009. Effects of colchicine on regeneration and polyploid induction from leaf *in vitro* of strawberry (*Fragaria × ananassa*). Journal of Plant Resources and Environment. 18: 69-73 (In Chinese with English abstract).
Zhao, M., J. Wang, Z. Wang, Y. Qian and W. Wu. 2014. GC-MS analysis of volatile components in chinese wild strawberry (*F. nilgerrensis* Schlecht.). Acta. Hort. 1049: 467-469.

Chapter 7

The Role of Polyploidization and Interspecific Hybridization in the Breeding of Ornamental Crops

Agnieszka Marasek-Ciolakowska[1,*], Paul Arens[2] and Jaap M. Van Tuyl[2]

ABSTRACT

In the history of ornamental plant breeding, polyploidization can be observed as one of the most important processes in domestication, with a large effect on crop improvement. In the ornamental crops (*Lilium*, *Tulipa*, *Narcissus* and *Begonia*) exemplified here, the origin of the assortment of varieties traces back to the diploid species, from which the modern, predominantly triploid or tetraploid cultivars were created through spontaneous interspecific crosses or applied polyploidization techniques. The use of meiotic (2*n* gametes) and mitotic polyploidization as methods to obtain polyploids are discussed. The three restitution mechanisms of 2*n* gamete formation (Second Division Restitution (SDR), First Division Restitution (FDR) and Intermediate Meiotic Restitution (IMR)) and their genetic impact as elucidated by molecular cytogenetic techniques (such as Genomic In-Situ Hybridization (GISH)) are presented. Through GISH analysis the presence of intergenomic recombination, which is desirable for introgression breeding, was clearly demonstrated to occur in 2*n* gamete formation.

Introduction

Polyploidization (genome doubling) and interspecific hybridization have played a major role in the evolution and the development of cultivated plants. Many important ornamental crops such as *Lilium* (Van Tuyl et al. 2002), *Tulipa* (Marasek et al. 2006), *Narcissus* (Brandham and Kirton 1987), *Gladiolus* (Van Tuyl 1997), *Crocus* (Ørgaard et al. 1995), *Iris* (Eikelboom and Van Eijk 1990), *Alstroemeria* (Ramanna 1992) and *Begonia* (Hvoslef-Eide and

[1] Laboratory of Biochemistry and Molecular Biology, Department of General and Molecular Biology, Research Institute of Horticulture, Konstytucji 3 Maja 1/3, 96-100 Skierniewice, Poland, E-mail: agnieszkamarasek@wp.pl

[2] Plant Breeding Wageningen University and Research Centre Wageningen, The Netherlands E-mail: paul.arens@wur.nl, jaap.vantuyl@wur.nl

* Corresponding author

Munster 2006; Dewitte 2010a) have been subjected to extensive interspecific or intergeneric hybridization followed by selection by breeders. All these genera include polyploid cultivars (Van Tuyl et al. 2002) which are characterized by the numerical change of complete copies of the nuclear chromosome set. For instance, triploids have three sets of chromosomes (genomes) ($2n = 3x$) whereas tetraploids have four sets of chromosomes ($2n = 4x$). Whole genome duplication may occur by somatic chromosome doubling (mitotic polyploidization) or sexually through gametic non-reduction (sexual polyploidization).

Harlan and De Wet (1975) showed that many plant species produce numerically unreduced ($2n$) gametes and discussed the importance of $2n$ gametes in the origin of polyploids. In ornamentals the spontaneous production of a small amount of $2n$ gametes by interspecific hybrids was observed, inter alia, in *Alstroemeria* (Ramanna and Jacobsen 2003), *Lilium* (Van Tuyl 1989; Lim et al. 2001), *Begonia* (Dewitte et al. 2009b) and *Tulipa* (Marasek-Ciolakowska et al. 2012a, 2012b). In many cases the functioning of $2n$ gametes has lead to formation of polyploids during interspecific hybridization. Triploids of spontaneous origins resulting from unreduced gametes have been recorded in *Lilium* (Noda 1986), *Crocus* (Ørgaard et al. 1995), *Narcissus* (Brandham 1986) and *Tulipa* (Van Scheepen 1996; Marasek et al. 2006). These polyploid genotypes were of great value for breeders because of their superior vigor, growth, flower size, sturdier stem, broader and thicker leaves or more compact plants, among other traits, as compared to their diploid progenitors (Van Tuyl et al. 2002; Ramanna et al. 2012).

This chapter focuses on breeding and polyploidization in some of the most important ornamental bulbous crops such as *Lilium, Tulipa, Begonia* and *Narcissus*. Some recent work on molecular cytogenetic technique such as genomic *in situ* hybridization (GISH) which has greatly facilitated our understanding of the modes of origins of polyploids and introgression is also reviewed in these crops.

Mode of Origin of Polyploids

It should be noted that neopolypoids are usually produced deliberately through artificial chromosome doubling (mitotic polyploidisation) or through the functioning of $2n$ gametes (meiotic doubling). Artificial chromosome doubling can be induced by treatment with chemicals such as colchicine, oryzalin or trifluralin during a stem regeneration process (Van Tuyl et al. 1992, 2002; Chauvin et al. 2006; Podwyszyńska 2012). This method is predominately used to restore the fertility of F_1 interspecific hybrids (Asano 1982; Van Tuyl et al. 1989). In ornamentals, mitotic polyploids have been introduced in many genera including *Lilium* (Van Tuyl et al. 1997), *Tulipa* (Eikelboom et al. 2001; Van Tuyl et al. 1992; Podwyszyńska 2012), *Alstroemeria* (Lu and Bridgen 1997), *Rosa* (Roberts et al. 1990; Zlesak et al. 2005), *Iris* (Yabuya 1985; Van Eijk and Eikelboom 1990) and *Hemerocallis* (Podwyszyńska et al. 2014).

Artificial chromosome doubling induces autopolyploids that may be fertile, but it does not result in introgression breeding, due to the absence of intergenomic recombination in the progenies (Lim et al. 2001, 2003; Xie et al. 2010a, 2010b). However, the phenotype of newly developed autopolyploids can sometimes be altered compared to the diploid progenitors, and autopolyploids can often be characterized by, among other traits, larger flowers, leaves, stomata and pollen grains (Podwyszyńska et al. 2014; Fig. 1).

Sexual polyploidization (meiotic polyploidization) is an alternative method to obtain polyploid cultivars. In this case, diploid genotypes that produce 2*n* gametes can be hybridized to give rise to polyploid progenies (Ramanna and Jacobsen 2003; Van Tuyl 1997).

Figure 1. Differences between unconverted diploid and tetraploid daylily plants (*Hemerocallis*) after *in vitro* polyploidisation with antimitotic agents: (A) flowers; (B) leaves; (C) flowering plants. For each photograph (A-C) the plant on the left is the diploid and the plant on the right is the tetraploid. Photos 1 A-C courtesy of Prof Małgorzata Podwyszyńska (D) Stomata in an unconverted diploid plant. (E) Stomata in a tetraploid plant. Bars represent 20 μm. Photos 1 D-E courtesy of Prof Barbara Dyki

The Origin of 2n Gametes

The mechanisms leading to formation of viable unreduced gametes include meiotic abnormalities such as omission of the first or second meiotic division, abnormal spindle morphology in the second division, and disturbed cytokinesis (Ramanna and Jacobsen 2003; Taschetto and Pagliarini 2003). Depending on the meiotic stage or mechanism by which the nucleus restitutes, three genetically different types of 2*n* gametes can be identified: those resulting from first division restitution (FDR), second division restitution (SDR), or from indeterminate meiotic restitution (IMR) (Lim et al. 2001, Barba-Gonzalez 2005, Barba-Gonzalez et al. 2008; Fig. 2). In FDR the whole chromosome complement divides equationally before telophase I leading to the formation of dyad without further division. GISH and FISH analysis of meiotic restitution mechanisms in *Alstroemeria* (Kamstra et al. 1999a, 1999b, 2004; Ramanna et al. 2003) and *Lilium* (Lim et al. 2000, 2001, 2003; Barba-Gonzalez et al. 2004, 2005a, 2005b, 2006c; Khan et al. 2009a, 2009b) have shown that there are two types of FDR gametes: those without homologous recombination between chromosomes and those with homologous recombination between chromosomes. The former may occur due to failure of homologous chromosome pairing, leading to a restitution nucleus at the end of meiosis I but with normal separation of sister chromatids after meiosis II. The latter may occur when meiosis I proceeds normally, but parallel spindles in meiosis II result in (recombined) homologous chromosomes reuniting.

In SDR the first meiotic division occurs normally; i.e. homologous chromosomes divide reductionally and cytokinesis occurs producing a dyad. The presence of cytokinesis and the formation of a cell wall after the telophase I is characteristic for most of the monocot plant having so called 'successive' type of microsporogenesis. In meiosis II, the centromeres of the half-bivalents divide, but the chromatids do not migrate to the poles. In SDR, chromosome assortment is random, which results in a very heterogeneous population of gametes. SDR is rare in hybrids because only in exceptional cases all chromosomes are paired as bivalents and tends to occurs in these hybrids in which the genomes are closely related (Ramanna and Jacobsen 2003).

The occurrence of IMR, which shows characteristics similar to both FDR and SDR, was reported for the first time in Longiflorum × Asiatic lily interspecific hybrids (Lim et al. 2001). In indeterminate meiotic restitution, both univalents and bivalents are formed during metaphase I. During the first meiotic division some bivalents disjoin reductionally as in SDR while some univalents divide equationally as in FDR, giving rise to 2*n* gametes with odd number of parental chromosomes (centromeres).

For breeding purposes, data have shown that 2*n* gametes of the FDR type are more advantageous then those obtained by SDR for transferring parental heterozygosity (Barcaccia et al. 2003). FDR is the basic mechanism of 2*n* pollen formation in *Lilium* (Lim at al. 2001; Barba-Gonzalez et al. 2008; Zhou et al. 2008a; Khan et al. 2010), *Alstroemeria* (Ramanna et al. 2003), *Begonia* (Dewitte et al. 2010a) and *Tulipa* (Marasek-Ciolakowska et al. 2014). The reason for this is probably that the chromosomal composition of FDR gametes is more balanced and thus gametes more viable than those of SDR and IMR. The occurrence of FDR with crossovers and of IMR during the origin of 2*n* gametes can also enlarge the genetic variation that occurs in polyploid progenies as well as the extent of introgressions (Okazaki 2005). GISH analysis confirmed the presence of recombinant chromosomes in meiotic polyploids in *Lilium* (Khan et al. 2009a, 2009b; Xie et al. 2010a) and *Tulipa* (Marasek-Ciolakowska et al. 2012a, 2012b, 2014).

The production of 2*n* gametes can be increased by manipulating environmental factors such as temperature (Bretagnolle and Thompson 1995; Ramsey and Schemske 1998). The formation of 2*n* gametes can be also induced by arresting the meiotic process with nitrous oxide (N_2O) gas. Akutsu et al. (2007) showed that effects were optimal when treatments

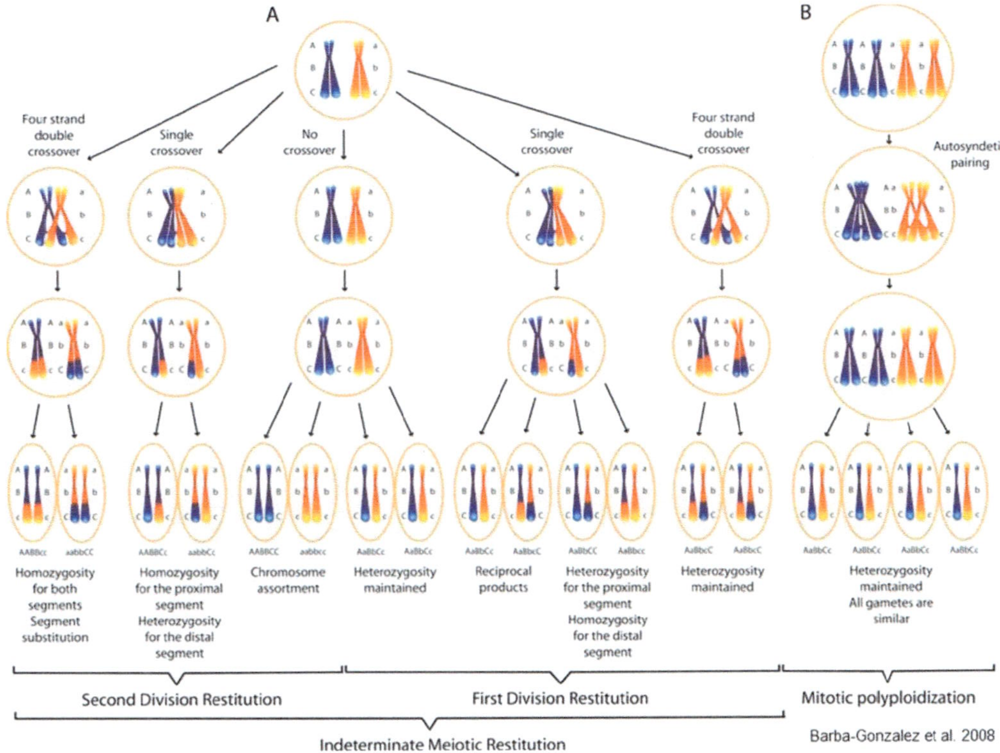

Figure 2. Schematic representation of three restitution mechanisms of 2*n* gamete formation in species with the monocotyledonous successive type of meiotic division and mitotic polyploidization. From Barba-Gonzalez R, Lim K.B. Zhou S. Ramanna M.S. and Van Tuyl J.M. (2008) Interspecific hybridization in lily: the use of 2*n* gametes in interspecific lily hybrids. In *Floriculture, ornamental and plant biotechnology vol. V*, ed. J.A. Teixeira da Silva, 138-145. Isleworth, UK: Global Science Books. (Reproduced with kind permission of Global Science Books).

started during pollen mother cell progression to metaphase I. Kitamura et al. (2009) found that N_2O mediates polyploidization by inhibiting microtubule polymerization, but not actin filament formation, during microsporocyte meiosis in *Lilium*. The treatments with N_2O to induce the formation of $2n$ pollen and to overcome hybrid sterility were successfully applied in lilies, tulips and begonia flower buds (Okazaki 2005; Okazaki et al. 2005; Barba-Gonzalez et al. 2006a, 2006b; Akutsu et al. 2007; Dewitte et al. 2010b; Nukui et al. 2011; Chung et al. 2013; Luo et al. 2013; Fig. 3) whereas it was used in young ovaries in tulips to induce tetraploid embryos (Zeilinga and Schouten 1968b).

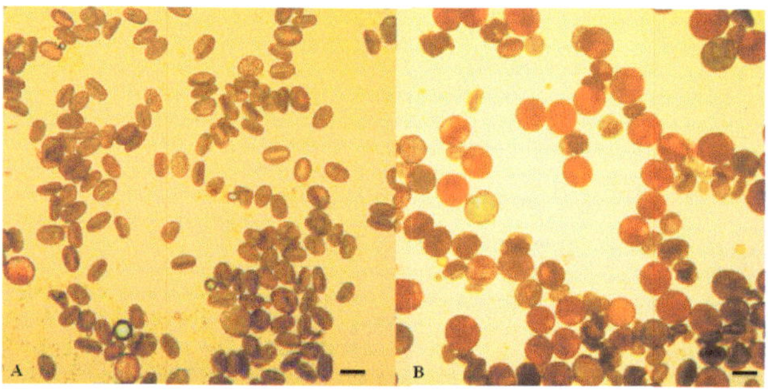

Figure 3. Overcoming hybrid sterility by treating sterile or partially sterile hybrid lilies with N_2O for 48 h: appearance of control pollen (A) and N_2O-induced pollen (B). Bar represents 50 μm. Photo courtesy of Dr Paul Arens.

Interploidy Crosses

Many polyploid cultivars have also been obtained as a result of interploidy crosses during the breeding process. Crossing tetraploids with diploids or vice versa resulted in triploids with vigorous growth in *Tulipa* (Van Scheepen 1996; Kroon and Van Eijk 1977) and *Lilium* (Lim et al. 2003; Zhou et al. 2008b). In triploid by tetraploid crosses ($3x \times 4x$) pentaploids have been found in *Lilium* (Lim et al. 2003), *Aloineae* (Brandham 1982), *Dactylis* (Jones and Borrill 1962), *Tulipa* (Marasek-Ciolakowska et al. 2014) and *Primula* (Hayashi et al. 2009).

In *Lilium*, when allotriploid BC_1 progeny of Oriental × Asiatic lily (AOA) was crossed with $2n$ gamete-producing Oriental × Asiatic lily hybrids (OA), aneuploid ($3x + 2$) and pentaploid ($2n = 5x = 60$) progenies were obtained and both progeny types possessed recombinant chromosomes (Barba-Gonzalez et al. 2006c). In comparison, in *Tulipa* in $3x \times 2x$ ($2n$) crosses aneuploids with 40–45 chromosomes were predominant, but tetraploids ($2n = 4x = 48$) and pentaploids ($2n = 5x = 60$) were also observed (Marasek-Ciolakowska et al. 2014). In the genus *Tulipa* diploid and aneuploid seedlings were obtained from $2x \times 3x$ and $3x \times 2x$ crosses (Upcott and Philip 1939; Bamford et al. 1939; Okazaki and Nishimura 2000; Lim et al. 2003). Similarly in *Lilium*, in $2x \times 3x$ or reciprocal crosses, progeny with nearly diploid chromosome numbers were recorded. According to Zhou et al. (2008b), when the triploid is used as the female parent most of the resulting genotypes are aneuploids, but when the same parent is used as a male parent the progenies are more likely to be diploid.

Lilium

Classification and Main Cultivated Groups

The genus *Lilium* belongs to the Liliaceae family, and comprises about 100 species (Comber 1949) which are taxonomically classified into following seven sections: Lilium, Martagon, Pseudolirium, Archelirion, Sinomartagon, Leucolirion and Oxypetalum (Comber 1949; De Jong 1974). The most important hybrid groups cultivated for cut flowers are Longiflorum (L), Asiatic (A) and Oriental hybrids (O) which originate from interspecific hybridization in the sections Leucolirion, Sinomartagon, and Archelirion, respectively.

Ploidy Level of Species

All native lilies (except for a triploid form of *Lilium tigrinum*) are diploid ($2n = 2x = 24$). In several species aneuploids have also been found (Stewart 1943). Supplementary B chromosomes have been reported in at least 17 *Lilium* species (Brandam 1967); for example in *L. callosum* (Kayano 1962) and in a tetraploid hybrid derived from chromosome doubling of a diploid hybrid between a cultivar of the Longiflorum (L) and the Trumpet (T) group (Xie et al. 2014).

Polyploidization in Lilies

Within 20 years more than 45% of the assortment became polyploid by application of mitotic and meiotic polyploidization techniques (Van Tuyl and Arens 2011). Chromosome doubling of lily was applied to many genotypes and resulted in thousands of auto- and allopolyploids (Van Tuyl al. 1992). However a more promising ploidization pathway in *Lilium* is via the use of $2n$ gametes. These unreduced gametes are (in different percentages) formed in intersectional hybrids. Crosses between the species or hybrids from different sections are only possible by application of embryo rescue techniques (Van Tuyl et al. 1991) which have become available in the last 35 years. These crosses between different hybrid groups (A = Asiatic Hybrids, L = Longiflorum hybrids, O = Oriental hybrids and T = trumpet hybrids) led to diploid F_1 hybrids such as LA, LO, OT and OA, which were mostly sterile, but in some cases $2n$ gametes were formed and backcrosses were produced (Fig. 4) (Lim and Van Tuyl 2006; Barba-Gonzalez et al. 2006a, 2006c; Van Tuyl and Arens 2011). These crosses led to triploid hybrids ($2n = 3x = 36$) like the LAA or ALA (mostly called LA-hybrids) and the OOT and OTO (mostly called OT-hybrids). These groups of triploid interspecific hybrids occupy nowadays more than 45% of the acreage for lily production. Khan et al. (2010) produced tetraploid progenies by using two separate F_1 LA hybrids as parents, one donating $2n$ eggs and the other $2n$ pollen.

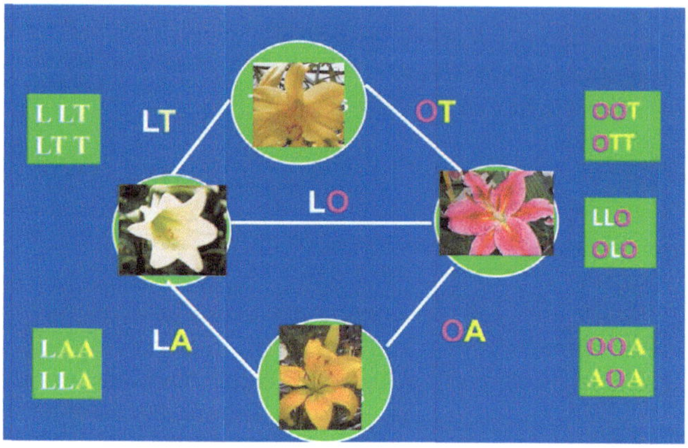

Figure 4. A lily crossing scheme between L (Longiflorum), A (Asiatic), O (Oriental) and T (Trumpet) lilies showing the diploid F_1 and triploid BC_1 progeny. Photos courtesy of Dr Jaap M. Van Tuyl.

Genome Differentiation Based on GISH Analysis

An extensive molecular cytogenetic approach has elucidated the genome composition of lily hybrids, the nature of the production of $2n$ gametes and the ability of the homoeologous chromosomes to pair and recombine. As the hybrids consist of distantly related species, the parental genomes can be easily discriminated by chromosome painting techniques. GISH has been used extensively in lily to recognize the three most used genomes of *Lilium, viz.* Longiflorum (L), Asiatic (A) and Oriental (O) genomes, and to study the recombinant chromosomes in the BC_1 and BC_2 progenies of LA and OA hybrids (Karlov et al. 1999; Lim et al. 2003; Barba-Gonzalez et al. 2004, 2005a; Zhou et al. 2008b). GISH was also applied to assess the intergenomic recombination and the nature of interspecific lily hybrids obtained from uni- and bilateral sexual polyploidization leading to allotriploid and allotetraploid formation (Khan et al. 2009a). It was found that in the populations of LA and OA hybrids obtained after unilateral sexual polyploidization, a majority of the different progenies had originated through functional $2n$ gametes via the First Division Restitution (FDR) mechanism, with or without cross overs. However, there were indications of Indeterminate Meiotic Restitution (IMR) mechanisms of $2n$ gamete formation as well. Most of the BC progenies exhibited recombination and the amount of recombination was dependent on the cross type. Intergenomic recombination was also determined cytologically in the plants of sib-mated LA hybrids where both parents had contributed $2n$ gametes (Khan et al. 2009a).

By using GISH techniques, it was shown that intergenomic recombination occurs at a high frequency in hybrids obtained by using unreduced gametes as compared to in hybrids obtained by chromosome doubling to overcome F_1 sterility (Lim et al. 2001; Barba-Gonzalez et al. 2005a, 2005b, 2006a, 2006c; Lim and Van Tuyl 2006; Xie et al 2010a; Fig. 5). Khan et al (2009a, 2009b) showed extensive intergenomic recombination among the chromosomes of diploid and triploid BC progenies of LA hybrids. Molecular markers also have the potential to identify recombination events at a very fine scale, as has been shown in an LA hybrid population (Shahin et al. 2012), and could be used to connect genetic data with GISH results.

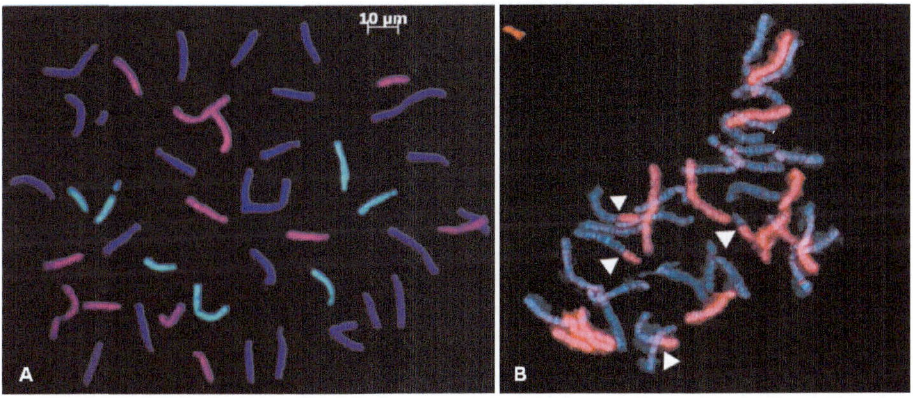

Figure 5. GISH results for *Lilium* hybrids. (a) LLO × LLTT hybrid ($2n = 4x - 5 = 43$) derived from crossing allotriploid LLO with allotetraploid LLTT (both resulted from mitotic polyploidization) showing no recombinant chromosomes. From Songlin Xie, Ramanna, M.S. and van Tuyl, J.M. (2010). Simultaneous identification of three different genomes in *Lilium* hybrids through multicolour GISH. Acta Hortic. 855, 299-304 DOI: 10.17660/ActaHortic.2010.855.45. (b) Triploid AOA hybrid ($2n = 3x = 36$) resulting from meiotic polyploidization and showing four recombinant chromosomes. From R. Barba-Gonzalez, K.-B. Lim, M.S. Ramanna, J.M. van Tuyl (2005b) Use of 2*n* gametes for inducing intergenomic recombination in lily hybrids. Acta Hort. 673, 161-166 DOI: 10.17660/ActaHortic.2005.673.18 (Reproduced with kind permission of ISHS).

Tulipa

Systematic, Species and Main Cultivated Groups in the Genus Tulipa

Tulips, *Tulipa* L. (Liliaceae), for centuries have been subjected to extensive interspecific hybridisation followed by selection, and more than 8,000 cultivars are available at present (Van Scheepen 1996). According to the classified list of the International Register of tulip cultivars (Van Scheepen 1996), tulips cultivars can be divided into 15 groups. The first nine groups belong to the species *T. gesneriana* (Single Early, Double Early, Triumph, Single Late, Lily-flowered, Fringed, Viridiflora, Parrot, and Double Late). At present it is not possible to determine the exact origin of these varieties (Killingback 1990). The other groups are: Darwin hybrid tulips, the Kaufmanniana group derived from *T. kaufmanniana*, the Fosteriana group from *T. fosteriana*, the Greigii derived from *T. greigii* and the Species group containing all other cultivated species. Of historical significance is the group of Rembrandt tulips, which are not commercially available.

Triumph tulips are currently the most important group of tulip varieties grown for cut flowers. The second most important commercial group is the Darwin hybrid tulips, which have been obtained from interspecific crosses between cultivars of *T. gesneriana* and *T. fosteriana* Hoog ex W. Irving genotypes (Van Tuyl and Van Creij 2006).

The Main Goals of Tulip Breeding

Modern tulip breeding is focused on the introgression of characteristics such as a short forcing period, good flower longevity, new flower colors and shapes and also the introgression of resistance against Tulip Breaking Virus (TBV), *Botrytis tulipae* and *Fusarium oxysporum* (bulb-rot) into the commercial assortment (Marasek-Ciolakowska et al. 2012a).

Interspecific crosses are usually made between genotypes of *T. gesneriana* and other *Tulipa* species (Van Eijk et al. 1991; Van Raamsdonk et al. 1995). Many successful crosses have been made especially between *T. gesneriana* cultivars and TBV resistant *T. fosteriana* cultivars, where some of the resulting hybrids (Darwin hybrid tulips) showed high resistance to TBV. In the past 50 years, more than 50 cultivars of Darwin hybrid tulips have been developed (Van Scheepen 1996). Several hybrids have been also obtained from crosses between *T. gesneriana* and species *T. kaufmanniana* Regel, *T. greigii* Regel, *T. eichleri* Regel, *T. ingens* Hoog, *T. albertii* Regel and *T. didieri* Jord (Marasek-Ciolakowska et al. 2012a).

Chromosome Numbers of Tulipa Species and Varieties

Most of the tulip species and cultivars are diploid ($2n = 2x = 24$), and a small number are triploid ($2n = 3x = 36$) and tetraploid ($2n = 4x = 48$) (Holitscher 1968; Kroon 1975; Zeilinga and Schouten 1968a, Kroon and Jongerius 1986; Van Scheepen 1996). The triploid tulip cultivars ($2n = 3x = 36$) mostly belong to the Darwin hybrids (Kho and Baër 1971; Kroon and Van Eijk 1997). Triploid Darwin hybrid cultivars such as 'Apeldoorn', 'Ad Rem', 'Pink Impression' and tetraploid ($2n = 4x = 48$) Darwin hybrids as 'Tender Beauty' were of unintentional origin as a result of $2n$ gametes that were spontaneously produced by their diploid progenitors (Van Scheepen 1996).

Polyploidization in Tulips

Since polyploid tulip hybrids are characterized by large flowers and sturdier leaves as compared to their diploid parents, many polyploidization attempts have been made (Okazaki et al. 2005). For instance, tetraploid 'Christmas Marvel' is derived from *in vitro* chromosome-doubling (Eikelboom et al.

2001). Highly fertile tetraploid cultivars of *T. gesneriana, T. fosteriana* and *T. kaufmanniana* were also obtained in the 90s as a result of laughing gas treatment of pollinated ovaries resulting in cultivars as 'Rambo', 'Zorro' and 'Hunter' (Zeilinga and Schouten 1968b; Straathof and Eikelboom 1997).

One of the main methods currently used to obtain polyploid tulip is meiotic polyploidization via crossing with $2n$ gamete-producing genotypes. Marasek-Ciolakowska et al. (2012b, 2014) demonstrated that some of the Darwin hybrid genotypes are fertile and could be backcrossed to *T. gesneriana.* Additionally, some F_1 Darwin hybrid tulips were found to produce both $2n$ and n gametes (Fig. 6). This provided unique opportunities to generate polyploid as well as diploid BC_1 progenies from backcrossing GF hybrids (Darwin hybrids) to *T. gesneriana* parents (Marasek-Ciolakowska et al. 2012a, 2014). In crosses between diploid *T. gesneriana* cultivars and diploid $2n$ gamete producers some triploid and tetraploid hybrids were obtained, where the tetraploids resulted from $2n$ gametes provided by both parents (Marasek-Ciolakowska et al. 2012a; 2014).

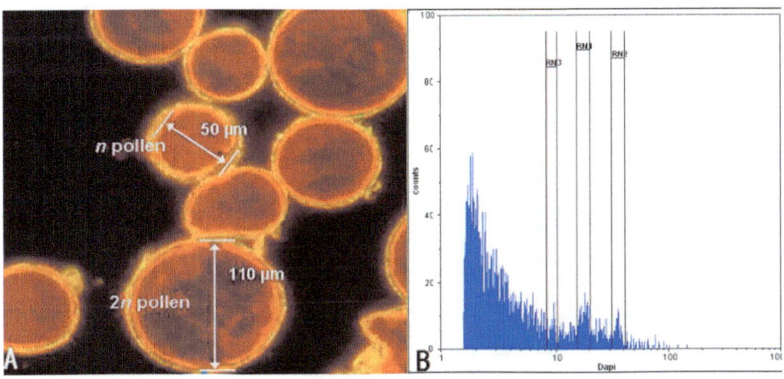

Figure 6. (A) Microscopic view of *Tulipa* F1 'Darwin' hybrid pollen stained with acetocarmine; (B) Flow cytometry analysis of pollen showing peaks located at 1C, 2C (*n* gamete) and 4C (*2n* gametes). From Marasek-Ciolakowska, A., Xie, S., Ramanna, M.S., Arens, P. and van Tuyl, J.M. (2012b). Ploidy manipulation and introgression breeding in Darwin hybrid tulips. Acta Hortic. 953, 187-192DOI: 10.17660/ActaHortic.2012.953.26 (Reproduced with kind permission of ISHS).

New polyploid tulips were also obtained as a result of interploidy crossing (Van Scheepen 1996; Straathof and Eikelboom 1997; Okazaki and Nishimura 2000; Marasek-Ciolakowska et al. 2014). Triploid cultivar 'World's Favourite' originated from a $4x \times 2x$ cross (Straathof and Eikelboom 1997). From $2x \times 4x$ crosses, triploid varieties such as 'Lady Margot', 'Benny Neyman' and 'Sun Child' and tetraploids such as 'Riant', 'Beauty of Canada' and 'Peerless Yellow' have been obtained (Kroon and Van Eijk 1977; Van Scheepen 1996). Polyploids were also obtained in crosses with $2n$ pollen induced via laughing gas treatment (Okazaki 2005; Okazaki et al. 2005; Barba-Gonzalez et al. 2006b).

Cytogenetic Analysis in Genus *Tulipa*

Genomic in situ hybridization (GISH) analyses have been used to elucidate the genome composition of diploid and polyploid Darwin hybrids and their BC progenies. In triploid Darwin hybrid cultivar 'Yellow Dover', GISH demonstrated the presence of two genomes of *T. gesneriana* and one genome of *T. fosteriana* (Marasek et al. 2006). GISH has also been successfully used for the elucidation of the genome composition of diploid Darwin hybrid 'Purissima' and its BC_1 and BC_2 hybrids (Marasek and Okazaki 2008; Marasek-Ciolakowska et al. 2009, 2011, 2012a). All 'Purissima' hybrids were diploid ($2n = 2x = 24$) with the exception of one triploid ($2n = 3x = 36$), one tetraploid BC_1 ($2n = 4x = 48$) and one aneuploid BC_2 ($2n = 2x = 25$) genotype. All of these progeny, except the triploid, possessed variable numbers of recombinant chromosomes that were clearly detectable by GISH (Marasek and Okazaki 2008; Marasek-Ciolakowska et al. 2011, 2012a). It is essential in tulip that genomic introgression can be also accomplished at the diploid level (Fig. 7). Through GISH analysis, the presence of intergenomic recombination, which is desirable for introgression breeding, was also observed in tetraploid cultivar 'Judith Leyster' ($2n = 4x = 48$) (Marasek and Okazaki 2008), and in tetraploid and pentaploid hybrids resulting from $3x \times 2x$ crosses where the paternal forms were $2n$ gamete-producers (Fig. 8) (Marasek-Ciolakowska et al. 2012a, 2014).

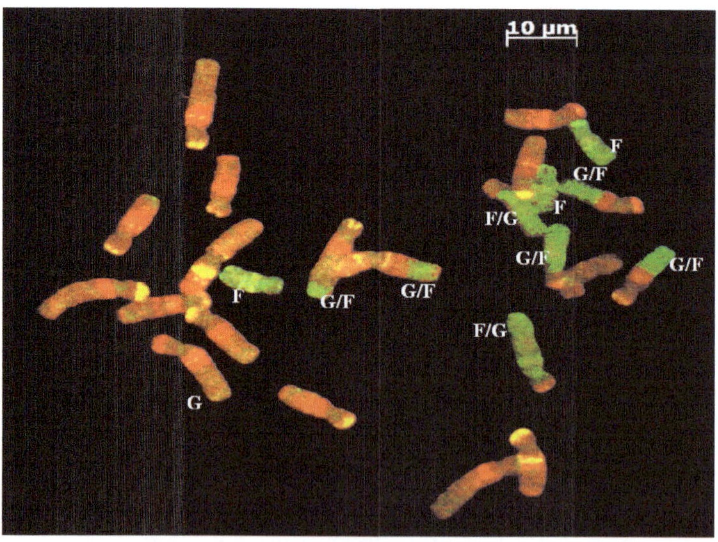

Figure 7. Chromosome differentiation in a diploid *Tulipa* BC_1 Darwin hybrid ($2n = 2x = 24$) derived from backcrossing 'Purissima' (GF) to *T. gesneriana* (G) cultivars using the GISH technique. Chromosome complement comprises 19 *T. gesneriana* chromosomes (red) and 5 *T. fosteriana* chromosomes (green) including 7 recombinant chromosomes. Recombinant chromosomes are defined as F/G and G/F, indicating a *T. fosteriana* centromere with a *T. gesneriana* chromosome segment(s) and a *T. gesneriana* centromere with a *T. fosteriana* chromosome segment(s), respectively. Bar 10 µm. Photo courtesy of Dr Agnieszka Marasek-Ciolakowska.

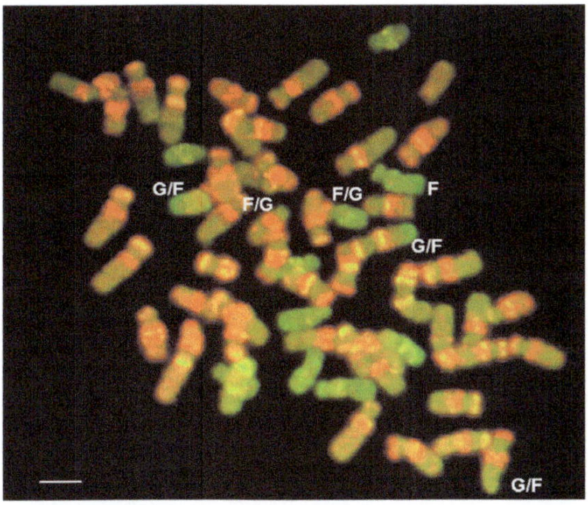

Figure 8. GISH results for a pentaploid *Tulipa* BC_1 hybrid ($2n = 5x = 60$) resulting from a $3x \times 2x$ cross showing 48 *T. gesneriana* (G) chromosomes (red fluorescence) and 12 *T. fosteriana* (F) chromosomes (green fluorescence) including 5 recombinant chromosomes. Bar represents 10 μm. From Marasek-Ciolakowska, A., Xie, S., Ramanna, M.S., Arens, P. and van Tuyl, J.M. (2012b). Ploidy manipulation and introgression breeding in Darwin hybrid tulips. Acta Hortic. 953, 187-192 DOI: 10.17660 /ActaHortic.2012.953.26 (Reproduced with kind permission of ISHS).

Begonia

Distribution and Classification

Begonia L. (Begoniaceae) is a genus distributed mainly in tropical and subtropical regions. It comprises more than 1500 species (Doorenbos et al. 1998; Wagner 1999). *Begonia* is an old crop: it was described for the first time in 1700, and by 1850 there were 200 species known in Europe. At present over 10,000 cultivars and hybrids are available (Cubey and Wesley 2006). According to Haegeman (1979) the genus *Begonia* can be divided into the following groups: Loraine begonias (*B. socotrana* × *B. dregei*), Tuberous begonias (*Begonia* × *tuberhybrida*), Elatior begonias (*B. socotrana* × tuberous hybrids), Semperflorens begonias (*B. semperflorens* × *B. schmidtiana*), Rex Begonias with ornamental foliage, which includes mainly cultivars of *B. rex*, *B. annulata*, *B. palmata*, *B. hatacoa*, *B. xanthina* and other species of the *Platycentrum* section, and a remaining group of other commercially grown Begonias that are not classified into any of the categories above.

The Main Cultivated Groups

The current commercial assortment is comprised mostly of 'Elatior-hybrids', which flower throughout the year and represent around 88% of the total *Begonia* production (Kroon 1993; Fig. 9). Most modern varieties of 'Elatior' begonias are the result of crosses between various tuberous *Begonia* species hybrids (*B.*

× *tuberhybrida* Voss) and *B. socotrana* Hook. F (Gleed 1961; Doorenbos 1973). Some varieties have also been obtained from backcrossing 'Elatior' begonias to *B. socotrana* or *B.* × *tuberhybrida* (Arends 1970). Recently the popularity of the Rex Cultorum group has increased: this group comprises over 500 cultivars characterized by beautiful leaf coloring (Cubey and Wesley 2006).

Figure 9. Elatior begonia hybrid 'Victoria Falls'. Photo courtesy of Dr. Jaap M. Van Tuyl.

Breeding Objectives

Begonia breeding focuses on the introgression of important agricultural traits to new cultivars, including winter flowering capacity, new flower shapes and colors, improved plant foliage and disease resistance. Interspecific crosses and polyploidization have played an important role in the development of cultivated *Begonia* ssp. (Horn 2004). A common way of creating new variation in begonia is via spontaneous mutations and mutations induced by radiation or chemicals (Hvoslef-Eide and Munster 2006). For instance, in Christmas begonias (*B.* × *cheimantha*), most of the cultivars on the Norwegian market are of spontaneous origin, whereas many cultivars of Elatior begonias resulted from induced mutations (Hvoslef-Eide and Munster 2006). According to Horn (2004), *B.* × *tuberhybrida* (Tuberous begonias) were developed from interspecific crosses in the 19th century, but mutation and polyploidization also played an important role in the production of horticulturally important characters such as double flowers and white and yellow flower color. New triploid semperflorens begonias and triploid *Begonia rex* have also been obtained as a result of interploidy crossing between diploid and tetraploid genotypes (Horn 2004; Van Tuyl pers. comm.)

Basic Chromosome Number and Karyology of Begonias

Chromosome numbers have been reported for many species of *Begonia*, ranging from $2n = 16$ in *B. rex* to $2n = 156$ in *B. acutifolia* (Okuno and Nagai 1954; Zeilinga 1962; Legro and Haegeman 1971; Nakata et al. 2007; Ye et al. 2004; Peng and Sue 2000; Dewitte et al. 2009a). Although no general basic chromosome number can be readily discerned (Dewitte et al. 2009a), the high variation in chromosome numbers suggests the occurrence of high levels of polyploidy in the genus. As an example of recent polyploidy, most of the 'Elatior' begonia hybrids are triploids, and a few are tetraploids (Mikkelsen 1976; Arends 1970; Hvoslef-Eide and Munster 2006; Marasek-Ciolakowska et al. 2010). *Begonia rex* from the Platycentrum section has a chromosome number of $2n = 22$, but the cultivated Rex begonias are mostly tetraploids ($2n = 44$).

According to Matsuura and Okuno (1936, 1943), the genus *Begonia* comprises different basic chromosome numbers of $x = 6$, 7 and 13, where the last may be of secondary origin derived from hybridization between the first two. Later, Legro and Haegeman (1971) postulated the basic chromosome numbers in *Begonia* to be $x = 13$ and 14. In general, in Taiwanese and Chinese *Begonia* the chromosome number $2n = 22$ is the most common (Oginuma and Peng 2002; Nakata et al. 2007), while in tuberous *Begonia* species chromosome numbers of $2n = 26$ and $2n = 28$ have been recorded (Legro and Haegeman 1971). *Begonia* is characterized by small chromosomes which range in length from 0.5 μm in *B. socotrana* (Arends 1970) to 5.43 μm in *B. coptidifolia* (Ye et al. 2004). Chromosomes are morphologically poorly differentiated, and centromeres are not distinguishable for many genotypes, such as 'Elatior' begonias (Arends, 1970) and tuberous hybrid Begonias (*B.* × *tuberhybrida* Voss) (Okuno and Nagai 1954). Chromosomes are often accompanied by satellites which can be mistaken for chromosomes (Okuno and Nagai 1953; Zeilinga 1962; Marasek-Ciolakowska et al. 2010).

Genome Differentiation by GISH

GISH analysis was used to assess the hybrid status of *Begonia* × *chungii* ($2n = 22$) (Kono et al. 2012) native to Taiwan. This study confirmed that *B.* × *chungii* is a natural F_1 hybrid between *B. longifolia* and *B. palmata,* and proved that parental genotypes have partially homologous genomes that may suggest that they originated from a common ancestral species (Kono et al. 2012). Genomic in situ hybridization (GISH) experiments were also reported in 'Elatior' begonia hybrids in which it was possible to distinguish between chromosomes derived from tuberous *Begonia* and *B. socotrana* (Marasek-Ciolakowska et al. 2010; Fig. 10). GISH analysis of the genome constitution of 'Elatior' begonia hybrids revealed the participation of unreduced gametes in the origin of some hybrids (Marasek-Ciolakowska et al. 2010).

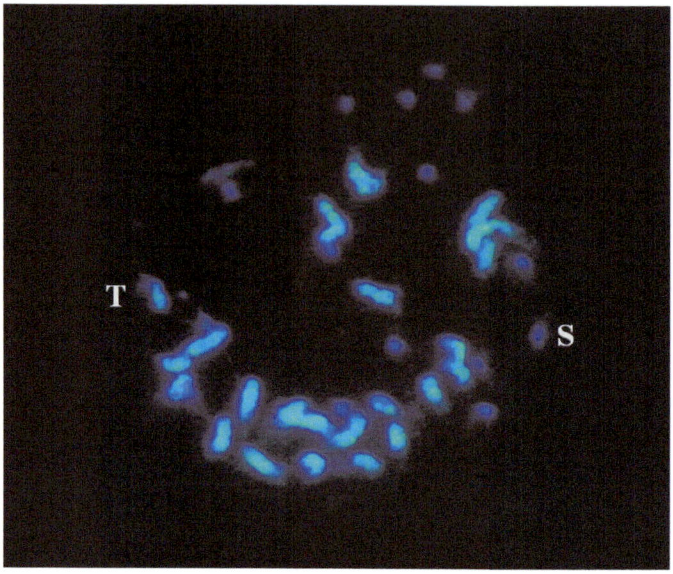

Figure 10. Genome differentiation in triploid Elatior begonia hybrids (28 T + 14 S) using the GISH technique. T defines genome of tuberous *Begonia* (green fluorescence); S indicates the genome of *B. socotrana* (blue fluorescence). Photo courtesy of Dr Agnieszka Marasek-Ciolakowska.

Unreduced Gametes

Dewitte et al. (2009b) proved based on pollen size measurements and flow cytometric analysis of pollen that occurrence of viable 2*n* pollen is a common phenomenon in *Begonia*. In their study, 14% of the investigated genotypes produced unreduced gametes. The frequency of large (2*n*) pollen depended on genotype, and varied from 1% in *Begonia* 'Rubaiyat' to 100% in 'Florence Rita' and B276. They also obtained successful crosses with genotypes producing 2*n* pollen, and showed via flow cytometry analysis the increase of DNA content in the progeny. Similarly, allotetraploid semperflorens begonias have been generated using unreduced gametes (Horn 2004). According to Dewitte et al. (2010a), FDR is the basic mechanism leading to 2*n* pollen formation in *Begonia*. Besides the spontaneous production of 2*n* gametes, the formation of 2*n* pollen was successfully induced by N_2O treatment of *Begonia* flower buds, and polyploid seedlings were obtained after crosses with N_2O-induced 2*n* pollen (Dewitte et al. 2010b).

Narcissus

All modern *Narcissus* cultivars are complex interspecific hybrids. Due to spontaneous (meiotic) polyploidization that occurred during domestication of this crop, a polyploid assortment was formed. For example, the old triploid *Narcissus* cultivars 'Emperor' and 'Empress' originated from diploid parents (Brandham 1992). Brandham (1986) reports that in the history of the origin of polyploids in this genus, before 1885 only a few diploid ($2n = 2x = 14$) or triploid ($2n = 3x = 21$) cultivars occurred, and that the first tetraploid was introduced in 1887. From the 1920s the introduction of new tetraploid cultivars accelerated rapidly, and at present nearly 75% of the commercial cultivars are tetraploids with $2n = 4x = 28$ chromosomes (Brandham 1986).

Cytological Analysis in Narcissus

Karyology of *Narcissus* is well-documented with respect to chromosome number (Brandham and Kirton 1987; Brandham 1986, 1992) and C-banding (De Domincis et al. 2002). The latter technique was used to analyze the origin of *N. biflorus* and to verify relationships with the parental species *N. tazetta* and *N. poëticus* (De Domincis et al. 2002). In the genus *Narcissus* there is variation in the basic number of chromosomes. The species in the subgenus *Narcissus* have basic chromosome numbers of $x = 7$, with series ranging from diploid ($2n = 2x = 14$) up to pentaploid ($2n = 5x = 35$), however the horticulturally optimal level of ploidy is considered to be $4x$. In the subgenus *Hermione*, species have been recorded with two basic chromosome numbers, $x = 10$ and 11 (Brandham 1986).).Hybridization can be made between members of these two subgenera; between plants with the same or different basic chromosome numbers; and between plants with the same or different polyploidy levels; which leads to great variability of chromosome numbers in resulting hybrids. According to Brandham (1992) the chromosome numbers of *Narcissus* cultivars range from $2n = 14$ to $2n = 46$.

The occurrence of B chromosomes has been reported in several cultivars: most cultivars have a single B chromosome; but four B chromosomes were observed in 'Ultimus' (Brandham 1992). The best known example of a cultivar with a B chromosome is the most popular *Narcissus* variety 'Tête-à-Tête', which is an allotriploid of spontaneous origin ($2n = 3x = 24 + 1B$) (Ramanna et al. 2012). Analysis of the genome constitution of this cultivar using both GISH and NBS (nucleotide-binding site) profiling confirmed the presence of two genomes of *N. cyclamineus* and one genome of *N. tazetta* together with a B chromosome (Wu et al. 2011).

Conclusions

A common feature of most of the ornamentals is that they include polyploid cultivars. In crops like lily and tulip, the change to a polyploid crop was manipulated by the breeder whereas in *Narcissus*, the assortment was developed through interspecific crosses and in a natural way it was followed by polyploidization. In the begonias interspecific crosses and mutation breeding have played an important role in creating new variation. Molecular cytogenetic techniques such as genomic *in situ* hybridization (GISH) played an important role in the elucidation of the modes of origins of polyploids and the nature of the production of $2n$ gametes and ability of the homoeologous chromosomes to pair and recombine. The use of 2*n* gametes has the big advantage that intergenomic recombination occurs which is favourable for introgression breeding.

References

Akutsu, M., S. Kitamura, R. Toda, I. Miyajima, K. Okazaki. 2007. Production of 2*n* pollen of Asiatic hybrid lilies by nitrous oxide treatment. Euphytica 155:143-152.

Arends, J.C. 1970. Somatic chromosome numbers in 'Elatior'-begonias. Meded Landbouwh Wageningen 70-20: 1-18.

Asano, Y. 1982. Overcoming interspecific hybrid sterility in *Lilium*. J Japan Soc Hort Sci. 51:75-81.

Bamford R., G.B. Reynard and J.M. Jr Bellows. 1939. Chromosome number in some tulip hybrids. Botanical Gazette 101: 482-490.

Barba-Gonzalez, R. 2005. The use of 2*n* gametes for introgression breeding in Oriental × Asiatic lilies. PhD thesis, Wageningen University, Wageningen, The Netherlands.

Barba-Gonzalez, R., K.B. Lim, M.S. Ramanna, R.G.F. Visser and J.M. Van Tuyl. 2005a. Occurrence of 2*n* gametes in the F1 hybrids of Oriental × Asiatic lilies (*Lilium*): Relevance to intergenomic recombination and backcrossing. Euphytica 143: 67-73.

Barba-Gonzalez, R., K.B. Lim, M.S. Ramanna, and J.M. Van Tuyl. 2005b. Use of 2*n* gametes for inducting intergenomic recombination in lily hybrids. Acta Hort 673: 161-166.

Barba-Gonzalez, R., K.B. Lim, S. Zhou, M.S. Ramanna and J.M. Van Tuyl. 2008. Interspecific hybridization in lily: the use of 2*n* gametes in interspecific lily hybrids. In *Floriculture, ornamental and plant biotechnology vol. V*, ed. J.A. Teixeira da Silva, 138-145. Isleworth, UK: Global Science Books.

Barba-Gonzalez, R., A.C. Lokker, K.B. Lim, M.S. Ramanna and J.M. Van Tuyl. 2004. Use of 2*n* gametes for the production of sexual polyploids from sterile Oriental × Asiatic hybrids of lilies (*Lilium*). Theor Appl Genet 109: 1125-1132.

Barba-Gonzalez, R., C.T. Mille, M.S. Ramanna and J.M. Van Tuyl. 2006a. Induction of 2*n* gametes for overcoming F1-sterility in lily and tulip. Acta Hort 714: 99-106.

Barba-Gonzalez, R., C.T. Miller, M.S. Ramanna and J.M. Van Tuyl. 2006b. Nitrous oxide (N_2O) induces 2*n* gametes in sterile F_1 hybrids between Oriental × Asiatic lily (*Lilium*) hybrids and leads to intergenomic recombination. Euphytica 148:303-309.

Barba-Gonzalez, R., A.A. Van Silfhout, M.S. Ramanna, R.G.F. Visser, J.M. Van Tuyl. 2006c. Progenies of allotriploids of Oriental × Asiatic lilies (*Lilium*) examined by GISH analysis. Euphytica 151: 243-250.

Barcaccia, G., S. Tavoletii, A. Mariani, F. Veronesi. 2003. Occurrence, inheritance and use of reproductive mutants in alfalfa improvements. Euphytica 133: 37-56.

Brandram, S. 1967. Cytogenetic studies of the Genus *Lilium*. M.Sc. thesis. The University of London.

Brandham, P.E. 1982. Inter-embryo competition in the progeny of autotriploid *Aloineae* (Liliaceae). Genetica 59: 29-42.

Brandham, P.E. 1986. Evolution of polyploidy in cultivated *Narcissus* subgenus *Narcissus*. Genetica 68:161-167

Brandham, P.E. 1992. Chromosome number in *Narcissus* cultivars and their significance to the plant breeder. The Plantsman. 14: 133-168

Brandham, P.E. and P.R. Kirton (1987) The chromosomes of species, hybrids and cultivars of *Narcissus* L. (Amaryllidaceae). Kew Bull 42: 65-102.

Bretagnolle, F. and J.D. Thompson. 1995. Gametes with the somatic chromosome number: mechanisms of their formation and role in the evolution of autoploid plants. New Phyt 129: 1-22.

Chauvin, J.E., A. Label and M.P. Kermarrec. 2006. In vitro chromosome-doubling in tulip (*Tulipa gesneriana* L.) Journal of Horticultural Science & Biotechnology 80: 693-698.

Chung, M.D., J.D. Chung, M.S. Ramanna, J.M. Van Tuyl and K.B. Lim. 2013. Production of polyploids and unreduced gametes in *Lilium auratum* × *L. henryi* hybrid. Int J Biol Sci 9: 693-701.

Comber, H.F. 1949. A new classification of genus *Lilium*. Lily Yearbook, Roy Hort Soc 13: 86-105.

Cubey, J. and W. Wesley. 2006. *Begonia* Rex Cultorum Group. RHS Plant Trials and Awards. Bulletin Number 16: 1-16.

De Dominicis, R.I., G.D'Amato and G.F. Tucci. 2002. On the hybrid origin of *Narcissus biflorus* (Amaryllidaceae): analysis of C-banding and rDNA structure. Caryologia 55: 129-134.

De Jong, P.C. 1974. Some notes on the evolution of lilies. Lily Yearbook, North Amer Lily Soc 27: 23-28.

Dewitte, A., T. Eeckhaut, J. Van Huylenbroeck and E. Van Bockstaele. 2010a. Meiotic aberrations during 2*n* pollen formation in *Begonia*. Heredity 104: 215-223.

Dewitte, A., T. Eeckhaut, J. Van Huylenbroeck and E. Van Bockstaele. 2010b. Induction of 2*n* pollen formation in *Begonia* by trifluralin and N2O treatments. Euphytica 171: 283-293.

Dewitte, A., L. Leus, T. Eeckhaut, I. Vanstechelman, J. Van Huylenbroeck and E. Van Bockstaele (2009a). Genome size variation in *Begonia*. Genome 52: 829-838.

Dewitte, A., T. Eeckhaut, J. Van Huylenbroeck and E. Van Bockstaele. 2009b. Occurrence of viable unreduced pollen in a *Begonia* collection. Euphytica 168: 81-94.

Doorenbos, J. 1973. Breeding 'Elatior'-begnia (*B.* × *hiemalis* Fotsch). Acta Hort 31: 127-131.

Doorenbos, J., M.S.M. Sosef and J.J.F.E. De Wilde. 1998. The sections of *Begonia*. Wageningen Agricultural University, 266 pp, Wageningen, The Netherlands.

Eikelboom, W. and J.P. Van Eijk. 1990. Prospects of interspecific hybridization in Dutch iris. Acta Hort 266: 353-356.

Eikelboom, W., T.P. Straathof and J.M. Van Tuyl. 2001. Tetraploïde "Christmas Marvel" Methoden om tetraploide tulpen te verkrijgen. Bloembollencultuur 112: 22-23.

Gleed, C.J. 1961. Hybrid winter-flowering Begonias. The story of raising some early varieties. J.R. Hort Soc 86: 319-322.

Haegeman, J. 1979. Tuberous Begonias. Origin and Development (268 pages). A.R. Gartner Verlag KG, Germany.

Harlan, J. and J. De Wet. 1975. The origins of polyploidy. Bot Rev 41: 361-390.

Hayashi, M., J. Kato, H. Ohashi and M. Mii. 2009. Unreduced 3*x* gamete formation of allotriploid hybrid derived from the cross of *Primula denticulata* (4*x*) × *P. rosea* (2*x*) as a causal factor for producing pentaploid hybrids in the backcross with pollen of tetraploid *P. denticulate*. Euphytica: 169(1): 123-131.

Holitscher, O. 1968. Pruhonicky sortiment tulipanu. Acta Pruhoniciana 18: 1-215.

Horn, W. 2004. The patterns of evolution and ornamental plant breeding. Acta Hort 651: 19-31.

Hvoslef-Eide, A.K. and C. Munster. 2006. *Begonia*. History & breeding. Chapter 9, Anderson N.O. (ed.), Flower Breeding and Genetics 241-275.

Hvoslef-Eide, A.K. and C. Munster. 2006. *Begonia*. History and breeding. In: Anderson NO (ed) Flower Breed Genet, Chapter 9. Springer, Heidelberg, pp. 241-275.

Jones, K. and M. Borrill. 1962. Chromosome status, gene exchange and evolution in Dactylis: 3, the role of the interploid hybrids. Genetica 32: 269-322.

Kamstra, S.A., J.H. De Jong, E. Jacobsen, M.S. Ramanna and A.G.J. Kuipers. 2004. Meiotic behaviour of individual chromosomes in allotriploid *Alstroemeria* hybrids. Heredity 93: 15-21.

Kamstra, S.A., M.S. Ramanna, M.J. De Jeu, G.J. Kuipers and E. Jacobsen. 1999a. Homoelogous chromosome pairing in the distant hybrid of *Alstroemeria aurea* × *A. indora* and the genome composition of its backcross derivatives determined by fluorescence in situ hybridization with species-specific probes. Heredity 82: 69-78.

Kamstra, S.A., A.G.J. Kuipers, M.J. De Jeu, M.S. Ramanna and E. Jacobsen. 1999b. The extent and position of homoelogous recombination in a distant hybrid of *Alstroemeria*: a molecular cytogenetic assessment of first generation backcross progenies. Chromosoma 108: 52-63.

Karlov, G.I., L.I. Khrustaleva, K.B. Lim and J.M. Van Tuyl. 1999. Homoeologous recombination in 2*n*-gamete producing interspecific hybrids of *Lilium* (Liliaceae) studied by genomic *in situ* hybridization (GISH). Genome 42: 681-686.

Kayano, H. 1962. Cytogenetic studies in *Lilium callosum*. V. Supernumerary B chromosomes in wild populations. Evolution 16: 246-253.
Khan, N., R. Barba-Gonzalez, M.S. Ramanna, P. Arens, R.G.F. Visser and J.M. Van Tuyl. 2010. Relevance of unilateral and bilateral sexual polyploidisation in relation to intergenomic recombination and introgression in *Lilium* species hybrids. Euphytica 171: 157-173.
Khan, N., R. Barba-Gonzalez, M.S. Ramanna, R.G.F. Visser and J.M. Van Tuyl. 2009a. Construction of chromosomal recombination maps of three genomes of lilies (*Lilium*) based on GISH analysis. Genome 52: 238-251.
Khan, N., Zhou, S., M.S. Ramanna, P. Arens, J. Herrera, R.G.F. Visser and J.M. Van Tuyl. 2009b. Potential for analytic breeding in allopolyploids: an illustration from Longiflorum × Asiatic hybrid lilies (*Lilium*). Euphytica 166: 399-409.
Kho, Y.O. and J. Baër. 1971. Incompatibility problems in species crosses of tulips. Euphytica 20: 30-35.
Killingback, S. 1990. Tulips — an illustrated identifier and guide to their cultivation. Apple Press, London, pp. 9-13.
Kitamura, S., M. Akutsu and K. Okazaki. 2009. Mechanism of action of nitrous oxide gas applied as a polyploidizing agent during meiosis in lilies. Sex Plant Rep 22: 9-14.
Kono, Y., M.C. Chung and C.I. Peng. 2012. Identification of genome constitutions in *Begonia × chungii* and its putative parents, *B. longifolia* and *B. palmata*, by genomic in situ hybridization (GISH). Plant Science 185-186: 156-160.
Kroon, G.H. 1975. Chromosome numbers of garden tulips. Acta Bot Neer 24: 489-490.
Kroon, G.H. 1993. Breeding research in *Begonia*. Acta Hort 337: 53-58.
Kroon, G.H. and M.C. Jongerius. 1986. Chromosome numbers of *Tulipa* species and the occurrence of hexaploidy. Euphytica 35: 73-76.
Kroon, G.H. and J.P. Van Eijk. 1977. Polyploidy in tulips (*Tulipa L.*). The occurrence of diploid gametes. Euphytica 26: 63-66.
Legro, R.A.H. and J.F.V. Haegeman. 1971. Chromosome number of hybrid tuberous Begonias. Euphytica 20: 1-13.
Lim, K.B., J.D. Chung, B.C.E. Van Kronenburg, M.S. Ramanna, J.H. De Jong and J.M. Van Tuyl. 2000. Introgression of *Lilium rubellum* Baker chromosomes into *L. longiflorum* Thunb.: a genome painting study of the F1 hybrid, BC1 and BC2 progenies. Chromosome Res 8: 119-125.
Lim, K.B., M.S. Ramanna, J.H. De Jong, E. Jacobsen and J.M. Van Tuyl. 2001. Indeterminate meiotic restitution (IMR): a novel type of meiotic nuclear restitution mechanism detected in interspecific lily hybrids by GISH. Theor Appl Genet 103: 219-230.
Lim, K.B., M.S. Ramanna, E. Jacobsen and J.M. Van Tuyl. 2003. Evaluation of BC2 progenies derived from 3*x*-2*x* and 3*x*-4*x* crosses of *Lilium* hybrids: a GISH analysis. Theor Appl Genet 106: 568-574.
Lim, K.B. and J.M. Van Tuyl. 2006. Lily, *Lilium* hybrids. Chapter 19 page 517-537 Flower breeding and genetics: Issues, challenges and opportunities for the 21st century. Kluwer Academic Publishers, Dordrecht (Ed. N.O. Anderson).
Lu, C. and M. Bridgen. 1997. Chromosome doubling and fertility study of *Alstroemeria aurea × A. caryophyllaea*. Euphytica 94: 75-81.
Luo, J.R., J.M. Van Tuyl, P. Arens and L.X. Niu. 2013. Cytogenetic studies on meiotic chromosome behaviors in sterile Oriental × Trumpet lily Genetics and molecular research: GMR 12(4): 6673-84.
Marasek, A. and K. Okazaki. 2008. Analysis of introgression of the *Tulipa fosteriana* genome into *Tulipa gesneriana* using GISH and FISH. Euphytica 160: 217-230.
Marasek, A., H. Mizuochi and K. Okazaki. 2006. The origin of Darwin hybrid tulips analyzed by flow cytometry, karyotype analyses and genomic *in situ* hybridization. Euphytica 151: 279-290.
Marasek-Ciolakowska, A., M.S. Ramanna, W.A. Ter Laak and J.M. Van Tuyl. 2010. Genome composition of 'Elatior'-begonias hybrids analyzed by genomic in situ hybridisation. Euphytica 171: 273-282.

Marasek-Ciolakowska, A., M.S. Ramanna and J.M. Van Tuyl. 2011. Introgression of chromosome segments of *Tulipa fosteriana* into *T. gesneriana* detected through GISH and its implications for breeding virus resistant tulips. Acta Hort 885: 175-182.

Marasek-Ciolakowska, A., M.S. Ramanna and J.M. Van Tuyl. 2009. Introgression Breeding in Genus *Tulipa* Analysed by GISH. Acta Hort 836: 105-110.

Marasek-Ciolakowska, A., H. He, P. Bijman, M.S. Ramanna, P. Arens and J.M. Van Tuyl. 2012a. Assessment of intergenomic recombination through GISH analysis of F1, BC1 and BC2 progenies of *Tulipa gesneriana* and *T. fosteriana*. Plant Syst Evol 298: 887-899.

Marasek-Ciolakowska, A., S. Xie, P. Arens and J.M. Van Tuyl. 2014. Ploidy manipulation and introgression breeding in Darwin hybrid tulips. Euphytica 198: 389-400.

Marasek-Ciolakowska, A., S. Xie, M.S. Ramanna, P. Arens and J.M. Van Tuyl. 2012b. Meiotic polyploidization in Darwin Hybrid Tulips. Acta Hort 953: 187-192.

Matsuura, H. and S. Okuno. 1936. Cytological study in *Begonia*. Jpn J Genet 12: 42-43.

Matsuura, H. and S. Okuno. 1943. Cytological study on *Begonia* (Preliminary survey). Cytologia 13: 1-18.

Mikkelsen, J.C. (1976) Enlarging the scope of flower crop improvement mutation breeding. Acta Hort 63: 197-202.

M.S. Ramanna, Z. Sawor, A. Mincione, A. Van de Steen, E. Jacobsen. 1998. Pollen markers for gene-centromere mapping in diploid potato. Theor Appl Genet 93: 1040-1047.

Nakata, M., K. Guan, J. Li, Y. Lu and H. Li. 2007. Cytotaxonomy of *Begonia rubropunctata* and *B. purpureofolia* (Begoniaceae). Bot J Lin Soc 155: 513-517.

Noda, S. 1986. Cytogenetic behavior, chromosomal differences, and geographic distribution in *L. lancifolium* (Liliaceae). Plant Species Biol. (Kyoto) 1: 69-78.

Nukui, S., S. Kitamura, T. Hioki, H. Ootsuka, K. Miyoshi, T. Satou, Y. Takatori, T. Oomiya and K. Okazaki. 2011. N_2O induces mitotic polyploidization in anther somatic cells and restores fertility in sterile interspecific hybrid lilies. Breed Science 61: 327-337.

Oginuma, K. and C.I. Peng. 2002. Karyomorphology of Taiwanese *Begonia*: taxonomic implications. J Plant Res 115: 225-235.

Okazaki, K. 2005. New aspects of tulip breeding: embryo culture and polyploid. Acta Hort 673: 127-140.

Okazaki, K. and M. Nishimura. 2000. Ploidy of progenies crossed between diploids, triploids and tetraploids in tulip. Acta Hort 522: 127-134.

Okazaki, K., K. Kurimoto and I. Miyajima. 2005. Induction of 2*n* pollen in tulips by arresting the meiotic process with nitrous oxide gas. Euphytica 143: 101-114.

Okuno, S. and S. Nagai. 1953. Cytological studies on *Begonia evansiana* Andr. with special reference to its meiotic chromosomes. Jpn J Genet 28: 132-136.

Okuno, S. and S. Nagai. 1954. Karyotypic polymorphism in *Begonia tuberohybrida* Voss. Jpn J Genet 29: 185-196.

Ørgaard, M., N. Jacobsen and I.S. Heslop-Harrison. 1995. The hybrid origin of two cultivars of *Crocus* (Iridaceae) analysed by molecular cytogenetics including genomic southern and *in situ* hybridization. Ann Bot 76: 253-262.

Peng, C.I. and C.Y. Sue. 2000. *Begonia × taipeiensis* (Begoniaceae), a new natural hybrid in Taiwan. Bot Bull Acad Sin 41: 151-158.

Podwyszyńska, M. 2012. *In vitro* tetraploid induction in tulip (*Tulipa gesneriana* L.) Acta Hort 961: 391-396.

Podwyszyńska, M., E. Gabryszewska, B. Dyki, A.A. Stępowska, A. Kowalski and A. Jasiński. 2014. Phenotypic and genome size changes (variation) in synthetic tetraploids of daylily (Hemerocallis) in relation to their diploid counterparts'. Euphytica 203: 1-16.

Ramanna, M.S. 1992. The role of sexual polyploidization in the origins of horticultural crops: Alstroemeria as an example. In A. Mariani and S. Tavoletti (eds). Proceedings of Workshop: Gametes with somatic chromosome number in the evolution and breeding of polyploidy polysomic species: Achievements and perspectives Tipolitographia Porzuincola-Assisi (PG) Italy pp. 83-89.

Ramanna, M.S. and E. Jacobsen (2003) Relevance of sexual polyploidization for crop improvement—a review. Euphytica 133: 3-18.

Ramanna, M.S., A.G.J. Kuipers and E. Jacobsen. 2003. Occurrence of numerically unreduced (2*n*) gametes in Alstroemeria interspecific hybrids and their significance for sexual polyploidisation. Euphytica 133: 95-106.

Ramanna, M.S., A. Marasek-Ciolakowska, S. Xie, N. Khan and J.M. Van Tuyl. 2012. The Significance of Polyploidy for Bulbous Ornamentals: A Molecular Cytogenetic Assessment. Chapter in: Floriculture and Ornamental Biotechnology. Special Issue: Bulbous Ornamentals. Editors: Jaap Van Tuyl and Paul Arens. Global Science Books. Ltd. UK Vol 1, pp. 116-121.

Ramsey, J. and D.W. Schemske. 1998. Pathways, mechanisms and rates of polyploid formation in flowering plants. An Rev Ecol Syst 29: 467-501.

Roberts, A., D. Lloyd D and K. Short. 1990. *In vitro* procedures for the induction of tetraploidy in a diploid rose. Euphytica 49: 33-38.

Shahin, A., M. Van Kaauwen, D. Esselink, J.W. Bargsten, J.M.Van Tuyl, R.G.F. Visser and P. Arens. 2012. Generation and analysis of expressed sequence tags in the extreme large genomes *Lilium* and *Tulipa*. BMC Genomics 13: 640.

Stewart, R.N. 1943. Occurrence of aneuploids in *Lilium*. Botanical Gazette 105: 620-626.

Straathof, Th.P. and W. Eikelboom. 1997. Tulip breeding at PRI. Daffodil and Tulip Yearbook 8: 27-33.

Taschetto, O.M. and M.S. Pagliarini. 2003. Occurrence of 2*n* and jumbo pollen in the Brazilian ginseng (*Pfaffia glomerata* and *P. tuberosa*). Euphytica 133: 139-145.

Upcott, M. and J. Philip. 1939. The genetic structure of *Tulipa*. IV. Balance, selection and fertility. J Genet 38: 91-123.

Van Eijk, J.P., L.W.D. Van Raamsdonk, W. Eikelboom and R.J. Bino. 1991. Interspecific crosses between *Tulipa gesneriana* cultivars and wild Tulipa species: a survey. Sexual Plant Reproduction 4: 1-5.

Van Eijk, J.P. and W. Eikelboom. 1990. Evaluation of breeding research on resistance to *Fusarium oxysporum* in tulip. Acta Hort 266: 357-364.

Van Raamsdonk, L.W.D., J.P. Van Eijk and W. Eikelboom. 1995. Crossability analysis in subgenus *Tulipa* of the genus *Tulipa* L. Botanical Journal of the Linnean Society 117: 147-158.

Van Scheepen, J. 1996. Classified list and international register of tulip names. Royal General Bulbgrowers' Association KAVB, Hillegom, The Netherlands.

Van Tuyl, J.M. 1989. Research on mitotic and meiotic polyploidization in lily breeding. Herbertia 45: 97-103.

Van Tuyl, J.M. 1997. Interspecific hybridization of flower bulbs: A review. Acta Hort 430: 465-476.

Van Tuyl, J.M., M.P. Van Diën, M.G.M. Van Creij, T.C.M. Van Kleinwee, J. Franken and R.J. Bino. 1991. Application of in vitro pollination, ovary culture, ovule culture and embryo rescue for overcoming Incongruity barriers in interspecific *Lilium* crosses. Plant Science 74: 115-126.

Van Tuyl, J.M. and P. Arens. 2011. *Lilium*: breeding history of the modern cultivar assortment. Acta Hort 900: 223-230.

Van Tuyl, J.M. and M.G.M. Creij. 2006. *Tulipa gesneriana* and *T.* hybrids. In: Flower breeding and genetics: Issues, challenges and opportunities for the 21st century. Anderson NO (Eds.) Chapter 23. Springer Verlag. Pp. 623-641.

Van Tuyl, J.M., J. N. de Vries, R.J. Bino and T.A.M. Kwakkenbos. 1989. Identification of 2npollen producing interspecific hybrids of *Lilium* using flow cytometry. Cytologia 54: 737-745.

Van Tuyl, J.M., B. Meijer and M.P. Van Diën. 1992. The use of oryzalin as an alternative for colchicine in *in-vitro* chromosome doubling of *Lilium* and *Nerine*. Acta Hort 352: 625-630.

Van Tuyl, J.M., B. Meijer and M.P. Van Diën. 1992. The use of oryzalin as an alternative for colchicine in *in-vitro* chromosome doubling of *Lilium*. The Lily Yearbook of North American Lily Society 43: 19-22.

Van Tuyl, J.M. 1997. Interspecific hybridization of flower bulbs: A review. Acta Hort 430: 465-476.

Van Tuyl, J.M., K.B. Kim and M.S. Ramanna. 2002. Interspecific hybridization and introgression. In: Breeding for ornamentals: Classical and molecular approaches, ed. A. Vainstein, 85-103. Dordrecht/Boston: Kluwer Academic Publishers.

Wu, H., M.S. Ramanna, P. Arens and J.M. Van Tuyl. 2011. Genome constitution of *Narcissus* variety, 'Tête-à-Tête', analysed through GISH and NBS profiling. Euphytica 181: 285-292.

Wagner, W.W. 1999. The French begonia society. The Begonian 66: 172-175.

Xie, S., N. Khan, M.S. Ramanna, L. Niu, A. Marasek-Ciolakowska, P. Arens and J.M. Van Tuyl. 2010a. An assessment of chromosomal rearrangements in neopolyploids of *Lilium* hybrids. Genome 53: 439-446.

Xie, S., M.S. Ramanna and J.M. Van Tuyl. 2010b. Simultaneous identification of three different genomes in *Lilium* hybrids through multicolour GISH. Acta Hortic. 855: 299-304.

Xie, S., A. Marasek-Ciolakowska, M.S. Ramanna, P. Arens, R.G.F. Visser and J.M. Van Tuyl. 2014. Characterization of B chromosomes in *Lilium* hybrids through GISH and FISH. Plant Syst Evol DOI 10.1007/s00606-014-1004-1.

Yabuya, T. 1985. Amphidiploids between *Iris laevigata* Fisch. and *I. ensata* Thunb. induced through *in vitro* culture of embryos treated with colchicine. Jap. J. Breed. 35: 136-144.

Ye, H.G., F.G. Wang, Y.S. Ye and C.I. Peng. 2004. *Begonia coptidifolia* (Begoniaceae), a new species from China. Bot Bull Acad Sin 45: 259-266.

Zeilinga, A.E. 1962. Cytological investigation of hybrid varieties of *Begonia semperflorens*. Euphytica 11: 126-136.

Zeilinga, A.E. and H.P. Schouten. 1968a. Polyploidy in garden tulips. I. Survey of *Tulipa* varieties for polyploids. Euphytica 17: 252-264.

Zeilinga, A.E. and H.P. Schouten. 1968b. Polyploidy in garden tulips. II. The production of tetraploids. Euphytica 17: 303-310.

Zhou, S., M.S. Ramanna, R.G.F. Visser and J.M. Van Tuyl. 2008a. Genome composition of triploid lily cultivars derived from sexual polyploidization of Longiflorum × Asiatic hybrids (*Lilium*). Euphytica 160: 207-215.

Zhou, S., K.B. Lim, R. Barba-Gonzalez, M.S. Ramanna and J.M. Van Tuyl. 2008b. Interspecific hybridization in lily (*Lilium*): Interploidy crosses involving interspecific F_1 hybrids and their progenies. In Floriculture, ornamental and plant biotechnology vol. V, ed. J.A. Teixeira da Silva, 152-156. Isleworth, UK: Global Science Books.

Zlesak, D.C., C.A. Thill and N.O. Anderson. 2005. Trifluralin-mediated polyploidization of *Rosa chinensis minima* (Sims) Voss seedlings. Euphytica 141: 281-290.

Chapter 8

Polyploidy in Maize: The Impact of Homozygosity and Hybridity on Phenotype

James A. Birchler[1,*] and Jacob D. Washburn[2]

ABSTRACT

In a maize polyploidy series, there is a decline in stature with increasing ploidy when the genome is homozygous. In contrast, when heterozygosity is maximized, the stature of plants increases with higher ploidy. Double cross tetraploids are more vigorous than double cross diploids. However, this increased vigor is difficult to capture for agricultural practices (grain yield) because the quadrivalent pairing of chromosomes and the phenomenon of double reduction lead to some sterility and variability in the progeny. These barriers to tetraploid corn production could in theory be overcome by introducing mechanisms of apomixis and/or chromosome level diploidization, but all attempts to do so in vivo have so far met with very limited success.

Introduction

Polyploidy is common in the plant kingdom and has been capitalized upon for plant cultivation. Indeed, cycles of polyploidization and gene copy reduction (fractionation) are very common in plants (Freeling 2009, Jiao et al. 2011) indicating that the actual level of polyploidy depends upon when one starts counting. The outcome of the fractionation process is not random with regard to gene function. Those genes that are involved in multisubunit complexes, which of note include transcriptional regulatory and signal transduction components, are retained longer following the whole genome duplication (WGD) than other general classes of genes (Freeling 2009). A loss of a member of a complex relative to its interactors is postulated to have a negative fitness effect similar to aneuploidy and therefore be selected against

[1,2] Division of Biological Sciences, 311 Tucker Hall, University of Missouri, Columbia, MO 65211,
* Corresponding author: E-mail; BirchlerJ@Missouri.edu

(Birchler et al. 2005, Birchler and Veitia 2012). This dosage balanced retention seems to decay eventually, but potentially lengthens the time for duplicate genes to gain new or split functions which seals the gene's retention by that mechanism thereafter (Birchler and Veitia 2012). These considerations need to be taken into account when contemplating the effects of ploidy in any one species. This is because a copy number change in chromosome complement that involves the whole genome might reflect different states of gene copy number, and the content of that number, in different taxa and species.

With these considerations in mind, the perception of polyploids has classically been that they are more robust than diploids or other lower ploidy counterparts. They are typically associated with larger flowers and seeds on the organismal level and with larger cells on the cytological level (Blakeslee 1934). The situation is actually more complicated than it appears on the surface and depends to some degree on the level of diversity of genomes that are present with increasing ploidy (Birchler et al. 2010, Birchler 2013). Indeed, this is the situation that occurs most commonly in natural populations with regard to allopolyploids, which are formed from the combination of different genomes of lower ploidy. This suggests an intersection of hybrid vigor, or heterosis, and ploidy with regard to plant stature (Chase 1980, Riddle et al. 2006, Riddle and Birchler 2008, Birchler et al. 2010, Birchler 2013).

Here, we will focus on the reaction of maize (*Zea mays* L.) to changes in ploidy and hybridity. Maize contains many duplicate genes because it experienced a whole genome duplication (WGD) about 5 million years ago (Schnable et al. 2011) with other WGDs preceding the emergence of the grass family. This fact should be kept in mind when considering how maize responds to ploidy change.

Monoploids of maize are less robust in phenotype than diploids but are generally healthy (Chase 1969, Chase 1980, Auger et al. 2004, Birchler and Veitia 2012). Because there is only a single set of chromosomes present, there are no homologues for pairing and thus there is very high sterility. There is, however, a certain level of spontaneous chromosome doubling in haploid maize that leads to diploid sectors in ears and tassels that provide viable haploid gametes. Consequently, many haploid individuals can be self-pollinated to produce totally homozygous progeny (Chase 1969). Various techniques have also been applied to cause chromosome doubling in developmental lineages, leading to flowers with an increased ability to self-pollinate (Kato and Geiger 2002). This approach is now used extensively in maize breeding to develop new homozygous lines in very few generations in contrast to producing inbred lines via inbreeding over many more generations.

Triploid maize derived from inbred lines appears to be similar to the progenitor diploid, or in some cases slightly less vigorous in stature (Auger et al. 2005, Yao et al. 2011, Yao et al. 2013). The fertility is low due to the distribution of chromosomes when three copies of each chromosome are present in meiosis. Two of the three centromeres segregate from each other and the third independently assorts although there are pairing switches along the chromosome arms. A range of kernel sizes are produced when pollen from a normal diploid is applied to the silks of triploid plants, due to variable numbers of chromosomes present and the impact that this situation has on endosperm development.

Tetraploid maize that is directly derived from an inbred line is routinely reduced in stature relative to the progenitor diploid (Randolph 1942, Levings et al. 1967, Kato and Birchler 2006). At first, this observation seemed in contrast to the conventional wisdom and previous observations on maize tetraploids that were produced by the meiotic mutation *elongate* (Rhoades and Dempsey 1966). However, in the latter case, it is likely that some degree of heterozygosity remains based on the observations on tetraploid hybridity that will be described below. The *elongate* method could not eliminate residual heterozygosity if this heterozygosity provided a selective advantage for maintenance of the stock, which appears likely. Tetraploid maize derived directly from inbred lines has not been known to be perpetuated in any of a variety of field conditions. Temperate greenhouse conditions are needed. As with most characteristics, there is considerable variation in maize lines for the response to ploidy (Riddle et al. 2006). In general, the cell, seed and pollen sizes in tetraploids are larger than those of their progenitor despite the fact that the overall plant size is reduced (Yao et al. 2011).

Pentaploid maize was described by Rhoades and Dempsey (1966), and in this case there was an observed decline in stature. As noted above, this material likely contains some residual heterozygosity but the negative impact of increased ploidy nevertheless becomes apparent at this ploidy level.

Hexaploid maize was described by Rhoades and Dempsey (1966) as well. In this case, there was a further decline in both vigor and female fertility. Yao et al. (2011) produced hexaploids that were direct descendants from a completely homozygous line and found a strongly defective stature, but cell size and pollen were still larger than the comparable tetraploids and diploids. The trend that strictly homozygous material is more severely defective appears to hold at this level of ploidy also.

Heptaploid maize is the highest ploidy level that was achieved using the *elongate* technique, and had the smallest stature of the ploidy series examined (Rhoades and Dempsey 1966). No fertility was documented.

Octoploid maize was generated by Randolph (1942) who described the plants as highly defective. Yao et al. (2011) found a single octoploid that was a depauperate seedling that died.

The trend from these studies is that with increasing ploidy of strictly homozygous materials, there is a decline in plant stature. A careful examination of the literature indicates that as a general rule, this trend is common across the plant kingdom (Abel and Becker 2007, Stupar et al. 2007, Miller et al. 2012).

Hybridity and Ploidy

In contrast, hybrid ploidy series, as far as they have been examined, behave in the opposite way (Riddle et al. 2010). Single cross hybrids of tetraploid maize are quite vigorous and can be cultivated in the field without difficulty. The degree of hybrid vigor is similar to that of the matched diploid hybrids (Riddle and Birchler 2008) (Fig 1).

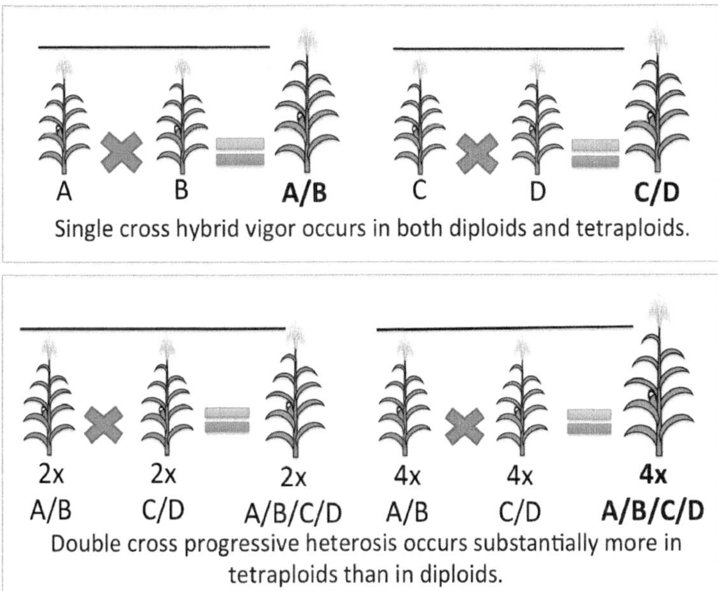

Figure 1. Heterosis can be seen in the progeny of both diploid and tetraploid maize single crosses. Double cross heterosis can also be seen in both diploid and tetraploid maize but is significantly stronger in tetraploids than diploids.

However, double cross hybrids that are produced by crossing together different hybrids exhibit an even greater boost of vigor that is significantly more in tetraploids than in diploids (Riddle et al. 2010, Washburn and Birchler 2014) (Fig 1). This phenomenon is referred to as "progressive heterosis". It was most thoroughly documented in alfalfa (Groose et al. 1989) and subsequently in maize (Alexander and Sonnemaker 1961, Sockness and Dudley 1989b, a, Riddle et al. 2010) and potato (Mok and Peloquin 1975).

Levings et al. (1967) analyzed diploid and tetraploid maize hybrids. They documented progressive heterosis as did Chase (1980). The tetraploids used likely contained some residual heterozygosity but the same phenomenon was recognized. Chase (1980) was able to further dissect the contributions of diversity of genotype in finding that the vigor relationship in stylized letter symbols was AAAA < AAAB < AABB < AABC < ABCD with the caveat that such genotypes cannot be perfectly established because of recombination among genomes in a tetraploid. It is of note that a related relationship occurs in wheat and its relatives. As a general rule the vigor of wheat species increases from diploid to allotetraploid to allohexaploid to allooctoploid in triticale. More interesting still, crosses of hexaploid wheat with tall wheat grass that is decaploid produced a plant that has 8 different genomes and spectacular vigor (Li et al. 2008). The generalized principle that can be gained from these observations seems to be that maximizing diversity of genomes with increasing ploidy will maximize the heterotic response.

Why Not Double Cross Tetraploid Maize in the Field?

Given this conclusion, one might ask why tetraploid maize is not adapted for agricultural production if even greater productivity could be achieved? There are several reasons why. First among them is that maize tetraploids behave as autotetraploids during meiosis, with associations of the four copies of each chromosome (Randolph 1935). The consequence of these associations is that 3: 1 or 4: 0 segregations can occasionally occur instead of the regular 2: 2 (Fig. 2).

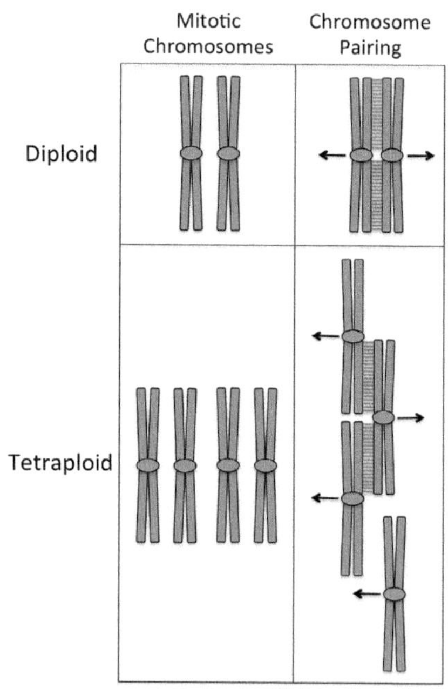

Figure 2. An example of one chromosome set (e.g., Chromosome 1 in maize) in both a diploid and autotetraploid state. One of many possible chromosome pairing scenarios is shown. Black arrows indicate the direction each chromosome will proceed after pairing. This particular pairing scenario would result in unbalanced gametes in the tetraploid and possible sterility.

This behavior leads to a low level of defective kernels in tetraploids resulting from the aneuploidy produced by this series of events (Birchler 2014). Thus, the fertility of hybrid tetraploid maize is somewhat reduced.

Further, the phenomenon of double reduction in tetraploids can produce homozygosity for chromosomal regions in gametes from a hybrid genotype (Welch 1962, Levings and Alexander 1966, Bingham et al. 1968). In other words, homologues from different parents with different alleles can cross over, and then because there are diploid gametes from a tetraploid, the same alleles can be present in the gametes rather than the different alleles segregating from each other. Thus, homogeneity in double cross maize tetraploids would be difficult to establish because individual progeny would not necessarily have the same genotype.

An attempt to overcome these issues was pursued by Doyle (1979; 1986). He selected for chromosomal behavior that might mimic an allotetraploid, such that the like chromosomes would preferentially pair with each other rather than undergo the random association that is typical in autotetraploids, by using structurally rearranged chromosomes from different parents (Doyle 1979, 1986). Limited success was achieved. If such a "homoeologous pairing" could be accomplished, then the issues of reduced fertility and double reduction might be overcome by such faithful chromosome pairing. With the ability to generate uniform genotypes in the field, optimization for field habit and harvest efficiency would still be needed.

Inbreeding Depression in Diploid and Tetraploid Maize

The converse of hybrid vigor is inbreeding depression. In an inbreeding scheme, the predictions for which alleles are made homozygous in diploids and tetraploids are very different from each other. In a diploid heterozygous for any one gene, a quarter of the zygotes from self-pollination will be homozygous for each of the two alleles present. However, in a tetraploid duplex hybrid of comparable genotype but with two alleles each (i.e. AABB rather than AB), self pollination will only generate 1/36 progeny homozygous for each of the two types of alleles, not considering the phenomenon of double reduction described above (Bennett 1976). Curiously, as first documented in alfalfa (Busbice and Wilsie 1966) and repeated in maize (Rice and Dudley 1974, Sockness and Dudley 1989b), the inbreeding depression in matched diploids and tetraploids is very similar. With regards to the simple assumption that making recessive detrimental mutations homozygous is the basis of inbreeding depression, the empirical results suggest that this is not the case, at least not in a major way.

What then does account for the similar rates of inbreeding depression in diploids and tetraploids? Busbice and Wilsie (1966), working with alfalfa, suggested that the dosage of alleles might contribute to these similar rates of inbreeding in diploids and tetraploids because the shifting of allele dosage is more similar at the two ploidy levels than the progression of homozygosity. An impact of allelic dosage in this case seems consistent with the results from different triploid hybrids noted below in which the dosage of different genomes produced a different heterotic response (Yao et al. 2013), and how the dosage of diverse alleles affects heterosis in tetraploids (Chase 1980). As noted above, genes contributing to multi-subunit complexes, including those involved in gene regulatory and signal transduction mechanisms, are the major classes that exhibit dosage effects. Thus, if dosage of alleles plays a role in inbreeding depression, it is likely that these classes of genes are preferentially involved, as is likely for any quantitative trait.

What does the Behavior of Hybrid Polyploids Contribute to an Understanding of Heterosis?

A popular idea to explain heterosis is that different slightly deleterious recessive alleles are present in the different parents (Charlesworth and Willis 2009). Then in the hybrid, different alleles from opposite parents will be complemented by a superior copy from the other parent. If a recessive allele has a detrimental effect on plant morphology, then that effect will be relieved. While this type of effect would occur, the behavior of heterosis in polyploids suggests that complementation is not the major factor. It seems likely that most alleles with such deleterious effects have been purged from most lines (East 1936) and that other genetic and physiological factors come into play.

Insight into heterosis was also gained from triploid hybrids (Yao et al. 2013). Reciprocal diploid hybrids are basically identical under most circumstances. In contrast, reciprocal triploid hybrids are different from each other and from homozygous triploids. Using stylized letter symbols, in diploids AA < AB = BA > BB, whereas in triploids AAA < AAB ≠ BBA > BBB (Yao et al. 2013). This finding indicates that the dosage of genomes produces a different heterotic response, which places constraints on genetic models of heterosis (Birchler 2013). The differential heterosis that occurs with reciprocal triploid hybrids is inconsistent with the concept of simple complementation as the major basis of heterosis (Yao et al. 2013).

The allele dosage concept is also consistent with the progression of inbreeding depression in autotetraploids as well. In addition, the complementation concept places unrealistic constraints on the genetic load of any line to explain progressive heterosis, i.e. the parents of single cross hybrids would have to differ in many homozygous alleles but share other detrimental alleles relative to the other single cross hybrid parents, which in turn must differ from each other and from the first pair in their complement of homozygous detrimental alleles. In a more general sense, heterosis tends to be greater in magnitude with increasing phylogenetic distance, as illustrated by the spectacular growth of many wide hybrids (Gravatt 1914, Li et al. 2008). It makes no sense that there would be an accumulation of slightly deleterious alleles in the *homozygous* state with advancing evolution, which would be required by the strict complementation model!

Dosage Component of Polyploidy Heterosis

The dosage component to heterosis is consistent with the broad experience that quantitative traits are typically multigenic and additive between parents of extreme phenotypes. Most mutations are completely recessive but genes identified as modulating quantitative traits tend to exhibit additive, i.e. semi-dominant, effects (Tanksley 1993, Lee et al. 1996). Cloning of example QTL and transformation of the responsible genes back into plants demonstrated a dosage effect on the phenotype (Liu et al. 2003). As noted in the Introduction, evolutionary genomics suggests that the relative dosages of genes involved with multi-subunit complexes have been subjected to selection. Many regulatory steps in gene expression involve such complexes, so this stoichiometric action likely impacts quantitative traits (Birchler and Veitia 2012). The dosage component might simply reflect the multigenic control of the process operating to produce heterosis. The relationship of the control of heterosis, quantitative traits and their intersection with a dosage component is unknown and an interesting avenue for research exploration.

Is there a Future for Polyploidy in Maize Cultivation?

A futuristic view might contemplate whether the progressive heterosis found in polyploids could be used for maize production. As noted above, polyploid maize has several disadvantages. However, there has been some research on developing apomictic maize that would bypass meiosis (Garcia-Aguilar et al. 2010, Singh et al. 2011), the root of the tetraploid liabilities. If apomixis could be combined with progressive heterosis, then this superior heterotic effect could be captured and perpetuated uniformly to progeny, thus overcoming the meiotic and non-uniformity issues of tetraploids described above. If a dominant apomictic property could be conferred into tetraploids, then pollen from such individuals potentially could be used onto sexual females to construct a tetraploid with diverse genomes in subsequent crosses. With the desired phenotype, the material could be perpetuated using the asexual kernel production. Alternatively, if maize tetraploids could be selected to act as allopolyploids, as noted above, some of the barriers to using tetraploids in field production might be overcome.

References

Abel, S. and H. C. Becker. 2007. The effect of autopolyploidy on biomass production in homozygous lines of *Brassica rapa* and *Brassica oleracea*. *Plant Breeding* 126: 642-643.

Alexander, D.E. and E.H. Sonnemaker. 1961. Inbreeding depression in autotetraploid maize. *Maize Genet. Coop. News Lett.* 35: 45.

Auger, D.L., T.S. Ream and J.A. Birchler. 2004. A test for a metastable epigenetic component of heterosis using haploid induction in maize. *Theor. Appl. Genet.* 108: 1017-1023.

Auger, D.L., A.D. Gray, T.S. Ream, A. Kato, E.H. Coe and J.A. Birchler. 2005. Nonadditive gene expression in diploid and triploid hybrids of maize. *Genetics* 169: 389-397.

Bennett, J.H. 1976. Expectations for inbreeding depression on self-fertilization of tetraploids. *Biometrics* 32: 449-452.

Bingham, E.T., C.R. Burnham and C.E. Gates. 1968. Double and single backcross linkage estimates in autotetraploid maize. *Genetics* 59: 399-410.

Birchler, J.A. 2013. Genetic rules of heterosis in plants, Polyploid and hybrid genomics 10.1002/9781118552872.ch19, 313-321. John Wiley & Sons, Inc.

Birchler, J.A. 2014. Interploidy hybridization barrier of endosperm as a dosage interaction. *Front. Plant Sci.* 5.

Birchler, J.A. and R.A. Veitia. 2012. Gene balance hypothesis: Connecting issues of dosage sensitivity across biological disciplines. *Proc Natl Acad Sci USA* 109: 14746-14753.

Birchler, J.A., N.C. Riddle, D.L. Auger and R.A. Veitia. 2005. Dosage balance in gene regulation: Biological implications. *Trends Genet.* 21: 219-226.

Birchler, J.A., H. Yao, S. Chudalayandi, D. Vaiman and R.A. Veitia. 2010. Heterosis. *Plant Cell* 22: 2105-2112.

Blakeslee, A.F. 1934. New jimson weeds from old chromosomes. *J. Hered.* 25: 81-108.

Busbice, T. and C.P. Wilsie. 1966. Inbreeding depression and heterosis in autotetraploids with application to *Medicago sativa* L. *Euphytica* 15: 52-67.

Charlesworth, D. and J.H. Willis. 2009. The genetics of inbreeding depression. *Nat. Rev. Genet.* 10: 783-796.

Chase, S.S. 1969. Monoploids and monoploid-derivatives of maize (*Zea mays* L.). *Bot. Rev.* 35: 117-168.

Chase, S.S. 1980. Studies of monoploid, diploid and tetraploids of maize in relation to heterosis and inbreeding depression. *Proc Arg Soc Gen.*

Doyle, G.G. 1979. The allotetraploidization of maize. 2. The theoretical basis – the cytogenetics of segmental allotetraploids. *Theor. Appl. Genet.* 54: 161-168.

Doyle, G.G. 1986. The allotetraploidization of maize. 4. Cytological and genetic evidence indicative of substantial progress. *Theor. Appl. Genet.* 71: 585-594.

East, E.M. 1936. Heterosis. *Genetics* 21: 375-397.

Freeling, M. 2009. Bias in plant gene content following different sorts of duplication: Tandem, whole-genome, segmental, or by transposition. *Annu. Rev. Plant Biol.* 60: 433-453.

Garcia-Aguilar, M., C. Michaud, O. Leblanc and D. Grimanelli. 2010. Inactivation of a DNA methylation pathway in maize reproductive organs results in apomixis-like phenotypes. *The Plant Cell* 22: 3249-3267.

Gravatt, F. 1914. A radish-cabbage hybrid: Cross between two genera shows extraordinary vigor but absolute sterility—pollen irregular both in size and in shape—two extra stamens present in some of the flowers. *J. Hered.* 5: 269-272.

Groose, R.W., L.E. Talbert, W.P. Kojis and E.T. Bingham. 1989. Progressive heterosis in autotetraploid alfalfa: Studies using two types of inbreds. *Crop Sci.* 29: 1173-1177.

Jiao, Y., N.J. Wickett, S. Ayyampalayam, A.S. Chanderbali, L. Landherr, P.E. Ralph, L.P. Tomsho, Y. Hu, H. Liang, P.S. Soltis, D.E. Soltis, S.W. Clifton, S.E. Schlarbaum, S.C. Schuster, H. Ma, J. Leebens-Mack and C.W. Depamphilis. 2011. Ancestral polyploidy in seed plants and angiosperms. *Nature* 473: 97-100.

Kato, A. and H.H. Geiger. 2002. Chromosome doubling of haploid maize seedlings using nitrous oxide gas at the flower primordial stage. *Plant Breeding* 121: 370-377.

Kato, A. and J.A. Birchler. 2006. Induction of tetraploid derivatives of maize inbred lines by nitrous oxide gas treatment. *J. Hered.* 97: 39-44.

Lee, E.A., L.L. Darrah and E.H. Coe. 1996. Dosage effects on morphological and quantitative traits in maize aneuploids. *Genome* 39: 898-908.

Levings, C.S. and D.E. Alexander. 1966. Double reduction in autotetraploid maize. *Genetics* 54: 1297-1305.

Levings, C.S., J.W. Dudley and D.E. Alexander. 1967. Inbreeding and crossing in autotetraploid maize. *Crop Sci.* 7: 72-73.

Li, Z., B. Li and Y. Tong. 2008. The contribution of distant hybridization with decaploid *Agropyron elongatum* to wheat improvement in China. *JGG* 35: 451-456.

Liu, J., B. Cong and S.D. Tanksley. 2003. Generation and analysis of an artificial gene dosage series in tomato to study the mechanisms by which the cloned quantitative trait locus *fw2.2* controls fruit size. *Plant Physiol.* 132: 292-299.

Miller, M., C. Zhang and Z.J. Chen. 2012. Ploidy and hybridity effects on growth vigor and gene expression in *Arabidopsis thaliana* hybrids and their parents. *G3* 2: 505-513.
Mok, D.W.S. and S.J. Peloquin. 1975. Breeding value of 2*n* pollen (diplandroids) in tetraploid *x* diploid crosses in potatoes. *Theor. Appl. Genet.* 46: 307-314.
Randolph, L.F. 1935. Cytogenetics of tetraploid maize. *J. Ag. Res.* 50: 591-605.
Randolph, L.F. 1942. The influence of heterozygosis on fertility and vigor in autotetraploid maize. *Genetics* 27: 163.
Rhoades, M.M. and E. Dempsey. 1966. Induction of chromosome doubling at meiosis by the elongate gene in maize. *Genetics* 54: 505-522.
Rice, J.S. and J.W. Dudley. 1974. Gene effects responsible for inbreeding depression in autotetraploid maize. *Crop Sci.* 14: 390-393.
Riddle, N. and J. Birchler. 2008. Comparative analysis of inbred and hybrid maize at the diploid and tetraploid levels. *Theor. Appl. Genet.* 116: 563.
Riddle, N., A. Kato and J. Birchler. 2006. Genetic variation for the response to ploidy change in *Zea mays* L. *Theor. Appl. Genet.* 114: 101.
Riddle, N.C., H. Jiang, L. An, R.W. Doerge and J.A. Birchler. 2010. Gene expression analysis at the intersection of ploidy and hybridity in maize. *Theor. Appl. Genet.* 120: 341-353.
Schnable, J.C., N.M. Springer and M. Freeling. 2011. Differentiation of the maize subgenomes by genome dominance and both ancient and ongoing gene loss. *Proc Natl Acad Sci U S A* 108: 4069-4074.
Singh, M., S. Goel, R.B. Meeley, C. Dantec, H. Parrinello, C. Michaud, O. Leblanc and D. Grimanelli. 2011. Production of viable gametes without meiosis in maize deficient for an argonaute protein. *The Plant Cell* 23: 443-458.
Sockness, B.A. and J.W. Dudley. 1989a. Morphology and yield of isogenic diploid and tetraploid maize inbreds and hybrids. *Crop Sci.* 29: 1029-1032.
Sockness, B.A. and J.W. Dudley. 1989b. Performance of single and double cross autotetraploid maize hybrids with different levels of inbreeding. *Crop Sci.* 29: 875-879.
Stupar, R.M., P.B. Bhaskar, B.S. Yandell, W.A. Rensink, A.L. Hart, S. Ouyang, R.E. Veilleux, J.S. Busse, R.J. Erhardt, C.R. Buell and J. Jiang. 2007. Phenotypic and transcriptomic changes associated with potato autopolyploidization. *Genetics* 176: 2055-2067.
Tanksley, S.D. 1993. Mapping polygenes. *Annu. Rev. Genet.* 27: 205-233.
Washburn, J.D. and J.A. Birchler. 2014. Polyploids as a "model system" for the study of heterosis. *Plant Reprod.* 27: 1-5.
Welch, J.E. 1962. Linkage in autotetraploid maize. *Genetics* 47: 367-396.
Yao, H., A. Kato, B. Mooney and J.A. Birchler. 2011. Phenotypic and gene expression analyses of a ploidy series of maize inbred oh43. *Plant Mol. Biol.* 75: 237-251.
Yao, H., A. Dogra Gray, D.L. Auger and J.A. Birchler. 2013. Genomic dosage effects on heterosis in triploid maize. *Proc Natl Acad Sci USA* 110: 2665-2669.

Chapter 9

Broadening the Genetic Basis for Crop Improvement: Interspecific Hybridization Within and Between Ploidy Levels in *Helianthus*

Michael Kantar[1,2,*], Sariel Hübner[1] and Loren H. Rieseberg[1,3]

ABSTRACT

The genus *Helianthus* is both economically important and genetically diverse. It contains two important crop species *H. annuus* and *H. tuberosus* in addition to being a model for evolutionary studies. The large number of species within the genus and the ability to hybridize makes the use of crop wild relatives as sources of novel phenotypes particularly promising as a way to introduce novel variation. Additionally, the promiscuity of the genus allows for an understanding of the genome dynamics within and between ploidy levels at many different evolutionary distances. *Helianthus* is an excellent system to study how hybridization can be used to explore the utilization of wild germplasm in crop improvement and how it led to the creation of hybrid sunflower industry. New mating designs and technology combined with the need to develop crops resilient to changing environments will increase the value of wild germplasm.

Introduction

Plant Genetic Resources: Crop Wild Relatives

Crop plants often are less resistant to biotic and abiotic stresses than their wild relatives. The loss of resistance accompanying crop domestication and improvement is hypothesized to be a by-product of selection for yield under ideal conditions. This hypothesis is based on reports of trade-offs

[1] Biodiversity Research Centre and Department of Botany, University of British Columbia, 3529-6270 University Boulevard, Vancouver, British Columbia V6T 1Z4, Canada
[2] Department of Agronomy and Plant Genetics, University of Minnesota 411 Borlaug Hall, 1991 Upper Buford Circle, St. Paul, MN 55108, USA
[3] Department of Biology, Indiana University, Bloomington, IN 47405, USA
* Corresponding author: E-mail; kant0063@umn.edu

between plant productivity and stress resistance (e.g. Mayrose et al. 2011). In addition, population bottlenecks during domestication and improvement have likely lead to the stochastic loss of resistance alleles in crop germplasm (Tanksley and McCouch 1997). The loss of stress resistance in crops, coupled with reduced diversity of crop gene pools, has become especially worrying recently as we attempt to increase crop productivity in the face of climate change, rapid population growth, and heightened competition for land and water (McCouch et al. 2013; Dempewolf et al. 2014; Gentzbittel et al. 2015).

One approach to minimize yield losses caused by environmental stress is to breed cultivars that combine high yield with resistance to biotic and abiotic stress. However, such breeding efforts require access to resistant germplasm, which may not exist in the cultivated gene pool. On the other hand, such resistance often resides in the wild relatives of crop plants, which often thrive under stressful conditions (Ricklefs and Jenkins, 2011; Thormann et al. 2012). While the genes and mutations underlying stress resistance can be obtained through genetic engineering, it is typically cheaper and less technologically challenging to obtain them through sexual hybridization and introgression. For hybridization to be useful to humans, crops must be cross-compatible with their wild relatives, hybrids must have non-zero fitness, and agronomically valuable alleles must be present in the wild background (Burke and Arnold, 2001; Rieseberg, 1997; Arias and Rieseberg, 1995). Indeed, many different crop wild relative species meet these criteria, and both intra- and interspecific hybridization has been employed to transfer useful traits into many different crop plants (reviewed in Hajjar and Hodgkin, 2007). In this chapter, we discuss how hybridization has been (or could be) employed to aid crop improvement in the genus *Helianthus*, which contains two crops: the diploid sunflower (*Helianthus annuus* L.) and the hexaploid *Helianthus tuberosus* L. ($2n = 6x = 102$).

Definitions

The term hybridization can be restricted to offspring formed by matings between species, or defined more broadly as the offspring of individuals from genetically differentiated populations. We prefer the second, broader definition for two reasons. First, it avoids issues with species definitions. Second, it is consistent with the plant breeding literature, in which hybridization is typically employed to describe crosses within the species, while wide hybridization is generally used to refer to crosses between different species. Likewise, introgression can be narrowly defined as the transfer of genes or traits between species via backcrossing, or more broadly as the sexual transfer of genetic material between genetically distinguishable populations. Again, we prefer the broader definition since it provides greater flexibility in usage.

Traditionally wild plant species related to cultivated species, i.e. crop wild relatives (CWR), have been placed in groups based on their crossing relationship with the crop. The primary germplasm have no crossing barriers with the crop producing fully fertile progeny, the secondary germplasm produce hybrids with some meiotic abnormalities (due to chromosome translocations etc. that influence F_1 meiosis but not viability or hybridization), the tertiary germplasm requires special techniques (e.g. embryo rescue, protoplast fusion) to produce hybrids, and within the quaternary germplasm hybrids cannot be produced via sexual or somatic means, although genetic introgressions can be achieved using recombinant DNA technology recombinant DNA technology (Harlan and De Wet, 1971; Harlan, 1976; Gepts 2000).

Helianthus

Helianthus is native to North America and contains 52 species (67 total taxa) (Marek and Seiler, 2011; Kane et al. 2013), occupying diverse ecological niches across much of North America including deserts, marshes and open plains. *Helianthus* is a promiscuous genus with many species readily hybridizing within and between ploidy levels. Hybridization and introgression appear to have facilitated colonization of diverse environments (Whitney et al. 2006, 2010; Thompson et al. 1981; Rogers et al. 1982). The two *Helianthus* crops originated from different sections within the genus. Sunflower arose within the annual section *Helianthus*, centered in central and western North America, while *H. tuberosus* is an autoallohexaploid from the perennial section *Divaracatus*, centered in northeastern North America.

Sunflower Domestication and Improvement

The process of domestication involves both the elimination of unwanted traits, and the development of traits that facilitate cultivation and improve yield. These two overlapping aspects of domestication transform plants from their natural forms to new and distinguishable types. Improvement refers to the phenotypic changes wrought by modern breeding, which may overlap with those sought by early farmers. Modern breeders typically target yield and quality traits, traits that confer resistance to biotic or abiotic stress, and traits that confer adaptation to local environments or that optimize crops for particular uses.

While it is clear that artificial selection (both conscious and unconscious) drives phenotypic evolution during domestication and improvement, the role of controlled crosses (both within and between species) in domestication is less clear. However, the use of controlled crosses is a critical component of modern breeding programs. Sunflower (*H. annuus*) offers an especially compelling example of the value of wide hybridization in crop improvement.

Sunflower ($2n = 2x = 34$) was domesticated in eastern North America approximately 4000 years ago (Harter et al. 2004, Blackman et al. 2011) in present day states of Arkansas, Kentucky, Illinois and Tennessee (Smith, 2006; Smith 2013). 'Wild' traits that were eliminated during sunflower domestication include branching, seed shattering, self-incompatibility and extended dormancy. Traits bred by early farmers to facilitate harvesting and enhance yield include increased seed size, increased oil content and adjustment of flowering time (Harlan et al. 1973; Burke et al. 2002a; Harter et al. 2004; Seiler and Jan, 2010; Blackman et al. 2011).

At the time of European contact, sunflower played a significant role in North American agriculture and was associated with many different Native American tribal nations (Sturtevant 1885; Jenks 1916; Jones et al. 1933; Thone 1936; Heiser, 1951; Heiser 1955; Kaplan 1963; Wasley 1962; Fritz 1990). Sunflower is often present in North American folklore (Heiser, 1951), and was of particular importance to the tribal nations of the American Southwest (Navajo, Apache, Pueblo and Hopi), figuring prominently in their folklore (Wallis 1936; Yarnell 1965; Minnis 1989). There were many regional and use-specific landraces (Nabhan and Richhardt, 1983); for example, the desert dwelling Hopi tribe bred sunflowers that produced specific dyes (Heiser 1951; Willis et al. 2010). Additionally, sunflower was part of traditional polycultures in the American Southeast (Scarry and Scarry 2005), with its complex agricultural systems continuing in modern times, particularly in organic farming systems (Jones and Gillett 2005).

Sunflower was introduced to Europe in the sixteenth century as an ornamental plant (Heiser 1955) and was further improved for oil content during the nineteenth century in Russia. Sunflower robustness in adverse environments was recognized, which enhanced its cultivation in wide range of environments around the globe (Hanna 1924, Shantz 1940). However, as alluded to above, domestication and improvement is typically accompanied bya reduction in genetic variation due to intensive inbreeding and selection, leading to increased vulnerability of cultivated varieties to environmental stresses, diseases and pests (Rieseberg et al. 1995; Harter et al. 2004). In sunflower, the cultivated genepool is estimated to include approximately 67% of the diversity present in wild populations of its progenitor, common sunflower (also *H. annuus*) (Lui and Burke 2006; Kolkman et al. 2007; Mandel et al. 2011). The relatively high proportion of genetic variation remaining in the cultivated gene pool is due in part to purposeful introgressions from the wild (Seiler, 1991a; Seiler, 1991b; Seiler, 1991c; Seiler and Marek, 2011; Baute et al. 2015).

Sunflower is currently cultivated on ~26 million hectares worldwide (FAO Stat), ranking second among hybrid crops in area harvested (Singh et al. 2007) and 13th among all crops. Commercial oilseed varieties dominate cultivation (75-90% of production) due to high demand for low trans-fat, high oleic vegetable oil (Berglund 2007). The biodiesel and direct consumption markets are smaller but still of economic value. Sunflower production in 2013 was 44.75 million metric tons worldwide and 0.92 million metric tons in U.S. (FAO stat). Over the last twenty years there has been an increase in sunflower production worldwide (Khoury et al. 2014) due to increases in both production area and productivity per hectare (Berglund 2007).

Domestication and Improvement of Jerusalem Artichoke

The second *Helianthus* crop is the hexaploid Jerusalem artichoke (*H. tuberosus*). *Helianthus tuberosus* displays a domestication syndrome consistent with tuber crops, where tuber number is reduced, individual tuber size increases and there is a more synchronous transition to reproduction. *Helianthus tuberosus* is native to central North America (Kays and Nottingham, 2008; Rogers et al. 1982) and was domesticated in the eastern United States prior to European contact. *Helianthus tuberosus* is an autoallohexaploid whose diploid progenitors are *H. divaricatus* and *H. grosseserratus* (Bock et al. 2014; Kostoff, 1934; Kostoff, 1939; Scibria, 1938). Levels of genetic diversity in the cultivated gene pool relative to that found in its wild progenitor (also *H. tuberosus*) are unknown.

The crop was first introduced to Europe in the early 17th century, where it was an immediate success among the royal court of France, quickly becoming an important food source among the European aristocracy: traveling in quick succession from France to Italy, to the Netherlands and then to England (Kays and Nottingham, 2008). In fact, extensive cultivation guides were published in the mid-18th century (Brookes, 1763). Jerusalem artichoke production continued to increase for ~200 years until it was largely replaced by potato production in the mid-19th century. However, the crop retained an important place in many European culinary traditions, with spikes in cultivation occurring at different times during different periods of history, for example during World War II (Kays and Nottingham, 2008).

The tubers have excellent nutritional properties and are a favorite in gourmet cooking, with some of the first recorded recipes emerging in England in the seventeenth century (Kays and Nottingham, 2008). In addition to food, proposed uses of JerusalemArtichoke include industrial (i.e. for rubber; Seiler et al. 1991), biofuel (Seiler and Campbell, 2006; Rodrigues et al. 2007), medicinal (i.e. inulin from tubers can be used treating diabetes; Kays and Nottingham, 2008), and as a forage crop (Seiler and Campbell, 2004). However, many of the desirable compounds are present at very low concentrations, making production economically unfeasible (Seiler et al. 1991).

When commercially produced, the crop is grown as a winter or summer annual. Production has fluctuated between food and forage production, with recent interest developing in biofuel production. Despite its widespread use there has been little information obtained on the genetics of wild *H. tuberosus*. Additionally, as far as we are aware, there has been little intentional interspecific introgression into domesticated *H. tuberosus*.

Gene Flow Between Wild and Cultivated Populations

Gene flow between wild and cultivated sunflowers occurs frequently as crop fields and wild *Helianthus* are often adjacent to one another (Burke et al. 2002b; Arias and Rieseberg, 1994). Most of the gene flow is with wild populations of *H. annuus*, but limited introgression has been observed with a related species, *H. petiolaris*, as well (Rieseberg and Kim, 1998). Gene flow is mainly from cultivated into wild populations, and cultivated alleles can persist for decades in wild or weedy populations (Snow et al. 1998; Whitton et al. 1997; Cummings et al. 2002). The mixed growth forms resulting from such admixture are sometimes observed in cultivated fields and can hurt production (Lu et al. 2013). With that said, significant care is taken to eliminate wild populations from areas of seed production. Thus, the influence of this unintended crop-wild gene flow on the genomic composition of the cultivar appears to be limited (Baute et al. 2015). Indeed, all domesticated sunflowers form a distinct lineage compared to wild relatives, with wild *H. annuus*making a fairly small direct contribution to domestic genomes (Mercer et al. 2006; Snow et al. 2003; Harter et al. 2004; Wills and Burke, 2006; Mandel et al. 2011; Baute et al. 2015).

How Can We Utilize Hybridization Within *Helianthus*?

Hybridization can be used in several ways. First, as discussed above, it offers a means for accessing agronomically valuable genetic variation, especially disease resistance alleles. Second, hybridization can reveal useful cryptic variation present in wild or cultivated germplasm. For many traits, individuals carry alleles with opposing effects (Tanksley, 1993). The existence of these alleles can be exposed by creating hybrid populations, potentially leading to extreme phenotypes, in a phenomenon referred to as transgressive segregation (Rieseberg et al. 2003; Nolte and Tautz, 2010; Mao et al. 2011; Dittrich-Reed and Fitzpatrick, 2013). Lastly, alleles derived through hybridization can mask deleterious alleles in cultivated lines, potentially leading to heterotic effects (Springer and Stupar, 2007; Birchler et al. 2010; Mezmouk and Ross-Ibarra, 2014). Below we describe some of these potential uses in *Helianthus*, drawing on examples from both natural and artificial hybridization experiments. We also describe potential barriers to introgression such as chromosomal rearrangements, and potential ways to overcome these barriers.

Hybridization in Nature

Interspecific hybridization and introgression permits large portions of the genome to change simultaneously, potentially facilitating rapid divergence or adaptation. Interspecific gene flow within *Helianthus* is common (Heiser 1947, 1951; Stebbins and Daly, 1961; Heiser, 1978), and the transfer of alleles between species has been shown to affect phenotype and fitness (Whitney et al. 2006, 2010). Studies of natural hybrids in the genus have provided insights into the genetics of adaptation, especially with respect to the roles of transgressive segregation and chromosomal rearrangements in ecological divergence (Strasburg et al. 2011; Sambatti and Rice, 2006; Kane and Rieseberg, 2007; Andrew et al. 2012; Scascitelli et al. 2010; Whitney et al. 2010).

Transgressive segregation appears to be common in interspecific *Helianthus* hybrids in both greenhouse and natural environments (Schwarzbach et al. 2001; Welch and Rieseberg, 2002a; Ludwig et al. 2004; Rieseberg et al. 2003). The most studied hybrids are natural derivatives of the two most widespread annual sunflowers, *H. annuus* and *H. petiolaris* (Heiser, 1947), as this hybrid combination created three different species: *H. paradoxus* (Rieseberg et al. 1990; Welch and Rieseberg, 2002), *H. deserticola* (Rieseberg, 1991a; Rieseberg, 1991b), and *H. anomalus* (Rieseberg, 1991a; Rieseberg, 1991b). The hybrid species have the same chromosome number as the parental species, and so represent examples of homoploid hybrid speciation. However, each of the hybrid species is comprised of a different combination of parental chromosomal segments (including rearrangements), which has resulted in distinctive transgressive phenotypes, as well as strong chromosomal sterility barriers (see below) relative to the parental species and to each other (Rieseberg et al. 1993, 2003; Gross et al. 2003; Rieseberg, 2001; Rieseberg et al. 1995; Lai et al. 2005). The extreme adaptations found in the natural hybrid species suggest that hybridization could be useful in adapting cultivars to abiotic stress.

Chromosomal Compatibility Within the Genus

Chromosomal rearrangements are commonly reported in progeny from both intra- and inter-specific crosses within *Helianthus*. Large-scale chromosomal translocations have been reported most frequently, mainly because the multivalent configurations they generate at meiosis are easily detected by conventional light microscopy. Large inversions have also been reported using the same approach. Both kinds of rearrangements were confirmed by initial low-density comparative genetic mapping studies (e.g., Rieseberg et al. 1995; Burke et al. 2004; Lai et al. 2005; Heesacker et al. 2009). Recent very high-density genetic maps have largely validated these initial mapping studies, and suggest that small-scale inversions and translocations are frequent as well (Barb et al. 2014).

Chromosomal translocations and inversions often cause reductions in the fertility of hybrids because recombinant gametes are frequently unbalanced (i.e., carry duplications or deletions). Because the unbalanced gametes are inviable, non-recombinant parental chromosomes will be over-represented in the gametes that survive, leading to an effective reduction of recombination rates in rearranged chromosomes. Hybrid fertility reduction is probably most important in preventing species' mergers following secondary contact (Noor et al. 2001; Rieseberg 2001). In some instances, mechanisms have evolved that suppress recombination in inversions prior to gamete development. In these situations, recombination suppression appears to be complete, although gene conversion rates can be surprisingly high (Gaut et al. 2007). Recombination suppression due to inversions has been shown to facilitate the accumulation of hybrid incompatibilities, as well as local adaptation in the presence of gene flow (Kirkpatrick and Barton, 2006; Lowry and Willis 2010).

From a breeding perspective, chromosomal structure impacts the success of intentional introgression with different species, with collinear portions of genomes being easier to introgressthan regions within or near rearrangements (Long, 1960; Whelan, 1978; Georgieva-Todorova, and Bohorova, 1980; Espinasse et al. 1995; Rieseberg et al. 1995; Rieseberg et al. 1996; Burke et al. 2004; Renaut et al. 2013; Barb et al. 2014). This is due both to the direct effects of the rearrangements and to linked hybrid incompatibilities (Orr et al. 1996; Lai et al. 2005). There is potential to utilize marker information to find rare recombinants through marker-assisted selection to eliminate these deleterious mutations (Robertson, 1960; Charlesworth, 2012).

Helianthus has a high rate of karyotypic evolution (Geisler, 1931; Seiler, 1992; Seiler and Rieseberg, 1997; Rieseberg et al. 1995; Fang et al. 2012; Feuk et al. 2006; Burke et al. 2004), estimated at 5.5-7.3 chromosomal rearrangements per million years (Chandler et al. 1986; Burke et al. 2004). This has led to the recognition of chromosomal subtypes among *Helianthus* species that predict crossing success (Schilling and Heiser, 1981; Chandler et al, 1986; Heiser et al. 1962; Sossey-Alaoui et al. 1998; Ceccarelli et al. 2007; Natali et al. 2008; Jan and Chandler, 1989). For example, the perennial polyploid species in *Helianthus* generally cross despite morphological differences (Long, 1955; Long 1960), different origins, and large variation in chromosome structure and pairing. Chromosomal subtypes have been identified; for example, homology has been reported between *H. ciliaris, H. tuberosus* and *H. annuus* as well as generally within the perennial diploid species' genomes (Espinassee et al. 1995). Different sections within the genus have shown differential abilities to hybridize with each other, with species within sections generally hybridizing better (Faure et al. 2002). The development of predictive chromosomal compatibility groups required that many different populations be tested, since there can be significant intraspecific variation in hybrid formation and vigor (Espinasse et al. 1995; Edmands, 2002). Understanding chromosomal structure provides an opportunity to better utilize wild populations in plant breeding, by providing insight into crossing success and what traits and genes are likely to be transferable. Chromosomal structure can often differ within species that are present in the primary, secondary and tertiary germplasm, and knowing which populations have structural variation could provide easier access to useful traits. While many crosses within *Helianthus* are possible, those involving more distant wild relatives sometimes require special techniques, such as embryo rescue and tissue culture or even the induction of an additional round of amphiploidy, with intraspecific variation in success (Jan and Chandler 1989, Feng and Jan 2008).

General Utility of Helianthus Crop Wild Relatives

The ability of the cultivated sunflower to readily hybridize with many other *Helianthus* species has been exploited extensively through the intentional introduction of genetic material from wild relatives (both annual and perennial) into the cultivated gene pool (Table 1). Wild relatives are sources of disease resistance genes (Feng et al. 2006), cytoplasmic male sterility (Seiler and Jan, 1994), quality traits, and yield traits. Many of these traits have been the target of mapping efforts (Bert et al. 2004; Qi et al. 2012; Yue et al. 2008; Yue et al. 2010), which have indicated that large chromosomal segments sometimes introgress with the traits of interest. Indeed, different interspecific hybridization events cover approximately 10% of the cultivated sunflower genome (Baute et al. 2015). Interspecific introgressions can be difficult to eliminate (or reduce in size if linked to a trait under selection) because of limited recombination between different chromosomal types (Livaja et al. 2013). As genotyping becomes cheaper and more efficient, the ability to utilize marker assisted selection to introgress precise genomic regions from crop wild relatives is an increasingly feasible option to limit the extent of donor parent contributions to cultivated material.

Cytoplasmic Male Sterility and the Formation of a Hybrid Seed Industry

Hybrid breeding has been used to improve performance in many crops, thereby making a fundamental contribution to the green revolution (Borlaug 2000). Hybrid vigor (heterosis) is formed by crossing different strains, varieties or species to produce offspring that outperform their parents in terms of biomass, growth rate, and fertility. This phenomenon was first described by Darwin in both natural and domesticated species (Darwin 1859). Hybrid production in crop species has been central to increasing crop production. The genetic basis of heterosis has been debated for over a century; however, a general consensus has been reached on three main models: dominance (Bruce 1910, Jones 1917), over-dominance (Shull 1908, East 1936, Crow 1948), and pseudo-overdominance (Crow 1952). The outcome of all three models is the same: increased performance in hybrid lines over their parents. Unfortunately, most crop plants bear anthers and stigmas in the same flower, or at closely associated flowers, so emasculation is required. In many plant species emasculation is tedious and requires a relatively high degree of technical training, reducing the economic potential of hybrid seed production (Kaya 2014).

In nature, several mechanisms have evolved to reduce self-fertilization and enhance outcrossing. One common mechanism is cytoplasmic male sterility (CMS), in which plants fail to produce functional pollen while maintaining female fertility. CMS is a maternally inherited trait and is thought to arise from an incompatibility between the nucleus and cytoplasm

Table 1. Interspecific hybridization within *Helianthus* with germplasm position defined by crossing relationships to the crop: the primary germplasm contains no crossing barriers, the secondary germplasm can be crossed but hybrids show meiotic abnormalities, and the tertiary germplasm requires special techniques such as embryo rescue to make the cross possible (Harlan, 1976).

Taxon	Position in Germplasm	Ploidy	Trait	Reference
Helianthus annuus	Primary	Diploid	Herbicide tolerance	Al-Khatib and Miller, 2000; Miller and Al-Khatib, 2002
Helianthus anomalus	Secondary	Diploid	Fertility restoration	Seiler, 1991a
Helianthus argophyllus	Secondary	Diploid	Downy mildew resistance, disease resistance, fertility restoration, salt tolerance, drought tolerance	Sieler, 1991a; Miller & Gulya, 1988; Jan et al. 2004; Hulke et al. 2010; Seiler, 1994
Helianthus arizonensis	Tertiary	Diploid	High linoleic acid concentrations in seed (potential)	Seiler, 1984
Helianthus atrorubens	Tertiary	Diploid	High linoleic acid concentrations in seed (potential)	Seiler, 1984
Helianthus bolanderi	Secondary	Diploid	Fertility restoration	Seiler, 1991a; Jan, 1992
Helianthus debilis	Secondary	Diploid	Powdery mildew resistance; fertility restoration	Jan & Chandler, 1988; Seiler, 1991a
*Helianthus debilis*subsp. *tariflorus*	Secondary	Diploid	Resistance to broomrape	Velasco et al. *2012*
Helianthus deserticola	Secondary	Diploid	Downy mildew resistance	Seiler, 1991b
Helianthus divaricatus	Tertiarty	Diploid	Broomrape resistance	Jan et al. 2002
Helianthus giganteus	Tertiary	Diploid	Fertility restoration; cytoplasmic male sterility	Whelan & Dedio, 1980; Seiler, 2000
Helianthus grosseserratus	Tertiary	Diploid	Broomrape resistance	Jan *et al*. 2002
Helianthus hirsutus	Secondary	Tetraploid	Fertility restoration	Seiler, 1991c; Seiler, 2000
Helianthus maximilianii	Tertiary	Diploid	Broomrape resistance; Cytoplasmic male sterility	Whelan & Dedio, 1980; Jan et al. 2002

cont.

Taxon	Position in Germplasm	Ploidy	Trait	Reference
Helianthus neglectus	Secondary	Diploid	Fertility restoration	Seiler, 1991a
Helianthus paradoxus	Secondary	Diploid	Salt tolerance; fertility restoration	Seiler, 1991a; Lexer et al. 2003; Lexer et al. 2004;
Helianthus pauciflorus	Tertiary	Tetraploid	Cytoplasmic male sterility; sclerotinia resistance	Jan et al. 2006
Helianthus petiolaris	Secondary	Diploid	Verticillium resistance; disease resistance ; cytoplasmic male sterility ; sunflower moth resistance; fertility restoration	Hoes et al. 1973; Rogers et al. 1984; Seiler, 1991a; Jan et al. 2004
Helianthus praecox	Secondary	Diploid	Downy mildew, rust, verticilliumwilt and broomrape resistance; fertility restoration; downy mildew resistance	Seiler, 1991a; Seiler, 1991b
Helianthus resinosus	Tertiary	Hexaploid	Fertility restoration	Seiler, 1991c
Helianthus salicifolius	Tertiary		High crude protein concentration in leaves (potential)	Seiler, 1983
Helianthus silphioides	Tertiary	Diploid	High oleic acid concentrations in seed (potential)	Seiler, 1984
Helianthus strumosus	Tertiary	Hexaploid	Fertility restoration	Seiler, 2000
Helianthus tuberosus	Secondary	Hexaploid	Broomrape resistance; sunflower moth resistance; fertility restoration	Putt, 1978; Rogers et al. 1984; Seiler, 2000

(Hanson and Conde 1985). Often, rearrangements in the mitochondrial DNA (mtDNA) have been associated with CMS, with collinear chloroplast DNA (cpDNA) in both male-fertile and male-sterile lines (Rieseberg and Seiler 1990). Although the molecular mechanism of CMS has been fully described in only a few species (Touzet and Meyer 2014), it provides an efficient mechanism to guide crossing in breeding programs and for production of hybrid seeds in many crop plants (Dewey et al. 1986, Makaroff et al. 1989, Bailey-Serres et al. 1986), including sunflower (Siculella and Palmer 1988). In parallel with the evolution of CMS in nature, a counteracting mechanism to this destructive mitochondrial effect has evolved to protect pollen functionality. This restoring mechanism is induced by nuclear genes that regulate the accumulation of transcripts or proteins associated with the CMS locus (Hanson and Bentolila 2004; Luo et al. 2013). Restorer genes typically belong to the penta-tricopeptide repeat (PPR) family (Brown et al. 2003, Wang et al. 2006), which is one of the largest gene families in plants. PPR genes occur in small clusters of closely related genes that have arisen through evolutionarily recent gene duplication and transposition, perhaps enabling them to respond quickly to the challenge posed by CMS (Schnable and Wise 1998).

Similar to in other species, CMS in sunflower is associated with genomic rearrangements in the mitochondria that lead to a chimeric open reading frame (ORF) (Leroy et al. 1985, Siculella and Palmer 1988, Köhler et al. 1991). The chimeric ORF in sunflower (ORF522) was shown to share sequence similarity with the ATP synthase subunit ORFB, which results in competition between the two proteins leading to decreased phosphorylation activity of the ATP synthase complex (Balk and Leaver 2001, Sabar et al. 2003). As the energetic demands increase during anther development, the expression of ORF522 compromises ATP complex activity and leads to developmental delay and pollen abortion. Although several nuclear restorers have been identified, the molecular mechanism of fertility restoration remains largely unknown. However, the most common locus (Rf-1) maps to a cluster of PPR genes on linkage group 13 (Baute 2015).

While many traits been introduced to cultivated sunflowers through interspecific hybridization, few have been as important as cytoplasmic male sterility (CMS). Today, commercial sunflower production is dominated by hybrid genotypes, made possible by the discovery of CMS in wild germplasm. However, commercial production of sunflower largely relies on a single cytoplasm, CMS PET-1, which originated from an interspecific cross of *Helianthus petiolaris* Nutt. With *H. annuus* L., and its corresponding fertility restoration gene, Rf-1 (Leclercq 1969, Gimenez and Fick 1975, Horn et al. 2003; Jan and Vick 2007). Hybrid production involves a female inbred parental line (CMS-HA), which carries the male-sterile cytoplasm (S-type cytoplasm) but not the fertility restorer allele (Rf) in the nucleus, and a male inbred parental line (RHA), which can carry a normal (N-type) or CMS (S-type) cytoplasm, but which carries the restorer allele (Rf). Therefore the female parent is a male-sterile inbred line (S-rfrf) and the male parent is male-fertile inbred line (S-RfRf or N-RfRf). The resulting hybrid is male-fertile, containing both the CMS and the restorer allele (S-Rfrf). This complex system of production is utilized because the increase in performance seen in hybrids greatly increases the economic value of the crop. Hybrid breeding has been the impetus for much of the gains seen in worldwide sunflower production over the past 40 years (FAO Stat).

More than 70 CMS sources have been identified in sunflower wild relatives (Serieys 2005), but the corresponding restorer genes are known for only about half (Jan and Seiler 2006). Most CMS systems are from wild relatives and show complete male sterility: e.g. *H. petiolaris* (PET1), *H. resinosus* (RES), *H. rigidus* (RIG), *H. giganteus* (GIG1), *H. maximiliani* (MAX1), and *H. argophyllus* (ARG3) (Kaya 2014). For the most commonly used CMS (PET1), two genes (Rf1 and Rf2) were identified to restore fertility and have been used extensively in sunflower hybrid production (Fick and Miller 1997). These restoring genes were found to be effective also for the ARG1 and ARG3 derived sterility (Christov 1991; Jan and Seiler 2006). However, other CMS sources are still not widely used for hybrid production at commercial scale (Kaya 2014).

Use of alternative CMS/Rf systems is recommended for sunflower hybrid seed production to reduce genetic vulnerability (e.g., susceptibility to diseases, pests, environmental stresses) associated with cytoplasmic uniformity. Breeders tend to avoid exploring new CMS systemsdue to the substantial effort involved in producing new CMS lines and the introgression of the corresponding Rf genes. Different cytoplasms may significantly impact drought tolerance and other traits (Sambatti et al. 2008). Perhaps a better understanding of the benefits of different cytoplasms will encourage expansion of the cultivated cytoplasm diversity.

Branching

One of the major domestication traits of sunflower is monocephaly (single head). A single head allows for more synchronous flowering and easier harvesting. However, after the initiation of hybrid sunflower production in the 1960's there was a need for male lines to flower for longer periods of time. This led to introgression of the recessive B-locus responsible for branching from wild *H. annuus* back into cultivated male lines (Mandel et al. 2013). This illustrates a difference in the needs of domestication and improvement phases of crop development and, like CMS (above), shows how wild diversity continues to contribute to sunflower improvement.

Introgression of Disease Resistance

Sunflower is one of a handful major crops that is widely cultivated at its center of origin, and is thereby exposed to the large number of pathogens that have coevolved with its wild progenitors. There are more than fifty diseases of *Helianthus*, although very few are of economic importance (Gulya et al. 1997). Genetic resistance is often the most cost effective way to deal with disease pressure (Talukder et al. 2014); therefore understanding the extent of both intra- and interspecific genetic resistance in germplasm collections is of great value. The large number of species within *Helianthus* that are adapted to a wide number of different pests and pathogens are an important resource for breeding efforts. As described above, most *Helianthus* species hybridize (Long et al. 1960; Chandler et al. 1986; Espinasse et al. 1995), and this ability has been extensively exploited with respect to the introgression of disease resistance (Table 1). At least twelve different species have been utilized to introgress disease resistance for eight different major diseases of cultivated sunflower (Table 1). The donor species occur across most of North America with the exception of the southeastern USA, where native *Helianthus* species are mainly part of the tertiary germplasm. The species that have been utilized are also well distributed across the secondary (67%) and tertiary (33%) germplasm. For example, members of the secondary germplasm, *H. argophyllus* and *H. praecox*, have been utilized for resistance to downy mildew (*Plasmopara halstedii*), a common disease in in the northern part of North America infecting seedlings and causing significant damage (up to 25% of a field). Resistance to the major fungal pathogen, *Sclerotinia sclerotiorum*, was identified in *Helianthus pauciflorus*; this disease has been a major disease in North America.

Crosses Between the Crops: Interspecific Hybridization Between H. tuberosus and H. annuus

Helianthus tuberosus diverged from *H. annuus* approximimently 1.7-8.2 million years ago (Schilling, 1997) and has been a major donor of useful genes for sunflower development (Hajjar and Hodgkin. 2007). The chromosomal interactions between the two crops have generated much interest and although the genomes are not the same, one of the sub-genomes of *H. tuberosus* pairs effectively with the *H. annuus* genome, such that strong viable hybrids are formed (Espinasse et al. 1995; Atlagic, 1993; Hulke and Wyse, 2008; Sujatha and Prabakaran, 2006). *Helianthus annuus* × *H. tuberosus* hybrids are tetraploid ($2n = 4x = 68$), with mixed bivalent and multivalent pairing (Sujatha and Prabakaran, 2006; Atlagic et al. 1995). Chromosomes from different populations of *H. tuberosus* pair differently with *H. annuus* during meiosis due to translocations and inversions, resulting in variable fertility in hybrid plants (Kostoff, 1939; Atlagic et al. 1995; Atlagic et al. 1993; Natali et al. 1998; Chandler et al. 1986; Atlagic et al. 1995; Kantar et al. 2014).

Interspecific *H. tuberosus* × *H. annuus* hybrids have been suggested to have agronomic value in their own right, including use as a forage crop, trait introgression-bridge, trap crop for blackbirds, and as a perennial oil seed (Seiler 1992; Atlagic, 2004; Kays and Nottingham, 2008; Kantar et al. 2014). Hybrid cultivars have been released in Russia and Sweden as a forage crop.

Development of hybrids as a perennial oil seed crop has been suggested as a way to increase sustainable agricultural and ecosystem service production (Glover et al. 2010). Exploration of such a possibility has been underway for two decades (Hulke and Wyse, 2008; Cox et al. 2010), with experimental tetraploid populations having been extensively phenotyped (Kantar et al. 2014). Ongoing work is exploring both neo-domestication of interspecific hybrid populations and ecosystem services available from an emerging perennial crop.

New Tools for Easier Use of Wild Relatives

Historically, exploration of crop wild relatives for use in plant breeding has utilized a biparental crossing approach, where diverse accessions (both intraspecific and interspecific) from *ex situ* collections were used to create populations in the search for useful phenotypic variation (Hajjar and Hodgkin, 2007). This method has been useful for the utilization of crop wild relatives (Khoury et al. 2015), but there are logistical limitations in the number of accessions that can be explored. Emerging sequencing technologies (2nd and 3rd generation), proteome data, metabolome data, and high-throughput phenotyping approaches provide a wealth of data for both basic and applied objectives and may help overcome logistical problems with conventional methods of exploring crop wild relative diversity (Mammadov et al. 2012; O'Driscoll et al. 2013). The cost, both per sample and per data point, of these emerging technologies is now affordable within standard laboratory budgets; this is particularly true for use of molecular markers (Edwards and Batley, 2010; Poland et al. 2012). Decreased costs have allowed for comprehensive assessments of genotypic diversity in crop wild relatives, and for implementation of marker-based selection schemes (e.g. marker assisted backcrossing, genomic selection) in many different species (Bernardo and Yu, 2007; Bernardo 2008; Storlie and Charmet, 2013). High resolution phenotyping also allows for increased precision and efficiencyin trait dissection in both greenhouse and field settings (Araus and Cairns, 2014). This combination of technologies leads to a better understanding of local adaptation and domestication,as well as a more targeted use of crop wild relatives in plant breeding (Fig. 1). Finally, leveraging data from multiple species may broaden the utility of wild germplasm by identifying specific genes with conserved function across species (Du et al. 2010; Monaco et al. 2013; Bolger et al. 2014).

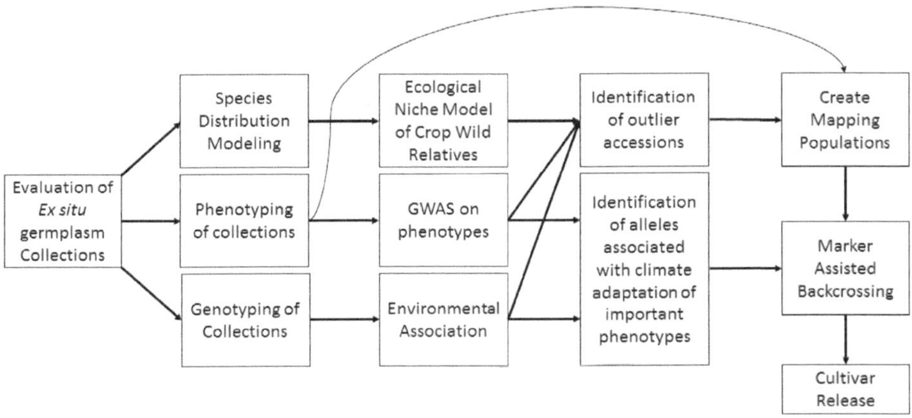

Figure 1. In order to fully utilize germplasm resources it is important to utilize both top-down and bottom-up approaches. These approaches are complementary, leveraging large databases (Germplasm, Genetic, Bioclimatic, and Biophysical), and local capacity to phenotype, to increase the use interspecific hybridization in plant breeding.

Next Generation Germplasm Resources

To further enhance utilization there has been interest in creating populations that will allow more efficient deployment of wild alleles in breeding programs. Novel mating designs are being used to enhance the number of recombinationevents, increase the number of wild alleles segregating in populations, and to allow the effects of wild alleles to be simultaneously tested across numerous genetic backgrounds. With biparental mating designs, there is limited resolution to identify the specific genes underlying quantitative variation. New mapping designs that leverage increased molecular marker density and increased numbers of recombination events include association mapping (individuals sampled directly from breeding or wild populations), nested association mapping (populations based on a hub parent design) and multi-parent intercross populations, in which multiple parents are randomly mated to increase the number of recombination events (Morrel et al. 2012). These mating designs provide easier access to and better knowledge of the effects of alleles present in the primary, secondary and tertiary germplasm of crop species.

Conclusions

Hybridization has been of central importance in the exploration and utilization of wild germplasm in sunflower improvement. It was central to the development of a hybrid sunflower industry, understanding chromosomal interactions among species, and the protection of yield through the provision of disease resistance. Hybridization is currently being used to develop new crops and to understand the physiological mechanisms and genetic basis of resistance to abiotic stress. New mating designs and technology which enable precision breeding, combined with the need to develop environmentally resilient crops, will enhance the value of wild germplasm in years to come.

References

Al-Khatib, K. and J.F. Miller. 2000. Registration of four genetic stocks of sunflower resistant to imidazolinone herbicides. Crop Sci., 40: 869-870.

Andrew, R.L., K.L. Ostevik, D.P. Ebert and L.H. Rieseberg. 2012. Adaptation with gene flow across the landscape in a dune sunflower. Mol. Ecol.; 21: 2078-91.

Araus J.L. and J.E. Cairns. 2014. Field high-throughput phenotyping: the new crop breeding frontier. Trends Plant. Sci. 19: 52-61.

Arias, D.M. and L.H. Rieseberg. 1994. Gene flow between cultivated and wild sunflowers. Theor. Appl. Genet. 89: 655-660.

Arias, D.M. and L.H. Rieseberg. 1995. Genetic relationships among domesticated and wild sunflowers (*Helianthus annuus*, Asteraceae). Econ. Bot. 49: 239-248.

Atlagić, J. 2004. Roles of interspecific hybridization and cytogenetic studies in sunflower breeding. Helia, 27: 1-24.

Atlagic, J., B. Dozet and D. Skoric. 1993. Meiosis and pollen viability in *Helianthus tuberosus L.* and its hybrids with cultivated sunflower. Plant Breeding 111: 318-324.

Atlagic, J., B. Dozet and D. Skoric. 1995. Meiosis and pollen viability in *Helianthus mollis*, *Helianthus salicifolius*, *Helianthus maximiliani* and F_1 hybrids with cultivated sunflower. Euphytica 81: 259-263.

Bailey-Serres, J., D.K. Hanson, T.D. Fox and C.J. Leaver. 1986. Mitochondrial genome rearrangement leads to extension and relocation of the cytochrome c oxidase subunit I gene in sorghum. Cell, 47: 567-576.

Balk, J. and C.J. Leaver. 2001. The PET1-CMS mitochondrial mutation in sunflower is associated with premature programmed cell death and cytochrome c release. Plant Cell, 13: 1803-1818.

Barb, J. G., J.E. Bowers, S. Renaut, J.I. Rey, S.J. Knapp, L.H. Rieseberg and J.M. Burke. 2014. Chromosomal evolution and patterns of introgression in *Helianthus*. Genetics 197: 969-979.

Baute, G.J. 2015. Exploration of the sunflower genome as it relates to domestication and prebreeding (Unpublished doctoral dissertation). University of British Columbia, Vancouver, Canada.

Baute, G.J., N.C. Kane, C.J. Grassa, Z. Lai and L.H. Rieseberg. 2015. Genome scans reveal candidate domestication and improvement genes in cultivated sunflower, as well as post-domestication introgression with wild relatives. New Phyt. 206: 830-838.

Berglund, D.R. (ed.). 2007. Sunflower production. Bull. A-1331 (EB 25 revised). North Dakota State Univ. Ext. Serv., Fargo.

Bernardo, R. 2008. Molecular markers and selection for complex traits in plants: learning from the last 20 years. Crop Sci., 48: 1649-1664.

Bernardo, R. and J. Yu. 2007. Prospects for genomewide selection for quantitative traits in maize. Crop Sci., 47: 1082-1090.

Bert, P.F., G. Dechamp-Guillaume, F. Serre, I. Jouan, D.T. de Labrouhe, P. Nicolas and F. Vear. 2004.Comparative genetic analysis of quantitative traits in sunflower (*Helianthus annuus* L.). Theor. Appl. Genet., 109: 865-74.

Birchler, J.A., H. Yao, S. Chudalayandi, D. Vaiman and R.A. Veitia. 2010. Heterosis. Plant Cell, 22: 2105-2112.

Blackman, B.K., M. Scascitelli, N.C. Kane, H.H. Luton, D.A. Rasmussen, R.A. Bye, D.L. Lentz and L.H. Rieseberg. 2011. Proc. Natl. Acad. Sci. U.S.A. 108: 14360-14365.

Bock, D.G., N.C. Kane, D.P. Ebert and L.H. Rieseberg. 2014. Genome skimming reveals the origin of the Jerusalem artichoke tuber crop species: neither from Jerusalem nor an Artichoke. New Phytol. 201: 1021-30.

Bolger, M.E., B. Weisshaar, U. Scholz, N. Stein, B. Usadel and K.F.X. Mayer. 2014. Plant genome sequencing—applications for crop improvement. Curr. Opin. Biotechnol. 26: 31-37.

Borlaug, N. 2000. We need biotech to feed the world. Wall Street Journal, 6.

Brookes, R. 1763. The natural history of vegetables. London.

Brown, G.G., N. Formanova, H. Jin, R. Wargachuk, C. Dendy, P. Patil, M. Laforest, J.F. Zhang, W.Y. Cheung and B.S. Landry. 2003. The radish *Rfo* restorer gene of Ogura cytoplasmic male sterility encodes a protein with multiple pentatricopeptide repeats. Plant J. 35: 262-272.

Bruce, A.B. 1910. The Mendelian theory of heredity and the augmentation of vigor. Science, 627-628.

Burke J.M., K.A. Gardner and L.H. Rieseberg. 2002b. The potential for gene flow between cultivated and wild sunflower (*Helianthus annuus*) in the United States. Am. J. Bot. 89: 1550-2.

Burke, J.M. and M.L. Arnold. 2001. Genetics and the Fitness of Hybrids. Annu. Rev. Genet 35: 31-52.

Burke, J.M., S. Tang, S.J. Knapp and L.H. Rieseberg. 2002a. Genetic analysis of sunflower domestication. *Genetics*. 161, 1257-1267.

Burke, J.M., Z. Lai, M. Salmaso, T. Nakazato, S.X. Tang, A. Heesacker, S.J. Knapp and L.H. Rieseberg. 2004. Comparative mapping and rapid karyotypic evolution in the genus *Helianthus*. Genetics 167: 449-57.

Ceccarelli, M., V. Sarri, L. Natali, T. Giordani, A. Cavallini, A. Zuccolo, I. Jurman, M. Morgante and P.G. Cionini. 2007. Characterization of the chromosome complement of *Helianthus annuus*by in situ hybridization of a tandemly repeated DNA sequence. Genome 50: 429-434.

Chandler, J.M., C.C. Jan and B.H. Beard. 1986. Chromosomal differentiation among the annual *Helianthus* species. Syst. Bot., 11: 354-371.

Charlesworth, B. 2012. The effects of deleterious mutations on evolution at linked sites. Genetics, 190: 5-22.

Christov, M. 1991. Possibilities and problems in the hybridization of cultivated sunflower with species of the genus *Helianthus* L. Helia 14: 35-40.

Crow, J.F. 1948. Alternative hypotheses of hybrid vigor. Genetics 33: 477-487.

Crow, J.F. 1952. Dominance and overdominance. Heterosis. Iowa State College Press, Ames, 282-297.

Cummings, C.L., H.M. Alexander, A.A. Snow, L.H. Rieseberg, M.J. Kim and T.M. Culley. 2002. Fecundity selection in a sunflower crop-wild study: can ecological data predict crop allele changes? Ecol. Appl. 12: 1661-71.

Darwin, C. 1859. On the origin of species by means of natural selection. London: Murray.

Dempewolf, H., R.J. Eastwood, L. Guarino, C.K. Khoury, J.V. Müller and J. Toll. 2014 Adapting agriculture to climate change: A global initiative to collect, conserve, and use crop wild relatives. Agroecology and Sustainable Food Systems, 38: 369-377.

Dewey, R.E., C.S. Levings and D.H. Timothy. 1986. Novel recombinations in the maize mitochondrial genome produce a unique transcriptional unit in the Texas male-sterile cytoplasm. Cell, 44: 439-449.

Dittrich-Reed, D.R. and B.M. Fitzpatrick. 2013. Transgressive hybrids as hopeful monsters. Evol. Biol. 40: 310-315.

Du, J.C., Z.X. Tian, C.S. Hans, H.M. Laten, S.B. Cannon, S.A. Jackson, R.C. Shoemaker and J. X. Ma. 2010. Evolutionary conservation, diversity and specificity of LTR-retrotransposons in flowering plants: insights from genome-wide analysis and multi-specific comparison. Plant J. 63: 584-598.

East, E.M. 1936. Heterosis. Genetics 21: 375-397.

Edmands S. 2002. Does parental divergence predict reproductive compatibility? Trends Ecol. Evol. 17: 520-7.

Edwards, D. and J. Batley 2010. Plant genome sequencing: applications for crop improvement. Plant Biotech. J., 8: 2-9.

Espinasse A., J. Foueillassar and G. Kimber. 1995. Cytogenetical analysis of hybrids between sunflower and four wild relatives. Euphytica 82: 65-72.

Fang, Z., T. Pyhäjärvi, A.L. Weber, R.K. Dawe, J.C. Glaubitz, J.d.J.S. González, C. Ross-Ibarra, J. Doebley, P.L. Morrell and J. Ross-Ibarra. 2012. Megabase-scale inversion polymorphism in the wild ancestor of maize. Genetics 191: 883-94.

FAOSTAT. Final Data 2013. Retrieved May, 2015. http: //faostat.fao.org.

Faure, N., H. Serieys, E. Cazaux, F. Kaan and A. Berville. 2002. Partial hybridization in wide crosses between cultivated sunflower and the perennial *Helianthus* species *H. mollis* and *H. orgyalis*. Ann.Bot., 89: 31-39.

Feng J., G. Seiler, T. Gulya and C. Jan. 2006. Development of Sclerotinia stem rot resistant germplasm utilizing hexaploid *Helianthus* species. In 28th Sunflower Research Workshop 11-2.

Feng, J. and C.C. Jan. 2008. Introgression and molecular tagging of Rf 4, a new male fertility restoration gene from wild sunflower *Helianthus maximiliani* L. Theor. Appl. Genet. 117: 241-249.

Feuk, L., A.R. Carson and S.W. Scherer. 2006. Structural variation in the human genome. Nat. Rev. Genet.7: 85-97.

Fick, G.N. and J.F. Miller. 1997. Sunflower Breeding. Agronomy Monograph, Sunflower Technology and Production 35: 395-439.

Fritz G.J. 1990. Multiple pathways to farming in precontact eastern north america. Journal of world prehistory 4(4), 387.

Gaut, B.S., S.I. Wright, C. Rizzon, J. Dvorak and L.K. Anderson. 2007. Recombination: an underappreciated factor in the evolution of plant genomes. Nat. Rev. Genet. 8: 77-84.

Geisler F. 1931. Chromosome Numbers in Certain Species of Helianthus, Vol. 2 Butler University Botanical Studies, Article 7.

Gentzbittel, L., S.U. Andersen, C. Ben, M. Rickauer, J. Stougaard and N.D. Young. 2015. Naturally occurring diversity helps to reveal genes of adaptive importance in legumes. Front. Plant Sci. 6: 269.

Georgieva Todorova, I.D., and Bohorova, N.E. 1980. Karyological investigation of the hybrid Helianthus annuus L.($2n = 34$) Helianthus hirsutus Ray ($2n = 68$). In Dokl. Bolg. Akad. Nauk (Vol. 33, No. 7, pp. 961-964).

Gepts, P. 2000. A Phylogenetic and Genomic Analysis of Crop Germplasm: A Necessary Condition for its Rational Conservation and Use. In *Genomes* (pp. 163-181). Springer US.

Gimenez, J.D. and G.N. Fick. 1975. Fertility restoration of male-sterile cytoplasm in wild sunflowers. Crop Sci. 15: 724-726.

Glover, J.D., J.P. Reganold, L.W. Bell, J. Borevitz, E.C. Brummer, E.S. Buckler, et al. 2010. Increased food and ecosystem security via perennial grains. Sci. 328:1638–1639.

Gross, B.L., A.E. Schwarzbach and L.H. Rieseberg. 2003. Origin (s) of the diploid hybrid species *Helianthus deserticola* (Asteraceae). Am. J. Bot 90: 170819.

Gulya, T., K.Y. Rashid and S.M. Masirevic. 1997. Sunflower diseases. In Sunflower technology and production. Crop Sci. Soc. Amer., Madison, Wis. pp. 263-379.

Hajjar, R. and T. Hodgkin. 2007. The use of wild relatives in crop improvement: a survey of developments over the last 20 years. Euphytica, 156: 1-13.

Hanna, W. F. 1924. Growth of corn and sunflowers in relation to climatic conditions. Botanical Gazette, 78: 200-214.

Hanson, M.R. and M.F. Conde. 1985. Functioning and variation of cytoplamsic genomes: Lessons from cytoplasmic-nuclear interactions affecting male fertility in plants. International Review of Cytology (USA) 94: 213-267.

Hanson, M.R. and S. Bentolila. 2004. Interactions of mitochondrial and nuclear genes that affect male gametophyte development. Plant Cell, 16: S154-S169.

Harlan, J.R. 1976. Genetic resources in wild relatives of crops. Crop Sci., 16: 329-333.

Harlan, J.R. and J.M.J. de Wet. 1971. Toward a rational classification of cultivated plants. Taxon, 20: 509-517.

Harlan, J.R., J.M.J. De Wet and E. G. Price. 1973. Comparative evolution of cereals. Evolution 27: 311-25.

Harter, A.V., K.A. Gardner, D. Falush, D.L. Lentz, R.A. Bye and L.H. Rieseberg. 2004. Origin of extant domesticated sunflowers in eastern North America. Nature 430: 201-205.

Heesacker, A., R.L. Brunick, L.H. Rieseberg, J.M. Burke, S.J. Knapp. 2009. Karyotypic evolution of the common and silverleaf sunflower genomes. Plant Genome 2: 233-246.

Heiser CB, Jr. 1947. Hybridization between the sunflower species *Helianthus annuus* and *H. petiolaris*. Evolution 1: 249-62.

Heiser Jr., C. B. 1951. The sunflower among the North American Indians. Proc. Am. Philos. Soc. 95: 432-448.

Heiser, C.B. 1978. Taxonomy of *Helianthus* and origin of domesticated sunflower. Sunflower science and technology, Agronomy Series, 19: 31-35.

Heiser, C.B., W.C. Martin and D. Smith. 1962. Species crosses in *Helianthus*: I. Diploid species. Brittonia 14: 137-47.

Heiser, Jr. C.B. 1955. The origin and development of the cultivated sunflower. The American Biology Teacher 17: 61-167.

Hoes, J.A., E.D. Putt and H. Enns. 1973. Resistance to Verticillium wilt in collections of wild *Helianthus* in North America. Phytopathology 63: 1517-1520.

Horn, R., B. Kusterer, E. Lazarescu, M. Prufe and W. Friedt. 2003. Molecular mapping of the *Rf1* gene restoring pollen fertility in PET1-based F1 hybrids in sunflower (*Helianthus annuus* L.). Theor. Appl. Genet. 106: 599-606.

Hulke, B.S. and D.L. Wyse. 2008. Using interspecific hybrids with *H. annuus* L. Proceedings of the 17th International Sunflower Conference, Cordoba, Spain; 729-34.

Hulke, B.S., J.F. Miller, T.J. Gulya and B.A. Vick. 2010. Registration of the oilseed sunflower genetic stocks HA 458, HA 459, and HA 460 possessing genes for resistance to downy mildew. *Journal of Plant Registrations* 4: 1-5.

Jan, C.C, J.M. Fernandez-Martinez, J. Ruso and J. Munoz-Ruz. 2002. Registration of four sunflower germplasms with resistance to *Orobanchecumana* Race F. Crop Sci. 42: 2217-2218.

Jan, C.C. 1992. Registration of sunflower amphiploid germplasm line, ANN-BOL-AMP1. Crop Sci. 32: 1513-1513.

Jan, C.C. and B.A. Vick. 2007. Inheritance and allelic relationships of fertility restoration genes for seven new sources of male-sterile cytoplasm in sunflower. Plant Breed. 126: 213-217.

Jan, C.C. and G. Seiler. 2006. Sunflower. In Genetic Resources, Chromosome Engineering, and Crop Improvement (103-165). *CRC Press*.

Jan, C.C. and J.M. Chandler. 1988. Registration of powdery mildew resistant sunflower germplasm pool, PM 1. Crop Sci. 28: 1039-1040.

Jan, C.C. and J.M. Chandler. 1989. Sunflower interspecific hybrids and amphiploids of *Helianthus annuus* × *H. bolanderi*. Crop Sci. 29: 643-645.

Jan, C.C., J.F. Miller, G.J. Seiler and G.N. Fick. 2006. Registration of one cytoplasmic male sterile and two fertility restoration sunflower genetic stocks. Crop Sci. 46: 1835-1835.

Jan, C.C., Z. Quresh and T.J. Gulya. 2004. Registration of seven rust resistant sunflower germplasms. CropSci. 44: 1887-1888.

Jenks, AE. 1916. Agriculture of the Hidatsa Indians. Science 44: 864-866.

Jones, D.F. 1917. Dominance of linked factors as a means of accounting for heterosis. Genetics 2: 466.

Jones, GA. and J.L. Gillett. 2005. Intercropping with sunflowers to attract beneficial insects in organic agriculture. The Florida Entomologist 88: 91-96.

Jones, W.B., P.A. Brannon, W. Hough et al. 1933. Archaeological Field Work in North America during 1932. American Anthropologist 35: 483-511.

Kane, N.C. and L.H. Rieseberg. 2007. Selective sweeps reveal candidate genes for adaptation to drought and salt tolerance incommon sunflower, *Helianthus annuus*. Genetics 175: 1823-34.

Kane, N.C., J.M. Burke, L. Marek, G. Seiler, F. Vear, G. Baute, S.J. Knapp, P. Vincourt and L.H. Rieseberg. 2013. Sunflower genetic, genomic and ecological resources. Mol. Ecol. Resour. 13: 10-20.

Kantar, M.B., K. Betts, J.M. Michno, J.J. Luby, P.L. Morrell, B.S. Hulke, R.M. Stupar and D.L. Wyse. 2014. Evaluating an interspecific *Helianthus annuus* × *Helianthus tuberosus* population for use in a perennial sunflower breeding program. Field Crops Res. 155: 254-64.

Kaplan, Lawrence. 1963. Archeoethnobotany of Cordova Cave, New Mexico. Econ. Bot. 17: 350-359.

Kaya, Y. 2014. Sunflower. In Alien Gene Transfer in Crop Plants, Volume 2 (pp. 281-315). Springer New York.

Kays, S.J. and S.F. Nottingham. 2008. Biology and chemistry of Jerusalem artichoke *Helianthus tuberosus* L. CRC Press, Boca Raton, FL.

Khoury, C.K., A.D. Bjorkman, H. Dempewolf, J. Ramirez-Villegas, L. Guarino, A. Jarvis, L.H. Rieseberg and P.C. Struik. 2014. Increasing homogeneity in global food supplies and the implications for food security. Proc. Natl. Acad. Sci. U.S.A. 111: 4001-4006.

Khoury, C.K., N.P. Castaneda-Alvarez, H.A. Achicanoy, C.C. Sosa, V. Bernau, M.T. Kassa, S.L. Norton, L.J.G. van der Maesen, H.D. Upadhyaya, J. Ramirez-Villegas, A. Jarvis and P.C. Struik. 2015. Crop wild relatives of pigeonpea [*Cajanuscajan* (L.) Millsp.]: Distributions, ex situ conservation status, and potential genetic resources for abiotic stress tolerance. Biological Conservation 184: 259-270.

Kirkpatrick M. and N. Barton. 2006. Chromosome inversions, local adaptation and speciation. Genetics 173: 419-34.

Long R.W. Biosystematics of two perennial species of Helianthus (Compositae). I. Crossing relationships and transplant studies. AmJ Bot 1960; 47: 729–35.

Köhler, R.H., R. Horn, A. Lossl and K. Zetsche. 1991. Cytoplasmic male sterility in sunflower is correlated with the co-transcription of a new open reading frame with the atpA gene. Mol. Gen. Genet. 227: 369-376.

Kolkman, J.M., S.T. Berry, A.J. Leon, M.B. Slabaugh, S. Tang, W.X. Gao, D.K. Shintani, J.M. Burke and S.J. Knapp. 2007. Single nucleotide polymorphisms and linkage disequilibrium in sunflower. Genetics 177: 457-68.

Kostoff, D. 1934. A contribution to the meiosis of *Helianthus tuberosus* L. Genetic Laboratory, Academy of sciences, Lenningrad USSR.

Kostoff, D. 1939. Autosyndesis and structural hybridity in F1-hybrid *Helianthus tuberosus* L. × *Helianthus annuus* L. and their sequences. Genetica 21: 285-299.

Lai, Z., T. Nakazato, M. Salmaso, J.M. Burke, S.X. Tang, S.J. Knapp and L.H. Rieseberg. 2005. Extensive chromosomal repatterning and the evolution of sterility barriers in hybrid sunflower species. Genetics 171: 291-303.

Leclercq, P. 1969. Cytoplasmic male sterility in sunflower. Ann. Amelior. Plant. 19: 99-106.

Leroy, P., S. Bazetoux, F. Quetier, J. Delbut and A. Berville. 1985. A comparison between mitochondrial DNA of an isogenic male-sterile (5) and male-fertile (F) couple (HA89) of sunflower. Curr. Genet. 9: 245-251.

Lexer, C., M.E. Welch, J.L. Durphy and L.H. Rieseberg. 2003. Natural selection for salt tolerance quantitative trait loci (QTLs) in wild sunflower hybrids: implications for the origin of *Helianthus paradoxus*, a diploid hybrid species. Mol. Ecol., 12: 1225-1235.

Lexer, C., Z. Lai and L.H. Rieseberg. 2004. Candidate gene polymorphisms associated with salt tolerance in wild sunflower hybrids: implications for the origin of *Helianthus paradoxus*, a diploid hybrid species. New Phyt. 161: 225-233.

Livaja, M., Y. Wang, S. Wieckhorst, G. Haseneyer, M. Seidel, V. Hahn, S.J. Knapp, S. Taudien, C.C. Schon and E. Bauer. 2013. BSTA: a targeted approach combines bulked segregant analysis with next generation sequencing and de novo transcriptome assembly for SNP discovery in sunflower. BMC Genomics 14: 628.

Long, R.W. 1960. Biosystematics of two perennial species of *Helianthus* (Compositae). I. Crossing relationships and transplant studies. Am. J. Bot. 47: 729-35.

Long, R.W., Jr. 1955. Hybridization in perennial sunflowers. Am. J. Bot. 42: 769-77.

Lowry, D.B. and J.H. Willis. 2010. A widespread chromosomal inversion polymorphism contributes to a major life-history transition, local adaptation, and reproductive isolation. PLoS Biol. 8: 2227.

Lu, B. 2013. Introgression of transgenic crop alleles: its evolutionary impacts on conserving genetic diversity of crop wild relatives. J. Syst. Evol. 51: 245-62.

Ludwig, F., D.M. Rosenthal, J.A. Johnston, N. Kane, B.L. Gross, C. Lexer, S.A. Dudley, L.H. Rieseberg and L.A. Donovan. 2004. Selection on leaf ecophysiological traits in a desert hybrid *Helianthus* species and early-generation hybrids. Evolution 58: 2682-92.

Lui, A. and J.M. Burke. 2006. Patterns of nucleotide diversity in wild and cultivated sunflower. Genetics 173: 321-330.

Luo, D.P., H. Xu, Z.L. Liu, J.X. Guo, H.Y. Li, L.T. Chen, C. Fang, Q.Y. Zhang, M. Bai, N. Yao, H. Wu, H. Wu, C.H. Ji, H.Q. Zheng, Y.L. Chen, S. Ye, X.Y. Li, X.C. Zhao, R.Q. Li and Y.G. Liu. 2013. A detrimental mitochondrial-nuclear interaction causes cytoplasmic male sterility in rice. Nat. Genet. 45: 573-577.

Makaroff, C.A., I.J. Apel and J.D. Palmer. 1989. The *atp6* coding region has been disrupted and a novel reading frame generated in the mitochondrial genome of cytoplasmic male-sterile radish. J. Biol. Chem. 264: 11706-11713.

Mammadov, J., R. Aggarwal, R. Buyyarapu and S. Kumpatla. 2012. SNP markers and their impact on plant breeding. Int. J. Plant Gen., 2012: 728398; doi: 10.1155/2012/728398.

Mandel, J., J. Dechaine, L. Marek and J. Burke. 2011. Genetic diversity and population structure in cultivated sunflower and comparison to its wild progenitor *Helianthus annuus* L. Theor. Appl. Genet. 123: 693-704.

Mandel, J.R., S. Nambeesan, J.E. Bowers, L.F. Marek, D. Ebert, L.H. Rieseberg, S.J. Knapp and J.M. Burke. 2013. Association Mapping and the Genomic Consequences of Selection in Sunflower. PLoS Genet 9: e1003378.

Mao, D.H., T.M. Liu, C.G. Xu, X.H. Li and Y.Z. Xing. 2011. Epistasis and complementary gene action adequately account for the genetic bases of transgressive segregation of kilo-grain weight in rice. Euphytica 180: 261-271.

Mayrose, M., N.C. Kane, I. Mayrose, K.M. Dlugosch and L.H. Rieseberg. 2011. Increased growth in sunflower correlates with reduced defences and altered gene expression in response to biotic and abiotic stress. Mol. Ecol. 20: 4683-4694.

McCouch, S., G.J. Baute, J. Bradeen, P. Bramel, P.K. Bretting, E. Buckler, J.M. Burke, D. Charest, S. Cloutier, G. Cole, H. Dempewolf, M. Dingkuhn, C. Feuillet, P. Gepts, D. Grattapaglia, L. Guarino, S. Jackson, S. Knapp, P. Langridge, A. Lawton-Rauh, Q. Lijua, C. Lusty, T. Michael, S. Myles, K. Naito, R.L. Nelson, R. Pontarollo, C.M. Richards, L. Rieseberg, J. Ross-Ibarra, S. Rounsley, R.S. Hamilton, U. Schurr, N. Stein, N. Tomooka, E. van der Knaap, D. van Tassel, J. Toll, J. Valls, R.K. Varshney, J. Ward, R. Waugh, P. Wenzl and D. Zamir. 2013. Agriculture: Feeding the future. Nature 499: 23-24.

Mercer, K.L., D.L. Wyse and R.G. Shaw. 2006. Effects of competition on the fitness of wild and crop-wild hybrid sunflower from a diversity of wild populations and crop lines. Evolution 60: 2044-2055.

Mezmouk, S. and J. Ross-Ibarra. 2014. The pattern and distribution of deleterious mutations in maize. G3 4: 163-171.

Miller, J.F. and K. Al-Khatib. 2002. Registration of imidazolinone herbicide-resistant sunflower maintainer (HA 425) and fertility restorer (RHA 426 and RHA 427) germplasms. *Crop Sci.* 42: 988-989.

Miller, J.F. and T.J. Gulya. 1988. Registration of six downy mildew resistant sunflower germplasm lines. Crop Sci. 28: 1040-1041.

Minnis, PE. 1989. Prehistoric Diet in the Northern Southwest: Macroplant Remains from Four Corners Feces. Am. Antiq. 54: 543-563.

Monaco, M.K., J. Stein, S. Naithani, S. Wei, P. Dharmawardhana, S. Kumari, V. Amarasinghe, K. Youens-Clark, J. Thomason, J. Preece, S. Pasternak, A. Olson, Y.P. Jiao, Z.Y. Lu, D. Bolser, A. Kerhornou, D. Staines, B. Walts, G.M. Wu, P. D'Eustachio, R. Haw, D. Croft, P.J. Kersey, L. Stein, P. Jaiswal and D. Ware. 2014. Gramene 2013: comparative plant genomics resources. Nucleic Acids Res. 42: D1193-D1199.

Morrell P.L., Buckler E.S., Ross-Ibarra J. 2012. Crop genomics: advances and applications. Nat Rev Genet 13: 85-96.

Nabhan, G.P. and K.L. Reichhardt. 1983. Hopi protection of *Helianthus anomalus*, a rare sunflower. The Southwestern Naturalist28: 231-235.

Natali, L., T. Giordani, E. Polizzi, C. Pugliesi, M. Fambrini and A. Cavallini. 1998. Genomic alterations in the interspecific hybrid *Helianthus annuus* × *Helianthus tuberosus*. Theor. Appl. Genet. 97: 1240-1247.

Natali, L., T. Giordani, E. Polizzi, C. Pugliesi, M. Fambrini and A. Cavallini. 2008. Phylogenetic relationships between annual and perennial species of *Helianthus*: evolution of a tandem repeated DNA sequence and cytological hybridization experiments. Genome 51: 1047-1053.

Nolte, A.W. and D. Tautz. 2010. Understanding the onset of hybrid speciation. Trends Genet. 26: 54-58.

Noor, M.A.F., K.L. Grams, L.A. Bertucci and J. Reiland. 2001. Chromosomal inversions and the reproductive isolation of species. Proc. Natl. Acad. Sci. U.S.A. 98: 12084-1208.

O'Driscoll, A., A. Daugelaite and R.D. Sleator. 2013. 'Big data,' Hadoop and cloud computing in genomics. J. Biomed. Inform. 46: 774-781. doi: 10.1016/j.jbi.2013.07.001

Orr, H.A. 1996. Dobzhansky, Bateson, and the genetics of speciation. Genetics 144: 1331.

Poland, J., J. Endelman, J. Dawson, J. Rutkoski, S.Y. Wu, Y. Manes, S. Dreisigacker, J. Crossa, H. Sanchez-Villeda, M. Sorrells and J.L. Jannink. 2012. Genomic selection in wheat breeding using genotyping-by-sequencing. Plant Gen. 5: 103-113.

Putt, E.D. 1978. History and present world status. In: *Sunflower and science and technology* (ed. Carter, J.P.). pp. 1-29. American Society of Agronomy, Madison, WI (USA).

Qi, L.L., G.J. Seiler, B.A. Vick and T. J. Gulya. 2012. Genetics and mapping of the R-11 gene conferring resistance to recently emerged rust races, tightly linked to male fertility restoration, in sunflower (*Helianthus annuus* L.). Theor. Appl. Genet. 125: 921-932.

Renaut, S., C. Grassa, S.Yeaman, B.T. Moyers, Z. Lai, N.C. Kane, J.E. Bowers, J.M. Burke and L.H. Rieseberg. 2013.Genomic islands of divergence are not affected by geography of speciation in sunflowers. Nat. Comm. 4: 1827.

Ricklefs, R.E. and D.G. Jenkins. 2011. Biogeography and ecology: towards the integration of two disciplines. Philos. Trans. R. Soc., B. 366: 2438-2448.

Rieseberg L.H. Hybridization in rare plants: insights from case studies in Helianthus and Cercocarpus. In: Falk DA, Holsinger KE (eds). Conservation of Rare Plants: Biology and Genetics. New York: Oxford University Press, Inc, 1991a: 171-81.

Rieseberg, L.H, R. Carter and S. Zona. 1990. Molecular tests of the hypothesized hybrid origin of two diploid *Helianthus* species (*Asteraceae*). Evolution 44: 1498-1511.

Rieseberg, L.H. 1991. Homoploid reticulate evolution in *Helianthus*: evidence from ribosomal genes. Am. J. Bot 78: 1218-37.

Rieseberg, L.H. 2001. Chromosomal rearrangements and speciation. Trends Ecol. Evol. 16: 351-8.

Rieseberg, L.H. and G.J. Seiler. 1990. Molecular evidence and the origin and development of the domesticated sunflower (*Helianthus annuus*, Asteraceae). Econ. Bot. 44, 3, Supplement: New Perspectives on the Origin and Evolution of New World Domesticated Plants, 79-91.

Rieseberg, L.H., A. Widmer, A.M. Arntz and J.M. Burke. 2003. The genetic architecture necessary for transgressive segregation is common in both natural and domesticated populations. Philos. Trans. R. Soc., B. 358: 1141-1147.

Rieseberg, L.H., C. Van Fossen and A.M. Desrochers. 1995. Hybrid speciation accompanied by genomic reorganization in wild sunflowers. Nature 375: 313-316.

Rieseberg, L.H., C.R. Linder and G. Seiler. 1995. Chromosomal and genic barriers to introgression in *Helianthus*. Genetics 141: 1163-71.

Rieseberg, L.H., H.C. Choi, R. Chan and C. Spore. 1993. Genomic map of a diploid hybrid species. Heredity 70: 285-93.

Rieseberg, L.H., M.J. Kim, and G.J. Seiler. 1999. Introgression between cultivated sunflowers and a sympatric wild relative, Helianthus petiolaris (Asteraceae). International Journal of Plant Sciences 160: 102-108.

Rieseberg, L.H., O. Raymond, D.M. Rosenthal, Z. Lai, K. Livingstone, T. Nakazato, J.L. Durphy, A.E. Schwarzbach, L.A. Donovan and C. Lexer. 2003. Major ecological transitions in wild sunflowers facilitated by hybridization. Science 301: 1211-1216.

Rieseberg, L.H., O. Raymond, D.M. Rosenthal, Z. Lai, K. Livingstone, T. Nakazato, J.L. Durphy, A.E. Schwarzbach, L.A. Donovan and C. Lexer. 1996. The effects of mating design on introgression between chromosomally divergent sunflower species. Theor. Appl. Genet. 93: 633-44.

Rieseberg, L.H. 2001. Chromosomal rearrangements and speciation. Trends Ecol. Evol. 16: 351-8.

Robertson A. 1960. A theory of limits in artificial selection. Philos. Trans. R. Soc., B. 153: 234-249.

Rodrigues, M.A., L. Sousa, J.E. Cabanas and M. Arrobas. 2007. Tuber yield and leaf mineral composition of Jerusalem artichoke (*Helianthus tuberosus* L.) grown under different cropping practices. Spanish Journal of Agricultural Research 5: 545-553.

Rogers, C.E., T.E. Thompson and G.J. Seiler. 1982. Sunflower species of the United States (pp. 1-75). Bismarck, ND: National Sunflower Association.

Rogers, C.E., T.E. Thompson and G.J. Seiler.1984. Registration of three *Helianthus* germplasms for resistance to the sunflower moth. Crop Sci. 24: 212-213.

Sabar, M., D. Gagliardi, J. Balk and C.J. Leaver. 2003. ORFB is a subunit of F1 FO-ATP synthase: insight into the basis of cytoplasmic male sterility in sunflower. EMBO reports, 4: 381-386.

Sambatti, J. and K.J. Rice. 2006. Local adaptation, patterns of selection, and gene flow in the Californian serpentine sunflower (*Helianthus exilis*). Evolution 60: 696-710.

Sambatti, J.B.M., D. Ortiz-Barrientos, E.J. Baack and L.H. Rieseberg. 2008. Ecological selection maintains cytonuclear incompatibilities in hybridizing sunflowers. Ecology Letters 11: 1082-1091.

Scarry, C.M. and J.F. Scarry. 2005. Native American 'Garden Agriculture' in Southeastern North America. World Archaeology 37: 259-274.

Scascitelli, M., K.D. Whitney, R.A. Randell, M. King, C.A. Buerkle and L.H. Rieseberg. 2010. Genome scan of hybridizing sunflowers from Texas (*Helianthus annuus* and *H. debilis*) reveals asymmetric patterns of introgression and small islands of genomic differentiation. Mol. Ecol.19: 521-41.

Schilling, E.E. and C.B. Heiser.1981. Infrageneric classification of *Helianthus* (Compositae). Taxon 30: 393-403.

Schilling, E.E. 1997. Phylogenetic analysis of Helianthus (Asteraceae) based on chloroplast DNA restriction site data. Theoretical and applied genetics, 94(6-7), 925-933.

Schnable, P.S. and R.P. Wise. 1998. The molecular basis of cytoplasmic male sterility and fertility restoration. Trends Plant Sci. 3: 175-180.

Schwarzbach, A.E., L.A. Donovan and L.H. Rieseberg. 2001. Transgressive character expression in a hybrid sunflower species. Am. J. Bot. 88: 270-277.

Scibria N. Hybrids between the Jerusalem Artichoke (*Helianthus tuberosus* L.) and the Sunflower (*Helianthus annuus* L.). C.R. Acad. Sci. URSS 1938; 2: 193-6.

Seiler G. and C.C. Jan. 2010. *Genetics, Genomics and Breeding of Sunflower.* (Hu J., Seiler G., Kole C., Eds.). New York: CRC Press (Taylor & Francis Group).

Seiler, G. J. 1994. Progress report of the working group of the evaluation of wild *Helianthus* species for the period 1991 to 1993. Helia, 17: 87-92.

Seiler, G. J. and L.G. Campbell. 2004. Genetic variability for mineral element concentrations of wild Jerusalem Artichokeforage. Crop Sci. 44: 289-292.

Seiler, G., and L.F. Marek. 2011. Germplasm resources for increasing the genetic diversity of global cultivated sunflower. Helia, 34: 1-20.

Seiler, G., and Marek, L. F. (2011). Germplasm resources for increasing the genetic diversity of global cultivated sunflower. Helia 34, 1–20.

Seiler, G.J. 1983. Protein and mineral concentrations of selected wild sunflower species. Crop Sci. 76: 289-294.

Seiler, G.J. 1984. Evaluation of seeds of sunflower species for several chemical and morphological characteristics. Crop Sci. 25: 183-187.

Seiler, G.J. 1991a. Registration of 15 interspecific sunflower germplasm lines derived from wild annual species. Crop Sci. 31: 1389-1390.

Seiler, G.J. 1991b. Registration of 13 downy mildew tolerant interspecific sunflower germplasm lines derived from wild annual species. Crop Sci. 31: 1714-1716.

Seiler, G.J. 1991c. Registration of six interspecific sunflower germplasm lines derived from wild perennial species. Crop Sci. 31: 1097-1098.

Seiler, G.J. 1992. Utilization of wild sunflower species for the improvement of cultivated sunflower. Field Crops Res. 30: 195-230.

Seiler, G.J. 2000. Registration of 10 interspecific germplasms derived from wild perennial sunflower. Crop Sci. 40: 587-588.

Seiler, G.J. and C. Jan. 1994. New fertility restoration genes from wild sunflowers for sunflower PET1 male-sterile cytoplasm. Crop Sci. 34: 1526-8.

Seiler, G.J. and L.G. Campbell. 2006. Genetic variability for mineral concentration in the forage of Jerusalem artichoke cultivars. Euphytica 150: 281-288.

Seiler, G.J. and L.H. Rieseberg. 1997. Systematics, origin, and germplasm resources of the wild and domesticated sunflower. Agronomy 35: 21-66.

Seiler, G.J., M.E. Carr and M.O. Bagby. 1991. Renewable resources from wild sunflowers (*Helianthus* spp., Asteraceae). Econ. Bot. 45: 4-15.

Serieys, H. 2005. Identification, study and utilization in breeding programs of new CMS sources, in the FAO Subnetwork. Proc. 2005 sunflower subnetwork progress report, 17-20.

Shantz, H. L.1940. Agricultural Regions of Africa. Part I – Basic Factors. Economic Geography, 16: 1-47.

Shull, G. H. 1908. The composition of a field of maize. J. Hered. 1: 296-301.

Siculella, L. and J.D. Palmer. 1988. Physical and gene organization of mitochondrial DNA in fertile and male sterile sunflower. CMS-associated alterations in structure and transcription of the atpA gene. Nucleic Acids Res. 16: 3787-3799.

Smith, B.D. 2006. Eastern North America as an independent center of plant domestication. Proc. Natl. Acad. Sci. U.S.A. 103: 12223-12228.

Smith, B.D. 2013. The domestication of *Helianthus annuus* L. (sunflower). Vegetation History and Archaeobotany, 1-18.

Snow, A.A., D. Pilson, L.H. Rieseberg, M.J. Paulsen, N. Pleskac, M.R. Reagon, D.E. Wolf and S.M. Selbo. 2003. A Bt Transgene Reduces Herbivory and Enhances Fecundity in Wild Sunflowers. Ecological Applications 13: 279-286.

Snow, A.A., P. Moran-Palma, L.H. Rieseberg, A. Wszelaki and G.J. Seiler. 1998. Fecundity, phenology, and seed dormancy of F1 wild-crop hybrids in sunflower (*Helianthus annuus*, Asteraceae). Am. J. Bot. 85: 794-801.

Sossey-Alaoui, K., H. Serieys, M. Tersac, P. Lambert, E. Schilling, Y. Griveau, F. Kaan and A. Berville. 1998. Evidence for several genomes in Helianthus. Theor. Appl. Genet. 97: 422-30.

Springer, N.M. and R.M. Stupar. 2007. Allelic variation and heterosis in maize: how do two halves make more than a whole? Genome Res. 17: 264-275.

Stebbins, G.L. and K. Daly. 1961. Changes in the variation pattern of a hybrid population of *Helianthus* over an eight-year period. Evolution 15: 60-71.

Storlie, E. and G. Charmet. 2013. Genomic selection accuracy using historical data generated in a wheat breeding program. The Plant Genome 6.doi: 10.3835/plantgenome2013.01.0001

Strasburg, J.L., N.C. Kane, A.R. Raduski, A. Bonin, R. Michelmore and L.H. Rieseberg. 2011. Effective population size is positively correlated with levels of adaptive divergence among annual sunflowers. Mol. Biol. Evol. 28: 1569-80.

Sturtevant, E.L. 1885. Indian Corn and the Indian. The American Naturalist 19: 225-234.

Sujatha, M. and A. Prabakaran. 2006. Ploidy manipulation and introgression of resistance to *Alternariahelianthi* from wild hexaploid *Helianthus* species to cultivated sunflower (*H. annuus* L.) aided by anther culture. Euphytica 152: 201-15.

Talukder, Z.I., L. Gong, B.S. Hulke, V. Pegadaraju, Q.J. Song, Q. Schultz and L.L. Qi. 2014. A high-density SNP map of sunflower derived from RAD-sequencing facilitating fine-mapping of the rust resistance gene R 12. PloS One 9: e98628.

Tanksley, S.D. 1993. QTL analysis of transgressive segregation in an interspecific tomato cross. Genetics 134: 585-596.

Tanksley, S.D. and S.R. McCouch. 1997. Seed banks and molecular maps: unlocking genetic potential from the wild. Science 277: 1063-1066.

Thompson, T.E., D.C. Zimmerman and C.E. Rogers. 1981. Wild *Helianthus* as a genetic resource. Field Crops Res. 4: 333-343.

Thone, F. 1936. First Farmers. The Science News-Letter 30: 314-316.
Thormann, I., H. Gaisberger, F. Mattei, L. Snook and E. Arnaud. 2012. Digitization and online availability of original collecting mission data to improve data quality and enhance the conservation and use of plant genetic resources. Genet. Resour. Crop Evol. 59: 635-644.
Touzet, P. and E.H. Meyer. 2014. Cytoplasmic male sterility and mitochondrial metabolism in plants. Mitochondrion 19: 166-171.
Velasco, L., Pérez-Vich, B., Yassein, A.A., Jan, C.C., and Fernández-Martínez, J.M. 2012. Inheritance of resistance to sunflower broomrape (*Orobanche cumana* Wallr.) in an interspecific cross between *Helianthus annuus* and *Helianthus debilis* subsp. *tardiflorus*. Plant Breeding, 131: 220-221.
Wallis, W.D. 1936. Folk Tales from Shumopovi, Second Mesa. The Journal of American Folklore 49: 1-68.
Wang, Z.H., Y.J. Zou, X.Y. Li, Q.Y. Zhang, L. Chen, H. Wu, D.H. Su, Y.L. Chen, J.X. Guo, D. Luo, Y.M. Long, Y. Zhong and Y.G. Liu. 2006. Cytoplasmic male sterility of rice with boro II cytoplasm is caused by a cytotoxic peptide and is restored by two related PPR motif genes via distinct modes of mRNA silencing. Plant Cell 18: 676-687.
Wasley, W.W. 1962. A Ceremonial Cave on Bonita Creek, Arizona. Am. Antiq. 27: 380-394.
Webb, L. Thompson, K.J. Edwards, S. Berry, A.J. Leon, M. Grondona, C. Olungu, N. Maes and S.L. Knapp. 2003. Towards a saturated molecular genetic linkage map for cultivated sunflower. Crop Sci. 43: 367-387.
Welch, M.E. and L.H. Rieseberg. 2002a. Habitat divergence between a homoploid hybrid sunflower species, *Helianthus paradoxus* (Asteraceae), and its progenitors. Am. J. Bot. 89: 472-478.
Whelan, E.D. 1978. Hybridization between annual and perennial diploid species of *Helianthus*. Can. J. Genet. Cytol. 20: 523-30.
Whelan, E.D.P. and W. Dedio. 1980. Registration of sunflower germplasm composite crosses CMG-1, CMG-2, and CMG-3. Crop Sci. 20: 832-832.
Whitney, K.D., R.A. Randell, and L.H. Rieseberg. 2006. Adaptive introgression of herbivore resistance traits in the weedy sunflower Helianthus annuus. American Naturalist 167:794-807.
Whitney, K.D., E.J. Baack, J.L. Hamrick, M.J. W. Godt, B.C. Barringer, M.D. Bennett, C.G. Eckert, C. Goodwillie, S. Kalisz, I. J. Leitch and J. Ross-Ibarra. 2010. A role for nonadaptive processes in plant genome size evolution? Evolution 64: 2097-109.
Whitton, J., D.E. Wolf, D.M. Arias, A.A. Snow and L.H. Rieseberg. 1997. The persistence of cultivar alleles in wild populations of sunflowers five generations after hybridization. Theor. Appl. Genet. 95: 33-40.
Wills, D.M. and J.M. Burke. 2006. Chloroplast DNA variation confirms a single origin of domesticated sunflower (*Helianthus annuus* L.). J. Hered. 97: 403-408.
Wills, D.M., H. Abdel-Haleem, S.J. Knapp and J.M. Burke. 2010. Genetic Architecture of Novel Traits in the Hopi Sunflower. J. Hered. 101: 727-36.
Yarnell, R.A. 1965. Implications of distinctive flora on Pueblo ruins. American Anthropologist 67: 662-674.
Yue, B., B.A. Vick, X. Cai and J. Hu. 2010. Genetic mapping for the *Rf1* (fertility restoration) gene in sunflower (*Helianthus annuus* L.) by SSR and TRAP markers. Plant Breed. 129: 24-28.
Yue, B., S.A. Radi, B.A. Vick, X. Cai, S. Tang, S.J. Knapp, T.J. Gulya, J.E. Miller and J. Hu. 2008. Identifying quantitative trait loci for resistance to *Sclerotinia* head rot in two USDA sunflower germplasms. Phytopathology 98: 926-31.

Chapter 10

Crop Improvement of *Phaseolus* spp. Through Interspecific and Intraspecific Hybridization

Elena Bitocchi[1,*], Domenico Rau[2], Monica Rodriguez[3] and Maria Leonarda Murgia[4]

ABSTRACT

This chapter aims to provide an up-to-date overview of the efforts that have been made in the pre-breeding and breeding of *Phaseolus* crop species, proceeding from a comprehensive summary of their evolutionary history, to an evaluation of their germplasm for traits of interest and its transfer into elite germplasm. In particular, this chapter is focused on common bean (*Phaseolus vulgaris*), as among the cultivated *Phaseolus* species, *P. vulgaris* is the most economically important, and thus the greater part of the research and breeding efforts have been aimed at its improvement. Here, we highlight the results obtained by using different souces of genetic diversity, including wild and domesticated forms of common bean from its two Mesoamerican and Andean gene pool and different *Phaseolus* species belonging to its secondary and tertiary gene pool, to improve biotic and abiotic resistance/tolerance, yields, and quality traits.

Introduction

The current level and organization of genetic diversity of crop species germplasm is the direct result of their evolution, from the appearance of the species as wild forms, with their co-evolution through their initial changing environments and interactions with humans during and after

[1] Department of Agricultural, Food and Environmental Sciences, Università Politecnica delle Marche, Via Brecce Bianche, 60131, Ancona, Italy, E-mail; e.bitocchi@univpm.it
[2] Department of Agriculture, Università degli Studi di Sassari, Via de Nicola, 07100 Sassari, Italy, E-mail: dmrau@uniss.it
[3] Department of Agriculture, Università degli Studi di Sassari, Via de Nicola, 07100 Sassari, Italy, Centro per la Conservazione e Valorizzazione della Biodiversità Vegetale, Università degli Studi di Sassari, Surigheddu, 07040 Alghero, Italy, E-mail: mrodrig@uniss.it
[4] Department of Agriculture, Università degli Studi di Sassari, Via de Nicola, 07100 Sassari, Italy, E-mail: mlmurgia@uniss.it
* Corresponding author

the domestication process, to their dispersal out of their centers of origin. They then continued to evolve and adapt under the continuous pressures of the different agro-ecosystem conditions. Crop breeding is based on the availability of this genetic diversity, without which no improvement is possible (Singh 2001). Thus, the knowledge of how variability originated and was shaped by evolutionary forces is crucial not only to better evaluate and use this diversity to improve crop varieties, but also to implement appropriate strategies of diversity conservation.

Among the approximately 70 *Phaseolus* species (Freytag and Debouck 2002), five were domesticated, four of which are included in the same secondary clade, known as the Vulgaris group: *P. vulgaris* L. (common bean), *P. dumosus* Macfad. (year bean), *P. coccineus* L. (runner bean), and *P. acutifolius* A. Gray (tepary bean). The fifth, *P. lunatus* L. (Lima bean), belongs to the Lunatus clade (Delgado-Salinas et al. 2006).

Origin, Domestication and Expansion of Phaseolus Crop Species

Common Bean

Wild forms of common bean grow under a variety of different climatic and environmental conditions, from northern Mexico to north-western Argentina (Toro et al. 1990; Fig. 1). Three eco-geographic gene pools characterize the wild germplasm of common bean: Mesoamerica, Andes, and northern Peru and Ecuador (see Bellucci et al. 2014a for review; Fig. 1). The Mesoamerica and Andes represent the major gene pools of this species, where both wild and domesticated materials are found. These are characterized by partial reproductive isolation (Gepts and Bliss 1985, Koinange and Gepts 1992), and their distinctiveness has been recognized in several studies based on morphology, agronomic traits, seed proteins, allozymes, and various molecular markers (see Bellucci et al. 2014a as review). The northern Peru and Ecuador gene pool includes wild populations that were discovered by Debouck et al. (1993) and which grow in a restricted geographic area on the western slopes of the Andes. On the basis of the analysis of the sequence of a portion of the gene coding for phaseolin (the main seed-storage protein), this gene pool carries a unique phaseolin type, type I ('Inca') (Kami et al. 1995).

Bitocchi et al. (2012) analyzed the nucleotide sequences of five gene fragments from a wide and representative sample of wild *P. vulgaris* accessions, through which they indicated that common bean originated in Mesoamerica (most likely in Mexico), and then became widespread into South America through subsequent migrations. The whole-genome sequencing analysis performed by Schmutz et al. (2014) confirmed this evolutionary pattern and further estimated the divergence time between the Andean and Mesoamerican wild gene pools at ~165,000 years ago. The major evidence of the Mesoamerican origin for *Phaseolus* is the occurrence of a bottleneck prior to domestication in the Andes, and the identification of four different genetic

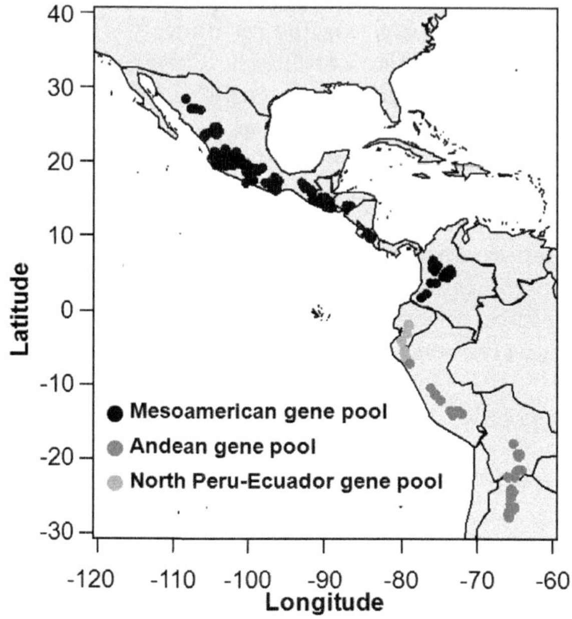

Figure 1. Geographic distribution of wild forms of *P. vulgaris*. The distribution was obtained by considering all of the wild *P. vulgaris* accessions with passport data present in the database of the International Centre for Tropical Agriculture.

groups in Mesoamerica, two of which are more closely related to the South American groups: one with Andean group, and the other with northern Peru and Ecuador accessions (Bitocchi et al. 2012).

The process of domestication is considered to be a milestone in human history, as it marked the transition of humans from hunter/gatherers to agricultural societies. This domestication involved plant and animal species that underwent several morphological and physiological modifications due to human selection. In particular, the main changes in common bean were related to growth habit (indeterminate *vs* determinate), appearance of a wide variety of seed sizes, shapes, and colors, loss of pod dehiscence and seed dormancy, and selection for photoperiod insensitivity. All of these structural and functional changes are known as the 'domestication syndrome', and distinguish the domesticated forms from their wild forms.

Domestication in common bean has been widely investigated in several studies (see Bellucci et al. 2014a for review), all of which agree with the occurrence of two independent events in the Americas that led to the formation of the two major domesticated gene pools (Bitocchi et al. 2013): one in Mesoamerica, and the other in the Andes. The major consequences of this evolutionary process were a reduction in genetic diversity (i.e., the domestication bottleneck) and increased divergence between the wild and domesticated populations. This arose from the reduction in population size (i.e., drift), which resulted in constriction of diversity at the genome-wide level, and from selection in target genomic regions (see Glémin and Bataillon 2009 for review). This selection also included neutral loci that were strictly linked to these target genomic regions (i.e., 'hitchhiking'; Smith and Haigh 1974). For common bean, bottlenecks due to domestication have clearly been identified in these two gene pools in several studies (e.g., Rossi et al. 2009, Kwak and Gepts 2009, Nanni et al. 2011, Mamidi et al. 2011, Bitocchi et al. 2013, Bellucci et al. 2014b, Schmutz et al. 2014). However, a difference was seen between these two gene pools, with the Mesoamerican gene pool showing a three-fold greater reduction in diversity than the Andes gene pool of the domesticated forms compared to the wild forms (Bitocchi et al. 2013). This particular scenario can be explained by the occurrence of a bottleneck prior to domestication in the Andes that strongly impoverished the genetic diversity of the Andean wild forms, and that subsequently affected the domestication bottleneck for this gene pool (Bitocchi et al. 2013). Recently, Bellucci et al. (2014b) investigated changes due to domestication at the transcriptome level in *P. vulgaris*. They analysed RNA sequencing data to compare the transcriptomes of a set of representative wild and domesticated accessions from Mesoamerica, and they found that domestication affected not only the genetic diversity (~60% reduction of diversity in the domesticated *versus* wild accessions at the level of expressed genomic regions), but also the gene expression patterns (18% reduction of diversity in domesticated *versus* wild forms). Moreover, the co-expression networks for the wild and domesticated accessions show distinct community structures that are enriched in different molecular functions (Bellucci et al. 2014b).

Making use of the availability of sequence data for ~27,000 genes, Bellucci et al. (2014b) also applied population genomics approaches to identify the genes involved in the domestication process of common bean. About ~9% of genes were found to be putatively under selection during domestication, with most of these showing evidence of positive selection associated with domestication; however, in a few cases (2.8% of loci), this selection increased the nucleotide diversity in the domesticated pool compared to the wild pool (Bellucci et al. 2014b). Functional analyses for loci putatively under selection during domestication of common bean in Mesoamerica, carried out with the aim to determine whether these loci are associated with the domestication process in other species, showed very interesting results (see Supplementary Methods 2 of Bellucci et al. 2014b for a detailed summary). Analyses that were aimed at the identification of candidate genes that control important domestication traits were also conducted by Schumtz et al. (2014), who identified sets of genes in the Mesoamerica and Andes gene pools that are putatively implicated in flowering time and seed size.

After domestication, cultivation of the domesticated materials was widespread in the Americas, and this led to the formation of the wide variability that can be observed among common bean landraces as a result of their evolution under the different agro-ecosystems conditions of the time. In particular, on the basis of morphological traits, agro-ecological adaptation, and biochemical and molecular markers, common bean landraces have been grouped into seven races (Singh et al. 1991a, Beebe et al. 2000a, 2001): *Durango, Jalisco, Mesoamerica,* and *Guatemala* from Mesoamerica, and *Nueva Granada, Peru* and *Chile* from South America. The landraces from Mesoamerica are small-seeded or medium-seeded (>25 g, 25-40 g per 100-seed weight, respectively), while those from the Andes are large seeded (>40 g per 100-seed weight). Controversy remained about their origins, with the suggestion that they derived from multiple domestication events within each gene pool (Singh et al. 1991a, b, c). This was resolved more recently by the work of Bitocchi et al. (2013), which indicated a single domestication event within each gene pool based on nucleotide sequences of five gene fragments combined with evidence from other studies based on multilocus molecular markers (Papa and Gepts 2003, Kwak and Gepts 2009, Kwak et al. 2009, Rossi et al. 2009, Nanni et al. 2011, Mamidi et al. 2011). This implies that the observed variability results from diversification under cultivation, both through gene flow and farmer selection. Seed variation in the different common bean gene pools for both wild and domesticated forms is shown in Fig. 2.

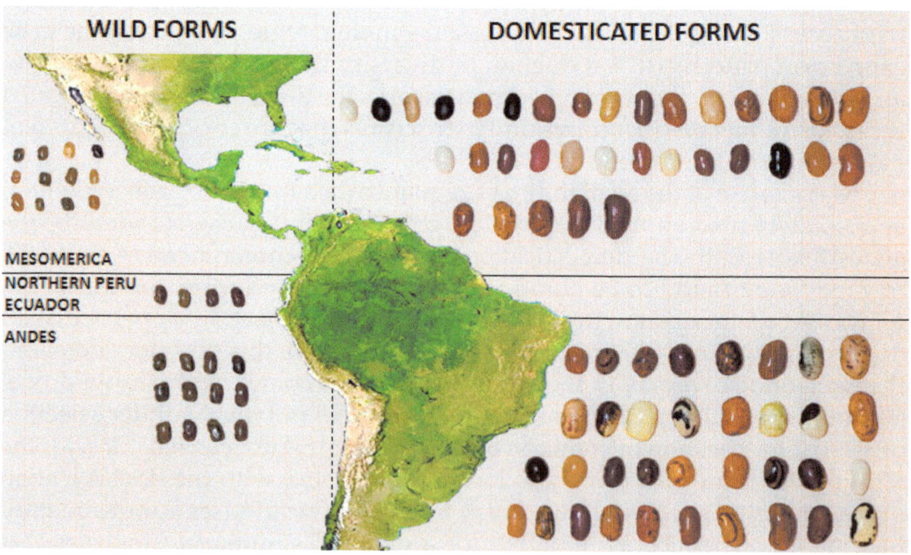

Figure 2. Seed diversity in common bean gene pools.

Gene flow and introgression between wild and domesticated common bean was investigated by Papa and Gepts (2003), using admixture proportion in Mesoamerican samples: introgression from domesticated to wild populations was about 3-4-fold higher than introgression from wild to domesticated populations. This appears to be because wild traits are usually dominant, or partially dominant, in common bean, and thus the F_1 and later generation hybrids will be phenotypically more similar to their wild than to their domesticated progenitors. This asymmetric migration can lead to genetic assimilation of the wild population (Papa and Gepts 2003). By contrast, the recessive nature of domestication alleles combined with the direct selection of farmers against wild phenotypes makes the genome-wide introgression of wild alleles into the domesticated background very unlikely. Moreover, Papa et al. (2005) showed how this scenario affected the diversity of the wild beans across the genome. In particular, selection acted as a barrier against introgression from the domesticated forms in the regions linked to domestication loci, while in the regions unlinked to domestication loci a strong reduction of genetic diversity was observed: this effect was due to unidirectional gene flow from the domesticated forms that were already depauperated from the domestication bottleneck. These results suggest that genomic regions linked to domestication loci in wild populations harbor the richest amount of diversity for use in pre-breeding programmes.

The introduction and spread of common bean out of the Americas appears to have been a very complex process. Different studies have investigated the genetic diversity of common bean germplasm on the different continents: several introductions from the New World were highlighted, along with exchanges between different continents and countries due to intensive commercial interactions (see Bellucci et al. 2014a for a review). Different proportions of Mesoamerican and Andean germplasm have been detected in common bean germplasm in the different continents. The Mesoamerican germplasm has been largely reported for Argentina (Burle et al. 2010) and China (Zhang et al. 2008), while the Andean germplasm appears predominant in Europe (Gepts and Bliss 1988, Lioi 1989, Logozzo et al. 2007, Angioi et al. 2010, Gioia et al. 2013). In Africa, the overall proportions of the Mesoamerican and Andean types are equal (50%), but remarkable differences between countries have been reported. Indeed, in Kenya, east Africa and southern Africa, the Andean type is predominant (Gepts and Bliss 1988, Asfaw et al. 2009), while the opposite was found for Ethiopia and central Africa (Asfaw et al. 2009, Blair et al. 2010a). Moreover, once out of the Americas, the spatial isolation between the Mesoamerican and Andean gene pools was removed, which led to greater possibilities for hybridization and introgression between these two different gene pools. This is a very important feature in terms of the dissemination of common bean around the world; indeed, novel genotypes and phenotypes (i.e., resistance to biotic and abiotic stress, nutritional quality; Angioi et al. 2010, Blair et al. 2010a, Santalla et al. 2010, Gioia et al. 2013) could then be produced that transgress the parental phenotypes (i.e., transgressive segregation; Allendorf and Luikart 2009). Common bean hybrids between gene pools have been detected in numerous studies that have focused on the germplasm from different continents (Angioi et al. 2010, Burle et al. 2010, Asfaw et al. 2009, Blair et al. 2010a, Zhang et al. 2008). Recently, through an analysis of a wide set of common bean accessions from the Americas and Europe, Gioia et al. (2013)

identified a very high level of hybridization between the Mesoamerican and Andean types, with a frequency that was almost four-fold greater in Europe (40.2%) than in America (12.3%). This indicates that the introduction of common bean into Europe was an exceptional evolutionary opportunity, such that a large number of landraces were developed and differentiated to adapt to the wide range of variable agro-ecological conditions.

Other Phaseolus *Crop Species*

As *P. dumosus* (year bean), *P. coccineus* (runner bean), *P. acutifolius* (tepary bean), and *P. lunatus* (Lima bean) are not as economically important as common bean, there is considerably less in the literature regarding their evolutionary history. Similar to common bean, the evolution of Lima bean appears to have been very complex. Its wild forms are distributed from central Mexico to northern Argentina (Allard 1960, Heiser 1965), with three main gene pools suggested (Serrano-Serrano et al. 2010, 2012, Andueza-Noh et al. 2013). One of these is in South America (Ecuador and northern Peru), and the other two are in Mexico, where their geographic distribution appears to be related to their adaptation to the different environments. However, a more complex genetic structure of wild Lima bean in Mexico was suggested by Martínez-Castillo et al. (2014), who proposed three wild gene pools in Mesoamerica. An Andean origin of Lima bean was also suggested (Serrano-Serrano et al. 2010), and this was indicated to have occurred recently (Delgado-Salinas et al. 2006), during the Pleistocene and after the major Andean orogeny that occurred ~2 Ma to 5 Ma ago (Gregory-Wodzicki 2000, Young et al. 2002).

Tepary bean, runner bean, and year bean are distributed in North and Central America, where they are thought to have originated. In particular, wild forms of tepary bean grow in the region from Central Mexico to southwestern USA (Blair et al. 2002). Wild forms of runner bean are distributed from Chihuahua to Panama (Delgado-Salinas 1988), while wild year bean is distributed in a very narrow area in Guatemala (Schmit and Debouck 1991).

A severe reduction in genetic diversity was found for both the Mesoamerican and Andean gene pools of Lima bean (Motta-Aldana et al. 2010, Serrano-Serrano et al. 2012, Andueza-Noh et al. 2013, 2015). Similarly, domestication led to loss of the genetic diversity of domesticated forms of tepary bean (Schinkel and Gepts 1988, 1989, Garvin and Weeden 1994, Munõz et al. 2006, Blair et al. 2012a) and year bean (Schmit and Debouk 1991). In contrast, the few studies on domestication of runner bean have shown no genetic erosion in its domesticated forms (Escalante et al. 1994, Spataro et al. 2011, Rodriguez et al. 2013). This indicates that the similar levels of genetic variation among wild and cultivated runner bean are mainly due to high gene flow between these two forms.

Evolution for Breeding

The domestication process and subsequent modern breeding have considerably reduced the genetic pools of elite germplasm in crop species, and consequently, there is an increased need to use new sources of genetic diversity to improve varieties. The studies described above represent a very useful starting point for the use of genetic diversity available from genetic resources, especially from wild forms and landraces.

For common bean such studies have been fundamental, inasmuch as they have provided breeders with information about both the available genetic diversity and the genetic control of important agronomic traits related to adaptation and domestication. An important outcome for common bean is the high genetic diversity in the Mesoamerican wild forms compared to those of the Andes (Bitocchi et al. 2012), which suggests that we have to look at these Mesoamerican materials for breeding purposes. On this basis, Rodriguez et al. (2015) recently analysed a wide sample of Mesoamerican wild common bean (417 accessions) to determine the role that demographic processes and selection for adaptation have had in the shaping of the current genetic structure of wild *P. vulgaris*. This analysis was based on 131 single nucleotide polymorphisms (SNPs), and they used spatial data in combination with genetic diversity data to uncover the effects of geography from those of ecology (Bradburd et al. 2013, Wang et al. 2013, Guillot et al. 2014, Kraft et al. 2014). By scanning these SNP markers, 26 loci were identified as under signatures of selection. Among these, different loci were shown to have compatible functions with adaptation features, such as chilling susceptibility, cold acclimation, and mechanisms related to drought stress. Moreover, studies on the domestication process allowed genes/genomic regions that are potentially involved in the domestication syndrome to be pinpointed, as well as the identification of genome regions that have been affected by selection (directly or from hitchhiking) during domestication (Bellucci et al. 2014b, Schumtz et al. 2014). This information is crucial to better exploit the genetic diversity in the wild relatives of a crop. In this respect, a large portion of the common bean genome (Papa et al. 2007, Bellucci et al. 2014b) was under selection during domestication, and this has probably been less exploited historically by farmers and breeders. However, these same studies have shown that these regions contain the highest diversity in wild bean progenitors. Moreover, particular interest must be focused on detected introgressions between the gene pools out of the Americas that might have created new interesting combinations of traits, such as greater adaptability to environmental stress, or to diseases and insects (Guzmán et al. 1995). This might also have helped, for example, to break the negative associations between seed weight and yield potential (Johnson and Gepts 1999). The exploration and use of this natural genetic variation will expand the genetic basis of the current elite germplasm, thus providing more flexibility for breeding (Tanskley and McCouch 1997, McCouch 2004). The potential applications are nearly infinite in terms of the construction of novel varieties in breeding programmes.

Intraspecific and Interspecific Hybridization in Phaseolus Species

Each of the five domesticated *Phaseolus* species is characterized by a primary gene pool that includes both its wild and domesticated forms, which are fully, or almost fully, sexually compatible (Smartt 1990). With particular focus on common bean (Fig. 3), its primary gene pool is characterized by the further subdivision into the Mesoamerican and Andean gene pools, each of which includes wild and domesticated forms, along with the wild populations from north Peru and Ecuador. However, hybridization between all common bean populations is not always possible, both for domesticated and wild forms, and incompatibility happens mostly for inter-gene pool crosses. One of the main causes of this incompatibility is the dose-dependent lethal (DL) gene system that leads to nonviable F_1 hybrids and to hybrid weakness visible in the dwarfing of F_1 plants and in lethal dwarfing that segregates in F_2 populations (Shii et al. 1980, Singh and Gutiérrez 1984, Gepts and Bliss 1985, White et al. 1991). The genetic control of this system is based on two semi-dominant alleles, DL1 and DL2, which are expressed at two complementary (epistatic) loci (Shii et al. 1980). The DL1 and DL2 alleles arose before domestication, and they have been found in some populations of the Mesoamerican and Andean gene pools, respectively (Koinange and Gepts 1992). At the same time, there are Mesoamerican and Andean forms that are not involved in this genetic barrier: these can easily be crossed and provide fully fertile progeny in the F_1 and subsequent generations (Koenig and Gepts 1989, Singh et al. 1995). This was clearly demonstrated also by hybrids found out of the Americas, where the geographic barriers between these two gene pools were removed (see Bellucci et al. 2014a for review). Both natural and human-mediated hybridization are very important to increase diversity and to transfer desirable traits into novel varieties.

The secondary gene pool for common bean includes *P. dumosus*, *P. coccineus* and *P. costaricensis*. Except for problems related to differences in flowering time and growth cycles, these species can easily cross with each other when *P. vulgaris* is the female parent (Mendel 1866, Wall 1970, Shii et al. 1982, Hucl and Scoles 1985, Camarena and Baudoin 1987, Singh et al. 1997). It is more difficult to obtain crosses when *P. vulgaris* is used as the pollen donor, and generally the progeny tend to revert back to the cytoplasm donor parent (Smartt 1970; Hucl and Scoles 1985; Debouck 1999). Crosses involving *P. dumosus* and the other two domesticated species of the Vulgaris group (i.e., *P. vulgaris, P. coccineus*) are easily achieved and can also occur naturally in areas of sympatry (Freytag and Debouck 2002), although *P. dumosus* × *P. coccineus* crosses are more frequent, and these progeny have a high level of fertility (Delgado-Salinas 1988, Freytag and Debouck 2002). This is in agreement with the hypothesis that *P. dumosus* originated from a cross of *P. vulgaris* as maternal parent and *P. coccineus* as the paternal parent,

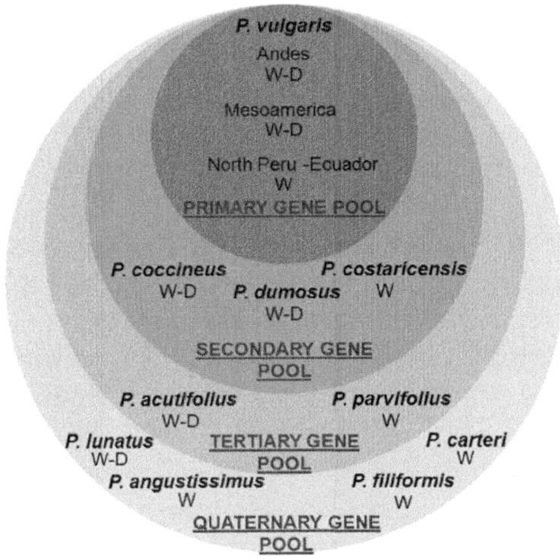

Figure 3. Primary, secondary, tertiary, and quaternary gene pools of *P. vulgaris*. W, wild; D, domesticated.

followed by successive backcrosses from *P. coccineus* as the paternal donor (Schmith et al. 1993, Llaca et al. 1994, Angioi et al. 2009). Indeed, *P. dumosus* is closer to *P. coccineus* according to nuclear DNA comparisons (Piñero and Eguiarte 1988, Delgado-Salinas et al. 1999), while according to chloroplast DNA comparisons, it appears to be more closely related to *P. vulgaris* (Llaca et al. 1994, Angioi et al. 2009). Successful interspecific hybrids have also been obtained between *P. vulgaris* and *P. costaricensis* by hand pollination (Singh et al. 1997).

The tertiary gene pool of common bean comprises *P. acutifolius* and *P. parvifolius*. As these species are more distantly related to common bean, their hybridization requires the use of *in vitro* techniques, such as embryo rescue. Their viable F_1 progeny are usually sterile (see, e.g. Smartt 1970, Mok et al. 1978, Pratt 1983, Thomas et al. 1983, Haghighi and Ascher 1988, Mejía-Jiménez et al. 1994). However, hybrid fertility can be restored through breeding methods, such as one or more generations of backcrossing with the recurrent common bean parent (Pratt et al. 1985, Thomas and Waines 1984), or by congruity backcrossing (recurrent backcrossing to each parent in alternate generations; Haghighi and Ascher 1988, Mejía-Jiménez et al. 1994). What is important to consider here is the choice of the parental genotypes, which can facilitate the hybridization and increase hybrid fitness (Federici and Waines 1988, Ferwerda et al. 2003).

Along with the wild species *P. carteri*, *P. filiformis* and *P. angustissimus*, *P. lunatus* is included in the quaternary gene pool of common bean. Unsuccessful attempts have been made to obtain interspecific hybrids between *P. vulgaris* and both *P. filiformis* and *P. angustissimus* (Balasubramanian et al. 2005). Similarly, it has still not been possible to obtain fertile hybrids from the reciprocal cross of *P. lunatus* × *P. vulgaris* (e.g., Mok et al. 1978, Kuboyama et al. 1991).

Another way to introgress useful genes into elite germplasm of beans to improve the current varieties is the use of genetic transformation. It is not easy to develop the appropriate transformation methods for common bean, considering that similar to other legumes, common bean is generally recalcitrant to *Agrobacterium*-mediated transformation due to its poor regeneration in tissue culture (Svetleva et al. 2003, Colpaert et al. 2008, Arellano et al. 2009). There are essentially two methods available for common bean transformation: indirect gene transfer using *Agrobacterium tumefaciens* or *Agrobacterium rhizogenes*; and direct gene transfer techniques that mainly include particle bombardment and electroporation (Dillen et al. 1995, Kim and Minamikawa 1997, Aragão et al. 2002, Rech et al. 2008, Amugune et al. 2011). The first examples of the production of stable and heritable transformation in *Phaseolus* crop species were those of Russell et al. (1993) and Aragão et al. (1998) for *P. vulgaris*, and of Dillen et al. (1997) for *P. acutifolius*. As part of their studies were oriented toward the design of functional genomics programs focused on root physiology, root metabolism, and root–microbe interactions, Estrada-Navarrete et al. (2006) reported the development of a transformation procedure with very high transformation efficiency rates (75%-90% frequency) using *A. rhizogenes* to generate transformed roots. Recently, genetic transformations of common bean were used to develop bean cultivars that are resistant to the bean golden yellow mosaic virus (BGYMV) through insertion of a mutated *rep* gene (Faria et al. 2006, Bonfim et al. 2007). The BGYMV-resistant engineered line that was developed for the common bean (Bonfim et al. 2007, Aragão and Faria 2009) was the first commercial variety of common bean, and it was evaluated at the molecular level by Aragão et al. (2013). The transgenes of this line were structurally stable for eight self-pollinated generations as well as after backcrosses with a non-transgenic commercial variety. Moreover, a bio-safety evaluation was conducted to demonstrate the safety of the transgenic line in terms of the environment and human and animal health. In their analysis of seeds cooked for 10 minutes, Aragão et al. (2013) did not find any small interfering (si)RNA signals, which indicated that this line is suitable for human consumption.

Biotic Stress

The yield and stability of common bean can be severely constrained by numerous biotic agents, such as bacteria, fungi and viruses, and also by pests that include several insects and nematodes. These can cause severe yield losses both quantitatively and qualitatively due to damage to roots, foliage, stems, pods, and/or seeds. Thus, effective control of diseases and pests is of paramount importance, as this contributes decisively to food security and crop sustainability. The role of plant breeding in this context is of primary relevance, as providing genetically resistant varieties allows reductions in the use of chemicals, and hence lowers the production costs and limits the impact of agriculture on the environment.

Legume breeding and more specifically common bean breeding for resistance to biotic stresses has been covered in several reviews (Miklas et al. 2006, Singh and Schwartz 2010a,b, Keneni et al. 2011, Rubiales et al. 2015), which have also focused in depth on the individual diseases (e.g., Liebenberg and Pretorius 2010, Schwartz and Singh 2013, Singh and Miklas 2015).

Moreover, looking at the breeding activities for biotic stress in common bean, the following consideration has emerged: as previously mentioned the evolutionary history of common bean has a strong impact on the breeding activities. Indeed, the two cultivated common bean gene pools (Mesoamerican and Andean) are genetically differentiated both overall and for their resistance to the fungi responsible for anthracnose (Sicard et al. 1997a,b), rust, (Steadman et al. 1995), and angular leaf spot (Guzman et al. 1995). In each of these plant–pathogen interactions, the plants of one cultivated gene pool were more resistant to the fungus coming from the other gene pool than to the fungus isolated from the same gene pool. This demonstrates that marker assisted selection (MAS) restricted to specific gene pools is a common occurrence for resistance-linked markers in bean; and that landraces have provided, and still provide, a 'bountiful harvest' of resistance genes (e.g., Gonçalves-Vidigal 2009, Ddamulira et al. 2014, Rana et al 2015, Souza et al. 2014, Sousa et al. 2015). Starting from these first two points, it can be anticipated that the screening of European germplasm for disease and pest resistance might provide some very interesting data due to the fact that frequency of hybrids is more than four-fold that in the Americas (Angioi et al. 2010), probably because of the breakdown of the geographic barriers and the co-occurrence of the two gene pools also in close sympatry in small gardens. Thus hybridization and selection might have produced genotypes that carry interesting combinations of different resistance genes from different gene pools with broad resistance. This expectation is reinforced by the observation that plant and pathogen co-evolution in metapopulations of landraces can also take place at relatively low spatial scales (Rau et al. 2015).

Fungal Pathogens

The most relevant fungal diseases of common bean are angular leaf spot, anthracnose, root rots, rust, web blight, and white mold. Individually, these have the potential to promote yield losses of between 50% (e.g., rust) and 100% (e.g., anthracnose, root rot) (see Table 1 of Singh and Schwartz 2010a).

Angular Leaf Spot (Causal Agent: Phaeoisariopsis griseola (Sacc.) Ferraris)

Angular leaf spot is particularly relevant in the tropical and sub-tropical areas of southern and central America and east Africa. This infection causes necrotic lesions in the aerial parts of the plant, with negative effects on the production and quality of the seeds. Results from studies aimed to provide an understanding of the pathogenic variations and to find sources of resistance have suggested co-evolution among the pathogen races and common bean

(Pastor-Corrales et al. 1998). In particular, despite the high genetic variability of *P. griseola* and the differences among different Latin American countries, such as for the common bean, *P. griseola* presents a clear population structure that is characterized by two main groups: Mesoamerican and Andean. Each of these groups is strongly virulent towards the common bean from their respective geographic area, although the Mesoamerican isolates have a broader virulence spectrum than the Andean isolates inasmuch as they also attack Andean beans (Pastor-Corrales et al. 1998). Screening for resistance to different angular leaf spot isolates has been carried out on wide collections of common bean accessions, and sources of resistance to diverse races of *P. griseola* have been found (Schwartz et al. 1982; Pastor-Corrales et al. 1998; Mahuku et al. 2003a). Several lines showing different levels of resistance to angular leaf spot were developed through traditional breeding; e.g., the A339, MAR1, MAR2 and MAR3 lines that were derived from single or interracial crosses between three Mesoamerican races (Singh et al. 2003), and the CAL143 line for the Andean gene pool, which represents the first Andean line that shows interesting levels of resistance to angular leaf spot (Aggarwal et al. 2004). Terán et al. (2013) performed simultaneous selections for resistance to five bacterial, fungal, and viral diseases, including angular leaf spot, in three Andean × Middle American inter-gene-pool double-cross populations, through which they obtained some interesting lines from all three of the populations that showed intermediate to high levels of resistance to the five diseases, when compared with the parents.

Along with materials from both of the common bean gene pools, Mahuku et al. (2003a) evaluated resistance to angular leaf spot in a core collection of plant materials from the secondary gene pools of common bean (i.e., *P. coccineus, P. dumosus*), plus about 1000 lines that were derived from interspecific crosses. They reported high levels of resistance in the secondary gene pool, which demonstrated that these materials are very important germoplasm for breeding. Moreover, 109 interspecific lines were highly resistant (Mahuku et al. 2003a).

From the evidence in the literature, plant resistance appears to be generally monogenic, and more often dominant than recessive (for review, see Miklas et al. 2006). However, several quantitative trait locus (QTL) studies have been carried out under both field and greenhouse conditions, and the genes with major effects have been identified, together with other minor genes (e.g., Miklas et al. 2006, Singh and Schwartz 2010a, Oblessuc et al. 2012, Keller et al. 2015). These studies have indicated a more complex inheritance towards resistance to angular leaf spot. QTL × environment interactions were observed, although some of the major QTLs identified in particular might be useful for bean breeding, as they are stable in all environments (Oblessuc et al. 2012, Keller et al. 2015). A major QTL was recently fine-mapped to a region containing 36 candidate genes, which include some promising representative candidate genes and SNP markers that have been developed for MAS (Keller et al. 2015). Co-segregation analysis also revealed that a gene for angular leaf spot resistance is inherited together with a gene for anthracnose resistance, and that these two genes are tightly linked in a *cis* configuration within a resistance gene cluster on chromosome Pv04 (Gonçalves-Vidigal et al. 2013). A molecular marker was used to simultaneously track both of these resistance genes, and this might be particularly useful for MAS applications, as gene pyramiding has been shown to be an effective breeding strategy against this pathogen (Ddamulira et al. 2015).

Moreover, high virulence diversity of *P. griseola* has been documented (Pastor-Corrales et al. 1998). There are different resistance genes that are effective against the same pathogen race, as well as against different pathogen races. Thus, the pyramiding of different resistant genes is needed to obtain broad resistance. Hence, classical breeding approaches, such as recurrent selection, might not be successful (de Oliveira Arantes et al. 2010).

Anthracnose (Causal Agent: Colletotrichum lindemuthianum *(Sacc. and Magn.) Lams.-Scrib.)*

Bean anthracnose is one of the most damaging of the diseases of common bean (Miklas et al. 2006, Singh and Schwartz 2010a). Screening of wild and domesticated common bean accessions for resistance to anthracnose has been carried out in various studies that included all of the common bean races and wild accessions from both Mesoamerica and Andes, and these have allowed the identification of sources of resistance (Schwartz et al. 1982, Vidigal Filho et al. 2007). Various resistance genes, which are known as Co-genes, can confer race-specific resistance to different strains of this highly variable fungus (Miklas et al. 2006, Singh and Schwartz 2010a). The large majority of these resistance genes are of Mesoamerican origin and are dominant, although a recessive gene and multiple allelisms have also been reported (Miklas et al. 2006). Several QTLs have also been identified (Miklas et al. 2006, Singh and Schwartz 2010a) that have uncovered the genetic architecture of *Colletotrichum lindemuthianum* resistance through the analysis of epistatic interactions (González et al. 2015) and through the identification of new chromosomal regions that are involved in resistance to anthracnose (Trabanco et al. 2015). Oblessuc et al. (2014) identified QTLs that are associated with different races of the anthracnose pathogen. At least 11 QTLs were identified, two of which show major effects on resistance, and can thus be regarded as valuable MAS targets. Molecular markers associated with Co-genes have been developed, and MAS can be applied (Miklas et al. 2006, Singh and Schwartz 2010a, Madakbas et al. 2013, Richard et al. 2014).

As anthracnose resistance genes have often been associated with resistance genes against other fungi, such as angular leaf spot and rust, MAS can allow simultaneous transfer of several genes (Miklas et al. 2006). However, to design effective gene pyramids and to provide durability of resistance, information on the pathogenic variability of *C. lindemuthianum* in any given area is needed (Miklas et al. 2006). Recently, the resistance gene

Co-x was fine-mapped to a 58-kb region at one end of chromosome Pv01. Comparative analysis between soybean and common bean has revealed that the *Co-x* syntenic region carries *Rhg1*, which is a major QTL that contributes to soybean cyst nematode resistance. The PCR-based markers generated should also be useful in MAS selection for pyramiding *Co-x* with other R genes (Richard et al. 2014). Moreover, it has been observed that common bean locus *Co-4*, which has traditionally been referred to as an anthracnose-resistant gene, contains a cluster of genes that are co-regulated with the basal immunity of common bean (Oblessuc et al. 2015).

Several lines have been released that carry single or combinations of different Mesoamerican and Andean resistant genes; e.g., cultivars 'Raven' (Kelly et al. 1994), 'Phantom' (Kelly et al. 2000), 'Chinook' 2000 (Kelly et al. 1999) and 'Jaguar' (Kelly et al. 2001). Other examples are the above-mentioned A339, MAR1, MAR2 and MAR3 lines that were derived from single and multiple crosses between Mesoamerican genotypes belonging to different races (i.e., Durango, Jalisco, Mesoamerica races) which are characterized not only by useful levels of resistance to angular leaf spot but also to antrachnose (Singh et al. 2003). Lines with intermediate to high levels of resistance to five bacterial, fungal, and viral diseases, including anthracnose, were developed from three inter-gene-pool (Andean × Mesoamerican) double-cross populations (Terán et al. 2013).

Mahuku et al. (2003b) evaluated a core collection of the secondary gene pool of *P. vulgaris* and interspecific lines derived from simple and complex crosses of primary and secondary genotypes for their resistance to antrachnose. Interestingly, they reported that none of the secondary gene pool genotypes was susceptible to *C. lindemuthianum*, and that *P. dumosus* has a higher level of resistance compared to *P. coccineus*. They identified several interspecific lines with high resistance that could be used in breeding programs as sources of resistance.

Root Rot (Causal Agents: Aphanomyces, Fusarium, Rhizoctonia, Pythium).

Root rot diseases are important in most bean production areas, and particularly in those with low soil fertility or with intensive cultivation and no crop rotation (Miklas et al. 2006). Breeding for root rot resistance is fundamental, as root health is fundamental for plant development and water and nutrient uptake. Moreover, root diseases can also aggravate problems of drought.

Fusarium solani is the predominant root rot pathogen in common bean (Miklas et al. 2006). Sources of resistance were identified in the Mesoamerican gene pool decades ago (Miklas et al. 2006, Singh and Schwartz 2010a). However, complex quantitative inheritance, coupled with difficulties in inter-gene-pool hybridization, have limited the incorporation of *Fusarium* root rot resistance into Andean bean cultivars (Miklas et al. 2006). The first QTL analysis with *F. solani* f. sp. *phaseoli* showed that the genetic resistance was multigenic, with individual QTLs explaining no more than 15% of the variance in resistance. Here, strong environmental effects and very low correlation between data obtained under field and greenhouse conditions were observed (Schneider et al. 2001). Minor QTLs have been detected in many other studies (see Miklas et al. 2006, Singh and Schwartz 2010a), although interestingly, a major QTL has also been detected that accounted for up to 53% of the phenotypic variation, and would probably be useful for MAS (Román-Avilés and Kelly 2005).

Marker assisted selection for *Fusarium* root rot resistance would be particularly useful, as selection under field conditions is very laborious and difficult. In this regard, an interesting resistance QTL to *Fusarium* root rot was reported by Kamfwa et al. (2013) that was confirmed in two populations. This might facilitate MAS for the transfer of resistance for *Fusarium* root rot into the highly susceptible Andean genotypes. Recently, a mapping population was evaluated for both *F. solani* f. sp. *phaseoli* root rot and *Aphanomyces euteiches* root rot over three and two seasons, respectively (Hagerty et al. 2015), and for a set of root architecture traits. Overall, relatively small effects were again observed with single QTLs, which explained between 5% and 22% of the total genetic variation. However, the QTL for resistance to the two diseases mapped to different genomic regions, which indicates different genetic control for these two pathogens. Moreover, the incidence of root rot disease did not correlate with root architecture.

Mukankusi et al. (2011) evaluated progenies of a 12 × 12 full di-allele mating scheme with reciprocal crosses using 12 domesticated lines from both the Mesoamerican and Andean gene pools (as intra- and inter-gene-pool crosses). They confirmed the very complex inheritance of resistance to *Fusarium* root rot, with additive gene actions and dominance, and various forms of epistasis and maternal and nonmaternal reciprocal effects involved. They showed that resistant parents that contained a number of different resistance genes can be combined with the expectation of producing strong and durable resistance, and they indicated four parental lines that contributed high levels of resistance in crosses and would thus be recommended for use in breeding programs. The use of resistant wild materials appeared to be very interesting for breeding. In particular, a cross between the Mexican cultivar 'Negro Tacaná' and a wild-type from Jalisco, 'G12947', gave F_1 resistant lines, with the resistance of the 'G12947' accessions tested and verified in field and greenhouse trials (Navarrete-Maya and Acosta-Gallegos 1999, Acosta-Gallegos et al. 2007). Similarly, 200 BC_2F_4 inbreed lines that were derived from intra-gene-pool crosses between the Mesomerican root rot resistant Durango race cultivar 'Pinto Villa' and a wild-type from Jalisco (Mexico) showed resistance under severe moisture and root rot stress (Acosta-Gallegos et al. 2007). Of note, the possibility to use *P. coccineus* as a source of root rot resistance in common bean breeding needs to be better exploited (Wallace and Wilkinson, 1965).

Rhizoctonia solani causes economically important root and hypocotyl damage throughout the world. An efficient screening method to evaluate damage from *R. solani* was developed and has been used to identify dry bean lines with some resistance to this disease, with the sources of resistance identified (Peña et al. 2013). The effects of co-inoculation of *R. solani* and *C. lindemuthianum* or *Uromyces appendiculatus* have also been studied (Paula Júnior et al. 2015). Very interestingly, antagonistic and synergistic interactions between *Rhizoctonia* root rot and these two common foliar bean diseases, respectively, have been observed.

Attempts at introgression of a *Pythium* root rot resistance gene into susceptible common bean cultivars have also been made, both considering this trait individually (Nzungize et al. 2011) and simultaneously with other disease-resistance traits (Kiryowa et al. 2015). Kiryowa et al. (2015) analyzed the effects of marker-aided pyramiding of anthracnose and *Pythium* root rot resistance genes on the agronomic characteristics among advanced breeding lines. They concluded that pyramiding higher numbers of resistance genes might result in a grain yield reduction through the number of seeds per plant. Therefore, they suggested that breeders should simultaneously select for a number of pyramided genes and the number of seeds per plant, and other highly associated traits.

Rust (Causal Agent: Uromyces appendiculatus *(Pers.: Pers.) Unger)*

As for other pathogens, *U. appendiculatus* co-evolved with the Andean and Mesoamerican common bean (Pastor-Corrales and Aime 2004). The evolutionary potential of *U. appendiculatus* is particularly high, and this has led to rapid breakdown of the major gene resistance in bean cultivars (Miklas et al. 2006, Singh and Schwartz 2010a, Liebenberg and Pretorius 2010). The first studies on the associations between molecular markers and leaf rust resistance genes were carried out more than 20 years ago (Miklas et al. 2006), and pyramiding different resistance genes that underlie the different resistance mechanisms (i.e., specific, adult plant, slow rusting, reduced pustule size, pubescence) has been considered a good strategy for obtaining more durable resistance (Miklas et al. 2006, Singh and Schwartz 2010a, Liebenberg and Pretorius 2010). High levels of resistance were found in both Andean and Mesoamerican cultivars and landraces (Singh and Schwartz 2010a). Nine major rust resistance genes have been identified: *Ur 3-7, Ur 9, Ur 11, Ur 12, Ur13,* and a further four unnamed genes. These rust resistance genes have been mapped, and associated markers that are useful for MAS have been developed (Miklas et al. 2006, Singh and Schwartz 2010a, Liebenberg and Pretorius 2010). The tendency to cluster appears to be stronger for these rust resistance genes than for anthracnose resistance genes (Miklas et al. 2006). Co-localization of rust and anthracnose resistance genes for positions across different chromosomes has also been observed (Miklas et al. 2006, Singh and Schwartz 2010a).

Given the very high virulence diversity of *U. appendiculatus* and its production of new virulent strains, the discovery of highly effective disease resistance genes is a very important step for the development of cultivars with effective rust resistance. For this reason, screening of plant genetic resources in this case appears strategic, as witnessed by the relatively recent discovery of new genes (de Souza et al. 2011) and of exceptional rust resistance that is conferred by a single dominant gene in a Mesoamerican common bean, which provides broad resistance (Pastor-Corrales et al. 2012). Interesting data have also been obtained by screening European germplasm from the Andean gene pool (Leitão et al. 2013). Screening for resistance to rust remains in progress in several parts of the world (e.g., Souza et al. 2013), and multiple disease-resistant lines have been registered, which will be useful as parents to enhance virus and rust resistance (Beaver et al. 2015). Introgression and pyramiding of different Mesoamerican and Andean rust resistant genes in new cultivars is the major breeding strategy that has been adopted (Miklas et al. 2006; Singh and Schwartz 2010a). Acevedo et al. (2013) tested different wild and domesticated accessions of both *P. vulgaris* and *P. coccineus* for rust resistance, and they reported higher levels of resistance in *P. coccineus* compared with *P. vulgaris*, which suggests that *P. coccineus* is a promising source of rust resistance.

White Mold (Causal Agent: Sclerotinia sclerotiorum *(Lib.) deBary)*

White mold development is highly influenced by the environment, although under favorable conditions (i.e., moist conditions in cool subtropical or temperate areas), this disease can be devastating, as it can result in severe loss of yield and quality. Physiological resistance and disease avoidance provided by plant-architecture-related traits have contributed to white mold field resistance. Thus, several attempts have been made to characterize white mold avoidance. As an example, Miklas et al. (2013) identified 13 QTLs where about half were strongly associated with disease avoidance traits measured under field conditions, such as lodging, canopy porosity, and height. In particular, resistance to lodging is extremely important for reducing disease severity. However, avoidance traits were less effective in reducing disease severity in trials where there was heavy disease infection. Hoyos-Villegas et al. (2015) reported on a major QTL that is associated with white mold avoidance, which is associated with disease avoidance traits such as canopy porosity, plant height, the stay green stem trait, and maturity (Hoyos-Villegas et al. 2015). However, breeding for disease avoidance is not recommended (Miklas et al. 2013); instead, selection for resistance to white mold in the field is preferable, in combination with high yield potential and acceptable maturity. As discussed by Miklas et al. (2006), recurrent selection solely for white mold resistance might result in lines showing other agronomically undesirable traits, such as lower yield and late maturation date. A strategy termed 'multi-trait bulking' might help overcome this limitation: QTLs are mapped by obtaining a bulk of individuals with physiological resistance, and at the same time, more desirable phenotypes (e.g., acceptable maturation date and yield) that can be contrasted with the bulk of the susceptible genotypes (Kolkman and Kelly 2003).

Resistance to white mold is inherited as a complex trait with low to moderate heritability, and multiple QTLs have been identified on the majority of chromosomes (Schwartz and Singh 2013). Major QTLs that have been confirmed in different genetic backgrounds have been reported (Schwartz and Singh 2013), and new resistance sources are under investigation (e.g., Lehner et al. 2015, Viteri and Singh 2015). The pyramiding of white mold resistance between and within common bean gene pools and with both wild and domesticated accessions has also been demonstrated to be a promising approach (Schwartz and Singh 2013, Singh et al. 2014). There have also been successful examples of interspecific hybridization of *P. vulgaris* with other *Phaseolus* species. Singh et al. (2009a) released two white mold resistant common bean lines, VCW54 and VCW55, that were developed using congruity backcrossing between a tropical small-seeded black common bean cultivar 'ICA Pijao' and the *P. coccineus* accession 'G35172'. Another example is the development of the wild mold resistant interspecific line 'VRW32', which was derived by recurrent backcrossing of 'ICA Pijao' with the wild *P. costaricensis* germplasm accession 'G40604' (Singh et al. 2009b, 2013). Populations derived form interspecific crosses between *P. vulgaris* and *P. coccineus* were also investigated to study inheritance of white mold resistance (Schwartz et al. 2006), which suggested that resistance of the *P. coccineus* accessions PI433246 and PI439534 are under the control of a single dominant gene.

Schwartz and Singh (2013) also proposed an integrated genetic improvement strategy for resistance to white mold, with germplasm enhancement and cultivar development using multiple-parent crosses and gamete selection methods of breeding. Partial resistance to white mold has been identified (e.g., Balasubramanian et al. 2014) and MAS backcrossing QTLs for partial resistance to *Sclerotinia* white mold in dry bean has also been attempted (Miklas 2007, Miklas et al. 2014a). It was concluded that MAS is an effective breeding tool for introgressing partial resistance to white mold, but it was also noted that because of the occurrence of linkage drag for yield, further selection for agronomic performance might be required to obtain lines that will be worth commercial production. Recently, next-generation sequencing approaches have been undertaken to identify a resistance gene to white mold (Mamidi et al. 2015).

Bacterial Pathogens

We consider here two relevant pathogens, common bacterial blight and Halo blight, both of which can cause yield losses of up to 45% (Singh and Schwartz 2010a).

Common Bacterial Blight (*Causal Agent:* Xanthomonas axonopodis *pv.* phaseoli)

Common bacterial blight is a seed-borne disease that constrains bean yields worldwide. The genetic basis of resistance to its causal agent, *Xanthomonas axonopodis*, is complex, with many minor QTLs scattered across the genome that are not stable across environments or across different plant organs (see Fig. 1 of Miklas et al. 2006). Major QTLs have also been detected as sequence characterized amplified region (SCAR) markers, and these have been developed for use in MAS (e.g., Mutlu et al. 2005a, b). As indicated by Miklas et al. 2006, "phenotypic selection is needed to retain minor effect QTLs and to select for epistatic interactions that contribute to improved resistance". Interestingly, genotyping with real-time PCR has revealed recessive epistasis between independent QTLs that confer resistance to common bacterial blight in dry bean (Vandemark et al. 2008). Thus, to obtain high resistance, it is better to select materials that are fixed for both of these QTLs. Moreover, interactions of common bacterial blight QTLs have been documented in a resistant inter-cross population of common bean (Durham et al. 2013). Using traditional breeding approaches, bean breeders have combined resistance sources from the primary and secondary gene pools to obtain cultivars and lines with improved resistance. Examples that have involved crosses between *P. vulgaris* and *P. coccineus* for the development of common bacterial blight resistant lines are numerous (Freytag et al. 1982, Miklas et al. 1999, Zapata et al. 2004). Similarly, successful introgressions of resistance to common bacterial blight from *P. acutifolius* are numerous (Scott and Michaels 1992, Singh and Muñoz 1999, Mutlu et al. 2005b, 2008). Singh and Muñoz (1999) released the 'VAX' common bean breeding lines (VAX1 to VAX6) that combine resistance genes from *P. vulgaris* and *P. acutifolius* and that show higher levels of common bacterial blight resistance than breeding lines derived from the combination of the resistance genes from *P. coccineus* and *P. vulgaris* (Miklas et al. 2006, Singh and Schwartz 2010a, Viteri et al. 2014). Beaver et al. (2012) released a common bacterial blight resistant line, PR0650-3 that they derived from crosses that involved a Mesoamerican domesticated genotype (BAT93), a wild genotype from Jalisco (Mexico), and the VAX 6 line. Interestingly, simultaneous selection for resistance to five bacterial (including common bacterial blight), fungal, and viral diseases in three Andean × Middle American inter-gene pool common bean populations was shown to be effective (Terán et al. 2013).

Halo Blight (*Causal Agent:* Pseudomonas syringae *pv.* phaseolicola) (*Psp*).

Halo blight can be particularly frequent and have high impact in humid and cool climates. Overall, five monogenic resistance genes have been reported, but these do not confer a wide spectrum of resistance (i.e., they are not effective against the five races of the pathogen *Pseudomonas syringae*). However, quantitative broad-resistance genes have also been identified (Miklas et al. 2006, Singh and Schwartz 2010a). QTLs for resistance to halo blight have been identified, and these were co-mapped with loci that determine resistance against the different races of the pathogen (Fourie et al. 2004). Recently, new resistant loci were also identified (Miklas et al. 2014b). Some sources of resistance to halo blight have been identified within *P. vulgaris* species, such as for 'Pinto US14', 'CAL143', 'GN Nebraska No.1 Sel. 27' and 'PI150414' (Taylor et al. 1978, 1996, Singh and Schwartz 2010a). In particular the 'PI150414' line has been used in different breeding programmes to develop common bean resistant lines (Singh and Schwartz 2010a).

Viral Pathogens

We have considered four viral pathogens here: bean common mosaic virus (BCMV), bean common mosaic necrosis virus (BCMNV), beet curly top virus (BCTV), and BGYMV, all of which can cause yield losses of up to 100% (Singh and Schwartz 2010a).

Bean Common Mosaic Virus and Bean Common Mosaic Necrosis Virus

Bean common mosaic virus and BCMNV are the most common and destructive potyviruses that infect common bean worldwide (Drijfhout 1978). Both of these viruses are seed-borne and can be transmitted by several aphid species in a non-persistent manner (Drijfhout 1978).

Genetic resistance to these potyviruses is determined by different genes. The *I* gene is dominant, and it confers hypersensitive resistance to several related potyviruses, while three different *bc* genes act in a recessive manner (for review, see Miklas, et al. 2006). As shown in several studies reviewed by Miklas et al. 2006, the *I* and *bc* genes subtend different mechanisms of resistance and are independent, which make them ideal for gene-pyramiding approaches. However, direct selection for the *I* gene is not possible, as the action of the dominant *I* gene is masked by the recessive *bc-3* gene. Thus, the adoption of MAS represents an important opportunity, and this has been used in several studies to develop enhanced germplasm with the *I* + *bc-3* gene combination (reviewed by Miklas et al. 2006). The efficiency of MAS is still hampered, however, due to the dominance, linkage phase, and loose linkage of previously developed markers (Bello et al. 2014). Thus, Bello et al. (2014) applied an *in-silico* bulked segregant analysis with SNPs for a very wide diversity panel, to develop co-dominant and tightly linked markers to the *I* gene that controls resistance to BCMV. Encouraging recent studies should be noted here: identification of genes for resistance to BCMV and BCMNV in snap bean breeding lines using conventional and molecular methods (Pasev et al. 2014); incorporation of resistance to angular leaf spot and BCMNV diseases into an adapted common bean genotype in Tanzania (Chilagane et al. 2013); and registration of lines with improved resistance to these viruses (Beaver et al. 2015).

Other attempts to improve common bean resistance to BCMV and BCMNV have been based on the development of cultivars derived from multiple-parent crosses (Terán et al. 2013). This method is based on intra- and inter-gene-pool crosses along with gamete selection, and it was proposed by Singh (1994) to develop breeding lines simultaneously resistant to more than one disease (Terán et al. 2013). Indeed, according to Singh and Schwartz (2010a), high levels of durable resistance can be achieved by introgression and pyramiding of resistant genes into landraces and cultivars from lines of the primary gene pool. An example was reported by Asensio-S. Manzanera et al. (2005, 2006), who used gamete selection in multiple-parent crosses to introgress the *I* gene subtending BCMV and BCMNV resistance from the commonly highly resistant Mesoamerican gene pool into the more susceptible Andean gene pool. On the other hand, a previous study allowed introgression of the *bc-3* gene from the Mesoamerican background into the Andean gene pool (Kornegay 1992, Johnson et al. 1997).

Interspecific breeding lines created to introgress biotic resistance from the secondary gene pool into the primary gene pool (e.g., see above for white mold and common bean blight) have also been shown to have increased resistance to BCMV and BCMNV (Singh and Muñoz 1999, Singh et al. 2001, 2013). Among others, the VAX-3 and VAX-6 interpecific breeding lines that were obtained from a cross between *P. vulgaris* and *P. acutifolius* were used as parents of the multiple-parent crosses that were developed to determine the effectiveness of simultaneous selection for resistance to five different diseases, including BCMV and BCMNV (Terán et al. 2013).

Beet Curly Top Virus

Beet curly top virus is a geminivirus that is vectored by the leafhopper *Circulifer tenellus* (Baker). BCTV occurs worldwide and it has a very wide host species range, which includes other relevant crops, such as tomato, pepper and sugar beet (Bennett 1971). Under field conditions, breeding for resistance to BCTV is difficult, because of the unpredictability of the attacks both in time and space, and also on the basis that under greenhouse conditions the present inoculation tests are very tedious (Miklas et al. 2006). Despite the difficulties of field and greenhouse screening, resistance genes have been identified for BCTV. A dominant resistant gene was identified, *Bct*, that is close to a region involved in viral (i.e., BGYMV), bacterial (i.e., common bacterial blight) and fungal (i.e., anthracnose, white mold) resistance (Larsen and Miklas 2004). A SCAR marker useful to track the inheritance of the *Bct* gene was developed specifically for BCTV resistance breeding of Andean bean lines, and several genotypes that are characterized by good levels of resistance were released following conventional plant breeding (Larsen and Miklas 2004, Miklas et al. 2006, Beaver and Osorno, 2009, Singh and Schwartz 2010a).

As observed for other diseases, small-seeded Middle American dry beans (e.g., 'T 39', 'Porrillo Sintetico', and 'Tio Canela 75') and cultivars of race Durango usually show higher levels of resistance to BCTV, when compared to Andean cultivars (Singh and Schwartz 2010a). Exploiting these differences, as also for the increased resistance of the secondary gene pool, would allow the broadening of the genetic bases of common bean for BCTV resistance. However, to the best of our knowledge, at present, no interspecific crosses have been developed to specifically address this objective (Singh and Schwartz 2010a).

Bean Golden Yellow Mosaic Virus

Bean golden yellow mosaic virus is a geminivirus and it is vectored by insects of *Bemisia* spp. (Gálvez and Morales 1989). To improve genetic resistance, phenotypic recurrent selection was used. Moreover, the breeding strategies have included combining resistance from different gene pools, as also across races and gene pools, along with transfer of resistance genes from *P. coccineus* (Singh and Schwartz 2010a). Several genes and QTLs have been reported, with SCAR co-dominant markers developed some 20 years ago that are useful for MAS (Urrea et al. 1996). Very recently, the adoption of genotyping-by-sequencing has enabled mapping and marker development for the *By-2* potyvirus resistance allele in common bean, and MAS of *By-2* (Hart and Griffiths 2015). MAS will have a relevant role in breeding for BGYMV resistance, because disease screening in the field is not predictable, and greenhouse screening is not efficient (Miklas et al. 2006). Moreover, bean lines (*P. vulgaris* L.) resistant to BGYMV and bean weevil (*Acanthoselides obtectus* Say) (González Vélez et al. 2012) have been developed, along with the registration of bean germplasm lines with resistance to BGYMV, BCMV, BCMNV and rust (Beaver et al. 2015).

Insects and Nematodes

Common bean is attacked by many insects, but relatively few of these are of major importance. Indeed, some of these pests are important worldwide, but others are important only in specific and more restricted areas. Yield losses can be severe, and have been estimated at between 35% and 100% (Singh and Schwartz 2010b).

The progress realized by common bean breeding has been different for different insects. The data have been satisfactory for pod weevil (*Apion godmani*), which damages pods and seeds. In this case, the availability of resistance (antibiosis) sources (albeit not very frequently; Garza et al. 1996, 2001) in landraces and wild forms (e.g., Garza et al. 2001) has allowed the transfer of resistant genes to breeding lines through mass pedigree selection. Similarly, the sources of resistance against the post-harvest storage bean weevil (*Zabrotres subfasciatus*) are relatively rare, and have been found only in wild common bean accessions from central Mexico (e.g., Sparvoli and Bollini 1998). In this case, resistance is due to the presence of a storage protein in the cotyledons, known as arceline. Using a recurrent backcross program, the resistance was transferred into different lines of cultivated bean (Cardona et al. 1990). For leafhoppers (*Empoasca kraemeri*), significant variations in antixenosis have been observed in the Mesoamerica race (e.g., Calderon and Backus 1992), albeit this is inherited quantitatively and with low heritability. The adoption of a method like recurring cycles of the bulk pedigree has lead to higher tolerance to leafhopper attacks in breeding lines (e.g., Kornegay and Cardona 1990). However, a source of resistance that has not yet been exploited is tepary bean (Cardona and Kornegay 1999), which can be crossed with common bean (e.g., Mejía-Jimenez et al. 1994). Interesting data have also been obtained for melon thrips (*Thrips palmi*), where there has been some evidence of antibiosis (Frei et al. 2003), and QTLs potentially exploitable with MAS have been detected (Frei et al. 2005).

Among the nematodes, the best data have been obtained for the root-knot nematode (*Meloidogyne* species), where several genotypes and breeding lines with good resistance levels have been found (Singh and Schwartz 2010b). Moreover, the genetic resistance to *Meloidogyne* was found to be due to a dominant marker (Me-1) with differences among breeding lines, and additionally, among different *Meloidogyne* species (see Singh and Schwartz 2010b for review). On the contrary, there is a paucity of information for lesion (*Pratylenchus* species) and soybean cyst (*Heterodera glycines*) nematodes. Also, as indicated by Singh and Schwartz (2010b), common bean cultivars with high levels of resistance to one or more insect pests and nematodes are rare.

Abiotic Stress

Wild *P. vulgaris* is distributed from northern Mexico to northwestern Argentina, and it occupies variable ecological niches, from the semi-arid regions and wet lowlands in Central America, to high altitude areas (>3,000 m a.s.l.) in the Andes (Toro et al. 1990, Gepts and Debouck 1991). At the same time, the domesticated form is grown worldwide under variable climate conditions (Singh 1989, Gepts and Debouck 1991, Broughton et al. 2003). This wide adaptation of *P. vulgaris* to different environments constitutes the basis to enhance the genetic diversity of the crop that can be exploited to cope with the pressure of different abiotic stresses (e.g., drought, heat, cold, aluminum) (Singh 2001). In addition, the other wild and domesticated *Phaseolus* species also represent sources of variable resistance that can be used for interspecific crosses with common bean (Miklas et al. 1994, Lin and Markhart 1996, Singh 2001, Muñoz et al. 2004, Rodiño et al. 2007, Butare et al. 2011). For the breeding efforts to improve grain yields, it has been estimated that up to 90% of the genetic variability in common bean and its sister species remains unutilized, or underutilized (Singh 2001).

Abiotic stress resistance is generally a polygenic trait, and its inheritance is often associated with different interacting mechanisms that render its investigation difficult at both the physiological and genetic levels. Despite this, breeding research for abiotic stress tolerance represents a compelling task, due to the negative effects that variable environments can have on crop growth and yield (Beebe et al. 2011). *P. vulgaris* is often exposed to the risk of drought, as it has been estimated that only 7% of its production occurs across well-watered areas (Broughton et al. 2003). Other yield limitations can occur in conjunction with drought when common bean is grown on acid soils with low phosphorous availability and/or aluminum toxicity; relevant constraints can also originate from low nitrogen availability and other edaphic stress (Broughton et al. 2003).

Drought

Drought represents the most limiting abiotic stress to crop productivity, and in 2009 the Crop Science Society of America included this constraint among the greatest of challenges that need to be addressed in the crop sciences (Lauer et al. 2012). A common definition of drought in an agricultural context refers to a period with declining soil moisture and the consequent crop failure (Mishra and Singh 2010). Plant responses to this low soil water content depend on their biological characteristics and stage of growth, and the properties of the soil (Mishra and Singh 2010). In particular, drought resistance refers to the different mechanisms developed by plants to survive in periods of drought stress, when responses can vary according to the pattern of drought encountered in the variable environments (Beebe et al. 2013). In common bean, early maturity usually follows in terminal drought stress environments, which are frequent in Central America and north-eastern Brazil, and in the African eastern highlands. Conversely, phenotypic plasticity might be more appropriate during the intermittent drought stress that is common in Mexico and southern Africa, and in much of Ethiopia (Schneider et al. 1997a; Frahm et al. 2004; Beebe et al. 2013; Asfaw and Blair 2014).

Although no definite numbers have been given in terms of yield losses due to drought, it is estimated that around 60% of common bean production worldwide is at risk, albeit with significant differences between countries determined by climate, agricultural management, availability of technical systems, and social factors (Broughton et al. 2003; Beebe et al. 2013). Yield limitations for common bean through drought stress are particularly significant in highland Mexico, Central America, north-eastern Brazil, and eastern and southern Africa, where this crop is often grown by small landholders and its consumption forms the basis of the diet of the population (Broughton et al. 2003, Beebe et al. 2011, 2013). As an example, it has been reported that in certain years in highland Mexico and north-eastern Brazil, yields can be around 0.4 t ha^{-1} to 0.45 t ha^{-1}, which is much less than the mean 1.8 t ha^{-1} of the USA, or 1.5 t ha^{-1} of Sao Paulo (Beebe et al. 2013).

Primary breeding efforts for drought resistance in *P. vulgaris* have concentrated on germplasm evaluation through the selection of lines that have shown higher yields when compared to other lines grown under the same drought conditions. Early trials started in the 1970s at the International Center for Tropical Agriculture in Colombia, followed by phenotype investigation in multiple environments (Ishitani et al. 2004). The races of common bean from the semi-arid regions of Mexico were believed to show some level of drought tolerance that could be used in breeding programs (Singh et al. 1991a). Indeed, preliminary studies showed that as well as some others, the BAT 477 breeding line (Mesoamerican race) was outstanding for its drought tolerance, which was mainly attributed to the root length density and its deeper soil moisture extraction (White et al. 1994). It was also demonstrated that local adaptation of genotypes was crucial in responses to drought stress. High levels of drought resistance were also found in the races Durango and Jalisco. Subsequent breeding efforts were based on the race Durango, which showed superior performance under drought stress, combined with early maturity, and high seed yield and 100-seed weight (Teràn and Singh 2002).

Lines obtained from inter-racial crosses of Mesoamerica × Durango and inter-gene pool crosses of Mesoamerica × Nueva Granada out-yielded cultivars from the races Durango and Jalisco, both in nonstressed and drought-stressed environments, with particular encouraging performances for the SEA 5 line (Teràn and Singh 2002). Consistently high performances under drought stress were observed for different lines derived from the race Durango which were evaluated for different phenological and physiological traits, yields and yield components in Michigan, USA (Ramirez-Valleyo and Kelly 1998), in the Rift Valley in east Africa (Abebe et al. 1998), and in the lowland tropics (Frahm et al. 2004). Furthermore, elite cultivars with satisfactory yields under drought stress and maximised yield potential under nonstressed environments were obtained through inter-racial crosses between the races Durango and Mesoamerica (Teràn and Singh 2002; Frahm et al. 2004; Beebe et al. 2008). On the contrary, less progress has been made with the Andean genotypes, even if germplasm evaluation and pre-breeding efforts have been carried out, and different drought resistant genotypes have been proposed among the Andean races Nueva Granada, Peru and Chile (Pérez-Vega et al. 2011; Beebe et al. 2013). An example of an efficient result was the release of the improved variety 'Pinto Villa', which includes the race Durango in its pedigree, albeit derived from a cross with an Andean cultivar (Acosta-Gallegos et al. 1995).

Breeding for drought tolerance in common bean has been based on different criteria, including phenological plasticity, photosynthate remobilization, biomass accumulation, seed yield traits, pod filling, and root architecture (White and Castillo 1992, Foster et al. 1995, Schneider et al. 1997a, Ramirez-Valleyo and Kelly 1998, Rao 2001, Frahm et al. 2004, Rosales-Serna et al. 2004). The effects of root genotype were shown to be more consistent than for shoots under water-deficit conditions (White and Castillo 1992), and nitrogen remobilization was indicated as an important adaptive trait under moderate or intermittent drought stress (Foster et al. 1995).

Selection for high yields followed by selection for low to moderate levels of drought susceptibility has been further suggested as the most effective approach in breeding for drought resistance (Ramirez-Valleyo and Kelly 1998). An example of this comes from the breeding program developed at the International Center for Tropical Agriculture, from which drought-tolerant lines were obtained that showed double or greater yields than commercial controls, as well as tolerance to low soil fertility, improved root traits and efficient mobilization of photosynthate to grain (Beebe et al. 2008). These lines were extracted from breeding populations created from inter-racial crosses within both Mesoamerican and Andean gene-pools using gamete and recurrent selection. Among the parental lines, genotypes from races Durango (e.g., progeny of SEA 5), Colombia (G21212), and Brazil (Carioca) were used (Beebe et al. 2008).

Drought tolerant lines in common bean have been mainly obtained from crosses within gene pools and based on traditional breeding, although some improvements have also been achieved with the application of molecular markers and MAS. Schneider et al. (1997b) demonstrated that selection based on markers from random amplification of polymorphic DNA (RAPD) that targeted QTLs for drought resistance was effective in extreme environments and not under moderate drought. Along this line, a Mesoamerican intra-gene pool genetic map (BAT 477 × DOR 364) (Blair et al. 2012b) and a recombinant inbred line (RIL) population (Mukeshimana et al. 2014) were developed to detect other QTLs associated with drought tolerance. The now available genome sequence and the improved molecular technologies have also been used to discover markers, genes and QTLs that are functionally associated with drought-stress responses that can be exploited in future breeding programs (Còrtes et al. 2011, Schmutz et al. 2014, Wu et al. 2014, Trapp et al. 2015).

At the intraspecific level, the use of common bean wild relatives for improvement of drought tolerance has been limited, although wild and domesticated forms can be easily crossed (Blair et al. 2006; Porch et al. 2013a). Some strategies have been proposed for future introgression of drought-related traits into cultivated forms by exploiting either newly discovered wild genetic resources or existing lines developed from sister species (Porch et al. 2013a). Córtes et al. (2012a,b) investigated both wild and domesticated common bean populations for nucleotide variations at the *Asr* and *DREB2* genes, which are implicated in plant responses to drought stress and are mediated and not mediated, respectively, by abscisic acid. They detected shared alleles between wild and domesticated genotypes, and proposed to focus on these alleles to transfer traits associated with drought tolerance into domesticated forms (Córtes et al. 2012a,b).

Along with the wild form, also the use of other *Phaseolus* species has been poor compared to common bean breeding for drought tolerance, and this use has been mainly confined to studies on tepary bean, as this species is extremely resistant to drought (Beebe et al. 2013). Common bean and tepary bean have been investigated and compared at the phenotypic, physiological and molecular levels under drought conditions, and various genotypes have been indicated for their outstanding drought resistance (Barrón and de Mejía 1998, Türkan et al. 2005, Rodriguez-Uribe and O'Connell 2006, Rao et al. 2013). Porch et al. (2013b) registered the two tepary bean genotypes TARS-Tep 22 (derived from hybridization and selection) and TARS-Tep 32 (a single plant selection from the landrace PI 477033) as improved germplasm. Rao et al. (2013) showed the superior performance of two accessions of tepary bean (G40159, G40068) and two elite lines of common bean (RAB 650, SEA 23), which were associated with increased efficiency in the remobilisation of photosynthates to the grain. Despite this, the interspecific lines evaluated in their study, along with the tepary and common bean genotypes, were not promising for their drought resistance (Rao et al. 2013). Possible explanations given by the authors were: lack of previous selection for drought, poor photosynthate remobilisation ability in the common bean parent, and the likely reduced introgression into the lines of the tepary bean genome. Indeed, previous attempts to transfer drought resistance traits from tepary bean into cultivated common bean showed low introgression also in interspecific hybrids obtained by congruity backcrossing, which has been shown to increase the rate of success of interspecific crosses (Mejía-Jiménez et al. 1994, Muñoz et al. 2004). Other studies have nonetheless shown that the yield potential of interspecific hybrids obtained from year beans and common beans can increase when they originate from drought-adapted common bean parents that also have high remobilisation capacity (Klaedtke et al. 2012).

Heat and Cold

Common bean is a short-day crop that usually grows at mean temperatures of 15°C to 25°C (Gepts 1998), and its production can be threatened by high and low temperatures. It has been shown that temperatures of >30°C during the day and/or >20°C at night can lead to yield reductions (Rainey and Griffiths 2005a, b). Common bean production can be particularly affected when high temperatures occur during the night and during reproductive development, such as macrosporogenesis and microsporogenesis, embryo-sac and pollen development, floral development, and pod and seed setting (Monterroso and Wien, 1990, Gross and Kigel 1994, Shonnard and Gepts 1994, Porch and Jahn 2001, Rainey and Griffiths 2005a). In addition, common bean suffers cold stress at temperatures below 10°C, and has little or no freezing tolerance at 0°C, when ice can be formed in the plant tissues (Balasubramanian et al. 2004).

With respect to drought stress, plant responses to high temperatures show relatively less variability across sites and seasons, and specific breeding for heat tolerance has given some interesting results (Shonnard and Gepts 1994, Rainey and Griffiths 2005a). A Mesoamerican small red seeded type

'Tio Canela' and other Mesoamerican lines, such as 'TARS-SR05' and 'TARS-MST1', were released as heat-resistant lines that also showed resistance to various diseases (Rosas et al. 1997, Smith et al. 2007, Porch et al. 2012). Examples of Andean lines obtained from a single cross between two rust-resistant lines are 'TARS-HT1' and 'TARS-HT2', which show tolerance to high daytime and night-time temperatures (Porch et al. 2010).

While interesting heat-tolerant lines have so far been developed mainly from the Mesoamerican gene pool, and mainly by classical breeding methods, more recently experimental data based on genetic studies have been provided to help in the development of new common bean varieties (Rainey and Griffiths 2005a, b). Dry common bean lines were evaluated during flower bud formation and pod filling, to determine the effects of heat stress on their reproductive development, and to understand the genetic basis of their heat tolerance (Shonnard and Gepts 1994). The experiments were performed on two sets of two heat-tolerant and two heat-susceptible parental lines and their corresponding progenies (including their F1 and F2 lines), which demonstrated that both traits show quantitative inheritance. These data revealed that a single gene controlling the growth habit was linked to heat tolerance during flower bud formation, and that additive genetic effects were observed for both traits, thus suggesting that a gain from selection is possible for improved heat tolerance (Shonnard and Gepts 1994). Subsequently, Rainey and Griffiths (2005a) identified an association between a flower abscission gene and genes that control pod number in a snap bean cross between heat-tolerant and heat-susceptible parental lines that can be exploited in breeding for heat tolerance in common bean.

Breeding for heat tolerant lines based on *P. vulgaris* sister species has produced limited data. Tepary beans show increased tolerance to high temperatures with respect to common bean, but the interspecific hybrids obtained so far need to be evaluated for heat resistance (Mejía-Jiménez et al. 1994, Ferwerda et al. 2003, Rainey and Griffiths 2005c). Alternative radical approaches have been proposed to overcome limits during reproduction and to achieve increased tolerant lines, such as apomixis (Hall 2004). As heat stress can have negative effects on pollination and other sexual reproduction stages, apomictic genotypes would assure reproduction without the need for pollination. Although still remote for common bean, the possibility to obtain facultative apomictic plants has been shown in nature, and apomictic soybean plants have been created that show increased hybrid vigour when compared to the parent (Taylor 2012).

Few advances have also been made in common bean for cold acclimation and freezing tolerance, even though the causes of freezing damage and plant responses to low temperatures have been investigated in detail (Thomashow 1999, 2001, Sinha et al. 2015). Early studies have demonstrated that it is possible to obtain common bean lines that are tolerant to low temperatures when the appropriate parents are chosen, but few studies have focused on breeding for tolerance to low temperatures (Zaiter et al. 1994, Otubo et al. 1996, Assefa et al. 2014). *P. angustissimus* has been identified as a possible source for common bean tolerance to low temperatures, but hybrids developed from crosses did not produce viable seeds, even with extensive backcrossing, and limited success has been achieved with *in-vivo* manipulation of pollination (Balasubramanian et al. 2004, Schryer et al. 2005, Gurusamy et al. 2007). In a recent study, Vijayan et al. (2011) developed expressed sequence tag (EST) sequences related to cold responses from *P. vulgaris* and *P. angustissimus* lines by subtraction suppression hybridization. The sequences obtained were also compared to an EST collection of *P. coccineus*, a species that also shows interesting levels of cold tolerance, thus overall supplying a valuable instrument for introgression into common bean of genes and alleles that target cold responses (Rodiño et al. 2007, Vijayan et al. 2011).

Phosphorus Availability

Phosphorus deficiency is among the most important abiotic constraint for common bean production in the tropics and subtropics (Smithson and Sanchez 2001, Lynch 2007). Phosphorus is a limiting nutrient for crop production, primarily because it is easily fixed in the soil and can be converted into forms that are no longer available to plants. Furthermore, phosphorus fertilisers might not be an adequate solution, especially in developing countries (Smithson and Sanchez 2001, Lynch 2007). A more straightforward solution is the development of plants with enhanced phosphorus efficiency, to achieve both improved uptake of phosphate from the soil (phosphorus acquisition efficiency) and improved productivity per unit phosphorus extracted from the soil (phosphorus use efficiency) (Richardson et al. 2011, Veneklaas et al. 2012). Phosphorus acquisition can be enhanced by improving root morphology, architecture and physiology, to also increase phosphorus mobilization in the rhizosphere, while phosphorus use efficiency can be optimised by more advantageous redistribution of phosphorus during plant development from tissues that no longer need it (e.g., senescing leaves) or where it is unfavorable (e.g., phytic acid in seeds), to those were it is useful at higher concentrations (Richardson et al. 2011, Veneklaas et al. 2012).

Common bean cultivars adapted to low phosphorus tolerance, such as the Brazilian 'Carioca', the Peruvian 'G19833', and the Mexican 'G2333', were obtained by not explicitly breeding for this trait (Lynch and Beebe 1995). A first extensive genetic study was performed on 364 common bean genotypes that included Mesoamerican and Andean cultivars and wild accessions. This revealed significant variability in phosphorus efficiency among the genotypes grown with and without phosphorus stress, with these results also associated with previous studies performed on root growth and architecture (Lynch and Beebe 1995, Beebe et al. 1997). The poor performance observed for wild beans suggested that phosphorus efficiency probably evolved after domestication, while the differences among the cultivars were related to the geographic origins of the genotypes (Beebe et al. 1997).

The lower performance of the wild genotypes is in agreement with data from Araújo et al. (1997), who showed higher root dry weight in cultivated common bean when similar shoot dry weight and phosphorus contents were detected. No correlation was found by Yan et al. (1995a,b, 1996) between genotype ranking for phosphorus efficiency and soil type, which indicated that genotype differences were related to plant root traits. These authors also detected superior performance in terms of yields of the Andean genotypes under low phosphorus availability, while the Mesoamerican genotypes were more responsive to added phosphorus.

Other efforts have been directed to design effective breeding methods for improving tolerance to low phosphorus. Genetic studies that have investigated different root traits and yield components in varying genotypes of different origins have shown complex inheritance under phosphorus stress (Fawole et al. 1982, Araújo et al. 2005, Kimani et al. 2007). Different mapping populations were subsequently investigated to map QTLs associated with phosphorus efficiency traits, and to use these for genotype selection. Yan et al. (2004) studied 86 $F_{5.7}$ recombinant inbred lines (RILs) that were obtained from a cross between G19833, the Andean landrace with high total phosphorus accumulation, and DOR 364, a Mesoamerican cultivar with low total phosphorus accumulation under low phosphorus conditions. These authors identified 19 QTLs that were associated with root hair, acid exudation, and phosphorus-uptake traits, and some of these root-trait QTLs were also closely linked with QTLs for phosphorus uptake in the field, thus suggesting the feasibility of MAS for screening of phenotypic root traits. Liao et al. (2004) further detected QTLs on the same RIL population that were associated with root gravitropism and root shallowness, which indicated that under phosphorus stress the basal root plasticity helps to increase the phosphorus acquisition efficiency of the superior genotypes. Beebe et al. (2006) identified 24 QTLs that were also associated with root traits, such as basal root development and root length and angle. These studies have contributed to a more detailed understanding of the root architecture; nonetheless, genetic improvements for phosphorus efficiency have been hampered by the complexity of this trait, while the use of other species through wide crossing or genetic transformation as sources of genes for successful breeding has not yet been documented.

Nitrogen Fixation

Nitrogen is the primary nutrient constraint for common bean production, and its intake and assimilation is almost as important for crop growth as the photosynthetic process (Broughton et al. 2003; Vance 2001). Although nitrogen is an abundant element on Earth, its input into agricultural systems is becoming critical for many reasons, including an imbalance in the global nitrogen cycle, the leaching of nitrogen into groundwater, volatilisation of nitrogen oxides, and the high cost of nitrogen fertilisers, especially in developing countries (Vance 2001).

High-quality protein-rich seeds can only be obtained through efficient acquisition and assimilation of nitrogen, which can be acquired from two principal sources: synthetic fertilisers, and/or mineralisation of organic matter from the soil and symbiotic nitrogen fixation (SNF) (Vance 2001). The most important source of biological fixed nitrogen in agricultural systems is the symbiotic associations between legumes and rhizobia, which provide 25% to 35% of the global intake (Herridge et al. 2008). These species have important roles in agriculture, because they are still intercropped or rotated with more than 50% of the crops grown in Africa, India and Latin America (Broughton et al. 2003, Mafongoya et al. 2009).

Symbiotic nitrogen fixation (SNF) derives from the symbiotic relationships between legumes and different bacteria of the family Rhizobiacae, which generates nitrogen fixing nodules in the legume root (Graham 2009). This process commonly starts with the release of metabolites from the legume roots, typically as flavonoids that trigger the release of *Rhizobium* Nod factors (e.g., lipochitooligosaccharides). When the plant detects the Nod factors, this induces the formation of the nodule (Gage 2009). The rhizobia within the nodule reduce atmospheric nitrogen to ammonium, which can then be assimilated by the plant. In return, the plant supplies the bacteria with carbon photo-assimilates (Vance 1997).

Breeding for common bean with improved SNF has been performed in the past, because with respect to other legumes, common bean is considered a poor nitrogen-fixing grain legume, and it is functional in unfavorable environments in developing countries (Bliss 1993, Herridge et al. 2008). Genetic variation linked to SNF and its correlated traits has been demonstrated, and bean lines with high potential for symbiotic nitrogen fixation have been released, such as 'Ouro Negro' in Brazil (Henson et al. 1993). A successful breeding program was conducted over a period of 13 years with the selection of lines for high yields in low nitrogen soils, which led to the release of five high nitrogen-fixing lines that were adapted to South American climates (Bliss 1993). Another successful example is the introgression of a QTL for nitrogen accumulation from BAT 477 into race 'Durango' cultivars using an associated microsatellite marker, which was performed to improve nitrogen fixation under conditions of water deficit (Miklas et al. 2006). Genetic improvements through breeding for increased nitrogen fixing capacity are therefore feasible in common bean, especially considering that no correlations have been reported between improved nitrogen fixation and insurgence of other unfavorable traits (Bliss 1993, Hardarson et al. 1993). Other studies have concentrated on investigations into segregating populations obtained from backcrosses between parental lines with high nitrogen fixing ability, such as the Mesoamerican landrace 'Puebla 152', and recurrent parents selected for other traits of interest, such as high yields (St. Clair and Bliss 1991, Barron et al. 1999). However, specific breeding for improved nitrogen fixation has not been carried out routinely, because it is still costly and time consuming to phenotype for SNF. However, molecular research is also underway to achieve more straightforward breeding methods (Miklas et al. 2006, Kamfwa et al. 2015).

Early studies to detect QTLs associated with SNF were performed under different nitrogen concentrations on the progeny of a cross between the Mesoamerican BAT 93 with the Andean Jalo EEP558, which were used to investigate rhizobium nodule number as an indirect measure of SNF, in association with resistance to common bacterial blight (Nodari et al. 1993, Tsai et al. 1998, Souza et al. 2000). A later study was performed on a RIL population that was obtained from an intergene pool cross between a Mesoamerican accession from Mexico and an Andean accession from Peru (G2333 × G19839) (Ramaekers et al. 2013). The progeny and the parents were grown under greenhouse and field conditions to evaluate the differences among the lines in terms of percentage and total nitrogen fixed. These authors succeeded in mapping QTLs associated with SNF traits on chromosomes Pv01, Pv04 and Pv10, partly confirming data from previous studies (Tsai et al. 1998, Souza et al. 2000, Ramaekers et al. 2013). An auxin-responsive transcription factor and an AP2/ERF-domain-containing transcription factor were also tentatively proposed as candidate genes that were implicated in SNF differences among these RILs (Ramaekers et al. 2013). A genome-wide association study that has helped in the gathering of genetic information on SNF and related traits was performed on a panel of 259 Andean common bean genotypes under greenhouse and field conditions (Kamfwa et al. 2015). SNPs associated with the percentage of nitrogen fixed were identified on chromosomes Pv03, Pv07 and Pv09 in the shoot at flowering, and for total nitrogen fixed in seed. While it was not possible to determine whether some of these QTL co-localised with those previously identified, two candidate genes, Phvul.007G050500 and Phvul.009G136200, were identified for future marker-assisted breeding; these code for leucine-rich repeat receptor-like protein kinase (LRRRLK) genes that have key roles in the signal transduction required for nodule formation.

Aluminum Tolerance

Aluminum toxicity is a major constraint for common bean production in acid soils (pH <5.0), which are common in tropical climates and are often associated with infertility and other mineral toxicities (Rao et al. 1993). It has been estimated that about 40% of the bean-growing area is affected by aluminum toxicity, with associated yield losses of between 30% and 60% relative to nonacid soils (Wortmann et al. 1998, Thung and Rao 1999). Crop productivity can be improved by addition of lime, but the subsoil is often unaffected by lime incorporation, and crops can still be susceptible to drought and have limited access to other macro and micro-nutrients (Rao et al. 1993). For these reasons, breeding for aluminum-tolerant cultivars represents a more valid economic alternative, especially in developing countries (Fageria et al. 1988; Rao et al. 1993).

Aluminum toxicity is primarily seen by reduction in root growth, and it can affect plant growth directly, e.g., by interfering with cell division in root tips and lateral roots and decreasing root respiration, or indirectly, e.g., by fixing phosphorous in forms that are not available to the plant or that interfere with the transport of other essential nutrients, such as calcium, magnesium, potassium, phosphate and iron (Fageria et al. 1988, Foy 1984, Barceló and Poschenrieder 2002). Plants can react to aluminum toxicity either through avoidance strategies, to exclude direct contact of aluminum ions with vital structures, and/or through mechanisms of internal tolerance, to create less toxic aluminum complexes in the plant (Barceló and Poschenrieder 2002). One common tolerance mechanism involves aluminum-induced secretion of organic acids and the subsequent chelation of aluminum. This mechanism was first reported in common bean by Miyasaka et al. (1991), who detected a 10-fold higher root exudation of citric acid in an aluminum-tolerant cultivar, compared to an aluminum-sensitive cultivar.

Genotype variation in response to aluminum toxicity was further reported by Rangel et al. (2007, 2009, 2010) in two common bean cultivars, the Mesoamerican 'Quimbaya' (aluminum resistant) and the Andean 'VAX-1' (aluminum sensitive). These authors observed inhibition of root elongation in both cultivars when aluminum was applied to the elongation zone of the root apex. After a lag phase, while continued recovery was observed for the tolerant genotype, only partial recovery with subsequent heavy damage was observed for the sensitive genotype (Rangel et al. 2007). The mechanisms of root inhibition were induced by apoplastic aluminum, and the recovery of the tolerant Andean genotype was mediated by an increase in the internal citrate pools and reduction of stably bound aluminum, which allowed cell elongation and division in roots (Rangel et al. 2009, 2010). This mechanism of aluminum resistance was subsequently associated with the expression of the citrate transporter *MATE* (multidrug and toxin extrusion family protein) gene, which was found to be crucial for citrate exudation, together with maintenance of the cytosolic citrate pool (Eticha et al. 2010).

Previous studies that have been conducted to improve bean production in low-fertility soils in Africa have led to the identification of cultivars that show promise in terms of their performances under different edaphic stresses, including aluminum toxicity, albeit no further news on the exploitation of these genotypes has become available (Wortmann et al. 1995). A more recent study that was conducted under hydroponic conditions allowed the identification of a group of Andean genotypes with superior aluminum resistance. This was shown by the analysis of different root morphological traits, which were suggested as selection criteria to distinguish between aluminum-tolerant and aluminum-resistant genotypes (Blair et al. 2009a).

Further studies have been carried out to identify other sources of resistance to aluminum toxicity among *P. vulgaris, P. coccineus* and *P. acutifolius* accessions under both hydroponic and field conditions (Butare et al. 2011). Overall, the Andean genotypes showed superior performance with respect to the Mesoamerican genotypes, as also confirmed by other studies (Blair et al. 2009a). However, VAX 1 showed superior performance to the Andean 'Quimbaya' under field conditions, due to its abundant and shallow root system that helps to avoid the aluminum toxicity (Butare et al. 2011). A greater level of aluminum resistance was detected in *P. coccineus* accessions (G35346, G35464), which also showed tolerance to combined aluminum and water stress factors (G35346). Accession G35346 was subsequently crossed with SER 16, a drought-resistant common bean line, to introduce aluminum resistance into common bean (Butare et al. 2012). Phenotypic evaluation of the 94 RILs obtained from the first backcross to SER 16 allowed the identification of lines with aluminum resistance combined with acid-soil adaptation, and these demonstrated the complex inheritance of aluminum resistance and acid-soil tolerance (Butare et al. 2012).

Other studies have been performed to gain further insights into aluminum resistance in common bean, and these provide the bases for future improvements in common bean breeding. López-Marín et al. (2009) developed a RIL population that was obtained from the cross of an Andean aluminum-resistant genotype (G19833) with a Mesoamerican genotype classified as intermediate in aluminum resistance (DOR364). These authors showed that root traits were under polygenic control, and they identified 24 QTLs through composite interval mapping analysis, of which 9 underlie the aluminum tolerance traits, while some were also associated to low phosphorous stress. Further contributions to make breeding for aluminum tolerance in common bean more straightforward have been made by the construction of a cDNA library from the G19833 Andean genotype, with subsequent EST sequencing (Blair et al. 2011a), and also by transcriptomic analysis of two common bean genotypes that showed differential gene expression under aluminum stress (Eticha et al. 2010).

Yield

Common bean is the most important food legume throughout the world for direct consumption, with an estimated production of ~12 million metric tons per year (http: //faostat3.fao.org/home/E). Although legume production has increased over the last 15 years at a greater rate than the growth rate of the world population, common bean is often grown under unfavorable conditions and further breeding efforts are needed to improve crop yields (Broughton et al. 2003, Akibode and Maredia 2011). To date, major enhancements in common bean yield have been achieved by breeders, from improved crop adaptation to biotic and abiotic stresses (see previous sections) to improved genetic yield potential per se (Rao 2001, Singh et al., 2007). Breeding for improved yields has exploited differences among the races and gene pools, while also including introgression of useful traits from wild genotypes (Kelly et al. 1998, Acosta-Gallegos et al. 2007, Wright and Kelly 2011, Beaver and Osorno 2009).

Successful results have been obtained by ideotype breeding, i.e., selection for a set of traits related to yield from intra-racial crosses of dry beans based on small-seeded Mesoamerican genotypes (Kelly et al. 1998). Ideotype breeding has also been used to evaluate inter-racial crosses of Durango × Mesoamerican and inter-gene pool crosses of Nueva Granada × Mesoamerica, and different genotypes with improved yields have been identified by recurrent selection (Kelly et al. 1998). The hypothesis that inter-racial and inter-gene pool crosses are more promising than intra-racial crosses for selection of lines with high yields was tested by Abreu et al. (1999). Indeed, different valuable common bean lines were obtained from inter-racial crosses, although the chances of obtaining lines that combine yield and the other traits preferred by consumers were reduced. Singh et al. (1999) also used recurrent selection to evaluate intra-racial and inter-racial crosses grown in three different environments, and they showed the usefulness of this method to improve seed yields in common bean.

Other breeding experiments have been based on exploitation of wild germplasm diversity. Inbred backcross methods have been used to select for interesting lines with superior yields, because in single crosses, this trait might be masked by traits related to the domestication syndrome (Acosta-Gallegos et al. 2007). The Mexican black bean cultivar 'Negro Tacaná' was crossed with the wild accession 'G24423' from Colombia, and five of the best $BC_2F_{4:7}$ lines were evaluated in different field trials from 2002 to 2005; these showed an average yield gain of 11% with respect to the recurrent parent 'Negro Tacaná' (Acosta-Gallegos et al. 2007). Improved yields were registered in subsequent trials based on different backcross populations obtained from wild and weedy parents and a domesticated parent. Inbred backcrossing has also been used to map QTLs on a mapping population of 157 lines that originated from a cross between the Andean genotype 'ICA Cerinza' and the wild genotype 'G24404' (Blair et al. 2006). Microsatellites, SCAR, and phaseolins were used to locate 41 QTLs on the 11 linkage groups. Thirteen QTLs were associated with plant height, yield and yield components, while there was a single QTL for seed size that showed positive alleles from the wild parent (Blair et al. 2006). Wright and Kelly (2011) also identified a QTL on Pv10 that accounted for ~28% of the variation in seed yield through analysis of a RIL population obtained from 'Jaguar', a black bean cultivar with acceptable canning quality, and the high-yielding 115M, a black bean IBL that was developed from the backcross of 'Tacana' with the wild accession G24423.

Results from breeding efforts conducted to improve common bean tolerance to different abiotic stresses (see previous sections) and/or biotic stresses have also produced relevant data for yield improvement. As an example, the common bean lines that originated from Mesoamerican × Andean inter-racial crosses and were selected by Beebe et al. (2008) for drought resistance showed greater harvest indices and increased seed yields in favorable environments. Other valuable data were obtained in the attempt to increase the yield potential of hybrid lines obtained from interspecific crosses of *P. vulgaris* × *P. dumosus* using drought-adapted *P. vulgaris* lines with enhanced ability to remobilize photosynthate from the shoots to the grain yield (Klaedtke et al. 2012). In their attempt to map QTLs associated with white mold resistance, Kolkman and Kelly (2003) also identified a QTL on chromosome Pv07 that was significantly associated with yield, seed size, lodging, and days to flower. More recently, in the progeny of an Andean intragene pool cross, a multi-environment QTL analysis allowed the identification of 59 QTLs that were implicated in genetic control of seed quality traits and seed yields (Yuste-Lisbona et al. 2014).

The development of broader studies aimed at developing genomic markers and QTLs related to other traits for local adaptation, such as the photoperiod response, would also help in the development of lines with improved abiotic stress resistance that might lead to improved seed yields in the future.

Nutritional Quality

Improved nutritional quality and health benefits of food are becoming increasingly important in current plant breeding programs. In this respect, the common bean is consumed by humans more than any other among the legumes. It is an important source of protein, especially in the poor countries of Latin America and Africa, where it complements cereals, which are primarily sources of carbohydrate; moreover, beans are also a good source of vitamins, minerals, and fiber (Broughton et al. 2003). Numerous studies have indicated that the health benefits from eating beans are numerous. This can be seen in reductions in blood cholesterol and sugar levels, which in turn can prevent or alleviate certain types of cancers, type 2 diabetes, and cardiovascular diseases (Anderson et al. 1984, Hernández-Saavedra et al. 2013). Diets rich in zinc and iron, two micronutrients that are abundant in beans, also delay onset of AIDS (Buys et al. 2002, Wong et al. 2006, Patrick and Ng 2004), and eating beans can result in significant reductions in the onset of breast cancer (Thompson et al. 2009), colon cancer (Hangen and Bennink 2002, Feregrino-Perez et al. 2008), and biomarkers for increased heart disease risk (Winham et al. 2007).

Proteins and Anti-Nutritional Compounds

In contrast to cereals, which are characterized by high starch and lower protein levels from the grain, legumes are rich in protein (Broughton et al. 2003); thus, legumes are a very important source of protein, especially in developing countries. However, the protein quality of beans is considered to be sub-optimal, especially in terms of deficiencies in essential amino acids such as methionine, cysteine and tryptophan, low protein digestibility, and the presence of anti-nutritional compounds, such as proteinase inhibitors, lectins, phytates, and polyphenols (see Krupa 2008 for review).

Vicilins (7S globulins) and legumins (11S globulins) are the two main seed-storage proteins in legumes. In particular, for common bean, up to 50% of the total seed protein is constituted by phaseolin, a glycoprotein that belongs to the 7S vicilin class (Bollini and Vitale 1981), while there are very low levels of the 11S globulin, legumin (Mühling et al. 1997). A further 5% to 10% of the total protein is represented by lectins (Vitale and Bollini 1995). Arcelin and α-amylase inhibitor, and the true lectin, phytohemagglutinin, belong to the lectin-related protein family.

Efforts in breeding to increase seed protein concentrations and to balance the composition of the essential amino acids in beans started from the 1970s (Bliss and Brown 1983). One of the strategies was to change the percentages of the protein fractions. Gepts and Bliss (1984) analysed the relationships between the available methionine concentrations and the levels of phaseolin using different genetic materials that were derived by (i) interspecific crosses between *P. vulgaris* cultivars and a *P. coccineus* subsp. *coccineus* line (cv. Mexican Red Runner) that has no detectable phaseolin; (ii) crosses between cultivated *P. vulgaris* lines and a Mexican wild bean accession (PI 325690-3) that carried a gene that produces a reduction in phaseolin content; and (iii) crosses between the Sanilac cultivar and two cultivated lines, Bush Blue Lake 240 and 15R 148. Here, Gepts and Bliss (1984) showed significant and positive correlations between phaseolin levels and available methionine concentrations. These data are consistent with the study of Chagas and Santoro (1997), where biochemical data on protein fractions indicated that methionine is associated with phaseolin. The same study also indicated that cysteine is present in non-phaseolin proteins. Osborn and Bliss (1985) investigated the effects of changing the amounts of lectin phytohemagglutinin. Phytohemagglutinin is the second most abundant storage protein in bean seeds, and it accounts for 5% to 10% of the total seed protein (Barker et al. 1976, Bollini and Chrispeels 1978). Furthermore, phytohemagglutinin does not contain sulfur amino acids, and it is toxic when consumed in raw bean. Osborn and Bliss (1985) showed that a single recessive gene controls for the absence of lectins in two common bean cultivars (Great Northern 'U.I. 1140', 'Pinto U.I. 111'). Here, analysis of backcross populations derived from crosses between 'Sanilac' cultivar (recurrent parent) were characterized by the presence of lectin, and the 'U.I. 1140' and 'Pinto U.I. 111' cultivars showed that backcross lines without lectin had substantially higher levels of phaseolin and slightly more total protein than lines with lectin. These data suggested that phaseolin overcompensated for the absence of lectin, to maintain a stable seed protein concentration. Burow et al. (1993) investigated changes in both phaseolin and phytohemagglutinin content in families that were segregating for phaseolin, lectin, and arcelin. These authors reported that genetic suppression of phaseolin and lectin accumulation increased the concentrations of other polypeptides in mature seeds; however, these did not contain greater concentrations of methionine than were seen for phaseolin. Following these results, and considering the positive relationship between phaseolin and total available methionine, Delaney and Bliss (1991a, b) investigated the possibility of increasing the frequency of favorable alleles that can increase the percentages of phaseolin through the use of recurrent selection methods. Recently, Campion et al. (2009a) developed and evaluated common bean lines with reduced anti-nutritional factors of the seeds (i.e., the APA proteins arcelin, phytohaemagglutinin, α-amylase inhibitor). Their lectin-free stable lines also included a white-seeded line, a trait that is often associated with reduced amounts of tannins and polyphenols in the seeds, and that thus represents a further improvement on the nutritional characteristics of the seeds. The common bean lines were evaluated in the field (two locations), and interestingly, there were no detrimental effects on the agronomic performances and germination of the lectin-free lines.

A study by Taylor et al. (2008) that represented the first report of a significant improvement in sulfur amino-acid content in seeds of common bean was based on natural genetic variations in seed-storage proteins. They analyzed three genetically related lines (SARC1, SMARC1-PN1, SMARC1N-PN1) that were developed by Osborn et al. (2003), and that had progressive deficiencies in phaseolin, phytohemagglutinin, and arcelin. These three lines shared ~85% of the background of their parental navy bean cultivar Sanilac. The SARC1 line integrated the arcelin-1 variant introduced from the wild accession G12882 into Sanilac. SARC1 contains the S-type phaseolin from Sanilac and phytohemagglutinin and arcelin from the wild parent, due to the tight linkage between the phytohemagglutinin and arcelin genes (Osborn et al. 1986, Kami et al. 2006). Alleles for phaseolin or phytohemagglutinin deficiency where introduced in the SARC1 background from a *P. coccineus* accession and the Great Northern U.I. 1140 cultivar, respectively. The derived lines were SMARC1-PN1, which is deficient in phaseolin, and SMARC1N-PN1, which is deficient in all three of these major seed-storage proteins. In particular, the SMARC1N-PN1 line showed 10% increase in methionine and 70% increase in cysteine in its soluble protein content, concomitant with 70% decrease in its *S*-methylcysteine concentration. Moreover, proteomic and transcriptomic profiles of the SMARC1N-PN1 line indicated increased levels of several sulfur-rich proteins, such as the 11S globulin legumin, albumin-2, defensin D1, Bowman-Birk-type proteinase inhibitor 2, albumin-1, basic 7S globulin, and Kunitz trypsin protease inhibitor (Marsolais et al. 2010, Yin et al. 2011, Liao et al. 2012).

Montoya et al. (2010) proposed a different strategy through conventional breeding to improve protein quality in common bean that was based on exploiting the wide diversity in phaseolin types and the increase in seeds of the types that have high digestibility. There have been different studies aimed at investigating the diversity of phaseolin in different common bean accessions, which also appeared to be a reliable evolutionary marker to investigate the domestication of common bean in both Central America and the Andes region (Gepts 1988). The domesticated beans have two main phaseolin types: the S (Sanilac) and T (Tendergreen) phaseolins (Gepts and Bliss 1986, Koenig et al. 1990), with Mesoamerican accessions mainly characterized by the S type, and Andean accessions by the T type (Gepts and Bliss 1986). However, other phaseolin types have been found in both wild and domesticated forms, such as for example the M (Middle America) and B (Boyacá) phaseolin types in Mesoamerica, the C (Contender), H (Huevo de Huanchaco), A (Ayacudo), and J (Jujuy) types in the Andes, and the I (Inca) type that is characteristic of wild populations from northern Peru and Ecuador (Kami et al. 1995, Koenig et al. 1990). On the basis of *in-vitro* protein digestibility corrected amino-acid scores, Montoya et al. (2010) showed that genotypes with highly digestible phaseolin types can increase their bio-available sulfur amino acids, as for their other essential amino acids. In particular, these authors reported that phaseolins with the highest degree of hydrolysis (DH) values provided 37% more sulfur-containing amino acids than those with the lowest DH values. This was also true for leucine, lysine, aromatic amino acids, and threonine. For example, beans containing the To1 and J1 phaseolins, which have DH values of 96%, provided 28% and 16% more sulfur-containing amino acids than beans that contained the S (DH, 58%) and T (DH, 71%) phaseolins, respectively (Montoya et al. 2010).

Another important anti-nutritional compound in beans is phytic acid, inasmuch as it represents a major constraint to micronutrient bioavailability. Phytic acid binds mineral cations, such as Fe^{2+}, Zn^{2+} and Ca^{2+}, to form insoluble complexes that are not digestible by humans and nonruminant animals. From a mutagenized population of common bean, Campion et al. (2009b) isolated and characterized a homozygous *lpa* mutant line (*lpa*-280-10) that compared to the wild type, showed a 90% reduction in phytic acid, a 25% reduction in raffinosaccharides, and a much higher amount of free or weakly bound iron cations in the seed. Moreover, they showed that the bean *lpa*-280-10 mutation line does not cause macroscopic negative effects on seed germination, plant growth, seed yield, and other traits of agronomic relevance.

Blair et al. (2012c) carried out a QTL analysis to investigate the inheritance of seed phytate in common bean that was based on the phenotypic and genotypic characterization of a segregant population derived from a cross between a Mesoamerican (DOR364) and an Andean (G19833) variety. They identified different QTLs on linkage groups b01, b03, b04, b06 and b10. In particular, one of these QTLs that was located on the linkage group b01 was associated with one of two paralogs of the *myo*-inositol (3)P_1 synthase gene family, and was expressed in common bean seed rather than in the vegetative tissues. Interestingly, Petry et al. (2013) showed that in a sample of 20 women, iron absorption from beans carrying the *lpa* mutation was significantly higher than that from their parent beans with normal phytic acid levels.

Micronutrients

Beans are an important source of micronutrients that have crucial roles in human health, such as iron, phosphorus, magnesium, and manganese, and to a lesser degree zinc, copper, and calcium (Welch et al. 2000). Increasing the concentrations of these micronutrients in common bean cultivars represents a goal for plant breeders, inasmuch as this will provide improvements to human nutrition, especially in developing countries (Frossard et al. 2000). The genetic and/or agronomic improvement of a staple crop that is used for nutritional value, and especially for its vitamin and mineral content, is known as bio-fortification. In particular, one of the major aims in common bean breeding is to increase the iron and zinc concentrations in seeds. Iron and zinc nutritional deficiencies affect over three billion human beings (Dwivedi et al. 2012), and especially women and young children, and these can lead to different diseases. Iron-deficiency anemia is a shortage of hemoglobin, which is the red pigment in red blood cells that carries the oxygen. Generally, the symptoms are weakness, shortness of breath, dizziness, and/or headaches; however, in severe cases, this can even cause death. Moreover, in childhood and adolescence, iron-deficiency anemia can cause impairment to mental and physical development (Pfeiffer and McClafferty 2007). Zinc is essential for multiple aspects of metabolism, and thus the physiological signs of zinc deficiency are linked to diverse biochemical functions, rather than to a specific function, which can lead to significant and diverse clinical problems (see Prasad 2013 for review).

Different studies have screened for iron and zinc content across wide and diverse genotypes of common bean. Islam et al. (2002) analyzed the levels of iron and zinc, along with other compositional characters and pathogen disease resistance levels, in seeds of a core collection of more than 1,000 domesticated accessions of common bean that were representative of both the Mesoamerican and Andean gene pools (Tohme et al. 1995). A similar evaluation was carried out by Blair et al. (2010a) on a sample of ~350 African bean landraces. Both of these studies reported high variability in the iron and zinc contents in the samples analyzed, which suggested that real improvements can be reached through breeding programs. The Andean beans showed a tendency for higher average seed iron concentration and significantly lower seed zinc concentration compared to the Mesoamerican beans. The genotypes that were derived by spontaneous inter-gene pool introgression appeared to be very interesting for breeding, as these showed the highest average seed iron concentration and similar seed zinc concentration to the Mesoamerican genotypes (Islam et al. 2002, Blair et al. 2010a). This was detected also by Islam et al. (2004), who analyzed a wide sample of cultivated germplasm from the Andean gene pool, and reported that the genotypes that showed introgression from the Mesoamerican gene pool had higher levels for all of the nutrient elements. Beebe et al. (2000b) also analyzed wild accessions of common bean, which showed only a narrow advantage in iron content. Guzmán-Maldonado et al. (2000) screened for iron and zinc content in a sample of ~70 wild and weedy accessions from Mesoamerica, in which they identified some wild and weedy genotypes from Jalisco and Durango Mexican states that showed very high levels of these nutrients in the seeds. Beebe et al. (2005) reported on an explorative evaluation of the mineral contents in seeds in core collections of the common bean sister species, *P. coccineus* and *P. dumosus,* with some of these genotypes promising for breeding programs based on interspecific crosses.

Information derived from germplasm screening has allowed the selection of parental genotypes to initiate crosses to study inheritance of iron and zinc accumulation in seeds. Diverse QTLs have been identified in studies based on inter-gene pool segregant populations (Blair et al. 2009b), including lines derived from crosses between wild and cultivated genotypes (Guzmán-Maldonado et al. 2003, Blair and Izquierdo 2012), and intra-gene pool segregant populations (Gelin et al. 2007, Cichy et al. 2009, Blair et al. 2010b, 2011b). All of these studies have indicated multigenic control of the micronutrient traits. Moreover, these have indicated that lines derived through inter-gene pool crosses have greater transgressive segregation for seed mineral content, compared to lines derived from intra-gene pool crosses. Identification of the genes and QTLs involved in the control of seed iron and zinc content is being pursued in common bean breeding so as to incorporate useful genomic regions into an elite germplasm background through MAS.

An essential issue to consider when trying to improve seed mineral content, as for any trait, is to what degree will this be stable across environments. For mineral content, it has been shown that the genotype × environment interaction is very important, with some lines showing large interactions with the environment, and other showing relative stability (Monserrate et al. 2007). Blair et al. (2010c) reported the first varietal release of nutritionally enhanced Andean lines (i.e., NUA35, NUA 56) that were selected from the ['CAL96' × (CAL96 × G14519)] backcross population, where CAL96 is a commercial cultivar in Colombia and Uganda, and G14519 is a high seed mineral germplasm accession from the International Center for Tropical Agriculture genebank. These were tested in multiple sites with varying climate, altitude and soil types in Central and South America. On average, these lines showed an iron content 18 mg/kg and 23 mg/kg higher for NUA35 and NUA56, respectively, than for CAL96, while the zinc content was 8 mg/kg and 7 mg/kg higher, respectively. Lines derived from interspecific crosses between common bean and *P. dumosus* and *P. coccineus* accessions showing high seed iron content are being developed and evaluated (Blair 2013).

Conclusions

The chapter presents a comprehensive overview of the genetic resources available, along with the major results obtained for the improvement of beans through pre-breeding and breeding. The available genetic resources to broaden the genetic background of common bean elite germplasm include: wild and domesticated materials from the two main gene pools, Mesoamerica and Andes; domesticated populations that are widespread out of the Americas and which are the result of evolution and adaptation to different environments as well as hybridization between Andean and Mesoamerican types due to the breakdown of geographical barriers; and finally, other *Phaseolus* species within the *P. vulgaris* secondary and tertiary gene pools. Common bean breeding based on the use of these sources of genetic diversity is well documented, and aims to improve biotic and abiotic resistance/ tolerance, yields, and quality traits. Major efforts have also been made especially towards obtaining varieties resistant to fungal, bacterial and viral pathogens. Thus, important lines have been released that were derived from both intra- and inter-gene-pool crosses and interspecific crosses; e.g., white mold resistant lines developed by introgression of resistance from *P. coccineus, P. acutifolius,* and *P. costaricensis* genotypes. Interesting results have also been obtained for yields and drought tolerance that have involved mostly intraspecific hybridization. For other traits, such as nitrogen fixation, aluminum tolerance, resistance to nematodes, and nutritional quality, most of the efforts are concentrated on pre-breeding: investigations into the inheritance of the trait and the evaluation of the genetic resources available. However, it is also well known that the majority of the genetic variability in common bean and its closely related species still remains unused, or at least underused, and thus major efforts are still needed to investigate this diversity and to implement appropriate breeding strategies. This especially applies to climate change effects, sustainability of production, and the challenges of the increasing human population.

References

Abebe, A., M.A. Brick and R.A. Kirkby. 1998. Comparison of selection indices to identify productive dry bean lines under diverse environmental conditions. Field Crop. Res. 58: 15-23.

Abreu, Â.D.F.B., M.A.P. Ramalho and D.F. Ferreira. 1999. Selection potential for seed yield from intra-and inter-racial populations in common bean. Euphytica 108: 121-127.

Acevedo, M., J.R. Steadman and J.C. Rosas. 2013. *Uromyces appendiculatus* in Honduras: Pathogen diversity and host resistance screening. Plant Dis. 97, 652-661.

Acosta-Gallegos, J.A., R. Ochoa-Marquez, M.P. Arrieta-Montiel, F. Ibarra-Perez, A. Pajarito-Ravelero and I. Sanchez-Valdez. 1995. Registration of 'Pinto Villa' common bean. Crop Sci. 35: 1211.

Acosta-Gallegos, J.A., J.D. Kelly and P. Gepts. 2007. Pre-breeding in common bean and use of genetic diversity from wild germplasm. Crop Sci. 47: S44-S59.

Akibode, S. and M. Maredia. 2011. Global and regional trends in production, trade and consumption of food legume crops. SPIA Report department of agricultural, food and resource economics, Michigan State University, East Lansing, MI.

Aggarwal, V.D., M.A. Pastor-Corrales, R.M. Chirwa and R.A. Buruchara. 2004. Andean beans (*Phaseolus vulgaris* L.) with resistance to the angular leaf spot pathogen (*Phaeoisariopsis griseola*) in southern and eastern Africa. Euphytica 136, 201-210.

Allard, R.W. 1960. Patterns of evolution of cultivated species. pp. 7-18. *In*: R.W. Allard (ed.). Principles of Plant Breeding. Wiley & Sons, New York.

Allendorf, F.W. and G. Luikart. 2009. Conservation and the genetics of populations. Wiley & Sons, New York.

Amugune, N.O., B. Anyango and T.K. Mukiama. 2011. *Agrobacterium*-mediated transformation of common bean. Afr. Crop Sci. J. 19: 137-147.

Anderson, J.W., L. Story, B. Sieling, W.J. Chen, M.S. Petro and J. Story. 1984. Hypocholesterolemic effects of oat-bran or bean intake for hypercholesterolemic men. Am. J. Clin. Nutr. 40: 1146-1155.

Andueza-Noh, R.H., M.L. Serrano-Serrano, M.I. Chacón, I. Sánchez del-Pino, L. Camacho-Pérez, J. Coello-Coello, J. Mijangos Cortes, D.G. Debouck and J. Martìnez-Castillo. 2013. Multiple domestications of the Mesoamerican gene pool of Lima bean (*Phaseolus lunatus* L.): evidence from chloroplast DNA sequences. Genet. Resour. Crop Evol. 60: 1069-1086.

Andueza-Noh, R.H., J. Martínez-Castillo and M.I. Chacón-Sánchez. 2015. Domestication of small-seeded lima bean (*Phaseolus lunatus* L.) landraces in Mesoamerica: evidence from microsatellite markers. Genetica 143, 657-669.

Angioi, S.A., F. Desiderio, D. Rau, E. Bitocchi, G. Attene and R. Papa. 2009. Development and use of chloroplast microsatellites in *Phaseolus* spp. and other legumes. Plant Biol. 11: 598-612.

Angioi, S.A., D. Rau, G. Attene, L. Nanni, E. Bellucci, G. Logozzo, V. Negri, P.L. Spagnoletti Zeuli and R. Papa. 2010. Beans in Europe: origin and structure of the European landraces of *Phaseolus vulgaris* L. Theor. Appl. Genet. 121: 829-843.

Aragão, F.J. L., S.G. Ribeiro, L.M.G. Barros, A.C. M. Brasileiro, D.P. Maxwell, E.L. Rech and J.C. Faria. 1998. Transgenic beans (*Phaseolus vulgaris* L.) engineered to express viral antisense RNAs show delayed and attenuated symptoms to bean golden mosaic geminivirus. Mol. Breed. 4: 491-499.

Aragão, F.J.L., G.R. Vianna, M.M.C. Albino and E.L. Rech. 2002. Transgenic dry bean tolerant to the herbicide glufosinate ammonium. Crop Sci. 42: 1298-1302.

Aragão, F.J.L. and J.C. Faria. 2009. First transgenic geminivirus-resistant plant in the field. Nat. Biotechnol. 27: 1086-1088.

Aragão, F.J.L., E.O.P.L. Nogueira, M.L.P. Tinoco and J.C. Faria. 2013. Molecular characterization of the first commercial transgenic common bean immune to the Bean golden mosaic virus. J. Biotechnol. 166: 42-50.

Araújo, A.P., M.G. Teixeira and D.L. De Almeida. 1997. Phosphorus efficiency of wild and cultivated genotypes of common bean (*Phaseolus vulgaris* L.) under biological nitrogen fixation. Soil Biol. Biochem. 29: 951-957.

Araújo, A.P., I.F. Antunes and M.G. Teixeira. 2005. Inheritance of root traits and phosphorus uptake in common bean (*Phaseolus vulgaris* L.) under limited soil phosphorus supply. Euphytica 145: 33-40.

Arellano, J., S.I. Fuentes, P. Castillo-Espana and G. Hernandez. 2009. Regeneration of different cultivars of common bean (*P. vulgaris* L.) via indirect organogenesis. Plant Cell Tissue and Organ Cult. 96: 11-18.

Asensio-S. Manzanera, M.C., C. Asensio and S.P. Singh. 2005. Introgressing resistance to bacterial and viral diseases from the Middle American to Andean common bean. Euphytica 143, 223-228.

Asensio-S. Manzanera, M.C., C. Asensio and S.P. Singh. 2006 Gamete selection for resistance to common and halo bacterial blights in dry bean intergene pool population. Crop Sci. 46, 131-135.

Asfaw, A., M.W. Blair and C. Almekinders. 2009. Genetic diversity and population structure of common bean (*Phaseolus vulgaris* L.) landraces from the East African highlands. Theor. Appl. Genet. 120: 1-12.

Asfaw, A. and M.W. Blair. 2014. Quantification of drought tolerance in Ethiopian common bean varieties. Agric. Sci. 5: 124-139.

Assefa, M., B. Shimelis, S. Punnuri, R. Sripathi, W. Whitehead and B. Singh. 2014. Common bean germplasm diversity study for cold tolerance in Ehtiopia. Am. J. Plant Sci. 5: 1842-1850.

Balasubramanian, P., A. Vandenberg, P. Hucl and L. Gusta. 2004. Resistance of *Phaseolus* species to ice crystallization at sub-zero temperatures. Physiol. Plant. 120: 451-457.

Balasubramanian, P., F. Ahmad, A. Vandenberg and P.J. Hucl. 2005. Barriers to interspecific hybridization of common bean with *Phaseolus angustissimus* A. Gray and *P. filiformis* Bentham. J. Genet. Breed. 59: 321-328.

Balasubramanian, P.M., R.L. Conner, D.L. McLaren, S. Chatterton and A. Hou. 2014. Partial resistance to white mould in dry bean. Can. J. Plant Sci. 94: 683-691.

Barceló, J. and C. Poschenrieder. 2002. Fast root growth responses, root exudates, and internal detoxification as clues to the mechanisms of aluminium toxicity and resistance: a review. Environ. Exp. Bot. 48: 75-92.

Barker, R.D.J., E. Derbyshire, A. Yarwood and D. Boulter. 1976. Purification and characterization of the major storage proteins of *Phaseolus vulgaris* seeds and their intracellular and cotyledonary distribution. Phytochemistry 15: 751-757.

Barrón, M.C. and E.G. de Mejía. 1998. Comparative study of enzymes related to proline metabolism in tepary bean (*Phaseolus acutifolius*) and common bean (*Phaseolus vulgaris*) under drought and irrigated conditions, and various urea concentrations. Plant Food. Human Nutr. 52: 119-132.

Barrón, J.E., Pasini, R.J., Davis, D.W., Stuthman, D.D. and Graham, P.H. 1999. Response to selection for seed yield and nitrogen (N2) fixation in common bean (*Phaseolus vulgaris* L.). Field Crop. Res. 62: 119-128.

Beaver, J.S. and J.M. Osorno. 2009. Achievements and limitations of contemporary common bean breeding using conventional and molecular approaches. Euphytica 168: 145-175.

Beaver, J.S., M. Zapata, M. Alameda and T.G. Porch. 2012. Registration of PR0401-259 and PR0650-31 dry bean germplasm lines. J. Plant Regist. 6, 81-84.

Beaver, J.S., J.C. Rosas, T.G. Porch, M.A. Pastor-Corrales, G. Godoy-Lutz and E.H. Prophete. 2015. Registration of PR0806-80 and PR0806-81 white bean germplasm lines with resistance to BGYMV, BCMV, BCMNV, and Rust. J. Plant Regist. 9: 208-211.

Beebe, S., J.B. Lynch, N. Galwey, J. Tohme and I. Ochoa. 1997. A geographical approach to identify phosphorus-efficient genotypes among landraces and wild ancestors of common bean. Euphytica 95: 325-336.

Beebe, S., P. Skroch, J. Tohme, M.C. Duque, F. Pedraza and J. Nienhuis. 2000a. Structure of genetic diversity among common bean landraces of middle-American origin based on correspondence analysis of RAPD. Crop Sci. 40: 264-273.

Beebe, S., A.V. Gonzalez and J. Rengifo. 2000b. Research on trace minerals in the common bean. Food Nutr. Bull. 21: 387-391.

Beebe, S., J. Rengifo, E. Gaitan, C. Duque and J. Tohme. 2001. Diversity and origin of Andean landraces of common bean. Crop Sci. 41: 854-862.

Beebe, S., C. Cajiao and O. Mosquera. 2005. Identification of high mineral accessions in sister species of common bean. p. 59-60. In: Centro Internacional de Agricultura Tropical. Project IP-1. Bean improvement for the tropics. Annu. Rep. CIAT, Cali, Colombia.

Beebe, S.E., M. Rojas-Pierce, X. Yan, M.W. Blair, F. Pedraza, F. Munoz, J. Tohme and J.P. Lynch. 2006. Quantitative trait loci for root architecture traits correlated with phosphorus acquisition in common bean. Crop Sci. 46: 413-423.

Beebe, S.E, I.M. Rao, C. Cajiao and M. Grajales. 2008. Selection for drought resistance in common bean also improves yield in phosphorus limited and favorable environments. Crop Sci. 48: 582 - 592.

Beebe, S., J. Ramirez, A. Jarvis, I.M. Rao, G. Mosquera, J.M. Bueno and M.W. Blair. 2011. Genetic improvement of common beans and the challenges of climate change. pp 356-369. *In:* Yadav, S.S., R.J. Redden, J.L. Hatfield, H. Lotze-Campen and A.E. Hall (eds.). Crop adaptation to climate change. John Wiley & Sons, Ltd., Published by Blackwell Publishing Ltd, Richmond, Australia.

Beebe, S.E., I.M. Rao, M.W. Blair and J.A. Acosta-Gallegos, J.A. 2013. Phenotyping common beans for adaptation to drought. Front. Physiol. 4: 35.

Bello, M.H., S.M. Moghaddam, M. Massoudi, P.E. McClean, P.B. Cregan and P.N. Miklas. 2014. Application of in silico bulked segregant analysis for rapid development of markers linked to Bean common mosaic virus resistance in common bean. BMC Genomics 15: 903.

Bellucci, E., E. Bitocchi, D. Rau, M. Rodriguez, E. Biagetti, A. Giardini, G. Attene, L. Nanni and R. Papa. 2014a. Genomics of origin, domestication and evolution of *Phaseolus vulgaris*. pp. 483-507. *In*: R. Tuberosa, A. Graner and E. Frison (eds.). Genomics of Plant Genetic Resources. Springer, Netherlands.

Bellucci, E., E. Bitocchi, A. Ferrarini, A. Benazzo, E. Biagetti, S. Klie, A. Minio, D. Rau, M. Rodriguez, A. Panziera, L. Venturini, G. Attene, E. Albertini, S.A. Jackson, L. Nanni, A.R. Fernie, Z. Nikoloski, G. Bertorelle, M. Delledonne and R. Papa. 2014b. Decreased nucleotide and expression diversity and modified coexpression patterns characterize domestication in the common bean. Plant Cell Online, tpc-114.

Bennett, C.W. 1971. The curly top disease of sugarbeet and other plants. Am. Phytopathol. Soc. Monogr. No. 7.

Bitocchi, E., L. Nanni, E. Bellucci, M. Rossi, A. Giardini, P. Spagnoletti Zeuli, G. Logozzo, J. Stougaard, P. McClean, G. Attene and R. Papa. 2012. Mesoamerican origin of the common bean (*Phaseolus vulgaris* L.) is revealed by sequence data. Proc. Natl. Acad. Sci. USA 109: E788-E796.

Bitocchi, E., E. Bellucci, A. Giardini, D. Rau, M. Rodriguez, E. Biagetti, R. Santilocchi, P. Spagnoletti Zeuli, T. Gioia, G. Logozzo, G. Attene, L. Nanni and R. Papa. 2013. Molecular analysis of the parallel domestication of the common bean (*Phaseolus vulgaris*) in Mesoamerica and the Andes. New Phytol. 197: 300-313.

Blair, M.W., L.C. Muñoz and D.G. Debouck. 2002. Tepary beans (*P. acutifolius*): molecular analysis of a forgotten genetic resource for dry land agriculture. Grain Legum. 36: 25-26.

Blair, M.W., G. Iriarte and S. Beebe. 2006. QTL analysis of yield traits in an advanced back cross population derived from a cultivated Andean × wild common bean (*Phaseolus vulgaris* L.) cross. Theor. Appl. Genet. 112: 1149-1163.

Blair, M. W., H.D. López-Marín and I.M. Rao. 2009a. Identification of aluminum resistant Andean common bean (*Phaseolus vulgaris* L.) genotypes. Braz. J. Plant Physiol. 21: 291-300.

Blair, M.W., C. Astudillo, M.A. Grusak, R. Graham and S.E. Beebe. 2009b. Inheritance of seed iron and zinc concentrations in common bean (*Phaseolus vulgaris* L.). Mol. Breed. 23: 197-207.

Blair, M.W., L.F. González, M. Kimani and L. Butare. 2010a. Genetic diversity, inter-gene pool introgression and nutritional quality of common beans (*Phaseolus vulgaris* L.) from Central Africa. Theor. Appl. Genet. 121: 237-248.

Blair, M.W., J.I. Medina, C. Astudillo, J. Rengifo, S.E. Beebe, G. Machado and R. Graham. (2010b). QTL for seed iron and zinc concentration and content in a Mesoamerican common bean (*Phaseolus vulgaris* L.) population. Theor. Appl. Genet. 121: 1059-1070.

Blair, M.W., F. Monserrate, S.E. Beebe, J. Restrepo and J.O. Flores, 2010c. Registration of high mineral common bean germplasm lines NUA35 and NUA56 from the red-mottled seed class. J. Plant Reg. 4: 55-59.

Blair, M.W., A.C. Fernandez, M. Ishitani, D. Moreta, M. Seki, S. Ayling and K. Shinozaki. 2011a. Construction and EST sequencing of full-length, drought stress cDNA libraries for common beans (*Phaseolus vulgaris* L.). BMC Plant Biol. 11: 171.

Blair, M.W., C. Astudillo, J. Rengifo, S.E. Beebe and R. Graham. 2011b. QTL for seed iron and zinc concentrations in a recombinant inbred line population of Andean common beans (*Phaseolus vulgaris* L.). Theor. Appl. Genet. 122, 511–523.

Blair, M.W., W. Pantoja and L.C. Muñoz. 2012a. First use of microsatellite markers in a large collection of cultivated and wild accessions of tepary bean (*Phaseolus acutifolius* A. Gray). Theor. Appl. Genet. 125: 1137-1147.

Blair, M.W., C.H. Galeano, E. Tovar, M.C.M. Torres, A.V. Castrillón, S.E. Beebe and I.M. Rao. 2012b. Development of a Mesoamerican intra-genepool genetic map for quantitative trait loci detection in a drought tolerant × susceptible common bean (*Phaseolus vulgaris* L.) cross. Mol. Breed. 29: 71-88.

Blair, M.W., A.L. Herrera, T.A. Sandoval, G.V. Caldas, M. Filleppi and F. Sparvoli. 2012c. Inheritance of seed phytate and phosphorus levels in common bean (*Phaseolus vulgaris* L.) and association with newly-mapped candidate genes. Mol. Breed. 30: 1265-1277.

Blair, M.W. and P. Izquierdo. 2012. Use of the advanced backcross-QTL method to transfer seed mineral accumulation nutrition traits from wild to Andean cultivated common beans. Theor. Appl. Genet. 125: 1015-1031.

Blair, M.W. 2013. Mineral biofortification strategies for food staples: the example of common bean. J. Agric. Food Chem. 61: 8287-8294.

Bliss, F.A. and J.W.S. Brown. 1983. Breeding common bean for improved quantity and quality of seed protein. pp. 59-102. *In*: J. Janick (ed.). Plant breeding reviews, vol 1. Wiley, New York.

Bliss, F.A. 1993. Breeding common bean for improved biological nitrogen fixation. Plant Soil 152: 71-79.

Bollini, R. and M.J. Chrispeels. 1978. Characterization and subcellular localization of vicilin and phytohemagglutinin, the two major reserve proteins of *Phaseolus vulgaris* L. Planta 142: 291-298.

Bollini, R. and A. Vitale. 1981. Genetic variability in charge microheterogeneity and polypeptide composition of phaseolin, the major storage protein of *Phaseolus vulgaris;* and peptide maps of its three major subunits. Physiol. Plant. 52: 96-100.

Bonfim, K., J.C. Faria, E.O.P.L. Nogueira, E.A. Mendes and F.J.L. Aragão. 2007. RNAi-mediated resistance to bean golden mosaic vírus in genetically engineered common bean (*Phaseolus vulgaris*). Mol. Plant Microbe Interact. 20: 717-726.

Bradburd, G.S., P.L. Ralph and G.M. Coop. 2013. Disentangling the effects of geographic and ecological isolation on genetic differentiation. Evolution 67: 3258-3273.

Broughton, W.J., G. Hernandez, M. Blair, S. Beebe, P. Gepts and J. Vanderleyden. 2003. Beans (*Phaseolus* spp.)-model food legumes. Plant Soil 252: 55-128.

Burle, M.L., J.R. Fonseca, J.A. Kami and P. Gepts. 2010. Microsatellite diversity and genetic structure among common bean (*Phaseolus vulgaris* L.) landraces in Brazil, a secondary center of diversity. Theor. Appl. Genet. 121: 801-813.

Burow, M.D., P.W. Ludden and F.A. Bliss. 1993. Suppression of phaseolin and lectin in seeds of common bean, Phaseolus vulgaris L.: increased accumulation of 54 kDa polypeptides is not associated with higher seed methionine concentrations. Mol. Gen. Genet. 241: 431-439.

Butare, L., I.M. Rao, P. Lepoivre, J. Polania, C. Cajiao, J. Cuasquer and S. Beebe. 2011. New genetic sources of resistance in the genus *Phaseolus* to individual and combined aluminium toxicity and progressive soil drying stresses. Euphytica 181: 385-404.

Butare, L., I. Rao, P. Lepoivre, C. Cajiao, J. Polania, J. Cuasquer and S. Beebe. 2012. Phenotypic evaluation of interspecific recombinant inbred lines (RILs) of *Phaseolus* species for aluminium resistance and shoot and root growth response to aluminium-toxic acid soil. Euphytica 186: 715-730.

Buys, H., M. Hendricks, B. Eley and G. Hussey. 2002. The role of nutrition and micronutrients in pediatric HIV infection. SADJ 57: 454-456.

Calderon, J.D. and E.A. Backus. 1992. Comparison of the probing behaviors of *Empoasca fabae* and *E. kraemeri* (Homoptera: Cicadellidae) on resistant and susceptible cultivars of common beans. J. Econ. Entomol. 85: 88-99.

Camarena, F. and J.P. Baudoin. 1987. Obtention des premiers hybrides interspécifiques entre *Phaseolus vulgaris* et *Phaseolus polyanthus* avec le cytoplasme de cette dernière forme. Bull. Rech. Agron. Gembloux 22: 43-55.

Campion, B., D. Perrone, I. Galasso and R. Bollini. 2009a. Common bean (*Phaseolus vulgaris* L.) lines devoid of major lectin proteins. Plant Breed. 128: 199-204.

Campion, B., F. Sparvoli, E. Doria, G. Tagliabue, I. Galasso, M. Fileppi, R. Bollini and E. Nielsen. 2009b. Isolation and characterisation of an *lpa* (low phytic acid) mutant in common bean (*Phaseolus vulgaris* L.). Theor. Appl. Genet. 118: 1211-1221.

Cardona, C., J. Kornegay, C.E. Posso, F. Morales and H. Ramirez. 1990. Comparative value of four arcelin variants in the development of dry bean lines resistant to the Mexican bean weevil. Entomol. Exp. Applic. 56: 197-206.

Cardona, C. and J. Kornegay. 1999. Bean germplasm resources for insect resistance. pp. 85-99. *In*: S.L. Clement and S.S. Quisenberry (eds.). Global plant genetic resources for insect resistance. CRC Press, Boca Raton, FL.

Chagas, E.P. and L.G. Santoro. 1997. Globulin and albumin proteins in dehulled seeds of three *Phaseolus vulgaris* cultivars. Plant Foods Hum. Nutr. 51: 17-26.

Chilagane, L.A., G.M. Tryphone, D. Protas, E. Kweka, P.M. Kusolwa and S. Nchimbi-Msolla. 2013. Incorporation of resistance to angular leaf spot and bean common mosaic necrosis virus diseases into adapted common bean (*Phaseolus vulgaris* L.) genotype in Tanzania. Afr. J. Biotechnol. 12: 4343-4350.

Cichy, K.A., G.V. Caldas, S.S. Snapp and M.W. Blair. 2009. QTL analysis of seed iron, zinc and phosphorus levels in an Andean bean population. Crop Sci. 49: 1742–1750.

Colpaert, N., S. Tilleman, M. van Montagu, G. Gheysen and N. Terryn. 2008. Composite *P. vulgaris* plants with transgenic roots as research tool. Afr. J. Biotechnol. 7: 404-408.

Cortés, A.J., M.C. Chavarro and M.W. Blair. 2011. SNP marker diversity in common bean (*Phaseolus vulgaris* L.). Theor. Appl. Genet. 123: 827-845.

Cortés, A.J., D. This, C. Chavarro, S. Madriñán and M.W. Blair. 2012a. Nucleotide diversity patterns at the drought-related *DREB2* encoding genes in wild and cultivated common bean (*Phaseolus vulgaris* L.). Theor. Appl. Genet. 125: 1069-1085.

Cortés, A.J., M.C. Chavarro, S. Madriñán, D. This and M.W. Blair. 2012b. Molecular ecology and selection in the drought-related *Asr* gene polymorphisms in wild and cultivated common bean (*Phaseolus vulgaris* L.). BMC Genet. 13: 58.

Ddamulira, G., C. Mukankusi, M. Ochwo-Ssemakula, R. Edema, P. Sseruwagi and P. Gepts. 2014. Identification of new sources of resistance to angular leaf spot among Uganda common bean landraces. Can. J. Plant Breed. 2: 55-65.

Ddamulira, G., C. Mukankusi, M. Ochwo-Ssemakula, R. Edema, P. Sseruwagi and P. Gepts. 2015. Gene pyramiding improved resistance to angular leaf spot in common bean. Am. J. Exp. Agric. 9: 2.

Debouck, D.G., O. Toro, O.M. Paredes, W.C. Johnson and P. Gepts. 1993. Genetic diversity and ecological distribution of *Phaseolus vulgaris* in northwestern South America. Econ. Bot. 47: 408-423.

Debouck, D.G. 1999. Diversity in *Phaseolus* species in relation to the common bean. pp. 25-52 *In*: S.P. Singh (ed.). Common Bean Improvement in the Twenty-First Century, Vol. 7. Kluwer Academic Publishers, Dordrecht, The Netherlands.

Delaney, D.E. and F.A. Bliss. 1991a. Selection for increased percentage phaseolin in common bean. 1. Comparison of selection for seed protein alleles and S1 family recurrent selection. Theor. Appl. Genet. 81: 301-305.

Delaney, D.E. and F.A. Bliss. 1991b. Selection for increased percentage phaseolin in common bean. 2. Changes in frequency of seed protein alleles with S1 family recurrent selection. Theor. Appl. Genet. 81: 306-311.

Delgado-Salinas, A. 1988. Variation, taxonomy, domestication and germplasm potentialities in *Phaseolus coccineus*. pp. 441-463. *In*: P. Gepts (ed.). Genetic Resources of Phaseolus Beans, 2nd edn. Kluwer Academic Publishers, Boston, MA.

Delgado-Salinas, A., T. Turley, A. Richman and M. Lavin. 1999. Phylogenetic analysis of the cultivated and wild species of *Phaseolus* (Fabaceae). Syst. Bot. 24: 438-460.

Delgado-Salinas, A., R. Bibler and M. Lavin. 2006. Phylogeny of the genus *Phaseolus* (Leguminosae): a recent diversification in an ancient landscape. Syst. Bot. 31: 779-791.

de Oliveira Arantes, L., Â.D.F.B. Abreu and M.A.P. Ramalho. 2010. Eight cycles of recurrent selection for resistance to angular leaf spot in common bean. Crop Breed. Appl. Biotechnol. 10: 232-237.

de Souza, T.L.P., S. N. Dessaune, D.A. Sanglard, M.A. Moreira and E.G. de Barros. 2011. Characterization of the rust resistance gene present in the common bean cultivar Ouro Negro, the main rust resistance source used in Brazil. Plant Pathol. 60: 839-845.

Dillen, W., G. Engler, M. van Montagu and G. Angenon. 1995. Electroporation-mediated DNA delivery to seedling tissues of *Phaseolus vulgaris* L. (common bean). Plant Cell Rep. 15: 119-124.

Dillen, W., J. De Clercq, A. Goossens, M. Van Montagu and G. Angenon. 1997. *Agrobacterium*-mediated transformation of *Phaseolus acutifolius* A. Gray. Theor. Appl. Genet.. 94: 151-158.

Drijfhout, E. 1978. Genetic interaction between *Phaseolus vulgaris* and bean common mosaic virus with implications for strain identification and breeding for resistance. Verslagen van Landbouwkundige Onderzoekingen 872: 1-89.

Durham, K.M., W. Xie, K. Yu, K.P. Pauls, E. Lee and A. Navabi. 2013. Interaction of common bacterial blight quantitative trait loci in a resistant inter-cross population of common bean. Plant Breed. 132: 658-666.

Dwivedi, S.L., K.L. Sahrawat, K.N. Rai, M.W. Blair, M. Andersson and W. Pfieffer. 2012. Nutritionally enhanced staple food crops. Plant Breed. Rev. 34: 169-262.

Escalante, A., G. Coello, L.E. Eguiarte and D. Piñero. 1994. Genetic structure and mating systems in wild and cultivated populations of *Phaseolus coccineus* and *P. vulgaris* (Fabaceae). Am. J. Bot. 81: 1096-1103.

Estrada-Navarrete, G., X. Alvarado-Affantranger, J.E. Olivares, C. Díaz-Camino, O. Santana, E. Murillo, G. Guillén, N. Sánchez-Guevara, J. Acosta, C. Quinto, D. Li, P. M. Gresshoff and F. Sánchez. 2006. *Agrobacterium rhizogenes* transformation of the *Phaseolus* spp.: a tool for functional genomics. Mol. Plant Microbe Interact. 19: 1385-1393.

Eticha, D., M. Zahn, M. Bremer, Z. Yang, A.F. Rangel, I.M. Rao and W.J. Horst. 2010. Transcriptomic analysis reveals differential gene expression in response to aluminium in common bean (*Phaseolus vulgaris*) genotypes. Ann. Bot. 105: 1119-1128.

Fageria, N.K., V.C. Ballgar and R.J. Wright. 1988. Aluminum toxicity in crop plants. J. Plant Nutr. 11: 303-319.

Faria, J.C., M.M.C. Albino, B.B.A. Dias, L.J. Cancado, N.B. da Cunha, L.M. Silva, G.R. Vianna and F.J.L. Aragão. 2006. Partial resistance to Bean Golden Mosaic Virus in a transgenic common bean (*Phaseolus vulgaris*) line expressing a mutated rep gene. Plant Sci. 171: 565-571.

Fawole, I., G.C. Gerloff, W.H. Gabelman and T. Nordheim. 1982. Genetic control of root development in beans (*Phaseolus vulgaris* L.) grown under phosphorus stress. J. Am. Soc. Hortic. Sci. 107: 98-100.

Federici, C.V.T. and J.G. Waines. 1988. Interspecific hybrid compatibility of selected *Phaseolus vulgaris* L. lines with *P. acutifolius* A. Gray, *P. lunatus* L. and *P. filiformis* Bentham. Annual report of the Bean Improvement Cooperative.

Feregrino-Perez, A.A., L.C. Berumen, G. Garcia-Alcocer, R.G. Guevara-Gonzalez RG, M. Ramos-gomez, R. Reynoso-Camach, J.A. Acosta-Gallegos and G. Loarca-Pina. 2008. Composition and chemopreventive effect of polysacchrides from common bean (*Phaseolus vulgaris* L.) on azoxymethane-induced colon cancer. J. Agric. Food Chem. 56: 8737-8744.

Ferwerda, F.H., M.J. Bassett and J.S. Beaver. 2003. Viability of seed of reciprocal interspecific crosses between *Phaseolus vulgaris* L. and *Phaseolus acutifolius* A. Grey. Annu. Rep. Bean Improv. Coop. USA 46: 29-30.

Foster, E.F., A. Pajarito and J. Acosta-Gallegos. 1995. Moisture stress impact on N partitioning, N remobilization and N-use efficiency in beans (*Phaseolus vulgaris* L). J. Agric. Sci. 124: 27-37.

Fourie, D., P. Miklas and H. Ariyaranthe. 2004. Genes conditioning halo blight resistance to races 1, 7, and 9 occur in a tight cluster. Annu. Rep. Bean Improv. Coop. 47: 103-104.

Foy, C.D. 1984. Physiological effects of hydrogen, aluminum, and manganese toxicities in acid soil. pp. 57-97. *In:* Adams, F. (ed.). Soil Acidity and Liming. Agronomy Monograph no. 12. ASA-CSSA-SSSA Publisher, Madison, WI.

Frahm, M.A., J.C. Rosas, N. Mayek-Pérez, E. López-Salinas, J.A. Acosta-Gallegos and J.D. Kelly. 2004. Breeding beans for resistance to terminal drought in the lowland tropics. Euphytica 136: 223-232.

Frei, A., H. Gu, J.M. Bueno, C. Cardona, and S. Dorn. 2003. Antixenosis and antibiosis of common beans to *Thrips palmi* Karny (Thysanoptera: Thripidae). J. Econ. Entomol. 96: 1577-1584.

Frei, A., M.W. Blair, C. Cardona, S.E. Beebe, H. Gu and S. Dorn. 2005. QTL mapping of resistance to karny in common bean. Crop Sci. 45: 379-387.

Freytag, G.F., M.J. Bassett and M. Zapata. 1982. Registration of XR-235-1-1 bean germplasm. Crop Sci. 22, 1268-1269.

Freytag, G.F. and D.G. Debouck. 2002. Taxonomy, distribution and ecology of the genus *Phaseolus* (Leguminosae-Papilionoideae) in North America, Mexico and Central America. SIDA Bot. Misc. 23: 1-300.

Frossard, E., M. Bucher, F. Mächler, A. Mozafar and R. Hurrell. 2000. Potential for increasing the content and bioavailability of Fe, Zn and Ca in plants for human nutrition. J. Sci. Food. Agric. 80: 861-879.

Gage, D.J. 2009. Nodule development in legumes. pp. 1-24. *In:* Emerich, D.W. and H.B. Krishnan (eds.). Nitrogen Fixation in Crop Production. Crop Science Society of America, Madison.

Gálvez, G.E. and F.J. Morales. 1989. Whitefly-transmitted viruses. pp. 379-408. *In:* Schwartz, H.F. and M.A. Pastor-Corrales (eds.). Bean Production Problems in the Tropics. CIAT, Cali, Colombia.

Garvin, D.F. and N.F. Weeden. 1994. Isozyme evidence supporting a single geographic origin for domesticated tepary bean. Crop Science 34: 1390-1395.

Garza, R., C. Cardona and S.P. Singh. 1996. Inheritance of resistance to the bean-pod weevil (*Apion godmani* Wagner) in common beans from Mexico. Theor. Appl. Genet. 92: 357-362.

Garza, R., J. Vera, C. Cardona, N. Barcenas and S.P. Singh. 2001. Hypersensitive response of beans to *Apion godmani* (Coleoptera: Curculionidae). J. Econ. Entomol. 94: 958-962.

Gelin, J.R., S. Forster, K.F. Grafton, P. McClean and G.A. Rojas-Cifuentes. 2007. Analysis of seed-zinc and other nutrients in a recombinant inbred population of navy bean (*Phaseolus vulgaris* L.). Crop Sci. 47: 1361-1366.

Gepts, P. and F.A. Bliss. 1984. Enhanced available methionine concentration associated with higher phaseolin levels in common bean seeds. Theor. Appl. Genet. 69: 7-53.

Gepts, P. and F.A. Bliss. 1985. F1 hybrid weakness in the common bean. Differential geographic origin suggests two gene pools in cultivated bean germplasm. J. Hered. 76: 447-450.

Gepts, P. and F.A. Bliss. 1986. Phaseolin variability among wild and cultivated common beans (*Phaseolus vulgaris*) from Colombia. Econ. Bot. 40: 469-78.

Gepts, P and F.A. Bliss. 1988. Dissemination pathways of common bean (*Phaseolus vulgaris*, Fabaceae) deduced from phaseolin electrophoretic variability. II Europe and Africa. Econ. Bot. 42: 86-104.

Gepts, P. 1988. Phaseolin as an evolutionary marker. pp. 215-241. *In*: P. Gepts (ed.). Genetic resources of *Phaseolus* beans. Kluwer, Dordrecht, the Netherlands.

Gepts, P. and D.G. Debouck. 1991. Origin, domestication, and evolution of the common bean, *Phaseolus vulgaris*. pp 7-53. *In:* Voysest, O. and A. Van Schoonhoven (eds.). Common Beans: Research for Crop Improvement. CAB, Oxon, UK.

Gepts, P. 1998. Origin and evolution of common bean, past event and recent trends. J. Am. Soc. Hortic. Sci. 33: 1124-1130.

Gioia, T., G. Logozzo, G. Attene, E. Bellucci, S. Benedettelli, V. Negri, R. Papa and P. Spagnoletti Zeuli. 2013. Evidence for introduction bottleneck and extensive inter-gene pool (Mesoamerica x Andes) hybridization in the European common bean (*Phaseolus vulgaris* L.) germplasm. PLoS ONE 8: e75974.

Glémin, S. and T. Bataillon. 2009. A comparative view of the evolution of grasses under domestication. New Phytol. 183: 273-290.

Gonçalves-Vidigal, M.C., S.V. Pedro Filho, A.F. Medeiros and M.A. Pastor-Corrales. 2009. Common bean landrace Jalo Listras Pretas is the source of a new Andean anthracnose resistance gene. Crop Sci. 49: 133-138.

Gonçalves-Vidigal, M. C., A.S. Cruz, G.F. Lacanallo, P.S. Vidigal Filho, L.L. Sousa, C.M.N.A. Pacheco, P. McClean, P. Gepts and M.A. Pastor-Corrales. 2013. Co-segregation analysis and mapping of the anthracnose Co-10 and angular leaf spot Phg-ON disease-resistance genes in the common bean cultivar Ouro Negro. Theor. Appl. Genet. 126: 2245-2255.

González Vélez, A., F. Ferwerda, E. Abreu and J.S. Beaver. 2012. Development of bean lines (*Phaseolus vulgaris* L.) resistant to BGYMV, BCMNV and bean weevil (*Acanthoselides obtectus* Say). Annu. Rep. Bean Improv. Coop. 55: 89-90.

González, A.M., F.J. Yuste-Lisbona, A.P. Rodiño, A.M. De Ron, C. Capel, M. García-Alcázar, M., R. Lozano and M. Santalla. 2015. Uncovering the genetic architecture of *Colletotrichum lindemuthianum* resistance through QTL mapping and epistatic interaction analysis in common bean. Front. Plant Sci. 6: 141.

Graham, P.H. 2009. Soil biology with an emphasis on symbiotic nitrogen fixation. pp 171-209. *In:* Emerich, D.W. and H.B. Krishnan (eds.). Nitrogen Fixation In Crop Production. Crop Science Society of America, Madison.

Gregory-Wodzicki, K.M. 2000. Uplift history of the central and northern Andes: A review. Geol. Soc. Am. Bull. 112: 1091-1105.

Gross, Y. and J. Kigel. 1994. Differential sensitivity to high temperature of stages in the reproductive development of common bean *Phaseolus vulgaris* L. Field Crop Res. 36: 201-212.

Guillot, G., R. Vitalis, A.l. Rouzic and M. Gautier. 2014. Detecting correlation between allele frequencies and environmental variables as a signature of selection. A fast computational approach for genome-wide studies. Spat. Stat. 8: 145-155.

Gurusamy, V., V. Vandenberg and K.E. Bett. 2007. Manipulation of in vivo pollination techniques to improve the fertilization efficiency of interspecies crosses in the genus *Phaseolus*. Plant Breed. 126: 120-124.

Guzmán, P., R.L. Gilbertson, R. Nodari, W.C. Johnson, S.R. Temple, D. Mandala, A.B.C. Mkandawire and P. Gepts. 1995. Characterization of variability in the fungus *Phaeoisariopsis griseola* subbests coevolution with the common bean (*Phaseolus vulgaris*). Phytopathol. 85: 600-607.

Guzmán-Maldonado, S.H., J. Acosta-Gallegos and O. Paredes-López. 2000. Protein and mineral content of a novel collection of wild and weedy common bean (*Phaseolus vulgaris* L.). J. Sci. Food Agric. 80: 1874-1881.

Guzmán-Maldonado, S.H., O. Martínez, J.A. Acosta, F. Guevara-Lara and O. Paredes-López. 2003. Putative quantitative trait loci for physical and chemical components of common bean. Crop Sci. 43: 1029-1035.

Hagerty, C.H., A. Cuesta-Marcos, P.B. Cregan, Q. Song, P. McClean, S. Noffsinger and J.R. Myers. 2015. Mapping *Fusarium solani* and *Aphanomyces euteiches* root rot resistance and root architecture quantitative trait loci in common bean. Crop Sci. 55: 1969-1977.

Haghighi, K. R. and P.D. Ascher. 1988. Fertile, intermediate hybrids between *Phaseolus vulgaris* and *P. acutifolius* from congruity backcrossing. Sex. Plant Reprod. 1: 51-58.

Hall, A.E. 2004. Comparative ecophysiology of cowpea, common bean and peanut. pp. 271-325. *In*: Nguyen, H.T. and A. Blum (eds.). Physiology and Biotechnology Integration For Plant Breeding, Marcel Dekker, New York.

Hangen, L. and M.R. Bennink. 2002. Consumption of black beans and navy beans (*Phaseolus vulgaris*) reduced azoxymethance-induced colon cancer. Nutr Cancer 44: 60-65.

Hardarson, G., F.A. Bliss, M.R. Cigales-Rivero, R.A. Henson, J.A. Kipe-Nolt, L. Longeri, A. Manrique, J.J. Peña-Cabriales, P.A.A. Pereira, C.A. Sanabria and S.M. Tsai. 1993. Genotypic variation in biological nitrogen fixation by common bean. Plant Soil 152: 59-70.

Hart, J.P. and P.D. Griffiths. 2015. Genotyping-by-Sequencing enabled mapping and marker development for the BY-2 potyvirus resistance allele in common bean. Plant Genome 8: 1.

Heiser, C.B. 1965. Cultivated plants and cultural diffusion in nuclear America. Am. Anthropol. 67: 930-949.

Henson, R.A., P.A.A. Pereira, J.E.S. Carneiro and F.A. Bliss. 1993. Registration of 'Ouro negro', a high dinitrogen-fixing, high-yielding common bean. Crop Sci. 33: 644.

Hernández-Saavedra, D., M. Mendoza-Sánchez, H.L. Hernández-Montiel, H.S. Guzmán-Maldonado, G. Loarca-Piña, L.M. Salgado and R. Reynoso-Camacho. 2013. Cooked common beans (*Phaseolus vulgaris*) protect against β-cell damage in streptozotocin-induced diabetic rats. Plant Foods Hum. Nutr. 68: 207-212.

Herridge, D.F., M.B. Peoples and R.M. Boddey. 2008. Global inputs of biological nitrogen fixation in agricultural systems. Plant Soil 311: 1-18.

Hoyos-Villegas, V., W. Mkwaila,P.B. Cregan and J.D. Kelly. 2015. Quantitative trait loci analysis of white mold avoidance in Pinto Bean. Crop Sci. 55: 2116-2129.

Hucl, P. and G.J. Scoles. 1985. Interspecific hybridization in the common bean: a review. Hort. Sci. 20: 352-357.

Ishitani, M., I. Rao, P. Wenzl, S. Beebe and J. Tohme. 2004. Integration of genomics approach with traditional breeding towards improving abiotic stress adaptation: drought and aluminum toxicity as case studies. Field Crop Res. 90: 35-45.

Islam, F.M.A., K.E. Basford, C. Jara, R.J. Redden and S.E. Beebe. 2002. Seed compositional and disease resistance differences among gene pools in cultivated common bean. Genet. Res. Crop Evol. 49: 285–293.

Islam, F.A., S. Beebe, M. Muñoz, J. Tohme, R.J. Redden and K.E. Basford. 2004. Using molecular markers to assess the effect of introgression on quantitative attributes of common bean in the Andean gene pool. Theor. Appl. Genet. 108: 243-252.

Johnson, W.C., P. Guzmán, D. Mandala, A.B.C. Mkandawire, S. Temple, R.L. Gilbertson and P. Gepts. 1997. Molecular tagging of the *bc-3* gene for introgression into Andean common bean. Crop Sci. 37, 248-254.

Johnson, W.C. and P. Gepts. 1999. Segregation for performance in recombinant inbred populations resulting from inter-gene pool crosses of common bean (*Phaseolus vulgaris* L.). Euphytica 106: 5-56.

Kamfwa, K., M. Mwala, P. Okori, P. Gibson and C. Mukankusi. 2013. Identification of QTL for *Fusarium* root rot resistance in common bean. J. Crop Improv. 27: 406-418.

Kamfwa, K., K.A. Cichy and J.D. Kelly. 2015. Genome-wide association analysis of symbiotic nitrogen fixation in common bean. Theor. Appl. Genet. DOI 10.1007/s00122-015-2562-5.

Kami, J., V.B. Velásquez, D.G. Debouck and P. Gepts. 1995. Identification of presumed ancestral DNA sequences of phaseolin in *Phaseolus vulgaris*. Proc. Natl. Acad. Sci. USA 92: 1101-1104.

Kami, J., V. Poncet, V. Geffroy and P. Gepts. 2006. Development of four phylogenetically-arrayed BAC libraries and sequence of the APA locus in *Phaseolus Vulgaris*. Theor. Appl. Genet. 112: 987-998.

Keller, B., C. Manzanares, C. Jara, J.D. Lobaton, B. Studer and B. Raatz. 2015. Fine-mapping of a major QTL controlling angular leaf spot resistance in common bean (*Phaseolus vulgaris* L.). Theor. Appl. Genet.128: 813-826.

Kelly, J.D., G.L. Hosfi eld, G.V. Varner, M.A. Uebersax, S.D. Haley and J. Taylor. 1994. Registration of 'Raven' black bean. Crop Sci. 34, 1406-1407.

Kelly, J. D., J.M. Kolkman and K. Schneider. 1998. Breeding for yield in dry bean (*Phaseolus vulgaris* L.). Euphytica 102: 343-356.

Kelly, J.D., G.L. Hosfield, G.V. Varner, M.A. Uebersax and J. Taylor. 1999. Registration of 'Chinook 2000' light red kidney bean. Crop Sci. 39, 293.

Kelly, J.D., G.L. Hosfield, G.V. Varner, M.A. Uebersax and J. Taylor. 2000. Registration of 'Phantom' black bean. Crop Sci. 40, 572.

Kelly, J.D., G.L. Hosfield, G.V. Varner, M.A. Uebersax and J. Taylor. 2001. Registration of 'Jaguar' black bean. Crop Sci. 41, 1447-1448.

Keneni, G., E. Bekele, E. Getu, M. Imtiaz, T. Damte, B. Mulatu and K. Dagne. 2011. Breeding food legumes for resistance to storage insect pests: potential and limitations. Sustainability 3: 1399-1415.

Kiryowa, M., S.T. Nkalubo, C. Mukankusi, H. Talwana, P. Gibson and P. Tukamuhabwa. 2015. Effect of marker aided pyramiding of anthracnose and pythium root rot resistance genes on plant agronomic characters among advanced common bean genotypes. J. Agricul. Sci. 7: 98.

Kim, J.W. and T. Minamikawa. 1997. Stable delivery of a canavalin promoter-β-glucuronidase gene fusion into French bean by particle bombardment. Plant Cell Physiol. 38: 70-75.

Kimani, J.M., P.M. Kimani, S.M. Githiri and J.W. Kimenju. 2007. Mode of inheritance of common bean (*Phaseolus vulgaris* L.) traits for tolerance to low soil phosphorus (P). Euphytica 155: 225-234.

Klaedtke, S.M., C. Cajiao, M. Grajales, J. Polanía, G. Borrero, A. Guerrero, M. Rivera, I. Rao, S.E. Beebe and J. León. 2012. Photosynthate remobilization capacity from drought-adapted common bean (*Phaseolus vulgaris* L.) lines can improve yield potential of interspecific populations within the secondary gene pool. J. Plant Breed. Crop Sci. 4: 49-61.

Koenig, R. and P. Gepts. 1989. Segregation and linkage of genes for seed proteins, isozymes and morphological traits in common bean (*Phaseolus vulgaris*). J Hered 80: 455-459.

Koenig, R., S.P. Singh and P. Gepts. (1990) Novel phaseolin types in wild and cultivated common bean (*Phaseolus vulgaris*, Fabaceae). Econ. Bot. 44: 50-60.

Koinange, E.M.K. and P. Gepts. 1992. Hybrid weakness in wild *Phaseolus vulgaris* L. J. Hered. 83: 135-139.

Kolkman, J.M. and J.D. Kelly. 2003. QTL conferring resistance and avoidance to white mold in common bean. Crop Sci. 43: 539-548.

Kornegay, J.L. and C. Cardona. 1990. Development of an appropriate breeding scheme for tolerance to *Empoasca kraemeri* in common bean. Euphytica 47: 223-231.

Kornegay, J. 1992. BCMV: CIAT's point of view. Ann. Rep. Bean Improv. Coop.

Kraft, K.H., C.H. Brown, G.P. Nabhan, E. Luedeling, J.D.J.L. Ruiz, G. Coppens d'Eeckenbrugge, R.J. Hijmans and P. Gepts. 2014. Multiple lines of evidence for the origin of domesticated chili pepper, *Capsicum annuum,* in Mexico. Proc. Natl. Acad. Sci. USA 111: 6165-6170.

Krupa, U. 2008. Main nutritional and antinutritional compounds of bean seeds-a review. Pol. J. Food Nutr. Sci. 58: 149-155.

Kuboyama, T., Y. Shintaku and G. Takeda. 1991. Hybrid plant of *Phaseolus vulgaris* L. and *P. lunatus* L. obtained by means of embryo rescue and confirmed by restriction endonuclease analysis of rDNA. Euphytica 54: 177-182.

Kwak, M. and P. Gepts. 2009. Structure of genetic diversity in the two major gene pools of common bean (*Phaseolus vulgaris* L, Fabaceae). Theor. Appl. Genet. 118: 979-992.

Kwak, M., J. Kami and P. Gepts. 2009. The putative Mesoamerican domestication center of *Phaseolus vulgaris* is located in the Lerma-Santiago Basin of Mexico. Crop Sci. 49: 554-563.

Larsen, R.C. and P.N. Miklas. 2004. Generation and molecular mapping of a sequence characterized amplified region marker linked with the *Bct* gene for resistance to Beet curly top virus in common bean. Phytopathol. 94: 320-325.

Lauer, J.G., C.G. Bijl, M.A. Grusak, P.S. Baenziger, K. Boote, S. Lingle, T. Carterg, S. Kaepplerh, R. Boermai, G. Eizengaj, P. Carterk, M. Goodmanl, E. Nafzigerm, K. Kidwelln, R. Mitchello, M.D. Edgertonp, K. Quesenberryq and M.C. Willcox. 2012. The scientific grand challenges of the 21st century for the Crop Science Society of America. Crop Sci. 52: 1003-1010.

Lehner, M.D.S., H. Teixeira, T.J. Paula Júnior, R.F. Vieira, R.C. Lima and J.E.S. Carneiro. 2015. Adaptation and resistance to diseases in brazil of putative sources of common bean resistance to white mold. Plant Dis. 99: 1098-1103.

Leitão, S.T., N.F. Almeida, A. Moral, D. Rubiales and M.C. Vaz Patto. 2013. Identification of resistance to rust (*Uromyces appendiculatus*) and powdery mildew (*Erysiphe diffusa*) in Portuguese common bean germplasm. Plant Breed. 132: 654-657.

Liao, H., X. Yan, G. Rubio, S.E. Beebe, M.W. Blair and J.P. Lynch. 2004. Genetic mapping of basal root gravitropism and phosphorus acquisition efficiency in common bean. Funct. Plant Biol. 31: 959-970.

Liao, D., A. Pajak, S.R. Karcz, B.P. Chapman, A.G. Sharpe, R.S. Austin, R. Datla, S. Dhaubhadel and F. Marsolais. 2012. Transcripts of sulphur metabolic genes are co-ordinately regulated in developing seeds of common bean lacking phaseolin and major lectins. J. Expt. Bot. 63: 6283-6295.

Liebenberg, M.M. and Z.A. Pretorius. 2010. 1 Common Bean Rust: Pathology and Control. Hortic. Rev. 37: 1.

Lin, T.Y. and A.H. Markhart. 1996. *Phaseolus acutifolius* A. Gray is more heat tolerant than *P. vulgaris* L. in the absence of water stress. Crop Sci. 36: 110-114.

Lioi, L. 1989. Geographical variation of phaseolin patterns in an old world collection of *Phaseolus vulgaris*. Seed Sci. Technol. 17: 317-324.

Llaca, V., S.A. Delgado and P. Gepts. 1994. Chloroplast DNA as an evolutionary marker in the *Phaseolus vulgaris* complex. Theor. Appl. Genet. 88: 646-652.

Logozzo, G., R. Donnoli, L. Macaluso, R. Papa, H. Knüpffer and P. Spagnoletti Zeuli. 2007. Analysis of the contribution of Mesoamerican and Andean gene pools to European common bean (*Phaseolus vulgaris* L.) germplasm and strategies to establish a core collection. Genet. Resour. Crop Evol. 54: 1763-1779.

López-Marín, H.D., I.M. Rao and M.W. Blair. 2009. Quantitative trait loci for root morphology traits under aluminum stress in common bean (*Phaseolus vulgaris* L.). Theor. Appl. Genet. 119: 449-458.

Lynch, J.P. and S. Beebe. 1995. Adaptation of beans (*Phaseolus vulgaris* L.) to low phosphorus availability. Hortic. Sci. 30: 1165-1171.

Lynch, J.P. 2007. Roots of the second green revolution. Turner review no. 14. Aust. J. Bot. 55: 493-512.

Madakbas, S.Y., M.C. Hiz, S. Küçükyan, and M.T. Sayar. 2013. Transfer of *Co-1* gene locus for anthracnose disease resistance to fresh bean (*Phaseolus vulgaris* L.) through hybridization and molecular marker-assisted selection (MAS). J. Agricul. Sci. 5: 94.

Mafongoya, P.L., S. Mpepereki and S. Mudyazhezha. 2009. The importance of biological nitrogen fixation in cropping systems in nonindustrialized nations. pp. 329-348. *In:* Emerich, D.W. and H.B. Krishnan (eds.). Nitrogen Fixation in Crop Production. Crop Science Society of America, Madison.

Mahuku, G.S., C. Jara, C. Cajiao and S. Beebe. 2003a. Sources of resistance to angular leaf spot (*Phaeoisariopsis griseola*) in common bean core collection, wild *Phaseolus vulgaris* and secondary gene pool. Euphytica 130, 303-313.

Mahuku, G.S., C.E. Jara, C. Cajiao and S. Beebe. 2003b. Sources of resistance to *Colletotrichum lindemuthianum* in the secondary gene pool of *Phaseolus vulgaris* and in crosses of primary and secondary gene pools. Plant Dis. 86, 1383-1387.

Mamidi, S., M. Rossi, D. Annam, S. Moghaddam, R. Lee, R. Papa and P. McClean. 2011. Investigation of the domestication of common bean (*Phaseolus vulgaris*) using multilocus sequence data. Funct. Plant Biol. 38: 953-967.

Mamidi, S., S.M. Moghaddam, R. Lee, J. Myers, P. Miklas and P. McClean. 2015. Next generation sequencing identifies regions of introgression and differentially expressed genes for *Sclerotinia* resistance in common bean. In: Plant and Animal Genome XXIII Conference, 10-14 January 2015, San Diego, CA.

Marsolais, F., A. Pajak, F. Yin, M. Taylor, M. Gabriel, D.M. Merino, V. Ma, A. Kameka, P. Vijayan, H. Pham, S. Huang, J. Rivoal, K. Bett, C. Hernández-Sebastià, Q. Liu, A. Bertrand and R. Chapman. 2010. Proteomic analysis of common bean seed with storage protein deficiency reveals up-regulation of sulfur-rich proteins and starch and raffinose metabolic enzymes and down-regulation of the secretory pathway. J. Proteomics 73: 1587-1600.

Martínez-Castillo, J., L. Camacho-Pérez, S. Villanueva-Viramontes, R.H. Andueza-Noh and M.I. Chacón-Sánchez. 2014. Genetic structure within the Mesoamerican gene pool of wild *Phaseolus lunatus* (Fabaceae) from Mexico as revealed by microsatellite markers: implications for conservation and the domestication of the species. Am. J. Bot. 101: 851-864.

McCouch, S. 2004. Diversifying selection in plant breeding. PLoS Biol. 2: e347.

Mejía-Jiménez, A., C. Muñoz, H.J. Jacobsen, W.M. Roca and S.P. Singh. 1994. Interspecific hybridization between common and tepary beans: increased hybrid embryo growth, fertility and efficiency of hybridization through recurrent and congruity backcrossing. Theor. Appl. Genet. 88: 324-331.

Mendel, G. 1866. Experiments on plant hybrids. pp. 1-48. *In:* C. Stern and E.R. Sherwood (eds.). The origin of genetics. W.H. Freeman & Company, S. Francisco.

Miklas, P.N., J.C. Rosas, L. Beaver, X. Telek and G.F. Freytag. 1994. Field performance of select tepary bean germplasm in the tropics. Crop Sci. 34: 1639-1644.

Miklas, P.N., M. Zapata, J.S. Beaver and K.F. Grafton. 1999. Registration of four dry bean germplasms resistant to common bacterial blight: ICB-3, ICB-6, ICB-8 and ICB-10. Crop Sci. 39, 594.

Miklas, P.N., J.D. Kelly, S.E. Beebe and M.W. Blair. 2006. Common bean breeding for resistance against biotic and abiotic stresses: from classical to MAS breeding. Euphytica 147: 105-131.

Miklas, P.N. 2007. Marker-assisted backcrossing QTL for partial resistance *to Sclerotinia* white mold in dry bean. Crop Sci. 47: 935-942.

Miklas, P.N., L.D. Porter, J.D. Kelly and J.R. Myers. 2013. Characterization of white mold disease avoidance in common bean. Eur. J. Plant Pathol. 135: 525-543.

Miklas, P.N., J.D. Kelly, J.R. Steadman and S. McCoy. 2014a. Registration of pinto bean germplasm line USPT-WM-12 with partial white mold resistance. J. Plant Registr. 8: 183-186.

Miklas, P.N., D. Fourie, J. Trapp, J. Davis and J.R. Myers. 2014b. New loci including conferring resistance to halo bacterial blight on chromosome Pv04 in common bean. Crop Sci. 54: 2099-2108.

Mishra, A.K. and V.P. Singh. 2010. A review of drought concepts. J. Hydrol. 391: 202-216.

Miyasaka, S.C., J.G. Buta, R.K. Howell and C.D. Foy. 1991. Mechanisms of aluminum tolerance in snapbeans. Root exudation of citric acid. Plant Physiol. 96: 737-743.

Mok, D.W.S., M.C. Mok and A. Rabakoarihanta. 1978. Interspecific hybridization of *Phaseolus vulgaris* with *P. lunatus* and *P. acutifolius*. Theor. Appl. Genet. 52: 209-215.

Monserrate, F., G. Hyman and M.W. Blair. 2007. Stability and geographic systems analysis of NUA advanced lines in Colombia and Bolivia. p. 270-272. *In*: Centro Internacional de Agricultura Tropical. Improved beans for the developing world. Outcome line SBA-1. Annu. Rep. CIAT, Cali, Colombia.

Monterroso, V.A. and H.C. Wien. 1990. Flower and pod abscission due to heat stress in beans. J. Amer. Soc. Hort. Sci. 115: 631-634.

Montoya, C.A., J.P. Lallès, S. Beebe and P. Leterme. 2010. Phaseolin diversity as a possible strategy to improve the nutritional value of common beans (*Phaseolus vulgaris*). Food Res. Int. 43: 443-449.

Motta-Aldana, .J, M.L. Serrano-Serrano, H.J. Torres, C.G. Villamizar, D.G. Debouck and M.I. Chacón. 2010. Multiple origins of Lima bean landraces in the Americas: evidence from chloroplast and nuclear DNA polymorphisms. Crop Sci. 50: 1773-1787.

Mühling, M., J. Gilroy and R.R.D. Croy. 1997. Legumin proteins from seeds of *Phaseolus vulgaris* L. J. Plant Physiol. 150: 489-492.

Mukankusi, C., J. Derera, R. Melis, P.T. Gibson and R. Buruchara. 2011. Genetic analysis of resistance to *Fusarium* root rot in common bean. Euphytica 182, 11-23.

Muñoz, L.C., M.W. Blair, M.C. Duque, J. Tohme and W. Roca. 2004. Introgression in common bean x tepary bean interspecific congruity-backcross lines as measured by AFLP markers. Crop Sci. 44: 637-645.

Muñoz, L.C., M.C. Duque, D.G. Debouck and M.W. Blair. 2006. Taxonomy of tepary bean and wild relatives as determined by amplified fragment length polymorphism (AFLP) markers. Crop Sci. 46: 1744-1754.

Mukeshimana, G., L. Butare, P.B. Cregan, M.W. Blair and J.D. Kelly. 2014, Identification of Quantitative Trait Loci associated with Drought Tolerance in Common Bean using SNP Markers, Crop Sci. 54: 923-938.

Mutlu, N., P.N. Miklas, J. Reiser, J. and D. Coyne. 2005a. Backcross breeding for improved resistance to common bacterial blight in pinto bean (*Phaseolus vulgaris* L.). Plant Breed. 124: 282-287.

Mutlu, N., P.N. Miklas, J.R. Steadman, A.M. Vidaver, D.T. Lindgren, J. Reiser, D.P. Coyne and M.A. Pator-Corrales. 2005b. Registration of common bacterial blight resistant pinto bean germplasm line ABCP-8. Crop Sci. 45, 805-806.

Mutlu, N., C.A. Urrea, P.N. Miklas, M.A. Pastor-Corrales, J.R. Steadman, D.T.J. Lindgren, J. Reiser, A.K. Vidaver and D.P. Coyne. 2008. Registration of common bacterial blight, rust and bean common mosaic resistant Great Northern common bean germplasm line ABC-Weihing. J. Plant Regist. 2, 53-55.

Nanni, L., E. Bitocchi, E. Bellucci, M. Rossi, D. Rau, G. Attene, P. Gepts and R. Papa. 2011. Nucleotide diversity of a genomic sequence similar to SHATTERPROOF (PvSHP1) in domesticated and wild common bean (*Phaseolus vulgaris* L.). Theor. Appl. Genet. 123: 1341-1357.

Navarrete-Maya, R. and J.A. Acosta-Gallegos. 1999. Reacción de variedades de frijol común a *Fusarium* spp. y *Rhizoctonia solani* en el Altiplano de México. (In Spanish, with English abstract.) Agron. Mesoamer. 10, 37-46.

Nodari, R.O., S.M. Tsai, P. Guzmán, R.L. Gilbertson and P. Gepts. 1993. Toward an integrated linkage map of common bean. III. Mapping genetic factors controlling host-bacteria interactions. Genetics 134: 341-350.

Nzungize, J., P. Gepts, R. Buruchara, A. Male, P. Ragama, J.P. Busogoro and J.P. Baudoin. 2011. Introgression of *Pythium* root rot resistance gene into Rwandan susceptible common bean cultivars. Afr. J. Plant Sci. 5: 193-200.

Oblessuc, P.R., R.M. Baroni, A.A.F. Garcia, A.F. Chioratto, S.A.M. Carbonell, L.E.A. Camargo and L.L. Benchimol. 2012. Mapping of angular leaf spot resistance QTL in common bean (*Phaseolus vulgaris* L.) under different environments. BMC Genet. 13: 50.

Oblessuc, P.R., R.M. Baroni, G. da Silva Pereira, A.F. Chiorato, S.A.M. Carbonell, B. Briñez, L. Da Costa E Silva, A.A.F. Garcia, L.E. Aranha Camargo, J.D. Kelly and L.L. Benchimol-Reis. 2014. Quantitative analysis of race-specific resistance to *Colletotrichum lindemuthianum* in common bean. Mol. Breed. 34: 1313-1329.

Oblessuc, P.R., C. Francisco and M. Melotto. 2015. The *Co-4* locus on chromosome Pv08 contains a unique cluster of 18 *COK-4* genes and is regulated by immune response in common bean. Theor. Appl. Genet. 128: 1193-1208.

Osborn, T.C. and F.A. Bliss. 1985. Effects of genetically removing lectin seed protein on horticultural and seed characteristics of common bean. J. Am. Soc. Hortic. Sci. 110: 484-488.

Osborn, T.C., T. Blake, P. Gepts and F.A. Bliss. 1986. Bean arcelin 2. Genetic variation inheritance and linkage relationships of a novel seed protein of *Phaseolus Vulgaris*. Theor. Appl. Genet. 71: 847-855.

Osborn, T.C., L.M. Hartweck, R.H. Harmsen, R.D. Vogelzang, K.A. Kmiecik and F.A. Bliss. 2003. Registration of *Phaseolus vulgaris* genetic stocks with altered seed protein compositions. Crop Sci. 43: 1570-1571.

Otubo, S.T., M.A.P. Ramalho, A.F.B. Abreu, J.B. dos Santos, B. Griffing, J.L. Jinks and H.S. Pooni. 1996. Genetic control of low temperature tolerance in germination of the common bean (*Phaseolus vulgaris* L.). Euphytica 89: 313-317.

Papa, R. and P. Gepts. 2003. Asymmetry of gene flow and differential geographical structure of molecular diversity in wild and domesticated common bean (*Phaseolus vulgaris* L.) from Mesoamerica. Theor. Appl. Genet. 106: 239-250.

Papa, R., E. Bellucci, M. Rossi, S. Leonardi, D. Rau, P. Gepts, L. Nanni and G. Attene. 2007. Tagging the signatures of domestication in common bean (*Phaseolus vulgaris*) by means of pooled DNA samples. Ann. Bot. 100: 1039-1051.

Pasev, G., D. Kostova and S. Sofkova, S. 2014. Identification of genes for resistance to Bean common mosaic virus and Bean common mosaic necrosis virus in snap bean (*Phaseolus vulgaris* L.) breeding lines using conventional and molecular methods. J. Phytopathol. 162: 19-25.

Pastor-Corrales, M.A., C. Jara and S.P. Singh. 1998. Pathogenic variation in, sources of, and breeding for resistance to *Phaeoisariopsis griseola* causing angular leaf spot in common bean. Euphytica 103: 161-171.

Pastor-Corrales, M.A. and M.C. Aime. 2004. Differential cultivars and molecular markers segregate isolates of *Uromyces appendiculatus* into two distinct groups that correspond to the gene pools of their common bean hosts. Phytopathol. 94: S82.

Pastor-Corrales, M.A., S.H. Shin and J. Wolf. 2012. Exceptional rust resistance in Mesoamerican common bean accession PI 310762. Annu. Rep. Bean Improv. Coop. 55: 147-148.

Patrick, H.K. and T.B. Ng. 2004. Coccicin, an antifungal peptide with antiproliferative and HIV-1 reverse transciptase inhibitory activities from large scarlet runner beans. Peptides 25: 2063-2068.

Paula Júnior, T.J., W.C. Jesus Junior, B. Admassu, M.D.S. Lehner and B. Hau. 2015. Interactions between rhizoctonia root rot and the foliar common bean diseases anthracnose and rust. J. Phytopathol. 163: 642-652.

Peña, P.A., J.R. Steadman, K.M. Eskridge and C.A. Urrea. 2013. Identification of sources of resistance to damping-off and early root/hypocotyl damage from *Rhizoctonia solani* in common bean (*Phaseolus vulgaris* L.). Crop Prot. 54: 92-99.

Pérez-Vega, J.C., M.W. Blair, F. Monserrate and M. Ligarreto. 2011. Evaluation of an Andean common bean reference collection under drought stress. Agron. Colomb. 29: 17-26.

Petry, N. and I. Egli, B. Campion, E. Nielsen and R. Hurrell. 2013. Genetic reduction of phytate in common bean (*Phaseolus vulgaris* L.) seeds increases iron absorption in young women. J. Nutr. 143: 1219-1224.

Pfeiffer, W.H. and B. McClafferty. 2007. HarvestPlus: breeding crops for better nutrition. Crop Sci. 47: S88–S105.

Piñero, D. and L. Eguiarte. 1988. The origin and biosystematic status of *Phaseolus coccineus* subsp. polyanthus: electrophoretic evidence. Euphytica 37: 199-203.

Porch, T.G. and M. Jahn. 2001. Effects of high temperature stress on microsporogenesis in heat-sensitive and heat-tolerant genotypes of *Phaseolus vulgaris*. Plant Cell Environ. 24: 723-731.
Porch, T.G., J.R. Smith, J.S. Beaver, P.D. Griffiths and C.H. Canaday. 2010. TARS-HT1 and TARS-HT2 heat-tolerant dry bean germplasm. Hort. Sci. 45: 1278-1280.
Porch, T.G., C.A. Urrea, J.S. Beaver, S. Valentin, P.A. Peña and J.R. Smith. 2012. Registration of TARS-MST1 and SB-DT1 multiple-stress-tolerant black bean germplasm. J. Plant Reg. 6: 75-80.
Porch, T.G., J.S. Beaver, D.G. Debouck, S.A. Jackson, J.D. Kelly and H. Dempewolf. 2013a. Use of wild relatives and closely related species to adapt common bean to climate change. Agronomy 3: 433-461.
Porch, T.G., J.S. Beaver and M.A. Brick. 2013b. Registration of tepary germplasm with multiple-stress tolerance, TARS-Tep 22 and TARS-Tep 32. J Plant Reg. 7: 358-364.
Prasad, A.S. 2013. Discovery of Human Zinc Deficiency: Its Impact on Human Health and Disease1-3. American Society for Nutrition. Adv. Nutr. 4: 176-190.
Pratt, R.C. 1983. Gene transfer between tepary and common beans. Desert Plants 5: 57-63.
Pratt, R.C., R.A. Bressan and P.M. Hasegawa. 1985. Genotypic diversity enhances recovery of hybrids and fertile backcrosses of *Phaseolus vulgaris* L.× *P. acutifolius* A. Gray. Euphytica 34: 329-334.
Rainey, K.M. and P.D. Griffiths. 2005a. Inheritance of heat tolerance during reproductive development in snap bean (*Phaseolus vulgaris* L.). J. Am. Soc. Hortic. Sci. 130: 700-706.
Rainey, K.M. and P.D. Griffiths. 2005b. Differential response of common bean genotypes to high temperature. J. Am. Soc. Hortic. Sci. 130: 18-23.
Rainey, K.M. and P.D. Griffiths. 2005c. Evaluation of *Phaseolus acutifolius* A. Gray plant introductions under high temperatures in a controlled environment. Genet. Resour. Crop Evol. 52: 117-120.
Ramirez-Vallejo, P. and J.D. Kelly. 1998. Traits related to drought resi stance in common bean. Euphytica 99: 127-136.
Ramaekers, L., C.H. Galeano, N. Garzon, J. Vanderleyden and M.W. Blair. 2013. Identifying quantitative trait loci for symbiotic nitrogen fixation capacity and related traits in common bean. Mol. Breed. 31: 163-180.
Rana, J.C., T.R. Sharma, R.K. Tyagi, R.K. Chahota, N.K. Gautam, M. Singh, P.N. Sharma and S.N. Ojha. 2015. Characterisation of 4274 accessions of common bean (*Phaseolus vulgaris* L.) germplasm conserved in the Indian gene bank for phenological, morphological and agricultural traits. Euphytica 205: 441-457.
Rangel, A.F., I.M. Rao and W.J. Horst. 2007. Spatial aluminium sensitivity of root apices of two common bean (*Phaseolus vulgaris* L.) genotypes with contrasting aluminium resistance. J. Exp. Bot. 58: 3895-3904.
Rangel, A.F., I.M. Rao and W.J. Horst. 2009. Intracellular distribution and binding state of aluminum in root apices of two common bean (*Phaseolus vulgaris*) genotypes in relation to Al toxicity. Physiol. Plant. 135: 162-173.
Rangel, A.F., I.M. Rao, H.P. Braum and W.J. Horst. 2010. Aluminum resistance in common bean (*Phaseolus vulgaris* L.) involves induction and maintenance of citrate exudation from root apices. Physiol. Plant. 138: 176-190.
Rao, I.M., R.S. Zeigler, R. Vera and S. Sarkarung. 1993. Selection and breeding for acid-soil tolerance in crops. BioScience 43: 454-465.
Rao, I.M. 2001. Role of physiology in improving crop adaptation to abiotic stresses in the tropics: the case of common bean and tropical forages. pp. 583-613. *In:* Pessarakli, M. (ed.) Handbook of Plant and Crop Physiology. Marcel Dekker, New York.
Rao, I., S. Beebe, J. Polania, J. Ricaurte, C. Cajiao, R. Garcia and M. Rivera. 2013. Can tepary bean be a model for improvement of drought resistance in common bean? Afr. Crop Sci. J. 21: 265-281.
Rau, D., M. Rodriguez, M.L. Murgia, V. Balmas, E. Bitocchi, E. Bellucci, L. Nanni, G. Attene and Papa, R. 2015. Co-evolution in a landrace meta-population: two closely related pathogens interacting with the same host can lead to different adaptive outcomes. Scientific reports, 5: 12834.

Rech, E.L., R.G. Vianna and F.J.L. Aragao. 2008. High efficiency transformation by biolistics of soybean, common bean and cotton transgenic plants. Nat. Protoc. 3: 410-418.

Richard, M.M., S. Pflieger, M. Sévignac, V. Thareau, S. Blanchet, Y. Li, S.A. Jackson and V. Geffroy. 2014. Fine mapping of *Co-x*, an anthracnose resistance gene to a highly virulent strain of *Colletotrichum lindemuthianum* in common bean. Theor. Appl. Genet. 127: 1653-166.

Richardson, A.E., J.P. Lynch, P.R. Ryan, E. Delhaize, F.A. Smith, S.E. Smith, PR. Harvey, M.H. Ryan, E.J. Veneklaas, H. Lambers, A. Oberson, R.A. Culvenor and R.J. Simpson. 2011. Plant and microbial strategies to improve the phosphorus efficiency of agriculture. Plant and soil, 349(1-2), 121-156.

Rodiño, A.P., M. Lema, M. Pérez-Barbeito, M. Santalla and A.M. De Ron. 2007. Assessment of runner bean (*Phaseolus coccineus* L.) germplasm for tolerance to low temperature during early seedling growth. Euphytica 155: 63-70.

Rodriguez, M., D. Rau, S.A. Angioi, E. Bellucci, E. Bitocchi, L. Nanni, H. Knüpffer, V. Negri, R. Papa and G. Attene. 2013. European *Phaseolus coccineus* L. landraces: population structure and adaptation, as revealed by cpSSRs and phenotypic analyses. PLoS ONE 8: e57337.

Rodriguez, M., D. Rau, E. Bitocchi, E. Bellucci, E. Biagetti, A. Carboni, P. Gepts, L. Nanni, R. Papa R and G. Attene. 2015. Landscape genetics, adaptive diversity, and population structure in *P. vulgaris*. New Phytol. *In press*

Rodriguez-Uribe, L. and M.A. O'Connell. 2006. A root-specific bZIP transcription factor is responsive to water deficit stress in tepary bean (*Phaseolus acutifolius*) and common bean (*P. vulgaris*). J. Exp. Bot. 57: 1391-1398.

Román-Avilés, B. and J.D. Kelly. 2005. Identification of quantitative trait loci conditioning resistance to *Fusarium* root rot in common bean. Crop Sci. 45: 1881-1890.

Rosales-Serna, R., J. Kohashi-Shibata, J.A. Acosta-Gallegos, C. Trejo-López, J. Ortiz-Cereceres and J.D. Kelly. 2004. Biomass distribution, maturity acceleration and yield in drought-stressed common bean cultivars. Field Crops Res. 85: 203-211.

Rosas, J.C., O.I. Varela and J.S. Beaver, J.S. 1997. Registration of 'Tío Canela-75' small red bean (race Mesoamerica). Crop Sci. 37: 1391.

Rossi, M., E. Bitocchi, E. Bellucci, L. Nanni, D. Rau, G. Attene and R. Papa. 2009. Linkage disequilibrium and population structure in wild and domesticated populations of *Phaseolus vulgaris* L. Evol. Appl. 2: 504-522.

Rubiales, D., S. Fondevilla, W. Chen, L. Gentzbittel, T.J. Higgins, M.A. Castillejo, K.B. Singh and N. Rispail. 2015. Achievements and challenges in legume breeding for pest and disease resistance. Crit. Rev. Plant Sci. 34: 195-236.

Russell, D.R., K.M. Wallace, J.H. Bathe, B.J. Martinell and D.E. McCabe. 1993. Stable transformation of *Phaseolus vulgaris* via electric-discharge mediated particle acceleration. Plant Cell Rep. 12: 165-169.

Santalla, M., A.M. De Ron and M. De La Fuente. 2010. Integration of genome and phenotypic scanning gives evidence of genetic structure in Mesoamerican common bean (*Phaseolus vulgaris* L.) landraces from the southwest of Europe. Theor. Appl. Genet. 120: 1635-1651.

Schinkel, C. and P. Gepts. 1988. Phaseolin diversity in the tepary bean *Phaseolus acutifolius* A. Gray. Plant Breed. 101: 292-301.

Schinkel, C. and P. Gepts. 1989. Allosyme variability in the tepary bean, *Phaseolus acutifolius* A. Gray. Plant Breed. 102: 182-195

Schneider, K.A., R. Rosales-Serna, F. Ibarra-Pérez, B. Cazares-Enriquez, J.A. Acosta-Gallegos, P. Ramirez-Vallejo, N. Wassimi and J.D. Kelly. 1997a. Improving common bean performance under drought stress. Crop Sci. 37: 43-50.

Schneider, K.A., M.E. Brothers and J.D. Kelly. 1997b. Marker-assisted selection to improve drought tolerance in common bean. Crop Sci. 37: 51-60.

Schneider, K.A., K.F. Grafton and J.D. Kelly. 2001. QTL analysis of resistance to *Fusarium* root rot in bean. Crop Sci. 41: 535-542.

Schmit, V. and D.G. Debouck. 1991. Observations on the origin of *Phaseolus polyanthus* Greenman. Econ. Bot. 45: 345-364.

Schmit, V., P. du Jardin, J.P. Baudoin and D.G. Debouck. 1993. Use of chloroplast DNA polymorphism for the phylogenetic study of seven *Phaseolus* taxa including *P. vulgaris* and *P. coccineus*. Theor. Appl. Genet. 87: 506-516.

Schmutz, J., P.E. McClean, S. Mamidi, G.A. Wu, S.B. Cannon, J. Grimwood, J. Jenkins, S. Shul, Q. Song, C. Chavarro, M. Torres-Torres, V. Geffroy, S.M. Moghaddam, D. Gao, B. Abernathy, K. Barry, M.W. Blair, M.A. Brick, M. Chovatia, P. Gepts, D.M. Goodstein, M. Gonzales, U. Hellsten, D.L. Hyten, G. Jia, J.D. Kelly, D. Kudrna, R. Lee, M.M.S. Richard, P.N. Miklas, J.M. Osorno, J. Rodrigues, V. Thareau, C.A. Urrea CA, M. Wang, Y. Yu, M. Zhang, R.A. Wing, P.B. Cregan, D.S. Rokhsar and S.A. Jackson. 2014. A reference genome for common bean and genome-wide analysis of dual domestications. Nature Genet. 46: 707-713.

Schryer, P.A., Q. Lu, A. Vandenberg and K.E. Bett. 2005. Rapid regeneration of *Phaseolus angustissimus* and *P. vulgaris* from very young zygotic embryos. Plant Cell Tissue Organ. Cult. 83: 67-74.

Schwartz, H.F., M.A. Pastor-Corrales and S.P. Singh. 1982. New sources of resistance to anthracnose and angular leaf spot of beans (*Phaseolus vulgaris* L.). Euphytica 31, 741-754.

Schwartz, H.F., K. Otto, H. Terán and M. Lema. 2006. Inheritance of white mold resistance in *Phaseolus vulgaris* × *P. coccineus* crosses. Plant Dis. 90, 1167-1170.

Schwartz, H.F. and S.P. Singh. 2013. Breeding common bean for resistance to white mold: a review. Crop Sci. 53: 1832-1844.

Scott, M.E. and T.E. Michaels. 1992. *Xanthomonas* resistance of *Phaseolus* interspecific cross selections confirmed by field performance. Hort Sci. 27, 348-350.

Serrano-Serrano, M.L., J. Hernandez-Torres, G. Castillo-Villamizar, D.G. Debouck and M.I. Chacón. 2010. Gene pools in wild Lima bean (*Phaseolus lunatus* L.) from the Americas: Evidences for an Andean origin and past migrations. Mol. Phylogenet. Evol. 54: 76-87.

Serrano-Serrano, M.L., R.H. Andueza-Noh, J. Martínez-Castillo, D.G. Debouck and M.I. Chacón. 2012. Evolution and domestication of Lima bean (*Phaseolus lunatus* L.) in Mexico: Evidence from ribosomal DNA. Crop Sci. 52: 1698-1712.

Shii, C.T., M.C. Mok, S.R. Temple and D.W.S. Mok. 1980. Expression of developmental abnormalities in hybrids of *Phaseolus vulgaris* L. J. Hered. 71: 218-222.

Shii, C.T., A. Rabakoarihanta, M.C. Mok, D.W.S. and Mok. 1982. Embryo development in reciprocal crosses of *Phaseolus vulgaris* and *Phaseolus coccineus*. Theor. Appl. Genet. 62: 59-64.

Shonnard, G.C. and P. Gepts. 1994. Genetics of heat tolerance during reproductive development in common bean. Crop Sci. 34: 1168-1175.

Sicard, D., Y. Michalakis, M. Dron and C. Neema. 1997a. Population structure of *C. lindemuthianum* in the three centres of diversity of the common bean. Phytopathol. 87: 807-813.

Sicard, D., S. Buchet, Y. Michalakis and C. Neema. 1997b. Genetic variability of *Colletotrichum lindemuthianum* in wild populations of common bean. Plant Pathol. 46: 355-365.

Singh, S.P. and A.J. Gutierrez. 1984. Geographical distribution of the DL1 and DL2 genes causing hybrid dwarfism in *Phaseolus vulgaris* L., their association with seed size and their significance to breeding. Euphytica 33: 337-345.

Singh, S.P. 1989. Patterns of variation in cultivated common bean (*Phaseolus vulgaris*, Fabaceae). Econ. Bot. 43: 39-57.

Singh, S.P., P. Gepts and D.G. Debouck. 1991a. Races of common bean (*Phaseolus vulgaris* L., Fabaceae).

Econ. Bot. 45: 379-396.

Singh, S.P., J.A. Gutiérrez, A. Molina, C. Urrea and P. Gepts. 1991b. Genetic diversity in cultivated common bean. II. Marker–based analysis of morphological and agronomic traits. Crop Sci. 31: 23-29.

Singh, S.P., R. Nodari and P. Gepts. 1991c. Genetic diversity in cultivated common bean. I. Allozymes. Crop Sci. 31: 19-23.

Singh, S.P. 1994. Gamete selection for simultaneous improvement of multiple traits in common bean. Crop Sci. 34, 352-355.

Singh, S.P., A. Molina and P. Gepts. 1995. Potential of wild common bean for seed yield improvement of cultivars in the tropics. Can. J. Plant Sci. 75: 807-813.

Singh, S.P. and H. Téran. 1995. Evaluating sources of water-stress tolerance in common bean. Annu. Rpt. Bean Improv. Coop. 38: 42-43.

Singh, S.P. and C.G. Muñoz. 1999. Resistance to common bacterial blight among *Phaseolus* species and common bean improvement. Crop Sci. 39, 80-89.

Singh, S.P., H. Terán, C.G. Muñoz and J.C. Takegami. 1999. Two cycles of recurrent selection for seed yield in common bean. Crop Sci. 39: 391-397.

Singh, S.P. 2001. Broadening the genetic base of common bean cultivars: A review. Crop Sci. 41: 1659-1675.

Singh, S.P., C.G. Muñoz and H. Terán. 2001. Registration of common bacterial blight resistant dry bean germplasm VAX1,VAX3, and VAX 4. Crop Sci. 41, 275-276.

Singh, S.P., H. Terán, J.A. Gutíerrez, M.A. Pastor-Corrales, H.F. Schwartz and F.J. Morales. 2003. Registration of A 339, MAR 1, MAR 2, and MAR 3 angular-leaf-spot and anthracnose-resistant germplasm. Crop Sci. 43, 1886-1887.

Singh, S.P., H. Terán, M. Lema, D.M. Webster, C.A. Strausbaugh, P.N. Miklas, H.F. Schwartz and M.A. Brick. 2007. Seventy-five years of breeding dry bean of the Western USA. Crop Sci. 47: 981-989.

Singh, S.P., H. Terán, H.F. Schwartz, K. Otto and M. Lema. 2009a. White mold–resistant interspecific common bean germplasm lines VCW 54 and VCW 55. J. Plant Regist. 3, 191-197.

Singh, S.P., H. Terán, H.F. Schwartz, K. Otto and M. Lema. 2009b. Introgressing white mold resistance from *Phaseolus* species of the secondary gene pool into common bean. Crop Sci. 49, 1629-1637.

Singh, S.P. and H.F. Schwartz. 2010a. Breeding common bean for resistance to diseases: a review. Crop Sci. 50: 2199-2223.

Singh, S.P. and H. Schwartz. 2010b. Review: Breeding common bean for resistance to insect pests and nematodes. Can. J. Plant Sci. 91: 239-250.

Singh, S.P., H. Terán, H.F. Schwartz, K. Otto, D.G. Debouck, W. Roca and M. Lema. 2013. White mold–resistant, interspecific common bean breeding line VRW 32 derived from *Phaseolus costaricensis*. J. Plant Registr. 7, 95-99.

Singh, S.P., H.F. Schwartz, D. Viteri, H. Terán and K. Otto. 2014. Introgressing white mold resistance from PI 439534 to common Pinto bean. Crop Sci. 54: 1026-1032.

Singh, S.P. and P.N. Miklas. 2015. Breeding common bean for resistance to common blight: a review. Crop Sci. 55: 971-98.

Sinha, S., B. Kukreja, P. Arora, M. Sharma, G.K. Pandey, M. Agarwal and V. Chinnusamy. 2015. The omics of cold stress responses in plants. pp 143-194. *In:* Pandey, G.K. (ed.). Elucidation of Abiotic Stress Signaling in Plants. Functional Genomics Perspectives, Vol. 2. Springer, New York.

Smartt, J. 1970. Interspecific hybridization between cultivated American species of the genus *Phaseolus*. Euphytica 19: 480-489.

Smartt, J. 1990. Grain legumes. Evolution and genetic resources. Cambridge University Press, Cambridge, England.

Smith, J.M. and J. Haigh. 1974. The hitchhiking effect of a favourable gene. Genet. Res. 23: 23-35.

Smith, J.R., S.J. Park, J.S. Beaver, P.N. Miklas, C.H. Canaday and M. Zapata. 2007. Registration of TARS-SR05 multiple disease-resistant dry bean germplasm. Crop Sci. 47: 457-458.

Smithson, P.C. and P.A. Sanchez. 2001. Plant nutritional problems in marginal soils of developing countries. pp32-68. *In:* Ae, N., J. Arihara, K. Okada and A. Srinivasan (eds.). Plant Nutrient Acquisition. Springer, Japan.

Sousa, L.L., A.O. Gonçalves, M.C. Gonçalves-Vidigal, G.F. Lacanallo, A.C. Fernandez, H. Awale and J.D. Kelly. 2015. Genetic characterization and mapping of anthracnose resistance of common bean landrace cultivar Corinthiano. Crop Sci. 55: 1900-1910.

Souza, A.A., R.L. Boscariol, D.H. Moon, L.E. Camargo and S.M. Tsai. 2000. Effects of *Phaseolus vulgaris* QTL in controlling host-bacteria interactions under two levels of nitrogen fertilization. Genet. Mol. Biol. 23: 155-161.

Souza, T.L.P.O., F.G. Faleiro, S.N. Dessaune, T.J.D. Paula-Junior, M.A., Moreira and E.G.D. Barros. 2013. Breeding for common bean (*Phaseolus vulgaris* L.) rust resistance in Brazil. Trop. Plant Pathol. 38: 361-374.

Souza, T.L.P.O., A. Wendland, M.S. Rodrigues, L.C.S. Almeida, F.R. Correia, L.A. Rodrigues, H.S. Pereira, L.C.Faria and L.C. Melo. 2014. Anthracnose resistance sources to be explored by the common bean breeding programs in Brazil. Annu. Rep. Bean Improv. Coop. 57: 217-218.

Sparvoli, F. and R. Bollini. 1998. Arcelin in wild bean (*Phaseolus vulgaris* L.) seeds: sequence of arcelin 6 shows it is a member of the arcelins 1 and 2 subfamily. Genet. Resour. Crop Evol. 45: 383-388.

Spataro, G., B. Tiranti, P. Arcaleni, E. Bellucci, G. Attene, R. Papa, P. Spagnoletti Zeuli and V. Negri. 2011. Genetic diversity and structure of a worldwide collection of *Phaseolus coccineus* L. Theor. Appl. Genet. 122: 1281-1291.

St. Clair, D.A. and F.A. Bliss. 1991. Intrapopulation recombination for 15N-determined dinitrogen fixation ability in common bean. Plant Breed. 106: 215-225.

Steadman, J. R., J. Beaver, M. Boudreau, D. Coyne, J. Groth, J. Elly, M. McMillan, R. McMillan, P. Miklas, M. Pastor Corrales, H. Schwartz and J. Stavely. 1995. Progress reported at the 2nd International Bean Rust Workshop. Annu. Rep. Bean Improv. Coop. 38: 1-10.

Svetleva, D.L., M.R. Veltcheva and G. Bhowmik. 2003. Biotechnology as a useful tool in common bean (*P. vulgaris* L.) improvement. Euphytica 131: 189-200.

Tanksley, S.D. and S.R. McCouch. 1997. Seed banks and molecular maps: unlocking genetic potential from the wild. Science 277: 1063-1066.

Taylor, J.H., N.L. Innes, C.L. Dudley and W.A. Griffiths. 1978. Sources and inheritance of resistance to halo-blight of *Phaseolus* beans. Ann. Appl. Biol. 90, 101-110.

Taylor, J.D., D.M. Teverson and J.H.C. Davis. 1996. Sources of resistance to *Pseudomonas syringae* pv. *phaseolicola* races in *Phaseolus vulgaris*. Plant Pathol. 45, 479-485.

Taylor, M., R. Chapman, R. Beyaert, C. Hernández-Sebastià and F.R. Marsolais. 2008. Seed storage protein deficiency improves sulfur amino acid content in common bean (*Phaseolus vulgaris* L.): redirection of sulfur from gamma-glutamyl-S-methyl-cysteine. J. Agric. Food Chem. 56: 5647-5654.

Taylor, G.R. 2012. Apomictic soybean plants and methods for producing. U.S. Patent Application 13/463,632.

Terán, H. and S.P. Singh. 2002. Comparison of sources and lines selected for drought resistance in common bean. Crop Sci. 42: 64-70.

Terán, H., C. Jara, G. Mahuku, S. Beebe and S.P. Singh. 2013. Simultaneous selection for resistance to five bacterial, fungal, and viral diseases in three Andean× Middle American inter-gene pool common bean populations. Euphytica 189: 283-292.

Thomas, C.V., R.M. Manshardt and J.G. Waines. 1983. Teparies as a source of useful traits for improving common beans. Desert plants 5: 43-48.

Thomas, C.V. and J.G. Waines. 1984. Fertile backcross and allotetraploid plants from crosses between tepary beans and common beans. J. Hered. 75: 93-98.

Thomashow, M.F. 1999. Plant cold acclimation: freezing tolerance genes and regulatory mechanisms. Annu. Rev. Plant Biol. 50: 571-599.

Thomashow, M.F. 2001. So what's new in the field of plant cold acclimation? Lots! Plant Physiol. 125: 89-93.

Thompson, M.D., M.A. Brick, J.N. McGinly and H.J. Thompson. 2009. Chemical composition and mammary cancer inhibitory activity of dry bean. Crop Sci. 49: 179-186.

Thung M. and I.M. Rao. 1999. Integrated management of abiotic stresses. pp. 331-370. *In:* Singh, S.P. (ed.). Common Bean Improvement in the Twenty-first Century. Springer, Netherlands.

Tohme, J., P. Jones, S. Beebe and M. Iwanaga. 1995. The combined use of agroecological and characterization data to establish the CIAT *Phaseolus vulgaris* core collection. Core Collections of Plant Genetic Resources. New York: John Wiley & Sons.

Toro, O., J. Tohme and D.G. Debouck. 1990. Wild bean (*Phaseolus vulgaris* L): Description and Distribution. Cali, Colombia: Centro Internacional de Agricultura Tropical.

Trabanco, N., A. Campa and J.J. Ferreira. 2015. Identification of a new chromosomal region involved in the genetic control of resistance to anthracnose in common bean. Plant Genome 8.

Trapp, J.J., C.A. Urrea, P.B. Cregan and P.N. Miklas. 2015. Quantitative trait loci for yield under multiple stress and drought conditions in a dry bean population. Crop Sci. 55: 1596-1607.

Tsai, S., R. Nodari, D. Moon, L. Camargo, R. Vencovsky and P. Gepts. 1998. QTL mapping for nodule number and common bacterial blight in *Phaseolus vulgaris* L. Plant Soil 204: 135-145.

Türkan, İ., M. Bor, F. Özdemir and H. Koca. 2005. Differential responses of lipid peroxidation and antioxidants in the leaves of drought-tolerant *P. acutifolius* Gray and drought-sensitive *P. vulgaris* L. subjected to polyethylene glycol mediated water stress. Plant Sci. 168: 223-231.

Urrea, C.A., P.N. Miklas, J.S. Beaver and R.H. Riley. 1996. A codominant randomly amplified polymorphic DNA (RAPD) marker useful for indirect selection of bean golden mosaic virus resistance in common bean. J. Am. Soc. Horticul. Sci. 121: 1035-1039.

Vance, C.P. 1997. The molecular biology of nitrogen metabolism. pp. 449-476. In: Dennis, D.T., D.H. Turpin, D.D. Lefebvre and D.B. Layzell (eds.). Plant Metabolism. Longman Scientific, Essex, UK.

Vance, C.P. 2001. Symbiotic nitrogen fixation and phosphorus acquisition. Plant nutrition in a world of declining renewable resources. Plant Physiol. 127: 390-397.

Vandemark, G.J., D. Fourie and P.N. Miklas. 2008. Genotyping with real-time PCR reveals recessive epistasis between independent QTL conferring resistance to common bacterial blight in dry bean. Theor. Appl. Genet. 117: 513-522.

Veneklaas, E. J., H. Lambers, J. Bragg, P.M. Finnegan, C.E. Lovelock, W.C. Plaxton, C.A. Price, W-R. Scheible, M.W. Shane, P.J. White and J.A. Raven. 2012. Opportunities for improving phosphorus-use efficiency in crop plants. New Phytol. 195: 306-320.

Vidigal Filho, P.S., M.C. Gonçalves-Vidigal, J. D. Kelly and W.W. Kirk. 2007. Sources of resistance to anthracnose in traditional common bean cultivars from Paraná, Brazil. J. Phytopathol. 155, 108-113.

Vijayan, P., I.A.P. Parkin, S.R. Karcz, K. McGowan, K. Vijayan, A. Vandenberg and K.E. Bett. 2011. Capturing cold-stress-related sequence diversity from a wild relative of common bean (*Phaseolus angustissimus*). Genome 54: 620-628.

Vitale, A. and R. Bollini. 1995. Legume storage proteins. pp. 73-102. *In*: J. Kigel and G. Galili (eds.). Seed Development and Germination. New York: Marcel Dekker.

Viteri, D.M., H. Terán, A.S. Manzanera, M. Carmen, C. Asensio, T.G. Porch, P.N. Miklas and S.P. Singh. 2014. Progress in breeding Andean common bean for resistance to common bacterial blight. Crop Sci. 54: 2084-2092.

Viteri, D.M. and S.P. Singh. 2015. Inheritance of white mold resistance in an Andean common bean A 195 and its relationship with Andean G 122. Crop Sci. 55: 44-49.

Wall, J.R. 1970. Experimental introgression in the genus *Phaseolus*. 1. Effect of mating systems on interspecific gene flow. Evolution 24: 356-366.

Wallace, D.H. and R.E. Wilkinson. 1965. Breeding for *Fusarium* root rot resistance in beans. Phytopathol. 55, 1227-1231.

Wang, I.J., R.E. Glor and J.B. Losos. 2013. Quantifying the roles of ecology and geography in spatial genetic divergence. Ecol. Lett. 16: 175-182.

Welch, R.M., W.A. House, S. Beebe and Z. Cheng. 2000. Genetic selection for enhanced bioavailable levels of iron in bean (*Phaseolus vulgaris* L.). J. Agric. Food Chem. 48: 3576–3580.

White, J.W., C. Montes and L.Y. Mendoza. 1991. Use of grafting to characterize and alleviate hybrid dwarfness in common bean. Euphytica 59: 19-25.

White, J.W. and J.A. Castillo. 1992. Evaluation of diverse shoot genotypes on selected root genotypes of common bean under soil water deficits. Crop Sci. 32: 762-765.

White, J.W., R. Ochoa, F. Ibarra and S.P. Singh. 1994. Inheritance of seed yield, maturity and seed weight of common bean (*Phaseolus vulgaris* L.) under semi-arid rainfed conditions. J. Agric. Sci.122: 265-273.

Winham, D.M., A.M. Hutchins and C.S. Johnston. 2007. Pinto bean consumption reduces biomarkers for heart disease risk. J. Am. Coll. Nutr. 26: 243-249.

Wong, J.H., X.Q. Zhang, H.X. Wang and T.B. Ng. 2006. A mitogenic defensin from white cloud beans (*Phaseolus vulgaris* L.). Peptides 27: 2075-2081.

Wortmann, C.S., L. Lunze, V.A. Ochwoh and J. Lynch. 1995. Bean improvement for low fertility soils in Africa. Afr. Crop Sci. J. 3: 469-477.

Wortmann, C.S., R.A. Kirkby, C.A. Eledu and D.J. Allen. 1998. Atlas of common bean (*Phaseolus vulgaris* L.) production in Africa. Cali, Colombia: CIAT.

Wright, E.M. and J.D. Kelly. 2011. Mapping QTL for seed yield and canning quality following processing of black bean (*Phaseolus vulgaris* L.). Euphytica 179: 471-484.

Wu, J., L. Wang, L. Li and S. Wang. 2014. De novo assembly of the common bean transcriptome using short reads for the discovery of drought-responsive genes. PLoS ONE 10: e0119369.

Yan, X., S.E. Beebe and J.P. Lynch. 1995a. Genetic variation for phosphorus efficiency of common bean in contrasting soil types. II. Yield response. Crop Sci. 35: 1094-1099.

Yan, X., J.P. Lynch and S.E. Beebe. 1995b. Genetic variation for phosphorus efficiency of common bean in contrasting soil types. I. Vegetative response. Crop Sci. 35: 1086-1093.

Yan, X.L., J.P. Lynch and S.E. Beebe. 1996. Utilization of phosphorus substrate by contrasting common bean genotypes. Crop Sci. 36: 936-941.

Yan, X., H. Liao, S.E. Beebe, M.W. Blair and J.P. Lynch. 2004. QTL mapping of root hair and acid exudation traits and their relationship to phosphorus uptake in common bean. Plant Soil. 265: 17-29.

Yin, F., A. Pajak, R. Chapman, A. Sharpe, S. Huang and F. Marsolais. 2011. Analysis of common bean expressed sequence tags identifies sulfur metabolic pathways active in seed and sulfur-rich proteins highly expressed in the absence of phaseolin and major lectins. BMC Genomics 12: 268.

Young, K.R., C. Ulloa Ulloa and J.L. Luteyn. 2002. Plant evolution and endemism in Andean South America: an introduction. Bot. Rev. 68: 4-21.

Yuste-Lisbona, F.J., A.M. González, C. Capel, M. García-Alcázar, J. Capel, A.M. De Ron, R. Lozano and M. Santalla. 2014. Genetic analysis of single-locus and epistatic QTLs for seed traits in an adapted × nuña RIL population of common bean (*Phaseolus vulgaris* L.). Theor. Appl. Genet. 127: 897-912.

Zaiter, H., E. Baydoun and M. Sayyed-Hallak. 1994. Genotypic variation in the germination of common bean in response to cold temperature stress. Plant Soil 163: 95-101.

Zapata, M., G. Freytag and R. Wilkinson. 2004. Release of five common bean germplasm lines resistant to common bacterial blight W-BB-11, W-BB-20-1, W-BB-35, W-BB-52, and W-BB-56. J. Agric. Univ. Puerto Rico 88, 91-95.

Zhang, X., M.W. Blair and S. Wang. 2008. Genetic diversity of Chinese common bean (*Phaseolus vulgaris* L.) landraces assessed with simple sequence repeats markers. Theor. Appl. Genet. 117: 629-640.

Chapter 11

Triticale

Sylwia Oleszczuk[1,*] and Zofia Banaszak[2]

ABSTRACT

In its 140 years of written history, triticale advanced from a botanical curiosity to an established crop; quite competitive with other cereals, especially under more demanding growing conditions. This advancement was not straight and simple; at times too much promise lead to quick disappointments and triticale faded from view for a while. Yet, triticale breeders worked quietly to solve all major issues of such a new crop and made it competitive. This is an impressive achievement when taking into account that all major cereals were domesticated some 10000 years ago and have undergone thousands more rounds of selection to make them more suitable for agriculture. Given the current rate of improvement, triticale may well have all its future still ahead. This chapter reviews some major points in triticale history, as well as some aspects of its genetics, breeding and usage.

Introduction

It was the year 1875 when the first report was published on a purposefully made wheat-rye hybrid. It was an F_1 hybrid, of course sterile, created by a Scottish breeder Wilson. In the 140 years since then, wheat-rye hybrids have advanced from a mere botanical curiosity to an established crop. The progress was slow, in spurts at times, often with overly inflated claims and rosy predictions, but few would question the statement that, at present,

[1] Institute of Plant Breeding and Acclimatization, National Research Institute, Radzików, 05-870 Błonie, Poland, E-mail: s.oleszczuk@ihar.edu.pl

[2] DANKO Plant Breeding Ltd. Choryń 27, 64-000 Kościan, Poland, E-mail: zofia.banaszak@danko.pl

* Corresponding author

triticale, as wheat-rye hybrids are called, is an established cereal. In 2013, the last year for which complete data are available, triticale was grown on over 4 million hectares (Fig. 1), almost the world over but with largest chunks of this acreage in Eastern Europe (Poland, Hungary, Belarus, Lithuania, Russian, France, Germany, Spain, Poland) and China. It is grown mostly for animal feed; human consumption is minimal even though it is quite suitable for this purpose, especially in cookies and flatbreads. Even leavened bread from triticale can be successfully made, but it requires changes in technology that few are willing to make given the abundance of wheat.

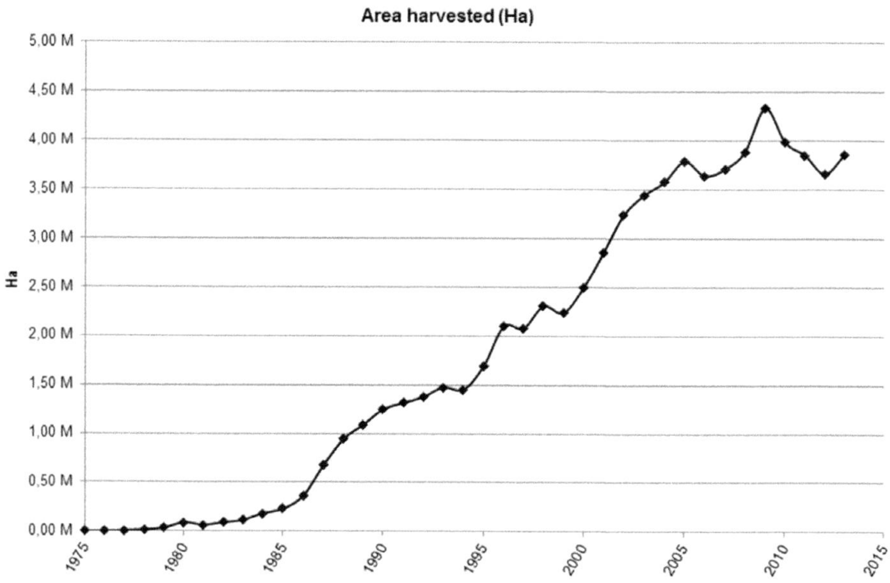

Figure 1. Changes in the worldwide triticale acreage between 1975 and 2014 (FAOSTAT, FAO).

Wilson's hybrid was of bread wheat and rye, so it had 28 chromosomes, seven from each of the four genomes (ABDR), and with no chromosome pairing in meiosis it was sterile. Chromosome doubling by colchicine had not yet been invented. Still, even before colchicine, the first octoploids became available (probably by spontaneous chromosome doubling via unreduced gametes) and the advantages of these hybrids became clear. They did combine, to some extent and not without problems, major advantages of the parents: productivity and quality of wheat with tolerance of harsh growing conditions of rye. And so the era of efforts with octoploid triticales begun, with Rimpau in Germany, A. Muntzing in Sweden, A.F. Shulyndin in Russia (cited after Muntzing 1979, Tarkowski 1989; for the less known aspects of early triticale history, see Miczyński 1905, Tajnsek and Kreft 1996). These days, essentially all effort in triticale breeding is focused on hexaploids, with basic genomic constitution of AABBRR.

Triticale acreage keeps increasing at a steady pace, from trace amounts in the 1970s, through ca. 635000 ha in 1984 (Proc. 3rd EUCARPIA Meeting of the Cereal Section on Triticale, Clermont-Ferrand, France), ca. 2500000 ha in 1991-92 (Varughese 1996), ca. 2700000 ha in 1998 (Proc. 4th Int. Triticale Symposium, Red Deer, Alberta, Canada), ca. 3000000 ha in 2004 (Mergoum and Gómez-Macpherson 2004) to over 4000000 ha at present (Fig. 1). The prospects for future growth have never looked better. An analysis of trends in official yield trials in Germany over the last 30 years clearly shows that triticale enjoys the highest genetic gain among twelve major crops tested (Laidig et al. 2014); in state run official yield trials in Poland over the last 30 years average triticale yields have increased by over 3% per year. Given the large biomass production of triticale and its relatively low harvest index (relative to wheat), significant future improvements are more than likely.

Apart from its potential as a new cereal, triticale also offers a unique opportunity to observe the evolution of a new amphipolyploid under domestication and strong human selection. Several instances of spontaneously formed allopolyploids have been noted and extensively studied in the last decades and many more have been made by artificial hybridization; triticale is the only recent amphipolyploid that is intensively managed and selected by humans, and on a fairly large scale. As is normally the case in polyploidization, the parental genomes in triticale go through the process of adaptation that includes rapid elimination of some DNA sequences, gene silencing and changes in gene expression (Feldman and Levy 2005, Ma and Gustafson 2005). Ma et al. (2004) reported that the wheat genomes were relatively highly conserved in triticales, whereas the rye genome consistently demonstrated a very high level of genomic sequence variation. Khalil et al. (2015) have shown that ca. 0.5% of genes highly expressed in rye are not expressed in triticale and conclude that the most likely to be silenced are the genes that are most dissimilar between the parents. As will be discussed later, some specific problems of triticale, such as reduced meiotic chromosome pairing, irregularities in the endosperm development and pre-harvest sprouting probably reflect some incompatibility, or lack of co-adaptation, of the constitutive genomes. These problems appear to have disappeared in allopolyploid wheat, but then, triticale's history goes back only ca. 140 years (and most stocks are much younger, especially the hexaploids); tetraploid wheat has ca. 0.5M years behind it; hexaploid wheat ca. 10,000 years.

This chapter will attempt to briefly outline the history and present status of triticale breeding and research. It is regrettable that much triticale research has dried out in recent decades, mostly because of limited funding available for new crops. Even breeding underwent a substantial change when the triticale program in CIMMYT was closed, as this was once the major supplier of germplasm, breeding lines and ready cultivars to the spring triticale world. Where triticale programs still exist, breeding efforts continue with essentially the same zeal but done quietly and away from the limelight, no longer promoting triticale as a world savior, but gradually improving it to the point where it often out-competes wheat. Because of limited research, old reviews of the subject, such as Gustafson (1982), Lukaszewski and Gustafson (1987), Tarkowski (1989), Lelley (1992) and the entire set of articles in Triticale Today and Tomorrow (1996) still stand today. Breeding and production aspects of triticale were comprehensively reviewed by Oettler (2005); Mergoum and Gómez-Macpherson (2004) present a good summary of the CIMMYT triticale program, its farewell song. Here we will refresh the old facts and add recent developments hopefully without personal biases.

Ploidy Levels and Chromosome Constitutions

All species in genus *Secale* are diploid with seven chromosomes per genome (designated R); the genus *Triticum* includes diploids, tetraploids and hexaploids, also with seven chromosomes per genome. Hybridization between wheat and rye can produce tetraploid, hexaploid or octoploid triticales (after chromosome doubling). Many such combinations have been made over the years; the most common are hybrids between cultivated tetraploid wheats (durum or macaroni wheat, genomic constitution AABB) or hexaploid wheat (bread wheat, genomic constitution AABBDD). Hybridization of these wheats with rye followed by chromosome doubling produces **hexaploid** and **octoploid** triticale, genomic constitution AABBRR and AABBDDRR, respectively. Attempts were made to cultivate octoploids but currently it appears that only hexaploid triticales are grown commercially.

Tetraploid triticales are quite rare; they can be created by direct hybridization of diploid species of *Triticum* with rye, or by selection of stable lines among progenies of ABRR hybrids, produced by a backcross of hexaploid triticale to rye, or a cross of durum wheat to tetraploid rye. Tetraploid triticales have one pair of wheat genomes, usually consisting of a mixture of A and B genome chromosomes; with some effort, D genome chromosomes can also be added to the mix. While tetraploid triticales can be very interesting from a cytogenetic or evolutionary point of view, at present they do not have any commercial value as a grain crop as their seed set is low. However, three cultivars with high biomass production have been registered in Spain (Ballesteros et al. 2007).

All early breeding efforts were centered on octoploid triticales. Octoploids are easier to make: while crossability of bread wheat with rye can be very low, most set seeds are viable. On the other hand, with few exceptions, seed produced by pollination of tetraploid wheats with rye is very badly shriveled and does not germinate. Hexaploid triticales started being produced on a large scale only when embryo rescue techniques and doubling of chromosome numbers by colchicine became available.

Triticales produced by direct hybridization of wheat with rye are called "primary". They tend to be chromosomally unstable (high proportion of aneuploids among progenies) and have low seed set. However, hybridization of such primary triticales (but also including crosses to bread wheat or diploid rye) followed by selection can lead to rapid improvement of their characteristics, and they quickly became the main and almost the only focus of triticale breeders. The authors are not aware of any breeding programs still focusing on octoploids even though they may be created regularly and used in crosses to hexaploids.

Materials obtained from intercrosses of primary triticale, or hybrids of primary triticales to wheat, rye or octoploid triticales, are referred to as "secondary" triticales (Kiss 1970). Secondary triticales are the main focus of breeding efforts these days; some breeders express the opinion that sufficient genetic variation now exists among secondary triticales that new primary amphiploids do not need to be created. This may a shortsighted view. Given that triticale is a very new creation and its genetic base is necessarily narrow, continuing efforts at creating new amphiploids, both hexaploid and octoploid, should be made.

While all primary **hexaploid triticales** are expected to have the standard AABBRR chromosome constitution (Fig. 2a), and hybridization within this pool is not expected to change it, chromosome constitutions deviating from that standard were quickly detected among secondary triticales. The first detected was the substitution 2D(2R) in spring materials developed in CIMMYT (Gustafson and Zillinsky 1973, Merker, 1975, Pilch 1981). This substitution most likely appeared following strong selection for daylength insensitivity (Lukaszewski and Gustafson 1984) and might have been a consequence of accidental pollination of a triticale line with wheat (Mergoum et al. 2004). The breeding program of triticale at CIMMYT, located in Mexico, started with materials imported from temperate zones, such as Canada, US and Europe (see Jenkins 1969), where they were adapted to much longer days. Chromosome 2D carries locus *Ppd1* controlling photoperiod response in wheat; in addition, the specific chromosome 2D introgressed into CIMMYT triticales also carried *Rht8*, a dwarfing gene from the Mediterranean wheats. As a consequence, all lines with this chromosome substitution were not only earlier but also had considerably shorter straw. This substitution was also found in two winter triticales developed in temperate regions of the globe, and it was probably selected from hybrids with CIMMYT materials widely distributed at that time (Lukaszewski 1988). The 2D(2R) substitution narrowed triticale's range of adaptation (Royo et al. 1993). Another substitution that appeared in CIMMYT materials was 6D(6A) (Lukaszewski 1988). Its origin and selective advantage are not clear; it might have been established under selection for tolerance of high soil acidity and perhaps the presence of soil aluminum. Similarly to 2D(2R), it also appeared in several winter triticales developed in Europe (Lukaszewski 1988) and it might have been transmitted from CIMMYT spring materials. There has been no systematic screening of triticale karyotypes for at least two decades and so it is not known if the two substitutions persist or if perhaps some new ones have established themselves among released cultivars and breeding lines.

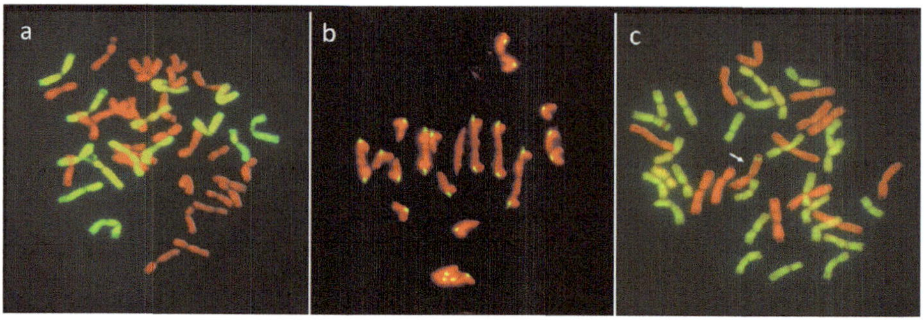

Figure 2. Chromosomes of wheat-rye hybrids a) Hexaploid triticale cv. Salvo: 14 chromosomes of rye labeled green and 28 chromosomes of wheat, labeled red; b) The effect of wheat chromosome arm 5BL with its *Ph1* locus on metaphase I chromosome pairing in a doubled haploid of tetraploid rye: 26 chromosomes of rye (with green-labeled centromeres) forming seven bivalents and 12 univalents. Chromosome arm 5BL forming a bivalent (in the middle, without the green label in the centromere); c) Hexaploid triticale cv. Presto with engineered chromosome 1R (arrowed). The engineered chromosome has two small inserts of wheat chromatin in the short arm and one larger one on the long arm. Here, wheat chromatin labeled green and rye chromatin labeled red. Photos courtesy of A.J. Lukaszewski.

Genetic variation in hexaploid triticale was greatly expanded when sets of chromosome substitutions from the D genome of wheat were systematically established in several winter triticales (Lukaszewski et al. 1987, Lukaszewski 1991, Kazman and Lelley 1994, Xu and Joppa 2000). Such substitutions are easily generated among progenies of hybrids between octoploids and tetraploids, as proposed by Krolow (1973); chromosome constitutions of tetraploids and the compensating ability within individual homoeologous groups determine the genomic location of the D-genome chromosomes. Among progenies of a hybrid AABBDDRR × AARR (F_1 hybrid genome complement AABDRR), chromosomes from the D genome replace B-genome chromosomes; if the starting cross is AABBDDRR × BBRR (F_1 hybrid genome complement ABBDRR), D-genome chromosomes will substitute for the A-genome chromosomes. Single D-genome substitutions were also produced by direct hybridization of lines of durum wheat with such substitutions (cv. Langdon) with rye. It was determined that in most cases, such substitutions had a negative impact on fertility (Xu and Joppa 2000). The agronomic value of individual D(A) or D(B) substitutions has never been firmly established but the existing sets are being tested for such characteristics as mineral uptake and efficiency as well as bread making quality. As expected, substitutions of chromosome 1D have a major impact on the parameters of bread making, as they introduce three gluten loci from bread wheat, including *Glu-D1* encoding high molecular weight glutenin, and increase the total gluten content in triticale. Several other D-genome substitutions, mainly those with the 4D chromosome, appear to increase nitrogen and phosphorus uptake and use efficiency (Oracka and Łapiński 2006) but their practical value remains unknown. Budzianowski and Woś (2004) have shown that lines with chromosome 4D substituted for 4A or 4B were among the most tolerant to high soil aluminum concentrations. Field trials of the 'Presto' D(A) and D(B) substitution set by Budak et al. (2004) showed yield reduction in all lines relative to unmodified 'Presto', thus no substitution line could serve as a commercial cultivar. However, the authors' conclusion, that no major benefit can be expected from D-genome substitutions, may not be unconditionally valid, as demonstrated by 2D(2R) and 6D(6A) substitutions in CIMMYT materials. To the best knowledge of the authors, hybrids between octoploids and tetraploids are never made in breeding programs so such substitutions have never been subjected to strict selection for agronomic performance.

The earlier review of triticale cytogenetics (Lukaszewski and Gustafson 1987) discussed several major cytogenetic problems of hexaploid triticale such as aneuploidy and kernel shriveling. Both appear to have been solved by triticale breeders, to a degree. Kernel shriveling does not appear to be an issue breeders complain much about; current releases have sufficiently high test weights. Originally, kernel shriveling was blamed on high telomeric heterochromatin content of rye chromosomes (Bennett and Gustafson 1982) but closer re-evaluation of the original materials showed that such a relationship was weak, if present at all (Gustafson et al. 1984, Gustafson and Lukaszewski 1985). There are no regular screenings of present-day triticale karyotypes but there are no indications that selection for higher test weight altered the heterochromatin content of rye chromosomes in any significant way, as the grain test weight kept increasing.

Aneuploidy still causes occasional problems with cultivar registration, violating the uniformity criterion. Aneuploids deviate morphologically from euploids and tend to stand out in trials. Aneuploidy is a direct consequence of reduced regularity of chromosome pairing in metaphase I (MI) of meiosis. In triticale this afflicts the rye genome to a greater extent than the wheat genomes. Several hypotheses attempting to explain this phenomenon were reviewed by Lukaszewski and Gustafson (1987). Now it appears that a conflict between the systems of chromosome pairing control of wheat and rye may the main culprit. This conflict was nicely illustrated in a series of experiments by T. Lelley (for a summary see Lelley 1992), who created primary triticales from defined genetic stocks of wheat and inbred lines of rye. It is typical for rye that the MI pairing indices drop upon inbreeding, and that normal pairing levels return upon hybridization of low-pairing inbreds. This suggests that in rye, some degree of chromosome polymorphism is required for normal MI pairing. In triticale, a crop treated as self-pollinating, rye chromosomes are in a state of extreme inbreeding and fail to pair regularly. At the same time, the presence of the rye genome (specifically chromosome 5R) appears to suppress to some degree the *Ph1* system of chromosome pairing of wheat, and occasionally wheat chromosomes also fail to pair. In wheat, *Ph1* does not tolerate heterozygosity; even in intervarietal hybrids MI pairing may drop significantly and some chromosomes or chromosome arms are unable to pair at all (Dvorak and McGuire 1981). In diploid rye, addition of a single copy of wheat chromosome 5B with the *Ph1* locus significantly reduces chromosome pairing (Schlegel et al. 1991). In tetraploid rye, the effect of *Ph1* is even clearer: it dramatically reduces the formation of multivalents and makes meiosis much more diploid-like (Fig. 2b), but selection of chromosomes for bivalent formation is not based on their similarity, as it is in wheat (Oleszczuk et al. 2014a). Chromosome pairing failure in standard triticale lines is, therefore, a consequence of contrasting requirements of wheat and rye pairing control systems: rye chromosomes fail to pair because they are too similar to pass their own requirement for pairing, while wheat chromosomes fail to pair because the *Ph1* system is partially suppressed by the rye genome.

Many cycles of strong selection pressure by triticale breeders for stability and uniformity mitigated the problem of chromosomal instability in conventional cultivars to a considerable degree and brought under control the problem of poor chromosome pairing and resulting aneuploidy. It will be interesting to watch how the problem will be solved in hybrid triticale breeding. Intervarietal triticale hybrids are notoriously meiotically unstable (Oleszczuk et al. 2011) and this affects both genomes: pairing indices for both wheat and rye chromosomes drop; those of rye to a far greater extent. Unlike in diploid rye (Lelley 1992), hybridization in triticale does not restore normal pairing of rye chromosomes. This is probably because of the *Ph1* system of wheat: it recognizes both wheat and rye homologues as different and prevents them from pairing (Lelley 1992, Lukaszewski and Kopecky 2010, Oleszczuk et al. 2014a). Perhaps pairing of both sets of chromosomes would improve in the absence of *Ph1*, or with its weaker version, such as *Ph2* on chromosome 3D. If this thinking is correct, the F_1 hybrids would perform better, but this solution does not address pairing problems of identical rye chromosomes in conventionally developed cultivars. They are too similar for their own good but again, breeder selection has made great strides in this area and only occasionally do some lines show unacceptable levels of aneuploidy that may disqualify them from registration.

Octoploids were the first triticales to be produced and the observations of their characteristics fed early enthusiasm for triticale. This enthusiasm was quickly quenched as octoploids are very unstable meiotically, with a high frequency of univalents in the first metaphase of meiosis. This generates aneuploid progeny with strange morphologies and reduced fertility. Because of chromosomal instability, affecting the rye genome to a far greater extent than the wheat genome, octoploid triticales are said to be capable of reverting to hexaploid wheat (Dou et al. 2006). Agronomically, octoploids show several desirable characteristics, such as good winter hardiness, high green biomass yields and good bread making quality, and this was the justification for early efforts in triticale. Once hexaploid triticales came along, especially the secondary hexaploids, interest in octoploids faded and few new ones are created today. This is unfortunate as octoploids are good crossing partners to hexaploids, and wider usage would expand genetic variation of hexaploids. Only in China octoploids have been released and grown commercially, but for an unknown purpose. The most recent report on a release of a cultivar of octoploid triticale was in 1998 (Yuanshu and Zengyuan 2004). Wang et al. (2010) still list three octoploids in cultivation in China in 2008 but their purpose, such as feed, biomass or grain, has not been stated.

Tetraploids remain relatively unexplored forms of wheat-rye hybrids. Based on limited experience it is difficult to state with any conviction whether they will remain a botanical curiosity, perhaps a model of genome evolution in polyploids via hybridization as postulated by Zohary and Feldman (1962), or whether they have commercial potential. Tetraploid triticale can be produced in several ways. Historically, the first was the approach of Krolow (1973), by hybridization of hexaploid triticale with diploid rye followed by self-pollination of the resulting ABRR hybrids. This process is slow and requires 6-7-8 generations to produce stable lines. It was greatly simplified by Baum and Lelley (1988) who demonstrated that a backcross of an ABRR hybrid as male to a hexaploid triticale (AABBRR) instantly fixes all chromosomes present in a 14-chromosome male gamete, and the extra seven wheat chromosomes present after the backcross (originating from the hexaploid) can be eliminated even in a single generation. If the terminology applied to hexaploids was to be used here such tetraploids deserve the term "secondary".

Tetraploids can be directly synthesized from diploid species of *Triticum* and *Secale* but isolation barriers make this approach difficult. Perhaps the largest collection of such amphidiploids was created by K.D. Krolow, and it included several S and D genome species (A.J. Lukaszewski, personal communication). A.J. Lukaszewski and B. Łapinski (personal communication) created an AARR tetraploid from *T. monococcum* and *S. cereale*, later exploited by Sodkiewicz (1984). Overall, only several centers around the world ever dabbled in tetraploids (Bulgaria, France, Germany, Poland and the former USSR and Russia now, recently Spain). Given that the standard approach of Krolow (1973) can produce 128 different chromosome constitutions from any starting hexaploid line, and that addition of the D-genome of wheat to the mix raises the number of possible combinations to 2187, the total number of chromosome combinations ever produced and described is but a tiny fraction of what is possible (and these numbers do not include A-B-D translocations and introgressions from or into the rye genome, and these do take place; Lukaszewski et al. 1984), and may not even be representative of all possible variation. While a majority of the A-B combinations have been produced, no one ever came close to exploring the ABD chromosome mixes because only several such lines have ever been produced (Hohmann and Krolow 1991). In this sense, the possible variation in chromosome structure at the tetraploid level has never been explored. It is interesting that regardless of the origin or the geographic location, chromosome constitutions combining more or less equal proportions of chromosomes from the original diploid wheat genomes appear to be favored by natural selection; tetraploids with more or less complete original diploid wheat genomes (A, B or D) seem weaker with very low seed set and at times are difficult to maintain. Among the 128 possible combinations of the A and B genome chromosomes creating the diploid wheat genome of a tetraploid triticale, some are favored in certain environments and genetic backgrounds, while quite the opposite combinations are favored elsewhere (Lukaszewski et al. 1984, Bernard et al. 1985, Łapiński and Apolinarska 1985, Sabeva 1985, Badaev et al. 1992). There is not enough data to draw any conclusions on the optimum composition of the wheat genome at this ploidy level, and the future prospects do not appear bright. The entire collection of lines created by Krolow was discarded after his retirement; a majority of the collection assembled by Lukaszewski perished in a freezer accident. The fate of lines created by Z. Sabeva is unknown. Judging from publication dates, some efforts with tetraploids still persist in Russia (Dubovets et al. 2008). Three germplasm lines have lately been registered in Spain but these represent new DDRR amphiploids (Ballesteros et al. 2007). The population of B. Łapinski in Poland still exists, built of a relatively narrow range of karyotypes with a 5AL.5RL translocation and propagated as bulks. Rye cytoplasm in these stocks appears to confer a higher growth rate in early spring (Łapiński 2002). So far, grain yield does not exceed 67% of hexaploid triticale standards (B. Łapiński, personal communication). Recently, these tetraploids were used as a bridge to transfer variation from rye hybrid breeding to hybrid triticale breeding (Łapiński 2014).

Triticale Breeding

Early efforts with triticale improvement, focusing on octoploids, were unsuccessful. Even A. Muntzing, after many years of struggle, admitted that if hexaploids had not come along, the history of triticale would have been a short one. Only China still reports octoploid triticale cultivars (Wang et al. 2010) but the purpose for which they are grown is unknown. Real progress in triticale breeding started only with hexaploids, and, specifically, when large-scale programs of secondary triticale production were undertaken in several places around the world. Early efforts led to the release of the first winter triticales in Hungary (Kiss 1966, 1970) and soon after, of cv. 'Rosner' in Canada (Salmon 2004). Canada seems to have been the very first country where triticale bread was regularly available on the shelves of a major grocery chain. However, those first cultivars were far from spectacular and disappointed more than satisfied. Their major problems were sterility, low yields, lodging, low grain test weight, and pre-harvest sprouting. However, they and their breeding relatives were the cornerstones for new programs that eventually dominated the world: at CIMMYT, Mexico, and at Poznanska Hodowla Roslin (currently DANKO) in Poland. The first grew to become the major creator and supplier of germplasm for spring triticale breeding and cultivation around the globe; the second dominated the winter triticale market in Europe for many years. Both used as their starting points germplasm from Hungary and Canada, among other sources, crossed to the locally produced germplasm.

For direct triticale breeding, crosses of hexaploid triticale × hexaploid triticale, octoploid triticale × hexaploid triticale and especially the three-way cross hexaploid triticale × bread wheat × hexaploid triticale made largely on a large scale in Polish winter triticale breeding programs, offer the best chances for quick progress. The selection of parental lines is based on the characteristics they represent and the breeding objectives. At this point, access to suitably variable starting material is critical, especially considering that triticale is not generous in transgressive segregation beyond the better parent (Lelley 1992). International exchange and production of new primary triticales is of primary importance. World-wide exchange of material under the auspices of CIMMYT left its mark everywhere; including winter triticales (see the part on chromosome constitutions above). A similar benefit was obtained from international trials under the EUCARPIA auspices in Europe. An additional approach to broadening the genetic base of triticale is hybridization with octoploids, with bread wheat and with rye. Various programs use these approaches to varying extents, even though in most cases such cross combinations may seem marginal in the overall effort. To use two Polish breeding programs as an illustration, of the 9415 cross combinations made at DANKO, 79.8% were crosses among hexaploid triticales, 10.9% were crosses of hexaploid triticales with bread wheat and 2.5%, crosses with diploid rye (Banaszak 2004). Overall, 32 813 crosses made at the DANKO triticale breeding program between 1973 and 2011 led to registration of a total of 28 cultivars, or an average of one cultivar per year. Their direct competitor, Plant Breeding Malyszyn IHAR (currently Strzelce), Poland, made ca. 25 000 crosses over 30 years, of which 90% were among hexaploid triticales while all other types of crosses amounted to only about 10%, and they registered 31 cultivars.

It is an open question if triticale will continue in its present form, with its standard AABBRR chromosome constitution, or if new chromosomal variants can be successful commercially. Substitutions and translocation of the D-genome chromosomes, as well as insertions of chromatin from a range of related species may affect important characteristics of triticale. After all, the triticale breeding program at CIMMYT became realistic only with the establishment of the 2D(2R) chromosome substitution (see above). It cannot be entirely discounted that some other introgressions will not have a similar effect. This area is relatively unexplored as research on triticale appears to be dying out and crosses that could potentially lead to such rearrangements are not routinely made in breeding programs.

For breeding purposes, hexaploid triticale is treated as a self-pollinating species and standard approaches are used with probably as many modifications as there are triticale breeders. As a general outline, F_1s are usually grown in the greenhouse or on small plots; no selection is done except that with heavy infection/infestation entire cross combinations may be eliminated. From the F_2 on, the material is advanced as bulks, either without selection, or with selection of individual heads and occasionally entire plants. For several generations the material may be head-rowed (that is, good-looking heads are selected and planted in individual rows) using standard seeding density to stimulate infections and to observe reaction to competition. From the F_5 on, selected heads can still be planted in individual rows while the rest of the seed harvested from the entire row (after head selection) may be planted on small plots in different locations (if possible) for a general evaluation of basic characteristics of the line such as winter hardiness, heading date, lodging resistance, height, disease resistance, and plot uniformity (Fig. 3). In head rows of each line, each of the critical characteristics is observed with special attention to those mandated under the International Union for the Protection of New Varieties of Plants (UPOV). In subsequent generations (F_6 through F_8) selections are planted in replicated trials in as many locations as feasible. Larger numbers of replications and locations provide greater credibility to results which are used to support applications for state trials.

Figure 3. Hexaploid winter triticale in tests and trials: a) Seedlings in flats prepared for tests of frost tolerance in a frost chamber; b) Frost damage in genotypes with good and poor frost tolerance; c) Plots of doubled haploids; d) Plots in a yield trial at the DANKO experiment station in Choryn, Poland.

To be registered and approved for the list of recommended cultivars, apart from meeting the criteria for the Value for Cultivation and Use (VCU) a new cultivar must pass tests of distinctness, uniformity and stability. Only then can it earn an official "report on technical examination" and "variety description". Because of these criteria, maintenance breeding is absolutely critical. It is conducted in parallel to all early trials and is based on head or single plant selection. Typical and uniform progenies are multiplied and bulked for the so-called "breeders material" from which further seed increases are made. Both in head and single plant increases strong negative selection must be carried out to eliminate off-types. These can be frequent in triticale, perhaps because of its tendency to outcross (Sowa and Krysiak 1996) or because of aneuploidy (Banaszak et al. 2007). Recent hybridization studies between triticale and parental species indicate crosses may produce viable seeds, although outcrossing under natural conditions seem unlikely (Kavanagh et al. 2010).

The most important yield-affecting trait in triticale is the number of productive heads per area unit (m^2) and the number of grains per head (Banaszak and Marciniak 2002, Butnaru et al. 2013). Two additional important characteristics are the 1000 kernel weight and test weight. The latter is probably more important as it correlates well with such characteristics as sprouting resistance, germination rate and resistance to fungal diseases. Yield structure is determined by both additive and non-additive effects; strong additive effects were observed for the 1000 kernel weight (Grzesik and Węgrzyn 2003). The overall yield is also affected by the heading and maturity dates, with earlier dates correlating with higher yields.

Breeding efforts in Poland resulted in significant grain yield increases. For winter triticales, the average yield between 1978 and 1995 increased by 2.44 t/ha. Using the average yields of cereals in official state trials in Poland in 1979 and 2003, triticale has shown the highest yield increase (93.9%), rye was somewhat lower at 56.9% and wheat came in third with only 26.4%, or just above 1% per year (vs. ca. 3% for triticale) (Banaszak 2004).

Frost Resistance

Inadequate winter hardiness limits triticale acreage in Europe, Canada and probably Russia as well. The cheapest and simplest methods for the evaluation of winter hardiness evaluation are field tests (Rajki 1980, Fowler and Limin 1997). However, with frequent mild winters, proxy tests are indispensable and these are done in frost chambers (Merker 1988) (Fig. 3 a, b) as frost tolerance is generally considered a good indicator of winter hardiness (Rajki 1980). This was confirmed in direct comparisons in two consecutive years of harsh winters, when the results of frost tolerance tests could be compared with field winter survival rates (Merker 1988, Banaszak et al. 1998). Frost tolerance can also be assessed using the chlorophyll fluorescence method but it requires optimal testing conditions (Fiust et al. 2015). Winter hardiness and frost tolerance are complex traits that can be improved by phenotypic selection, and three major QTL for both traits were identified on chromosomes 5A, 1B and 5R (Liu et al. 2014).

Lodging Resistance

The first triticale lines showed an unpleasant tendency to lodge, a combination of relatively long straw and heavy heads. The tendency is exacerbated by heavy rains during ripening and by high levels of nitrogen fertilization. The degree of lodging depends on straw length and strength, as well as on the susceptibility to crown diseases (Maćkowiak et al. 1998). In general, reduction in straw length reduces lodging (Berzonsky and Lafever 1993, McCaig and Clarke 1994). Straw length can be reduced by dwarfing genes of both wheat and rye origin. Haesert (1994) reports that *Rht1*, a dwarfing gene on chromosome 4B of wheat, is common in European triticales; CIMMYT early triticales were most likely double dwarfs with *Rht1* and *Rht8* (the latter on chromosome 2D in the 2D(2R) substitutions). Dwarfing genes from wheat *Rht1*, *Rht 2* and *Rht3* were tested in standard cv. 'Presto' but they never produced satisfactory yield levels (Kraska and Tarkowski 1998, Kociuba and Kowalczyk 2000). New perspectives for the straw length reduction appeared with the rye mutant EM1 carrying a dominant dwarfing gene *Ddw1*, developed by Kobylyanski (1972). Using this dwarfing source, Wolski (1991) and Wolski et al. (1998) initiated breeding of semidwarf winter triticales and in 1997 registered the first Polish cultivar with reduced straw length–Fidelio. Earlier, Zilinsky and Borlaug (1971) initiated crosses of triticale with a dwarf rye carrying a recessive dwarfing gene but the agronomic value of the resulting material was below expectations.

Pre-Harvest Sprouting

Pre-harvest sprouting is a serious problem of triticale in all regions of the world with high humidity at harvest, such as Brazil, Canada, Northern and Central Europe and East Africa (Skovmand et al. 1984, Mergoum et al. 1998). The sprouting disaster in Germany in 2002 confirmed that triticale has the least sprouting resistance of all cereals (F. Hartmann, personal communication). Despite many efforts focused on this problem, the overall level of sprouting resistance in triticale is unsatisfactory and requires further serious attention. Tests of sprouting resistance are more difficult in triticale than in rye or wheat, as some incompatibility between the two parental systems introduces a degree of disorder to the enzymatic control of the germination process and confuses standard testing methods (Oettler 1990). Tests for the α-amylase activity, such as the falling number, are not reliable in triticale and show considerable variation in readings between growing seasons. Not everyone even agrees that α-amylase activity is tightly associated with sprouting tendencies (Haesert and De Baets 1996). Combination of two tests, the falling number and sprouting in the head, offer a better chance of selecting the right genotypes. However, given the time and effort required, both tests can only be performed on advanced breeding materials, perhaps missing the best genotypes. Triticale breeders urgently need a testing method capable of producing reliable results regardless of the specific field conditions (Haesaert and De Baets 1996).

Fungal Diseases

Just as all other cereals, triticale is susceptible to a range of diseases of stem, leaf and head. For many years the symptoms of these diseases were clearly milder than in parental species (Sowa and Maćkowiak 1990, Zamorski and Schollenberger 1995), but already in the 1980s predictions were being made that the resistance of triticale may break down and yields will suffer (Kronberga 2002). Audenaert et al. (2014) suggest that the recent surge in triticale susceptibility to pathogens is a consequence of the host range expansion: typical wheat pathogens acquired the ability to colonize triticale. At present there are no clear indications that rye pathogens are doing the same. A relatively narrow genetic base of hexaploid triticale may make breeding for resistance difficult. However, introduction of genetic variation from related species, via chromosome engineering or other means, is just as feasible in triticale as it is in wheat.

New pathogens appear to emerge and gain importance; on winter triticales in Poland these are *Rynchosporium secalis, Pyrenopora tritici-repentis, Stagonospora nodorum;* in Australia it is leaf rust; in North West US it is stripe rust. For many years the symptoms of powdery mildew caused by *Blumeria graminis* on triticale were relatively mild and did not cause undue concern. However, its severity has increased dramatically since 2000. Cv. Lamberto, probably the most widely grown winter triticale in Europe (in 2003 it was registered in Belgium, Czech Republic, Estonia, France, Germany, Hungary, Lithuania, Germany, Norway, Sweden, Slovenia, Switzerland, and UK) was free of mildew at release but soon became so susceptible that it had to be withdrawn from the market, with the symptoms appearing in all stages, from seedling to adult plant. Triticale breeders are of the opinion that the seedling stage susceptibility to powdery mildew is less threatening, unless the infection is lethal. The adult stage resistance appears to be polygenic and hence less likely to break down (Bouguennec et al. 2013), but when it breaks down, it is more damaging as it reduces the leaf area. In recent years, a serious threat to triticales in Europe is from stripe rust.

Aluminium Tolerance

Acidic soils and associated toxic effect of aluminum ions on the root system are one of the main factors limiting triticale productivity (Anioł 1981, 1985). Triticale tends to have lower tolerance to toxic effects of aluminum relative to rye, but greater tolerance than that of wheat, although large variation has been noted (Zhang and Jessop 1998). Triticales with complete AABBRR genomes tend to have the highest Al tolerance, at times even exceeding the levels observed in rye (Kim et al. 2001). The presence of a complete rye genome (especially chromosome 3R) is necessary for triticale tolerance to aluminum (Budzianowski and Woś 2004). Although improved Al tolerance levels are coming from genes present in the rye genome (Kim et al. 2001), the expression of rye Al-tolerance genes is thought to be partially suppressed by the interaction with the wheat genetic background (Aniol and Gustafson 1984). The loci associated with the trait were mapped to chromosomes 3R, 4R, 6R and 7R (Ma et al. 2000, Budzianowski and Woś 2004, Niedziela et al. 2012). It was shown that the major gene coding for the aluminum-activated malate transporter (ALMT) is located on chromosome 7R, and it explains up to 36% of phenotypic variation (Niedziela et al. 2014). Recent observations suggest that Al tolerance may be co-regulated by epigenetic processes identified at the DNA methylation level (Niedziela et al. 2015).

Hybrid Triticale

As in many other crops, the development of hybrids in triticale is a hot topic in some programs. Heterosis in general refers to better levels of biomass and grain yield, growth rate, stature and fertility in the hybrid offspring than in the parents (Chen 2010). The level of hybrid vigor is of primary interest as only sufficient yield increases would justify to a farmer the extra cost for seed purchase. Darvey and Roake (2002) claim up to 20% grain yield increases for some hybrids but this may be overly optimistic. Mühleisen et al. (2015) report a maximum 11.5% yield increase relative to the mid parent but no hybrid performed better than the best inbred (parental line). However, hybrids demonstrated better yield stability (Mühleisen et al. 2014). Modestly higher grain yield increases were reported in earlier tests (Oettler et al. 2003, 2005), ranging from 10.1% to 22.4% over the mid-parent and up to 17.4% over the better parent, but without data on best standard cultivars at that time, complete evaluation of data is difficult. Accurate prediction of the hybrid effect based on the performance of parents would reduce the costs and field trials required, e.g. by focusing on the most promising hybrid combinations and/or estimation of parental genetic distance. The importance of specific combining ability and general combining ability in triticale are discussed by Grzesik and Węgrzyn (1998) and Oettler et al. (2003).

Hybrid breeding in triticale is based primarily on cytoplasmic male sterility (cms), using the cytoplasm of *Triticum timopheevi*. Some lines are male sterile in this cytoplasm but their fertility can be restored by genotypes carrying restorer genes. Tests of wheat-rye addition lines indicated that diploid rye (*Secale cereale*) carries two restorers for the *timopheevi* cytoplasm, a weak one on chromosome 4R and a potent one on chromosome 6R (Curtis and Lukaszewski 1993). All tested additions had the 6R restorer but allelic variation must exist as some triticales were male sterile with the *timopheevi* cytoplasm. With two restorer genes frequent in the rye genome in triticale, cms maintainer lines are rare and require serious efforts to identify (M. Pojmaj, H. Woś, personal communication). On the other hand, it would appear that with the high frequency of restorers, fertility restoration in triticale would not be a problem. However, breeders admit that really good restoring lines are in fact rare (H. Woś, personal communication) creating some room for speculation that restorer genes in the wheat genomes are also required. This would make the cms system of hybrid triticale even more complex than in wheat, with only one advantage, that of good anther extrusion and large amounts of pollen.

Recent results of Warzecha et al. (2014) suggest that hybrid rye maintainers could be used as a source of variation for building stable male sterility in the *timopheevi* cytoplasm-based hybrid triticale. An interesting alternative is the cms system based on the rye 'Pampa' cytoplasm (Łapiński 2014). The main expected advantage of this system over that based on the *timopheevi* cytoplasm is a link with complementary gene pools of rye hybrid breeding. The system of mass cross-pollination control works well in winter triticale. Since 2003, levels of 100% sterility of female lines and over 90% fertility restoration in the hybrids have been achieved in the Polish hexaploid triticale gene pool. A 100% fertility restoration has been reached in the majority of F_1s produced after the first cycle of enrichment of male lines with rye restorer genes. However, the heterosis level remains low and yields do not exceed those of pure cultivar standard lines.

An alternative to the cms-based hybrids is making use of chemical hybridizing agents (CHA – even if "chemical sterilizing agent" or CSA is more precise). Triticale hybrids generated via chemical sterilization showed an average mid-parent heterosis of around 10% (Fischer et al. 2010), while the cms-generated hybrids averaged only about 2% (Gowda et al. 2013). This may be an indication of a negative effect of the *timopheevi* cytoplasm on plant vigor. Warzecha et al. (1998) has shown that six F_1 hybrids with the standard (*aestivum*) cytoplasm had 11.7% higher grain yield than reciprocal hybrids with the *timopheevi* cytoplasm. The effect is not necessarily general as one cultivar ('Vero') showed a 6.2% yield increase in the *timopheevi* cytoplasm. 'Vero' reacted similarly to the 'Pampa' cms system (Łapiński 2014). While limited in scope, these tests suggest that the issue of triticale sterilization (*timopheevi*, Pampa, CHA) is still open and requires further tests.

Regardless of which system is eventually adopted, quick progress in hybrid triticale breeding may be difficult. As of now, no clear heterotic groups have been identified (Tams et al. 2005, 2006) which means that the hit-and-miss approach is the only viable option for selecting parental lines. This absence of heterotic groups may well be a consequence of wide germplasm exchange among companies and programs. A study by Tams et al. (2004) has shown that almost 85% of the detected genetic variation among winter triticales in Europe is within individual breeding programs and only ca. 15% among them. While this wide germplasm exchange clearly advanced triticale breeding and led to its commercial success, it also homogenized the gene pool to a considerable extent.

An issue asking for serious attention in hybrid triticale breeding is meiotic stability of the F_1 hybrids. In as much as breeders managed to solve the issues of meiotic instability and hence high frequency of aneuploids in conventionally bred materials, it still manifests itself in hybrids. Among four F_1 hybrids used to generate mapping populations in triticale, the proportions of meiocytes with complete chromosome pairing ranged from barely 13 to 46% (Oleszczuk et al. 2011). This means that in the most meiotically stable hybrid a majority (54%) of meiocytes had incomplete chromosome pairing and it could not possibly have produced a majority of normal gametes.

The first registered triticale hybrid cultivar 'Hybridel' (Hans Winzeler from Delley Samen Pflanzen, Switzerland) was released in 1998; it was based on a CHA (or, rather, a chemical sterilizing agent). The first cms-based commercial hybrid 'HYT Prime' (breeder E. Weissmann from Saatzucht Dr. Hege GbR, Germany) was officially registered in 2010 in France (Longin et al. 2012) for grain use with relatively high protein content. There are some hybrid triticale cultivars (e.g. Kulula, HYTMax, HYTGamma; S. Wiessmnn, personal communication) and lines in the pipeline (H. Woś, M. Pojmaj, personal communication)

Doubled Haploids of Triticale

As mentioned above, most breeding programs treat triticale as a self-pollinating species and the breeding schemes rest on the assumption that self-pollination predominates. This is not necessarily a fully correct assumption as triticale shows a fair degree of outcrossing (Sowa and Krysiak 1996). Still, standard breeding procedures typical for self-pollinators eventually lead to increased levels of homozygosity among selected materials. Even if an assumption is made of complete self-fertilization, it takes 6-7 generations to produce acceptably uniform material. As in many other crops, this time to homozygosity in triticale can be drastically reduced through the use of doubled haploid (DH) technology. At least in theory, the method is capable of generating within one year sufficient quantities of homozygous grain samples for replicated field trials. Doubled haploids can be produced from existing lines/cultivars for purification purposes, or from F_1 hybrids to accelerate the breeding process. Apart from their breeding application, they can and are being used in genetic mapping and, especially, in identification of quantitative trait loci for subsequent marker assisted selection (Kalih et al. 2014). They can also be used to generate male sterile lines, maintainers and restorers required for hybrid seed production (Warzecha et al. 2005). Triticale is quite amenable to the DH approach.

The value of the approach depends primarily on the efficiency of the system (the number of plants produced vs. effort required) and appropriate assessment of the lines' value in practical breeding. Making use of DH lines dramatically changes the emphasis in a breeding program. Traditionally, breeders start from a very large number of early generation hybrids which receive only cursory evaluation. As selection progresses and the numbers drop, the precision, and the cost, of the evaluation process increases. With the DH approach, a single step is made directly from F_1 hybrids to replicated plot tests, with all expenses associated with it. It has never been established with any precision how many DH lines are required to properly evaluate a given cross combination. In winter triticale breeding programs in Poland, 50-200 lines per cross combination are taken as satisfactory, but this number may change depending on the direction of breeding, the number of segregating traits and available resources.

Of several systems of DH production, two are used in triticale with some success: chromosome elimination and androgenesis. The former is by maize pollination (Laurie and Bennett 1988, P. Bowater, J. Braune, personal communication), exploiting the phenomenon of maize chromosome elimination in early embryo/zygote divisions leaving only the haploid triticale genome. Androgenesis is via anther or microspore culture, where under specific conditions immature cells of the male gametophytic pathway (microspores) are induced to switch their development to sporophytic, creating embryos that eventually can be induced to grow into full-size green plants (Fig. 4).

Figure 4. Androgenesis and regeneration of haploids in hexaploid triticale: a) anthers with microspore-derived androgenic structures after 6 weeks of *in vitro* culture on the induction medium; b) germinating embryos on the regeneration medium; c) haploid/doubled haploid regenerants on the rooting medium.

Relatively little work has been done on the maize pollination system in triticale; in wheat this system is used routinely, essentially on a commercial scale. Rogalska and Mikulski (1995, 1996), and Pratap et al. (2006) published some results on the efficiency of the method; Węgzony et al. (1998) have shown that the efficiency can be improved by modifications of the embryo culture media and concentration of growth regulators. In a direct comparison, Pratap et al. (2006) have shown that the chromosome elimination method was far more effective than androgenesis, but it needs to be pointed out, discreetly, that their androgenesis was surprisingly inefficient. The generally perceived low productivity of this method led to a lack of wider acceptance in triticale. Production of haploids via maize pollination remains an alternative for hybrids recalcitrant to the anther/microspore culture; the method of choice in triticale is androgenesis.

The first DH triticale lines produced via androgenesis were publicized some 40 years ago, and since then the protocols for triticale have been improved, optimized, and applied in large scale production both in genetic studies and breeding (Bernard et al. 1996, Fossati et al. 1996, Kalih et al. 2015). This eventually led to the release of the first commercial cultivars developed entirely via androgenesis (Z. Banaszak, H. Woś, F. Eudes, personal communication) even though the numbers are not nearly as impressive as those for barley or rapeseed. Anther culture has been applied to all triticale ploidy levels, from tetraploid to octoploid, and their hybrids (Lehmann and Krolow 1991, Thiemt and Oettler 2008, Lantos et al. 2014).

In recent years, the production of DH lines through the culture of isolated microspores has progressed considerably in cereals, and for some genotypes it is more efficient than anther culture. In triticale, isolated microspore culture can be quite effective (Pauk et al. 2000, Oleszczuk et al. 2004, Eudes and Amundsen 2005, Asif et al. 2013, Würschum et al. 2014a). Apart from the commercial/breeding applications, large quantities of isolated microspores that represent a single-cell system are a precious tool for embryological studies (monitoring of the earliest phases of embryogenesis), genetic studies and various biotechnological applications (*in vitro* selection, mutagenesis and transformation) (Eudes et al. 2014). However, the prerequisite for the widespread use of this system in breeding programs is an efficient method of plant regeneration. The factors affecting the efficiency of the isolated microspore culture have been reviewed by Ryöppy (1997) and Eudes and Chug (2009). These include the growth conditions of the donor plants, the appropriate developmental stage of the microspores/immature pollen grains, various stress-like pretreatments, media composition, modifications of the culture conditions and assorted media additives. A typical protocol for anther and microspore cultures waspublished in a manual by Maluszynski et al. (2003).

The androgenic response of triticale appears to be under genetic control, with some lines and hybrids being recalcitrant while others are superbly amenable to the approach. Various combinations of genetic and cytoplasmic factors may regulate the androgenic potential (Charmet and Bernard 1984); a positive effect of the *timopheevi* cytoplasm has also been observed (Arzani and Darvey 1996). Both the embryo induction and plant regeneration phases appear to be controlled by additive effects, indicating the absence of significant epistasis (Balatero et al. 1995, Gonzalez et al. 1997). Several QTLs have been identified as responsible for the androgenic response (Martinez et al. 1994, Gonzalez et al. 2005, Krzewska et al. 2012). So, while triticale shows considerable inter-line variation for the androgenic response, in most cases sufficiently large populations of DH lines can be obtained for any purpose desired (Tuvesson et al. 2000).

For the development of DH lines for breeding purposes, the average efficiency is 20 green plants per spike (range of 1.8-54.8) – 'spike' is understood here as a quantitative measure of the number of anthers plated (usually 130-160) (S. Oleszczuk, based on 45 Polish standard breeding hybrids of winter triticale, unpublished data). Many reports indicate a wide range in the number of green regenerants obtained per 100 anthers: 0.3-5.8 (Schumann 1990), 0-29 (Tuvesson et al. 2003), 1.6-31.4 (Ponitka and Ślusarkiewicz-Jarzina 2007), 2.48-20.88 (Lantos et al. 2014) in view of dependence of the androgenesis process on genotype. On the other end of the spectrum, as many as 150-170 green plants per 100 anthers plated (220-240 green regenerants per spike) have been produced in some genotypes (S. Oleszczuk, unpublished data). Given that plating of perhaps 500 anthers per hour is a modest accomplishment for trained personnel, the efficiency of the process is satisfactory indeed. As triticale is still a minor crop and is bred on a very limited scale, perhaps only several thousand DH lines are developed worldwide annually.

The basic assumption of androgenesis is that each derived line is unique, as each one is supposed to be generated from a single microspore. Therefore, the total number of lines produced should equal the number of gametes sampled. Unfortunately, this is not quite the case. A recent study (Oleszczuk et al. 2014b) has clearly illustrated that clones are frequent among populations of DH triticale lines, at times in frightening proportions. The study also identified critical points in the procedures when such clones are generated (secondary embryogenesis in the callus stage and formation of twin and polyembryos). Hopefully, with these points identified, steps can be taken to limit or eliminate the problem. It needs to be pointed out that the presence of clones (identical lines) among sets of DH lines produced for breeding purposes severely limits the variation available for selection/breeding and may even hinder proper assessment of a hybrid's value. In genetic experiments such clones can be identified upon genotyping, and eliminated, only driving the cost up. Among breeding lines they go unnoticed as these are never genotyped.

The DH approach to triticale breeding still has to deal with several restrictive factors related to genetics and culture conditions. The key concerns are a low rate of chromosome doubling, insufficient regeneration from androgenic structures, a high frequency of albinos and aneuploidy.

Triticale shows a relatively low rate of spontaneous diploidization of the microspore-derived plants, usually below 20% (Charmet et al. 1986, Ślusarkiewicz-Jarzina and Ponitka 1997, Pauk et al. 2000). The highest frequencies of spontaneous doubling (ca. 57.5%) were reported by Ślusarkiewicz-Jarzina and Ponitka (2003). Several mechanisms are known to result in doubling of the chromosome numbers in microspore-generated plants (e.g. cell fusion in the very early stages of the microspore divisions, endoreduplication, c-mitosis) (Seguí-Simarro and Nuez 2008). The frequency of spontaneous diploidization can at times be increased e.g. by optimization of the pretreatment and stress conditions (Indrianto et al. 1999, Immonen and Robinson 2000) but it varies among procedures and genotypes. The ploidy levels of regenerated plants can be determined genetically (Arzani and Darvey 2001).

As the average rate of spontaneous chromosome doubling among androgenic progenies in triticale is low, some antimitotic agents must be applied to enhance chromosome doubling, and these are added either to the culture media during plantlet regeneration, or to rooted plants (Würschum et al. 2012). Various efficiencies of the chromosome doubling methods have been reported, from 41% (Tuvesson et al. 2003), to 82.3% (Arzani and Darvey 2001) and to essentially 100% if skillfully applied. It has been observed that making use of parental lines created via standard DH production methods clearly increases the rate of spontaneous chromosome doubling among their androgenic progeny (S. Oleszczuk, unpublished data), as if via some imprinting mechanism. Application of antimitotics is viewed by many as cumbersome. Perhaps for that reason, genetic means of chromosome doubling should be considered, such as making use of meiotic restitution (Oleszczuk and Lukaszewski 2014).

Albinism is one of the major barriers that must be overcome to make androgenesis a truly routine procedure in triticale breeding. At times the proportions of albino regenerants are extremely high, and this of course severely reduces the efficiency of the process, especially because it is not discovered until very late in the procedure, when it is too late to compensate for its effects by increasing sample sizes (Schumann 1990, González et al. 1997, Ślusarkiewicz-Jarzina and Ponitka 1997). The issues explaining the inability of proplastids to transform into chloroplasts, resulting in albinism, focused on cytology, plastid and nuclear genome studies and other mechanisms of this phenomenon, as well as external factors which may reduce the number of albino plants, have been discussed (Ankele et al. 2005, Makowska and Oleszczuk 2014).

Any deviation caused by chromosome aberrations among regenerants has a clear adverse effect on the efficiency of the process, as it disqualifies the produced material from research and breeding. Unfortunately, aneuploids are frequently recovered during triticale androgenesis, at times reaching unpleasant proportions (up to 60% in extreme cases; Immonen and Robinson 2000, Oleszczuk et al. 2011). This does not appear to be a consequence of chromosome rearrangements during the *in vitro* phase but rather, it is dictated by the donor material itself (Jouve and Soler 1996). As was explained earlier, triticale in general, at all ploidy levels, does have issues with meiotic regularity and univalents are almost always present, especially in the F_1 hybrids. Univalency afflicts rye chromosomes more often than those of wheat, and as a consequence, haploids and doubled haploids with unusual constitutions (nullisomy, tetrasomy and various combinations of both) for rye chromosomes are frequent (Charmet et al. 1986, Oleszczuk et al. 2011). Since it is the F_1 hybrids which are most often used for haploid production, aneuploidy further reduces the overall efficiency of the system. On the positive side, many aneuploids are less vigorous and have altered morphology even at the tillering stage when antimitotics are used to double chromosome numbers, and thus they can be removed from the populations (Fig. 5). Aneuploids may also be more recalcitrant to chromosome doubling (Charmet et al. 1986).

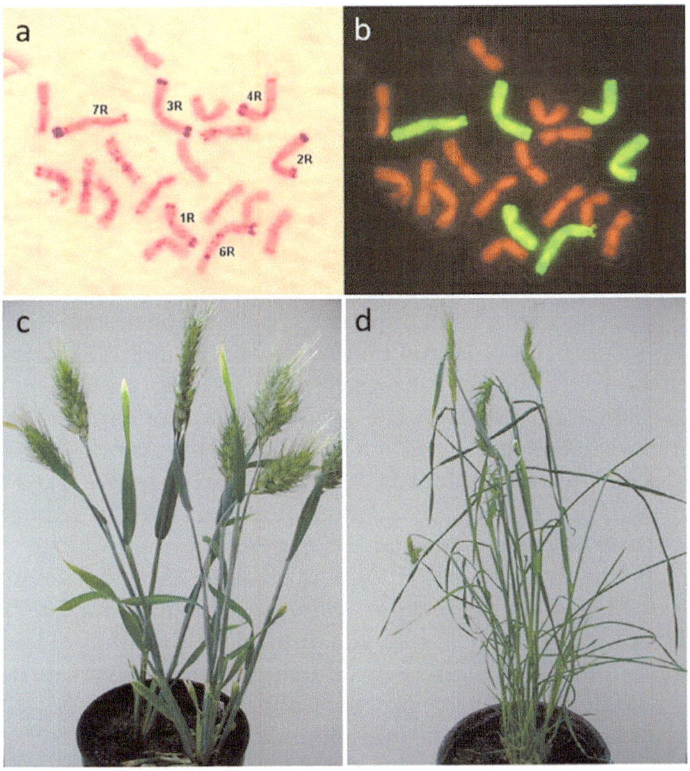

Figure 5. Aneuploidy among androgenic haploids of hexaploid triticale: a, b) sequential C-banding (a) and *in situ* probing with labeled DNA (b) of a nullisomic 5R. Rye chromatin labeled green; wheat chromatin is red. Two most frequent phenotypes among aneuploids; c) nullisomic 2R; d) nullisomic 5R.

Some of the issues in DH production of triticale listed above can be mitigated by proper choice of parental lines. This works well in experimental systems. In breeding, however, the choice of parental lines is not dictated by the efficiency requirements of the laboratory protocols but by breeding objectives, and for this very reason, additional tests of the haploid regeneration systems and handling need to be devised.

Utilization of Triticale

At present, triticale is grown primarily for feed, either as grain or hay/silage. In feeding, similarly to other cereals, triticale grain can constitute the basic ingredient in feed mixtures for monogastric animals. Modern cultivars of triticale are a suitable, high energy source for all classes of animals, with growth performance parameters comparable to or even better than wheat (Boros 2006, Widodo et al. 2015). Variation in feed quality is due primarily to differences in the nutrient contents and the presence of anti-nutritional compounds such as nonstarch polysaccharides and their propensity to create a viscous environment in the small intestine. The primary use of triticale at present is in poultry and hog feed, usually in mixtures of grain. However, in Australia, South Africa, Canada, Plains of the USA, Portugal, Brazil, triticale is also grown as double-use fodder: grazed in the spring by cattle and then allowed to regenerate and produce grain (Macas et al. 2002). In several places, such as Northwest US, Northern Mexico, Spain, triticale is used primarily for green biomass and hay. In such cases, awnlessness is a highly desirable characteristic.

At the beginning of its career as a commercial crop, the protein content in triticale grain was at a level similar to that of wheat, sometimes higher, and with good nutritional quality (Hulse and Laing 1974, Cygankiewicz 1995). Over the years, with great progress in breeding, triticale has revealed a negative correlation between the grain protein content and grain yield (Schlenker et al. 1988). Strong selection for kernel plumpness and high test weight reduced protein content in triticale grain (Amaya and Peña 1990). However, it still shows high lysine content and can be fed to animals without supplementation of this amino acid (Myer 1998).

An interesting aspect of triticale utilization is as an energy source (Karpenstein-Machan and Scheffer 1998, Jorgensen et al. 2007). Triticale seems to be superior to wheat in traits desired by the ethanol fermentation plants and is superior for biomass production, which reflects its potential as an ethanol feedstock in most regions of Canada (Beres et al. 2013). The main advantages of using triticale for bio-ethanol include a higher biomass and starch content, softer grain, lower requirements of nitrogen, and improved conversion of starch to alcohol compared to other cereals (Goyal et al. 2011). Obuchowski et al. (2010) claim that many related factors affect the effectiveness of bioethanol production from triticale, but the nitrogen fertilization level had the most noticeable effect on the grain starch content, while grain yield per hectare had the most significant effect on the ethanol output. The straw of triticale is expected to fit into different industrial purposes in view of its properties and its polymer composite properties (Mihai and Ton-That 2014).

Triticale would likely gain in stature if it became a staple for human consumption (Pattison et al. 2014). Many modern cultivars of triticale contain significantly higher amounts of valuable grain components compared to bread wheat (Boros 2006). However, at present, with its feed-crop status, triticale is looked upon as a "poor boy" on the block with a rather limited appeal. High quality bread can be made from triticale flour but it requires changes in handling and baking procedures that few are willing to make. On the other hand, alterations to the standard chromosome constitution of hexaploid triticale, via whole chromosome substitutions and engineered chromosomes (Fig. 2c) (Lukaszewski and Curtis 1992, 1994; Kazman and Lelley 1994, Lukaszewski 2006, Rabiza-Swider et al. 2010) introduce into hexaploid triticale all pertinent gluten loci from the D-genome of wheat, producing triticales with essentially the same load of gluten loci as bread wheat. All these alterations increase the parameters of bread making quality to the levels typical for wheat, including loaf volume and texture. The effect is not uniform and probably depends to a large extent on the background effects; triticale is not usually selected for bread making quality. Additional problems that negatively impact bread making quality are high sprouting that ruins flour quality, and inadequate kernel hardness that affects the milling process. The latest research shows that standard methods of quality assessment of wheat cannot be applied to triticale as the test results do not correlate well with bread making quality. The most important quality parameter for triticale is the falling number; it should be above 150s (Woś and Brzeziński 2015). As to kernel hardness, a study of Gasparis et al. (2013) clearly shows that its regulation in triticale is different from that in wheat. Perhaps triticale breeders will have to find their own solutions to testing problems rather than following the lead of wheat. There is a clear need for classification of cultivars into end- use classes to facilitate production and marketing of quality triticale (Pattison and Trethowan 2013). Current uses of triticale and the environmental effects on its grain quality were revived by McGoverin et al. (2011).

Recent Developments

Androgenesis and rapid production of doubled haploids was certainly one of the important milestones in recent triticale history. The offshoots were mapping populations which offer a chance to locate in the genome all pertinent loci and perhaps even facilitate their assembly into beneficial combinations. Careful studies of such mapping populations would also offer a chance to uncover at least some of the intergenomic interactions that complicate breeding. Unfortunately, molecular methods are expensive and with little funding trickling down to triticale research, progress in this area is minimal. However, the first genetic maps of triticale have been created, using several DNA marker systems such as SSR, AFLP and DArT (Tams at al. 2004, Alheit et al. 2011, Tyrka et al. 2011). These genetic maps do not appear to differ in any significant way from maps created separately for wheat and rye, indicating that there are no major differences in genetic recombination among these species (Bolibok-Bragoszewska et al. 2009, Milczarski et al. 2011). The existing mapping populations of triticale can be used to study not only the genetics and map locations of agronomically important loci, as these can probably be deduced from genetic maps of rye and wheat, but of specific interactions between individual genomes in polyploids and of unique triticale problems, such as grain sprouting and aneuploidy.

QTL mapping in triticale became evident only recently (Alheit et al. 2013, Stojałowski et al. 2013, Hahn et al. 2014, Würschum et al. 2014b) but at this rate it will be decades before we accumulate sufficient knowledge of the triticale genomes, and their interactions, to contemplate genomic selection. At the same time, a consensus map was developed (Alheit et al. 2011) and a framework for association mapping has been created, and used for identification of loci related to aluminum tolerance (Niedziela et al. 2012, 2014), winter hardiness and frost tolerance (Liu et al. 2014). However, the wide material exchange among breeders and geneticists, and the presence of chromosome rearrangements, may offer serious obstacles to genomic selection in triticale.

Manipulation of triticale by genetic engineering is still in its infancy. The initial gene transfer methods developed for triticale were restricted to direct delivery of DNA into protoplasts (Stolarz and Lörz 1988). Subsequent stable transgenic events relied on the microprojectile bombardment of the embryo scutella (Zimny et al. 1995, Doshi et al. 2007). Recent years have witnessed serious advances in triticale transformation via *Agrobacterium* (Nadolska-Orczyk et al. 2005, Chugh et al. 2009, Hensel et al. 2012). Efforts are underway to knock out the pathway responsible for preharvest sprouting (A. Linkiewicz, personal communication) or lignin accumulation (F. Eudes, personal communication).

Triticale is a tempting target for far reaching genetic modifications: it is a cultivated crop with a very high yield potential. As such, it will be productive and manageable in commercial agriculture. At present it is not used for human consumption so it would appear that separation of two streams, the genetically modified and natural, would not be a major issue. There seems to be no real danger of gene escape as hybrids with its progenitors (parents) have at best only marginal fertility and no fertile hybrids have been noted with its wild relatives, and so rare hybrids do not pose a substantial risk (Kavanagh et al. 2013). Still, the development of transgenic triticale as a source of bio-products would require serious environmental risk assessment (Hills et al. 2007). Caution is urged before any serious modifications are contemplated lest the Starlink history repeats itself (Bucchini and Goldman 2002). With genetically modified organisms not scoring any points with the general public, having triticale tainted by such modifications could once and for all remove its chances of being developed and used for direct human consumption. Perhaps this would not even be missed now; many years

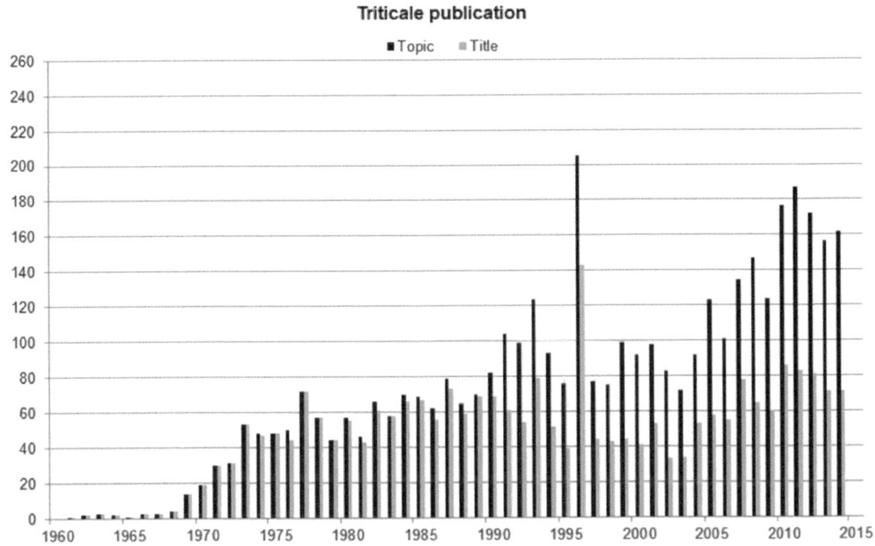

Figure 6. Research activity in triticale over the last 50 years in numbers of technical publications per year. Topic: triticale as one of the research objects; title: triticale listed in the title.

from now the story may be quite different. Triticale clearly demonstrates a far greater yield potential than its two parents; if its grain yield increases continue at the same pace for the next ten or twenty years, wheat will be left far behind. After almost two decades of clear stagnation in triticale research, it seems to be gaining momentum again (Fig. 6) so it may very soon become an established and respectable member among basic cereals.

Acknowledgements

The authors thank Prof. A. Lukaszewski for professional guidance, insightful and constructive suggestions in writing this chapter.

References

Alheit, K.V, J.C. Reif, H.P. Maurer, V. Hahn, E.A. Weissmann, T. Miedaner and T. Würschum. 2011. Detection of segregation distortion loci in triticale (× *Triticosecale* Wittmack) based on a high-density DArT marker consensus genetic linkage map. BMC Genomics 12: 380.

Alheit, K.V., L. Busemeyer, W. Liu, H.P. Maurer, M. Gowda, V. Hahn, S. Weissmann, A. Ruckelshausen, J.C. Reif and T. Würschum. 2013. Multiple line cross QTL mapping for biomass yield and plant height in triticale (× *Triticosecale* Wittmack). Theor. Appl. Genet. 127: 251-260.

Amaya, A. and R.J. Peña. 1990. Triticale industrial quality improvement at CIMMYT: Past, present and future. CIMMYT Proc. 2nd Int. Triticale Symp. Passo Fundo, Brasil. 412-421.

Anioł, A. 1981. The laboratory methods for testing cereals for Al-tolerance. Biul. IHAR 143: 3-14. [in Polish with English summary].

Anioł, A. 1985. Principles of breeding plants tolerant to aluminium toxicity. Biul. IHAR 156: 185-194. [in Polish with English summary].

Aniol, A. and J.P. Gustafson. 1984. Chromosomal location of genes controlling aluminum tolerance in wheat, rye and triticale. Can. J. Genet. Cytol. 26: 701-705.

Ankele, E., E. Heberle-Bors, M.F. Pfosser and B.J. Hofinger. 2005. Searching for mechanisms leading to albino plant formation in cereals. Acta Physiol. Plant. 27: 651-665.

Arzani, A. and N.L. Darvey. 1996. Positive effect of *timopheevii* cytoplasm on anther culture response of triticale. pp. 391-393. *In:* H. Guedes-Pinto H, N. Darvey and V.P. Carnide (eds.) Triticale: today and tomorrow. Kluwer Academic Publishers, Dordrecht, The Netherlands.

Arzani, A. and N.L. Darvey. 2001. The effect of colchicine on triticale anther-derived plants: microspore pre-treatment and haploid plant treatment using a hydroponic recovery system. Euphytica 122: 235-241.

Asif, M., F. Eudes, A. Goyal, E. Amundsen, H. Randhawa and D. Spaner. 2013. Organelle antioxidants improve microspore embryogenesis in wheat and triticale. *In Vitro* Cell. Dev. Biol.-Plant 49: 489-497.

Audenaert, K., V. Troch, S. Landschoot and G. Haesaert. 2014. Biotic stresses in the anthropogenic hybrid triticale (× *Triticosecale* Wittmack): current knowledge and breeding challenges. Eur. J. Plant Pathol. 140: 615-630.

Badaev, N.S., E.D. Badaev, N.I. Dubovets, N.L. Bolsheva, V.E. Bormotov and V. Zelenina. 1992. Formation of a synthetic karyotype of tetraploid triticale. Genome 35: 311-317.

Balatero, C.H., N.L. Darvey and D.J. Luckett. 1995. Genetic analysis of anther-culture response in 6*x* triticale. Theor. Appl. Genet. 90: 279-284.

Ballesteros, J., A. Cabrera, A. Aardse, M.C. Ramírez, S.G. Atienza and A. Martín. 2007. Registration of TS1, TS10 and TS41, Three High Biomass Production Tetraploid Triticale Germplasm Lines. J. Plant Registrations 1: 71-72.

Banaszak, Z. 2004. Analysis of breeding progress in winter triticale (× *Triticosecale* Wittm.). PhD Dissertation, IGR PAN Poznań, Poland.

Banaszak, Z. and K. Marciniak. 2002. Wide adaptation of DANKO triticale varieties. *In*: A. Aniol et al. (eds.), Proc. 5th Int. Triticale Symp., June 30–July 5, Radzików, Poland. 1: 217-222.

Banaszak, Z., K. Marciniak and L. Brykczyńska.1998. Evaluation of frost hardiness examination methods with reference to triticale in DANKO. Bull. IHAR 205/206: 213-218. [in Polish with English summary].

Banaszak, Z., P. Kaźmierczak, A. Trąbka, B. Apolinarska and J. Bocianowski. 2007. The influence of some factors on uniformity of triticale cultivars. Bull. IHAR 244: 145-154. [in Polish with English summary].

Baum, M. and T. Lelley. 1988. A new method to produce 4*x* triticales and their application in studying the development of a new polyploid plant. Plant Breed. 100: 260-267.

Bennett, M.D. and J.P. Gustafson. 1982. The effect of telomeric heterochromatin from *Secale cereale* L. on triticale (× *Triticosecale* Wittmack). II. The presence or absence of block of heterochromatin in isogenic backgrounds. Can. J. Genet. Cytol. 24: 93-100.

Beres, B.L., C.J. Pozniak, F. Eudes, R.J. Graf, H.S. Randhawa,. D.F Salmon, G.J. McLeod, Y. Dion., R.B. Irvine, H.D. Voldeng, R.A. Martin, D. Pageau, A. Comeau, R.M. DePauw, S.M. Phelps and D.M. Spaner. 2013. A Canadian ethanol feedstock study to benchmark the relative performance of Triticale: I. Agronomics. Agron. J. 105:1695-1706.

Bernard, M., S. Bernard and B. Saigue. 1985. Tetraploid triticales: investigation of their genome and chromosome constitution. *In*: M. Bernard and S. Bernard (eds.). Proc. 3rd EUCARPIA Meeting of Cereal Section on Triticale, Genetics and Breeding of Triticale. Clermont-Ferrand, July 2-5, 1984. INRA, France. 277-287.

Bernard, M., S. Bernard, H. Bonhomme, C. Faurie, G. Gay and L. Jestin. 1996. Triticale research and breeding programmes in France: recent developments. pp. 643-664. *In:* H. Guedes-Pinto, N. Darvey and V.P. Carnide (eds.). Triticale: today and tomorrow. Kluwer Academic Publishers, Dordrecht, The Netherlands.

Berzonsky, W.A. and H.N. Lafaver. 1993. Progress in Ohio soft red winter wheat breeding: grain yield and agronomic traits of cultivar released from 1871 to 1987. Crop Sci. 33: 1382-1386.

Bolibok-Bragoszewska, H., K. Heller-Uszyńska, P. Wenzl, G. Uszyński, A. Kilian and M. Rakoczy Trojanowska. 2009. DArT markers for the rye genome-genetic diversity and mapping. BMC Genomics. 10: 578.

Boros, D. 2006. Triticale of high end-use quality enhances opportunities to increase its value in world cereals market. *In*: W.C. Botes, D. Boros, N. Darvey, P. Gustafson, R. Jessop, G.F. Marais, G. Oettler and D. Salmon (eds.). Proc. of the 6th Inter. Triticale Symposium, Stellenbosch, South Africa.© 2006 ITA & SU-PBL – Faculty of AgriSciences, Stellenbosch. 119-125.

Bouguennec, A., M. Trottet, P. Cheyron and P. Lonnet. 2013. Triticale powdery mildew: population characterization and wheat gene efficiency. Comm. in Agricultural and Applied Biological Sci. Ghent Univ. 8th Int. Triticale Symp. June 10-14, 106-113.

Bucchini, L. and L.R. Goldman. 2002. Starlink corn: a risk analysis. Environ. Health Perspect. 110: 5-13.

Budak, H., P.S. Baenzinger, B.S. Beecher, R.A. Graybosch, T. Campell, J. Shipman., M. Erayman. and K.M. Eskridge. 2004. The effect of introgressions of wheat D-genome chromosomes into 'Presto' triticale. Euphytica 137: 261-270.

Budzianowski, G. and H. Woś. 2004. The effect of single D-genome chromosomes on aluminium tolerance of triticale. Euphytica 137: 165-172.

Butnaru, G., I. Sarac and S. Ciulca. 2013. Relationship among yield and plant specific traits on triticale Romanian varieties in Timisoara environment. *In*: Communications in Agricultural and Applied Biological Sciences. Ghent Univ., Belgium, Proc. 8th Int. Triticale Symp. June 10-14, 2013. 201-210.

Charmet, G. and S. Bernard. 1984. Diallel analysis of androgenetic plant production in hexaploid *Triticale* (× *Triticosecale* Wittmack). Theor. Appl. Genet. 69: 55-61.

Charmet, G., S. Bernard and M. Bernard. 1986. Origin of aneuploid plants obtained by anther culture in triticale. Can. J. Genet. Cytol. 28: 444-452.

Chen, Z.J. 2010. Molecular mechanisms of polyploidy and hybrid vigor. Trends Plant Sci. 15: 57-71.

Chugh, A., E. Amundsen and F. Eudes. 2009. Translocation of cell-penetrating peptides and delivery of their cargoes in triticale microspores. Plant Cell Rep. 28: 801-810.

Curtis, C.A. and A.J. Lukaszewski. 1993. Localization of genes in rye that restore male fertility to hexaploid wheat with *timopheevi* cytoplasm. Plant Breed. 111: 106-112.

Cygankiewicz, A. 1995. Quality estimation of new winter and spring triticale breeding strains. Bull. IHAR.195/196: 123-129. [in Polish with English summary].

Darvey, N. and J. Roake. 2002. Development of spring hybrid triticale. *In*: A. Aniol et al. (eds.). Proc.5th Int. Triticale Symp., June 30–July 5, Radzików, Poland. 1:207-209.

Doshi, K.M., E. Eudes, A. Laroche and D. Gaudet. 2007. Anthocyanin expression in marker free transgenic wheat and triticale embryos. *In Vitro* Cell Dev. Biol. Plant 43: 429-435.

Dou, Q.W., H. Tanaka, N. Nakata and H. Tsujimoto. 2006. Molecular cytogenetic analyses of hexaploid lines spontaneously appearing in octoploid Triticale. Theor. Appl. Genet. 114: 41-47.

Dubovets, N.I., E.A. Sycheva, L.A. Solovey, T.I. Shtyk and E.B. Bondarevich. 2008. Recombinant genome of cereals: the pattern of formation and the role in evolution of polyploid species. Russ. J. Genetics 44: 44-50.

Dvorak, J. and P. McGuire. 1981. Nonstructural chromosome differentiation among wheat cultivars with special reference to differentiation of chromosomes in related species. Genetics 97: 391-414.

Eudes, F. and A. Chug. 2009. An overview of triticale doubled haploids. *In*: A. Touraev, B.P. Forster and S.M. Jain (eds). Advances in haploid production in higher plants.Springer, Heidelberg. 87-96.

Eudes, F. and E. Amundsen. 2005. Isolated microspore culture of Canadian 6*x* triticale cultivars. Plant Cell Tissue Organ Cult. 82: 233-241.

Eudes, F., Y.-S.Shim and F. Jiang. 2014. Engineering the haploid genome of microspores. Biocatal. Agric. Biotechnology 3(1): 20-23.

FAOSTAT, FAO http://faostat3.fao.org/home/E

Feldman, M. and A.A. Levy. 2005. Allopolyploidy – a shaping force in the evolution of wheat genomes. Cytogenet. Genome Res. 109: 250-258.

Fischer, S., H.P. Maurer, T. Würschum, J. Möhring, H.P. Piepho, C.C. Schön, E.M. Thiemt, B.S. Dhillon, E.A. Weissmann, A.E. Melchinger and J.C. Reif. 2010. Development of heterotic groups in triticale. Crop Sci. 50: 584-590.

Fiust, A., M. Rapacz, M. Wójcik-Jagła, K. Śniegowska-Świerk, K. Żmuda, Z. Banaszak, M. Niewińska, M. Pojmaj, P. Kaźmierczak, E. Czerwińska., J. Haremza., Z. Nita and W. Orłowska-Job. 2015. Usefulness of chlorophyll parameter fluorescence measurement in freezing tolerance evaluation of Polish triticale genotypes depends on winter conditions. Proc. Ogólnopolska Konferencja Naukowa: Nauka dla Hodowli i Nasiennictwa Roślin Uprawnych. Zakopane 2-6 Feb 2015. 311-312.

Fossati, A., D. Fossati and G. Kleijer. 1996. Triticale breeding in Switzerland. pp. 649-653 *In:* H. Guedes-Pinto H, N. Darvey and V.P. Carnide (eds) Triticale: today and tomorrow. Kluwer Academic Publishers, Dordrecht, The Netherlands.

Fowler, D.B. and A.E. Limin. 1997. Breeding for winter hardiness in cereals. Proc. Int. Symp. on Cereal Adaptation to Low Temperature Stress in Controlled Environments, June 2 – 4. 1997, Martonvasar, Hungary. 169-176.

Gasparis, S., W. Orczyk and A. Nadolska-Orczyk. 2013. Sina and Sinb genes in triticale do not determine grain hardness contrary to their orthologs Pina and Pinb in wheat. BMC Plant Biol. 13:190.

González, L., M. Muniz and N. Jouve. 2005. Mapping of QTLs for androgenetic response based on a molecular genetic map of × *Triticosecale* Wittmack. Genome 48: 999-1009.

González, M., I. Hernandez and N. Jouve. 1997. Analysis of anther culture response in hexaploid triticale. Plant Breed. 116: 302-304.

Gowda, M, Y. Zhao, H.P. Maurer, E.A. Weissmann, T. Würschum and J.C. Reif. 2013. Best linear unbiased prediction of triticale hybrid performance. Euphytica 191: 223-230.

Goyal, A., B.L Beres, H.S. Randhawa, A. Navabi, D.F. Salmon and F. Eudes. 2011. Yield stability analysis of broadly adaptive triticale germplasm in southern and central Alberta, Canada, for industrial end-use suitability. Can. J. Plant Sci. 91:125-135.

Grzesik, H. and S. Węgrzyn. 2003, Heritability of some winter triticale yield components. Bull. IHAR.226/227/1: 191-195. [in Polish with English summary].

Grzesik, H., and S. Węgrzyn. 1998. Heterosis and combining ability in some varieties of triticale. *In*: P. Juskiw (ed). Proc 4th Int. Triticale Symp. Red Deer, July 26-31, AB, Canada.

Gustafson, J.P. 1982. Cytogenetic of triticale. pp. 228-250. *In*: M.S. Swaminathan, P.K. Gupta and U. Sinha (eds.). Cytogenetics of Crop Plants, Macmillan India Ltd.

Gustafson, J.P. and A.J. Lukaszewski. 1985. Early seed development in *Triticum-Secale* amphiploids. Can. J. Genet. Cytol. 27: 542-548.

Gustafson, J.P. and F. Zillinsky. 1973. Identification of D-genome chromosomes from hexaploid wheat in 42-chromosome triticale. *In*: E.R. Sears and L.M. Sears (eds). Proc 4th Int. Wheat Genet. Symp, MO Agri. Exp. Sta, Columbia, MO. pp. 225-232.

Gustafson, J.P., A.J. Lukaszewski and B. Skovmand. 1984. Heterochromatin content and early endosperm development in 42-chromosome spring triticale. Can. J. Genet. Cytol. 26: 85-90.

Haesaert, G. and A.E.G. De Baets. 1996. Preharvest sprouting resistance in triticale. pp. 616-622. *In:* H. Guedes-Pinto. N. Darvey and V.P. Carnide (eds). Triticale: Today and Tomorrow. Kluwer Academic Publishers, Dordrecht, The Netherlands.

Haesert, G. 1994. Triticale breeding: problems and possibilities. Acta. Hort. 355: 121-134.

Hahn, V., A. Ruckelshausen and H.P. Maurer. 2014. Mapping dynamic QTL for plant height in triticale. BMC Genetics 15: 59.

Hensel, G., S. Oleszczuk, D.E. Daghma, J. Zimny, M. Melzer and J. Kumlehn. 2012. Analysis of T-DNA integration and generative segregation in transgenic winter triticale (× *Triticosecale* Wittmack). BMC Plant Biol. 25, 12: 171.

Hills, M.J., L.M. Hall, D.F. Messenger, R.J. Graf, B.L. Beres and F. Eudes. 2007. Evaluation of crossability between triticale (× *Triticosecale* Wittmack) and common wheat, durum wheat and rye. Environ. Biosafety Res. 6: 249-257.

Hohmann, U. and K.D. Krolow. 1991.Introduction of D-genome chromosomes from *Aegilops squarrosa* L. into tetraploid triticale (AB) (AB) RR ($2n = 28$). Theor. Appl. Genet. 82: 777-783.
Hulse, J.H. and E.M. Laing. 1974. Nutritive value of triticale protein (and the proteins of wheat and rye.) Pub. IDRC, Ottawa, Canada.
Immonen, S. and J. Robinson. 2000. Stress treatment and ficoll for improving green plant regeneration in triticale anther culture. Plant. Sci. 150: 77-84.
Indrianto, A., E. Heberle-Bors and A. Touraev. 1999. Assessment of various stresses and carbohydrates for their effect on the induction of embryogenesis in isolated wheat microspores. Plant. Sci. 143: 71-79.
Jenkins, B.C. 1969. History of the development of some presently promising hexaploid Triticales. Wheat Inf. Serv. 28:18-19.
Jorgensen, J.R., L.C. Deleuran and B. Wollenweber. 2007. Prospects of whole grain crops of wheat, rye and triticale under different fertilizer regimes for energy production. Biomass and Bioenergy 31: 308-317.
Jouve, N. and C. Soler. 1996. Triticale genomic and chromosomes' history. pp. 91-103. *In*: H. Guedes-Pinto, N. Darvey and V.P. Carnide (eds), Triticale: today and tomorrow, Dordrecht: Kluwer Academic Publishers.
Kalih, R., H.P. Maurer and T. Midaner. 2015. Genetic architecture of fusarium head blight resistance in four winter triticale populations. Phytopathology 105: 334-41.
Kalih, R., H.P. Maurer, B. Hackauf and T. Midaner 2014. Effect of a rye dwarfing gene on plant height, heading stage, and Fusarium head blight in triticale (× *Triticosecale* Wittmack). Theor. Appl. Genet. 127: 1527-1536.
Karpenstein-Machan, M. and K. Scheffer. 1998. Triticale for industrial uses, produced in a sustainable cropping system. *In*: P. Juskiw (ed). Proc. 4th Int. Triticale Symp. Red Deer, Alberta, Canada. 273-283.
Kavanagh, V.B., L.M. Hall and J.C. Hall. 2010. Potential hybridization of genetically engineered triticale with wild and weedy relatives in Canada. Crop Sci. 50: 1128-1140.
Kavanagh, V.B., M. J. Hills, A. Goyal, H.S. Randhawa, A.K. Topinka, F. Eudes and L.M. Hall. 2013. Molecular markers as a complementary tool in risk assessments: quantifying interspecific gene flow from triticale to spring wheat and durum wheat. Transgenic Res. 22: 767-778.
Kazman, E. and T. Lelley. 1994. Rapid incorporation of D-genome chromosomes into A and/or B genomes of hexaploid triticale. Plant Breed. 113: 89-98.
Khalil, H.B., M.R. Ehdaeivand, Y. Xu, A. Laroche and P.J. Gulick. 2015. Identification and characterization of rye genes not expressed in allohexaploid triticale. BMC Genomics 16: 281.
Kim, B.Y., A.C. Baier, D.J. Somers and J.P. Gustafson. 2001. Aluminum tolerance in triticale, wheat, and rye. Euphytica 120: 329-337.
Kiss, A. 1966. Experiments with hexaploid triticale. Novenytermeles. 15: 311-328. [in Hungarian with English summary].
Kiss, A. 1970. Development of short secondary hexaploid triticale by crossing triticale with wheat. Wheat Inf. Service 31: 27-28.
Kobylyanski, V. D. 1972. On genetics of the dominant factor of short-strawed rye. Genetika 8: 12-17. [in Russian].
Kociuba, W. and K. Kowalczyk K. 2000. Influence of dwarfing genes insensitive to exogenous gibberelic acid on selected yield components of winter triticale. Folia Univ. Agric. Stetin. 206 Agricultura 82: 125-132. [in Polish].
Kraska, P. and Cz. Tarkowski. 1998. The influence of genes Rht1, Rht2, Rht3 on yield components of winter triticale. *In*: P. Juskiw (ed). 4th Int. Triticale Symp. Red Deer, 26-31 July, Alberta, Canada. pp. 24-26.
Krolow, K.D. 1973. 4*x* triticale production and use in triticale breeding. *In*: E.R. Sears and L.M. Sears (eds). Proc 4th Int. Wheat Genet. Symp, MO Agri Exp Sta, Columbia, MO. pp. 691-696.
Kronberga, A. 2002. Achievements and problems involvin triticale ideotype for Latvia growing conditions. *In*: A. Anioł et al. (eds.), Proc. 5th Int. Triticale Symp., June 30 – July 5, Radzików, Poland, 1: 375-379.

Krzewska M., I. Czyczyło-Mysza, E.Dubas, G. Gołębiowska-Pikania, E. Golemiec, S. Stojałowski, M. Chrupek and I. Żur. 2012. Quantitative trait loci associated with androgenic responsiveness in triticale (× *Triticosecale* Wittm.) anther culture, Plant Cell Rep. 31(11): 2099-2108.

Laidig, F., H.P. Piepho, T. Drobek and U. Meyer. 2014. Genetic and non-genetic long-term trends of 12 different crops in German official variety performance trials and on-farm yield trends. Theor. Appl. Genet. 127: 2599-2617.

Lantos, C., L. Bona, K. Boda and J. Pauk. 2014. Comparative analysis of *in vitro* anther- and isolated microspore culture in hexaploid Triticale (× *Triticosecale* Wittmack) for androgenic parameters. Euphytica 197: 27-37.

Łapiński, B. 2002.A new source of earliness in tetraploid Secalotriticum. *In*: A. Anioł et al. (eds.), Proc. 5th Int. Triticale Symp., June 30 – July 5, Radzików, Poland, pp. 49-53.

Łapiński, B. 2014. Hybrid breeding of Pampa-cytoplasmic triticale based on crosses with rye. Proc. of 8th Int. Triticale Symp. Ghent, Belgium. June 10-14, 2013. *In:* Communications in Agricultural and Applied Biological Sciences. Ghent Univ. 37-48.

Łapiński, B. and B. Apolinarska. 1985. Polish work upon 4*x* triticale. *In*: M. Bernard and S. Bernard(eds.). Proc 3rd EUCARPIA Meeting of Cereal Section on Triticale, Genetics and Breeding of Triticale. Clermont-Ferrand, July 2-5, 1984. INRA, France. 261-264.

Laurie, D.A. and M.D. Bennett. 1988. The production of haploid wheat plants from wheat *x* maize crosses. Theor. Appl. Genet. 76: 393-397.

Lehmann, C. and K.D. Krolow. 1991. Experiments on haploid production from tetraploid triticales by the *Hordeum bulbosum* system and anther culture. Cereal Res. Commun. 19: 283-290.

Lelley, T. 1992. Triticale, still a promise? Plant Breed. 109: 1-17.

Liu, W., H.P. Maurer, G. Li, M.R. Tucker, M. Gowda, E.A. Weissmann, V. Hahn and T. Würschum. 2014. Genetic architecture of winter hardiness and frost tolerance in triticale. PLoS One. 13:9 (6): e99848.

Longin, C.F., J. Mühleisen, H.P. Maurer, H. Zhang, M. Gowda and J.C. Reif. 2012. Hybrid breeding in autogamous cereals. Theor. Appl. Genet. 125: 1087-96.

Lukaszewski, A.J. 1988. Chromosome constitution of hexaploid triticale lines in the recent international yield trials. Z. Pflanzenzuchtg 100: 268-272.

Lukaszewski, A.J. 1991. Development of aneuploid series in hexaploid triticale.CIMMYT. Proc. 2nd Int. Triticale Symp. Passo Fundo, Brasil pp. 397-400.

Lukaszewski, A.J. 2006. Cytogenetically engineered rye chromosomes 1R to improve bread-making quality of hexaploid triticale. Crop Sci. 46: 2183-2194.

Lukaszewski, A.J. and C.A. Curtis.1992. Transfer of the *Glu-D1* gene from chromosome 1D of breadwheat to chromosome 1R in hexaploid triticale. Plant Breed. 109: 203-210.

Lukaszewski, A.J. and C.A. Curtis. 1994. Transfer of the *Glu-D1* gene from chromosome 1D to chromosome 1A in hexaploid triticale. Plant Breed. 112: 177-182.

Lukaszewski, A.J. and D. Kopecky. 2010. The *Ph1* Locus from wheat controls meiotic chromosome pairing in autotetraploid rye (*Secale cereale* L.). Cytogenet. Genome Res. 129: 124-132.

Lukaszewski, A.J. and J.P. Gustafson. 1984. The effect of rye chromosomes on heading date of triticale *x* wheat hybrids. Z. Pflanzenzuchtg 93: 246-250.

Lukaszewski, A.J. and J.P. Gustafson. 1987. Cytogenetics of triticale. vol. 5, pp. 41–93. *In*: J. Janick (ed.). Plant Breeding Reviews. AVI Publishing: New York, NY, USA.

Lukaszewski, A.J., B. Apolinarska and J.P. Gustafson. 1987. Introduction of the D-genome chromosomes from bread wheat into hexaploid triticale with a complete rye genome. Genome 29: 425-430.

Lukaszewski, A.J., B. Apolinarska, J.P. Gustafson and K.D. Krolow. 1984. Chromosome constitution of tetraploid triticale. Z. Pflanzenzuchtg 93: 222-236.

Ma, J.F., S. Taketa, and Z.M. Yang. 2000. Aluminium tolerance genes on the short arm of chromosome 3R are linked to organic acid release in triticale. Plant Physiol. 122: 687-694.

Ma, X.F. and J.P. Gustafson. 2005. Genome evolution of allopolyploids: a process of cytological and genetic diploidization. Cytogenet. Genome Res. 109: 236-249.

Ma, X-F., P. Fang and J. P. Gustafson. 2004. Polyploidization-induced genome variation in Triticale. Genome 47: 839-848.

Macas B., J. Coutinho and A. Costa A. 2002. Breeding of triticales and oats for dual purpose management systems. pp. 383-390. *In*: A. Aniol et al. (eds.). Proc. 5th Int. Triticale Symp., June 30 – July 5, Radzików, Poland.

Maćkowiak, W., G. Budzianowski, A. Cicha, H. Cichy, L. Mazurkiewicz G. Milewski, K. Paizert, B. Szeląg., J. Szeląg and H. Woś. 1998. Triticale breeding program in the Plant Breeding and Acclimatization Experiment Station Małyszyn. Bull. IHAR 205/206: 303-319 [in Polish with English summary].

Makowska, K. and S. Oleszczuk. 2014. Albinism in barley androgenesis. Plant Cell Rep. 33(3): 385-92.

Maluszynski, M., K.J. Kasha, B.P. Forster and I. Szarejko (eds). 2003. Doubled haploid production in crop plants. Kluwer Academic Publishers, Dordrecht, The Netherlands.

Martinez, I., M. Bernard, P. Nicolas and S. Bernard. 1994. Study of androgenetic performance and molecular characterization of a set of wheat-rye addition lines. Theor Appl. Genet., 89: 982-990.

McCaig, T.N. and J.M. Clarke. 1994. Breeding durum wheat in western Canada: Historical trends in yield and related variables. Can. J. Plant Sci. 75: 55-60.

McGoverin, C.M., F. Snyders, N. Muller, W. Botes, G. Fox and M. Manley. 2011. A review of triticale uses and the effect of growth environment on grain quality. J. Sci. Food. Agric. 91: 1155-1165.

Mergoum, M. and H. Gómez-Macpherson (eds). 2004. Triticale improvement and production. FAO Plant Production and Protection Paper, Food and Agriculture Organization of the United Nations. Rome.

Mergoum, M., W. Pfeiffer, S. Rajaram and R.J. Pena. 1998. Triticale at CIMMYT: improvement and adaptation. *In*: Juskiw P (ed.). Proc. 4th Int. Triticale Symp. Red Deer, Alberta, Canada. pp. 58-64.

Mergoum, M., W.H. Pfeiffer, J.R. Peña, K. Ammar and S. Rajaram S. 2004. Triticale crop improvement: the CIMMYT programme. pp. 11-26. *In*: M. Mergoum and H. Gomez-Macpherson (eds). Triticale improvement and production. Food and Agriculture Organization of the United Nations. Rome.

Merker, A. 1975. Chromosome composition of hexaploid triticale. Hereditas 80: 41-52.

Merker, A. 1988. Triticale improvement in Svalov. Proc. EUCARPIA Triticale Meeting at Schwerin pp. 385-392.

Miczyński, K. 1905. *Sur la création des nouvelles variétiés des plantes par l'hybridation*. Kosmos 30:130-147 [in Polish].

Mihai, M. and M.-T. Ton-That. 2014. Novel polylactide/triticale straw biocomposites: processing, formulation, and properties. Polym. Eng. Sci. 54: 446-458.

Milczarski, P., H. Bolibok-Brągoszewska and B. Myśków, S.Stojałowski, K. Heller-Uszyńska, M. Góralska, P. Brągoszewski, G. Uszyński, A. Kilian, and M. Rakoczy-Trojanowska. 2011. A high density consensus map of rye (*Secale cereale* L.) based on DArT markers. PLoS One 6:e28495. doi: 10.1371/journal.pone.0028495.

Mühleisen, J., H.P. Piepho, H.P. Maurer and J.C. Reif. 2015.Yield performance and stability of CMS-based triticale hybrids. Theor. Appl. Genet. 128: 291-301.

Mühleisen, J., H.P. Piepho, H.P. Maurer, C.F.H. Longin and C.F. Reif. 2014. Yield stability of hybrids versus lines in wheat, barley, and triticale. Theor Appl. Genet. 127: 309-316.

Muntzing, A. 1979. Triticale: results and problems. Advances in Plant Breeding. Z. Pflanzenzuchtg (supplement) 10: 1-103.

Myer, R.O. 1998. Evaluation of triticale in nursery diets for early weaned pigs. pp. 196-200 *In*: Proc. 4th Int. Triticale Symp. Vol.1: Oral presentations. Int. Triticale Assoc., Red Deer and Lacombe, Canada.

Nadolska-Orczyk, A., A. Przetakiewicz, K. Kopera, A. Binka and W. Orczyk. 2005. Efficient method of Agrobacterium mediated transformation for triticale (× *Triticosecale* Wittmack). J. Plant Growth Regul. 24: 2-10.

Niedziela, A, P.T. Bednarek, H. Cichy, G. Budzianowski, A. Kilian and A. Anioł. 2012. Aluminum tolerance association mapping in triticale. BMC Genomics 13: 67.

Niedziela, A., P.T. Bednarek, M. Labudda, D. R. Mańkowski and A. Anioł.2014. Genetic mapping of a 7R Al tolerance QTL in triticale (× *Triticosecale* Wittmack). J. Appl. Genet. 55: 1-14.

Niedziela, A., W. Jarska, D Mańkowski, P.T. Bednarek. 2015. HPLC and metAFLP analysis of triticale lines exposed to aluminum stress. EWAC-EUCARPIA Cereals Section International Conference, 24-29 May, Lublin, Poland (book of abstracts).

Obuchowski, W., Z. Banaszak, A. Makowsk and M. Łuczak. 2010. Factors affecting usefulness of triticale grain for bioethanol production. J. Sci. Food Agric. 90: 2506-2511.

Oettler, G. 1990. Alpha-amylaze activity and falling number during grain development in primary triticales and their parents. Proc. of 2nd Int. Int. Triticale Symp., Passo Fundo, Brazil. 483-485.

Oettler, G. 2005. The fortune of a botanical curiosity – Triticale: past, present and future. J. Agr. Sci. 143: 329-346.

Oettler, G., H. Burger and A.E. Melchinger. 2003. Heterosis and combining ability for grain yield and other agronomic traits in winter triticale. Plant Breed. 122: 318-321.

Oettler, G., S.H. Tams, H.F. Utz, E. Bauer and A.E. Melchinger. 2005. Prospects for hybrid breeding in winter triticale: I. Heterosis and combining ability for agronomic traits in European elite germplasm. Crop Sci. 45: 1476-1482.

Oleszczuk, S. and A.J. Lukaszewski. 2014. The origin of unusual chromosome constitutions among newly formed allopolyploids. Am. J. Bot. 101(2): 318-26.

Oleszczuk, S., J. Rabiza-Swider, J. Zimny and A.J. Lukaszewski. 2011. Aneuploidy among androgenic progeny of hexaploid triticale (× *Triticosecale* Wittmack). Plant Cell Rep. 30: 575-586.

Oleszczuk, S., M. Tyrka and A.J. Lukaszewski. 2014a. The *Ph1* locus of wheat does not discriminate between identical and non-identical homologues in rye. Cytogenet. Genome Res. 142: 293-298.

Oleszczuk, S., M. Tyrka and J. Zimny J. 2014b. The origin of clones among androgenic regenerants of hexaploid triticale. Euphytica 198: 325-336.

Oleszczuk, S., S. Sowa and J. Zimny. 2004. Direct embryogenesis and green plants regeneration from isolated microspores of hexaploid triticale (× *Triticosecale* Wittmack) cv. Bogo. Plant Cell Rep. 22: 885-893.

Oracka, T. and B. Łapiński. 2006. Nitrogen and phosphorus uptake and utilization efficiency in D(R) substitution lines of hexaploid triticale. Plant Breed. 125: 221-224.

Pattison, A.L. and R.M. Trethowan. 2013. Characteristics of modern triticale quality: commercially significant flour traits and cookie quality. Crop Pasture Sci. 64: 874-880.

Pattison, A.L., M. Appelbee and R.M. Trethowan. 2014. Characteristics of modern triticale quality: glutenin and secalin subunit composition and mixograph properties. J. Agric. Food Chem. 62: 4924-31.

Pauk, J., M.Puolimatka, K.L. Tóth and T. Monostori.2000. *In vitro* androgenesis of triticale in isolated microspore culture. Plant Cell Tiss. Org. Cult. 61(3): 221-229.

Pilch J. 1981. Analysis of the chromosome constitution and the amount of telomeric heterochromatin in the widely and narrowly adapted hexaploid triticales. Theor. Appl. Genet. 60: 145-149.

Ponitka, A. and A. Ślusarkiewicz-Jarzina.2007. The effect of liquid and solid medium on production of winter triticale (× *Triticosecale* Wittm.) anther-derived embryos and plants. Cereal Res. Commun. 35: 15-22.

Pratap, A., G.S. Sethi and H.K. Chaudhary. 2006. Relative efficiency of another culture and chromosome elimination techniques for haploid induction in triticale × wheat and triticale × triticale hybrids. Euphytica 150: 339-345.

Rabiza-Swider, J., W. Brzezinski and A.J. Lukaszewski. 2010. Breeding behavior of chromosomes 1R cytogenetically engineered for breadmaking quality in hexaploid triticale. Crop Sci. 50: 808-814.

Rajki, E. 1980. Winter hardiness. Frost resistance. Acta Agron. Acad. Sci. Hung 29: 451-468.

Rogalska, S.M. and W. Mikulski. 1995. Induction of haploids in triticale (× *Triticosecale* Witt.) by means of crossing with maize (*Zea mays*). Bull. IHAR 195/196: 55-64 [in Polish with English summary].

Rogalska, S.M. and W. Mikulski.1996. Induction of haploid in triticale (× *Triticosecale* Wittm.) by crossing it with maize (*Zea mays* L.). pp. 379-382 *In:* H. Guedes-Pinto. N. Darvey and V.P. Carnide (eds). Triticale: Today and Tomorrow. Kluwer Academic Publishers, Dordrecht, The Netherlands.

Royo, C., A. Rodrigues and I. Romagosa.1993. Differential adaptation of complete and substituted triticale. Plant Breed. 111: 113-119.

Ryöppy, P.H. 1997. Haploidy in triticale. Vol. 4, pp. 117-131 *In:* S. Mohad Jain, SK Sopory and R.E. Veilleux (eds), *In vitro* haploid production in higher plants. Kluwer Academic Publishers, Dordrecht.

Sabeva, Z. 1985. Obtaining and investigation of tetraploid triticale forms. Cereal Res. Commun. 13: 71-76.

Salmon, D.F. 2004. Production of triticale on the Canadian Prairies. pp. 103-108. *In*: M. Mergoum and H. Gómez-Macpherson (eds.). Triticale improvement and production. FAO plant production and protection paper, 179. Food and Agriculture Organization of the United Nations, Rome.

Schlegel, R., A. Boerner, V. Thiele and G. Melz. 1991. The effect of the *Ph1* gene in diploid rye, *Secale cereale* L. Genome 34: 913-917.

Schlenker, R., E. Prodoehl, E. Schuetzler and M. Kummer 1988.Aims and results of triticale breeding in the German Democratic Republic. Proc. 4th EUCARPIA Meeting of Cereal Section on Triticale, Schwerin, Germany. 393-399.

Schumann, G. 1990. *In vitro* production of haploids in triticale. pp. 382-402, *In*: Y.P.S. Bajaj (ed). Biotechnology in Agriculture and Forestry.Vol. 13, Wheat, Springer-Verlag, Berlin, Heidelberg.

Seguí-Simarro, J.M. and F. Nuez. 2008. Pathways to doubled haploidy: chromosome doubling during androgenesis. Cytogenet. Genome Res. 120: 358-369.

Skovmand, B., P.N. Fox and R.L. Villareal. 1984. Triticale in commercial agriculture: Progress and promise. Advances in Agronomy. 37: 1-45.

Ślusarkiewicz-Jarzina, A. and A. Ponitka A. 2003. Efficient production of spontaneous and induced doubled haploid triticale plants derived from anther culture. Cereal Res. Commun. 31: 289-296.

Ślusarkiewicz-Jarzina, A. and A. Ponitka. 1997. Effect of genotype and media composition on embryo induction and plant regeneration from anther culture in triticale. J. Appl. Genet. 38: 253-258.

Sodkiewicz, W. 1984. Amphiploid *Triticum monococcum* L. *x Secale cereale* L. (AARR) – a new form of tetraploid triticale. Cereal Res. Commun. 12: 35-40.

Sowa, W. and H. Krysiak. 1996. Outcrossing in winter triticale. pp. 593-596. *In:* H. Guedes-Pinto, N. Darvey and V.P. Carnide (eds). Triticale: today and tomorrow. Kluwer Academic Publishers, Dordrecht, The Netherlands.

Sowa, W. and W. Maćkowiak. 1990. Breeding Triticale that is well adapted to variable growing conditions at IHAR. CIMMYT. Proc. 2nd Int. Triticale Symp. Passo Fundo, Brasil. 136-139.

Stojałowski, S., A. Bobrowska, M. Hanek and B. Myśków. 2013. The importance of chromosomes from the sixth homeologic group in the restoration of male fertility in winter triticale with *Triticum timopheevii* cytoplasm. J. Appl. Genet. 54: 179-184.

Stolarz, A. and H. Lörz. 1988. Protoplast culture and transformation studies of triticale (× *Triticosecale* Wittmack). Plant Cell Tiss. Org. Cult. 12: 227-230.

Tajnsek, A. and I. Kreft. 1996. Less known studies on triticale in Central and Eastern Europe. pp. 83-87. *In:* H. Guedes-Pinto. N. Darvey and V.P. Carnide (eds). Triticale: Today and Tomorrow. Kluwer Academic Publishers, Dordrecht, The Netherlands.

Tams, S.H., A.E. Melchinger and E. Bauer. 2005. Genetic similarity among European winter triticale elite germplasms assessed with AFLP and comparisons with SSR and pedigree data. Plant Breed. 124: 154-160.

Tams, S.H., E. Bauer, G. Oettler and A.E. Melchinger. 2004. Genetic diversity in European winter triticale determined with SSR markers and coancestry coefficient. Theor. Appl. Genet. 108: 1385-1391.

Tams, S.H., E. Bauer, G. Oettler, A.E. Melchinger and C.C. Scho. 2006. Prospects for hybrid breeding in winter triticale: II. Relationship between parental genetic distance and specific combining ability. Plant Breed. 125: 331-336.

Tarkowski, C. 1989. Biology of Triticale Państwowe Wydawnictwo Naukowe, Warsaw, Poland [in Polish].

Thiemt, E.M. and G. Oettler. 2008. Agronomic performance of anther-derived doubled haploid and single seed descent lines in crosses between primary and secondary winter triticale. Plant Breed. 127: 476-479.

Tuvesson, S., A. Ljungberg, N. Johanson, K.E. Karlsson, W. Suijs and J.P. Josset. 2000. Large-scale production of wheat and triticale doubled haploids through theuse of a single-anther culture method. Plant Breed. 119: 455-459.

Tuvesson, S., R. von Post and A. Ljungberg. 2003. Triticale anther culture. pp 117-121. *In*: M. Maluszynski et al. (eds). Doubled haploid production in crop plants. Kluwer Academic Publishers, Dordrecht, The Netherlands.

Tyrka, M., P.T. Bednarek, A. Kilian, M. Wędzony, T. Hura and E. Bauer. 2011. Genetic map of triticale compiling DArT, SSR, and AFLP markers. Genome 54: 391-401.

Varughese, G. 1996. Triticale: present status and challenges ahead. pp. 13-20. *In:* H. Guedes-Pinto, N. Darvey and V.P. Carnide (eds). Triticale: today and tomorrow. Kluwer Academic Publishers, Dordrecht, The Netherlands.

Wang, Z., Y. Sun, K. Ammar, A.P.A. Bonjean, X. Chen and F. Sun. 2010. Development of Triticale in China. pp. 79-86, *In*: Z. He and A.P.A. Bonjean (eds). Cereals in China, Mexico, D.F.: CIMMYT.

Warzecha, R., K. Warzecha and Z. Staszewski. 1998. Development and use of triticale CMS system in hybrid breeding. Proc 4th Int. Triticale Symp. Red Deer, Canada. 1:79-85.

Warzecha, R., S. Sowa, K. Salak-Warzecha, S. Oleszczuk, E. Śliwińska and J. Zimny. 2005. Doubled haploids in production of male sterility maintaining triticale (*Triticosecale* Wittmack) lines. Acta Physiol. Plant. 27: 245-250.

Warzecha, T., A. Sutkowska and H. Góral. 2014. Male sterility of triticale lines generated through recombination of triticale and rye maintainers. Spanish Journal of Agricultural Research 12: 1124-1130.

Wędzony, M., I. Marcińska, A. Ponitka, A. Ślusarkiewicz-Jarzina and J. Woźna. 1998. Production of doubled haploids in triticale (× *Triticosecale* Wittm.) by means of crosses with maize (*Zea mays* L.) using picloram and dicamba. Plant Breed. 177: 211-215.

Widodo, A.E, J.V. Nolan and P. Iji. 2015. The nutritional value of new varieties of high-yielding triticale: Feeding value of triticale for broiler chickens. South African J. Anim. Sci. 45: 74-81.

Wolski T., J. Gryka, B. Jarząbek, E. Czerwińska. 1998. Semidraft triticale breeding in DANKO. Bull. IHAR.205/206: 299-302 [in Polish with English summary].

Wolski, T. 1991. Impact of semidwarf rye germplasm on triticale improvement. CIMMYT, Proc. 2nd Int. Triticale Symp. Passo Fundo, Brasil. 144-149.

Woś, H. and W. Brzeziński. 2015. Triticale for food- the quality driver. pp 213-232. *In:* F. Eudes (ed.). Triticale. Springer International Publishing Switzerland 2015. DOI 10.1007/978-3-319-22551-7_11.

Würschum, T., M. R. Tucker and H.P. Maurer. 2014a. Stress treatments influence efficiency of microspore embryogenesis and green plant regeneration in hexaploid triticale (× *Triticosecale* Wittmack L.). *In vitro* Cell Dev. Biol. – Plant. 50: 143-148.

Würschum, T., M.R. Tucker, J.C. Reif and H.P. Maurer. 2012. Improved efficiency of doubled haploid generation in hexaploid triticale by *in vitro* chromosome doubling. BMC Plant Biol. 12: 109.

Würschum, T., W. Liu, L. Busemeyer, M.R. Tucker, J.C. Reif, E.A. Weissmann, V. Hahn, A. Ruckelshausen and H.P. Maurer. 2014b. Mapping dynamic QTL for plant height in triticale BMC Genetics 15:59.

Xu, S.J. and L.R. Joppa. 2000. Hexaploid triticales from hybrids of Langdon durum D genome substitutions with Gazelle rye. Plant Breed. 119: 223-226.

Yuanshu, S. and W. Zengyuan. 2004. Triticale development in China. pp. 103-108. *In:* M. Mergoum and H. Gómez-Macpherson (eds.) Triticale improvement and production. FAO plant production and protection paper, 179. Food and Agriculture Organization of the United Nations, Rome.

Zamorski, C. and M. Schollenberger. 1995. The occurrence of triticale diseases in Poland. Bull. IHAR 195/196: 197-207 [in Polish with English summary].

Zhang, X. and R.S. Jessop. 1998. Analysis of genetic variability of aluminum tolerance response in triticale. Euphytica 102: 177-182.

Zilinsky, F. and N. Borlaug. 1971. Progress in developing triticale. CIMMYT Res. Bull. 7:1-27.

Zimny, J., D. Becker, R. Brettschneider and H. Lörz. 1995. Fertile transgenic triticale (× *Triticosecale* Wittmack). Mol Breed. 1: 155-164.

Zohary, D. and M. Feldman. 1962. Hybridization between amphidiploids and the evolution of polyploids in the wheat (*Aegilops-Triticum*) group. Evolution 16: 44-61.

Chapter 12

Polyploidy and Interspecific Hybridization in *Cynodon*, *Paspalum*, *Pennisetum*, and *Zoysia*

Wayne W. Hanna[1,*], Byron L. Burson[2], and Brian M. Schwartz[1]

ABSTRACT

Desirable traits such as pest and stress resistances/tolerances, quality traits etc. needed to improve cultivated species are often found in their wild relatives. However, differences in their ploidy levels and genomic compositions can prevent hybridization, and when hybrids are recovered various levels and forms of sterility occur. Germplasm from wild species can be used with various levels of success by using colchicine to double chromosome numbers, developing bridging hybrids, and using genetic mechanisms that control chromosome/genome segregation in *Pennisetum* species. When vegetative propagation is a possibility, as is the case with *Cynodon* and *Zoysia* hybrids, the chances of successfully using germplasm from polyploid wild species in a breeding program has a much higher potential because sterility issues are circumvented. Germplasm from wild species have been successfully used to incorporate valuable traits into commercial hybrids and to improve germplasm of *Pennisetum* and *Cynodon* species. Interspecific *Paspalum* hybrids have been used to establish the phylogenetic relationships of several *Paspalum* species, including determining the origin of apomictic common dallisgrass.

Introduction

Harlan and de Wet (1971) categorized germplasm into primary, secondary, and tertiary gene pools, based on the difficulty of crossing with cultivated species. The primary gene pool includes the cultivated species and 'wild

[1] Professor and Associate Professor, respectively; University of Georgia, Tifton Campus Tifton, GA 31793, E-mail: whanna@uga.edu; tifturf@uga.edu

[2] Research Geneticist, U.S. Department of Agriculture, Agricultural Research Service, College Station, TX 77843, E-mail: byron.burson@ars.usda.gov

* Corresponding author

relatives' that cross easily with it and usually produce fertile hybrids. Some difficulty is usually encountered when crossing secondary gene pool species with the cultivated species. Some sterility can be encountered and the secondary gene pool species may have a genome in common with the cultivated species. Members of the tertiary gene pool cross with difficulty, if at all, with the cultivated species and any hybrids produced have non-viable pollen and sterile seed.

There are a number of ways that 'wild' germplasm can be introgressed into the cultivated species. These include backcrossing a trait or cytoplasm to the cultivated species, transferring a related genome/chromosome(s) to the cultivated species, or using the cultivated × wild species as a cultivar. Using polyploidy and interspecific hybrids to identify genomes in apomictic polyploid species is also an avenue for improving the apomictic species by recreating it through crossing species with contributing genomes.

Male and female sterility is usually encountered in diverse crosses which can be a problem in crop cultivars where grain production is the main product. However, in forage and ornamental crops, seed and pollen sterility are advantageous if the cultivar can be vegetatively propagated. Our goal is to provide examples of how manipulating ploidy and backcrossing have been used to diversify cytoplasm and genes in cultivated species of four genera: *Cynodon, Pennisetum, Paspalum, and Zoysia.*

Cynodon Germplasm

The genus *Cynodon* Rich. is composed of about eight species (Harlan et al. 1970). *Cynodon dactylon* (L.) Pers. var. *dactylon* or bermudagrass (the main cultivated species) is the most genetically diverse and morphologically variable species. Chromosome numbers for this species are either diploid ($2n = 2x = 18$) or tetraploid ($2n = 4x = 36$). The 36-chromosome cytotype is most commonly used in forage and turf hybrids and cultivar development. *Cynodon dactylon* is cosmopolitan and grows under a broad range of soil and climatic conditions, as far north as 53°N latitude and at elevations 3000 m above sea level (Taliaferro, 2003). It produces deep growing rhizomes. Stargrass or *Cynodon nlemfuensis* Vanderyst var. *nlemfuensis* is reported to be either diploid ($2n = 4x = 18$) or tetraploid ($2n = 4x = 36$). This species tends to be robust, coarser-textured than bermudagrass, and produces no rhizomes. *Cynodon transvaalenis* Burtt-Davy is a diploid ($2n = 2x = 18$) species. It is characterized by fine-textured leaves and a low growth habit.

Turf Hybrids

The turf industry was revolutionized with the development and release of 'Tifgreen' bermudagrass, a triploid ($2n = 3x = 27$) interspecific hybrid resulting from the cross *C. transvaalensis* ($2x = 18$) × *C. dactylon* ($4x = 36$). The triploid combined the toughness, wear-tolerance, cold tolerance, and disease resistance of *C. dactylon* with the fine texture of *C. transvaalensis*. The triploid hybrids produce dense stolons and shallow rhizomes which favor wear tolerance and regeneration from close mowing. Tifgreen allowed golfers to putt on a dense grass mowed at 6.25 mm instead of sand or coarse common bermudagrass. Since the release of Tifgreen, there have been a number of other high turf quality triploid hybrids such as 'Tifdwarf', 'Tifway', 'Midiron', 'Midlawn', 'TifSport', 'TifEagle', and 'TifGrand' released that have made a significant contribution to the turf industry (Taliaferro 2003). More recently, another interspecific triploid hybrid with more cold tolerance, 'Northbridge', was released (Wu et al. 2013). These triploid hybrids are used on major golf courses and athletic fields around the world up to 38° N and S of the equator. Although most *C. transvaalensis* × *C. dactylon* crosses are triploids, Taliaferro et al. (2006) patented a tetraploid ($2n = 4x = 36$) interspecific hybrid, released as 'Patriot' (Taliaferro et al. 2006), by crossing *C. transvaalensis* with 'Tifton 10' (Hanna et al. 1990), a $2n = 6x = 54$ chromosomes hexaploid. The resulting hybrid was partially fertile.

The triploid hybrids are pollen and seed sterile. Therefore, the hybrids are increased and commercially planted by vegetative propagation. Vegetative propagation and seed sterility make possible desirable morphologically uniform sports turf (Hanna and Anderson 2008).

Forage Hybrids

The development and release of 'Coastal' (an intraspecific hybrid within $4x = 36$ *C. dactylon*) bermudagrass (Burton 1943) revolutionized forage production, which impacted beef and milk production in the southern U.S. An even greater impact on the forage and cattle industry resulted from crossing a *C. dactylon* introduction ($2n = 4x = 36$) with 'Tifton 68' [*C. nlemfuensis* ($2n = 4x = 36$)] to produce 'Tifton 85' bermudagrass (Burton et al. 1993). Tifton 85 combined desirable characteristics from both species: ease of vegetative propagation, high palatability, high dry matter yield, and critical late season forage production. Although it lacks some of the rhizome development which enhances cold tolerance, it has a rapid growth rate to help it recover from cold damage. It produced 26% more dry matter than Coastal.

Pennisetum Germplasm

The genus *Pennisetum* has more than 140 species (Jauhar 1981). This may be a little misleading since some taxonomists have assigned species names to races. However, this genus has base chromosome numbers of $x = 5$, 7, 8, and 9, with ploidy levels ranging from diploid to octaploid and can reproduce sexually or by apomixis (Jauhar 1981). Cultivated pearl millet ($2n = 2x = 14$), *P. glaucum* (L.) R. Br., is grown on over 20 million ha for food, feed, forage, and fuel, mainly in India and Africa. It is a high quality summer annual forage crop in the United States. *Pennisetum glaucum* subsp. *monodii* (Maire) Brunken is a weedy invasive subspecies ($2n = 2x = 14$) of pearl millet that grows wild along roadsides and around farmers' fields in Africa. Napiergrass,

P. purpureum Schumach. ($2n = 4x = 28$), is a forage crop mainly grown in India, Africa, and South America. It has a genome (A′) in common with pearl millet (A genome). The perennial nature of napiergrass, the vegetative propagation of this species, and the dominant nature of the B genome over the A′ genome makes this species a valuable source of 'stored' germplasm for the genus *Pennisetum* (to be discussed later). *Pennisetum squamulatum* Fresen. ($2n = 8x = 56$) is a wild apomictic species that appears to be part of the secondary gene pool but was previously assigned to the tertiary gene pool (Yukio et al. 2006). Tetraploid *P. glaucum* ($2n = 4x = 28$) × *P. squamulatum* crosses result in partially fertile (pollen and seed) interspecific hybrids.

Interspecific Forage Hybrids

The pearl millet ($2n = 2x = 14$, AA genome) × napiergrass ($2n = 4x = 28$, A′A′BB genomes) triploid interspecific hybrid ($2n = 3x = 21$, AA′B genomes) has been widely made at numerous research locations around the world (Hanna et al. 2004, Muldoon and Pearson 1979); it is used mainly in India but also in Africa and South America. The hybrid is pollen- and seed-sterile and combines the forage quality of pearl millet with the vigor and yield potential of napiergrass. It produces no seed, so a triploid hybrid has to be vegetatively propagated.

Crosses are usually hand-made and superior hybrids are selected from the progeny. However, it is possible to produce commercial seed of the triploid hybrid by alternating rows of a cytoplasmic-nuclear male sterile (CMS) pearl millet genotype with rows of napiergrass (Osgood et al. 1997). This approach for making interspecific hybrids has to be done in a frost-free area because napiergrass is short-day sensitive and does not flower before frost. Seeds produced on the CMS maternal parent are triploid. However, because napiergrass is highly heterozygous, the resulting progeny can be quite variable, consisting of poor to outstanding hybrids (Hanna and Monson 1980). Early work has shown that crosses using some napiergrass clones produced more uniform progeny than other clones. It has also been reported that tall pearl millet inbreds produced higher yielding triploid hybrids than dwarf pearl millet inbreds when crossed with various napiergrass clones. Pearl millet × napiergrass crosses can be made in either direction, but using pearl millet as the female works the best.

Trispecific Ornamental Hybrids

Red tetraploid ($2n = 4x = 28$) pearl millet was crossed with a napiergrass ($2n = 4x = 28$) × *P. squamulatum* ($2n = 8x = 56$) interspecific hybrid. One red vigorous plant was selected and used to pollinate red 'Princess' napiergrass ($2n = 4x = 28$). Ten unique trispecific reddish/purple hybrids were selected, tested, released, patented (two are in the pending stage), and licensed for commercial production. Tift 17 marketed as 'Princess Caroline' was the first to be registered (Hanna et al. 2010).

Genome (or Partial) Transfer

Although the pearl millet × napiergrass triploid hybrids are pollen and seed sterile, pollen and seed fertility can be restored by treating the triploids with colchicine to produce the hexaploid ($2n = 6x = 42$, AAA'A'BB genomes) (Gonzales and Hanna 1984). Efforts have been made to commercialize seed production of the hexaploids, but with little success because the hexaploids are not persistent or sufficiently vigorous to justify the cost of seed production (summarized by Hanna et al. 2004).

The B genome of napiergrass masks the morphological traits of the A' genome. In a study where a number of different hexaploid genotypes were used to pollinate a cytoplasmic-nuclear male pearl millet inbred, two major types of progeny were produced (Hanna 1990). Most progenies from the crosses showed morphological characteristics from the B genome of napiergrass (probably $2n = 4x = 28$, AAA'B) and were pollen and seed sterile (as expected). However, certain crosses produced fertile, pollen shedding pearl millet progeny, quite different from the CMS female inbred parent. These progenies were variable for inflorescence length, CMS restoration, seed characteristics, height, and other phenotypic traits. Two inbreds were developed and released from these progenies. 'Tift 93' (Hanna, 2000a) was used to produce a CMS F_1 single cross hybrid, 'Tift 8593', (Hanna 2000b) that is used to produce commercial pearl millet forage hybrids. The CMS single cross female parent more than doubles the production of commercial hybrid seed compared to an inbred female parent. One of the leading pearl millet forage hybrids in the U.S. is 'Tifleaf 3' (Hanna 2000c), which uses Tift 8593 as the female parent in commercial hybrid seed production. Another inbred derived from the A' genome of napiergrass was 'Tift 454', the pollinator or restorer for a pearl millet grain hybrid 'TifGrain' (Hanna et al. 2005).

New Stable Cytoplasm

In 1980, a grassy pearl millet-like plant was received from Dr. A. Lambert, an ICRISAT scientist in Senegal, who wanted the accession identified (Hanna 1989). We counted 14 chromosomes, determined it flowered only in the greenhouse during the short days of winter, and found that it set good seed when crossed with a cultivated pearl millet. We identified it as *P. glaucum* subsp. *monodii*. The hybrids with *monodii* were vigorous and immune to rust (*Puccinia substriata* Ell. & Barth. var. *indica* Ramachar & Cumm.), the most serious disease on pearl millet in the U.S. in the field, due to a dominant gene for resistance. This resistance was rapidly backcrossed (BC) into inbreds to produce 'Tifleaf 2', a rust-immune commercial forage hybrid. However, after 3 years, the rust immunity was lost. During the development of the inbreds to produce 'Tifleaf 2', the *monodii* cytoplasm was transferred through backcrossing into our best pearl millet inbred ('Tift 23'). We found that the *monodii* cytoplasm, later named A_4, was stable and did not produce fertile revertants like the standard A_1 cytoplasm used to produce commercial hybrids. The A_4 cytoplasm is now the standard for producing forage hybrids, but because it is more difficult to restore pollen fertility in hybrids (male sterility is advantageous in forage hybrids and especially in annual species) using this cytoplasm, more research is needed to identify male parents that give pollen fertility restoration in grain hybrids.

Transfer of Apomixis

Hanna and Bashaw (1987) discussed the advantages of apomixis in plant breeding and cultivar development. Apomixis is a mechanism that eliminates the need for a sperm cell to fertilize an egg to produce an embryo. It allows cloning or vegetative propagation through the seed to produce true-breeding genotypes or hybrids. In the major cultivated crops, apomixis is usually found in polyploid species belonging to the tertiary gene pool, which makes transfer and use difficult.

In the 1970s, a program was initiated to transfer apomixis from a wild *Pennisetum* species to cultivated pearl millet. *Pennisetum squamulatum* was selected as the donor species for apomixis, after experimenting with a number of *Pennisetum* species crosses with cultivated pearl millet. *Pennisetum squamulatum* was originally thought to belong to the tertiary gene pool but later was determined to belong to the secondary gene pool (Akiyama et al. 2006). For the original cross to be successful, colchicine was used to double the chromosomes of diploid, sexually-cultivated pearl millet to produce a sexual, tetraploid pearl millet ($2n = 4x = 28$). Tetraploid pearl millet crossed freely with *P. squamulatum*, but all crosses using diploid pearl millet as the female parent were unsuccessful. The BC_2 progeny appeared to be a 'road block' in continuing the backcrossing program, but by manipulating ploidy levels and using bridging hybrids, we progressed to the BC_3 generation (Dujardin and Hanna 1984, Dujardin and Hanna 1989). The BC program has progressed to the BC_8 generation with apomictic tetraploid pearl millet-like plants (Singh et al. 2010) that have only one chromosome from *P. squamulatum* with the apomictic genetic mechanism. Backcrossing has only been successful to tetraploid pearl millet. The main problem at this generation is maintaining seed development by preventing endosperm abortion. It appears that the endosperm develops if only one polar nucleus is fertilized.

We have reported on successful uses of polyploidy to produce cultivars and germplasm. In reality, not all efforts at using polyploidy are successful. A good example is our effort to transfer the apomictic trait from *Pennisetum orientale* Rich. (tertiary gene pool) to *P. glaucum*. Interspecific hybrids and backcrosses could be produced, but seed and pollen sterility required other techniques for transferring germplasm (Dujardin and Hanna 1987).

***Paspalum* Germplasm**

The genus *Paspalum* L. consists of nearly 400 species (Chase 1929). This number may be inflated because apomixis is prevalent within the genus and taxonomists have classified different apomictic ecotypes within a species as different species. Regardless, it is a large genus consisting of cytologically and morphologically diverse members. Two different base chromosome numbers ($x = 6$ and 10) are reported for species in the genus, with ten being the most prevalent. Reported chromosome numbers range from a low of $2n = 2x = 12$ for *Paspalum hexastachyum* Parodi (Quarin 1974) to a high of $2n = 16x = 160$ or $17x = 170$ for *Paspalum floridanum* Michx. (Gould 1975). Some species have a range of ploidy levels. For example bahiagrass, *Paspalum notatum* Flügge, has cytotypes with $2n = 2x = 20$, $3x = 30$, $4x = 40$, and $5x = 50$ chromosomes (Tischler and Burson 1995), and dallisgrass, *Paspalum dilatatum* Poir., has biotypes with $2n = 4x = 40$, $5x = 50$, and $6x = 60$ chromosomes (Bashaw and Forbes 1958, Moraes Fernandes et al. 1968, Burson et al. 1991). Because apomixis is prevalent within the genus, fertilization of unreduced eggs ($2n + n$) has increased the ploidy levels of many species. This reproductive phenomenon is discussed in detail later in this chapter.

A wide range of morphological diversity exists among the species within the genus. Some species are low growing, spreading, stoloniferous and/or rhizomatous types that seldom grow more than 20 to 30 cm in height. Seashore paspalum, *Paspalum vaginatum* Swartz, is one such species and is widely used as a turfgrass, especially on saline soils because of its salt tolerance. Members of the Quadrifara and Virgata groups have coarse leaves and are robust bunchgrasses that can grow in excess of 3 m in height. These have received attention as potential ornamental grasses. Most species are smaller bunchgrasses that are used for forage, especially where they are indigenous. Two species that are widely grown, have become naturalized throughout much of the world, and are used for forage and pasture purposes are bahiagrass and dallisgrass. Most forage cultivars are naturally occurring apomictic ecotypes that were collected in the wild. These were evaluated for desirable forage characteristics and the more productive accessions were released as new cultivars. Cultivars of sexual ecotypes with superior turf or forage characteristics were either selected from naturally occurring types and released or improved using traditional breeding methods involving intraspecific hybridization. Numerous interspecific *Paspalum* hybrids have been produced, but none have been released as improved cultivars primarily because of sterility issues. Essentially all were produced for phylogenetic investigations.

Dallisgrass Phylogenetics

Common dallisgrass is an important forage grass in the tropics and subtropics. However, apomixis and irregular meiosis prevents the possibility of improving this grass using traditional breeding methods. Research was conducted to identify the genomes of this species in an attempt to introduce genetic variation by resynthesizing this natural hybrid. Several different biotypes of *P. dilatatum* are native to an area consisting of eastern Argentina, southern Brazil, and Uruguay (Evers and Burson 2004). Common dallisgrass is the most widely distributed and economically important because of its superior forage attributes. This biotype is an obligate apomict and a natural hybrid with 50 chromosomes that associate as 20 bivalents and 10 univalents during metaphase I of meiosis. Because of the absence of homologues to pair with the 10 univalents, these chromosomes lag behind the members of the dividing bivalents during anaphase I and some are not incorporated into the developing microspores. Consequently, its pollen grains are lacking some of the univalents and are low in viability. This significantly reduces the utility of this biotype as a pollen parent in crosses. Even though the common biotype is meiotically irregular and produces aneuploid gametes, it maintains a constant chromosome number of $2n = 5x = 50$ for successive generations because of its apomictic method of reproduction. The grass is highly susceptible to ergot (*Claviceps paspali* Stevens & Hall), which reduces seed production and can be lethal to animals that ingest infected inflorescences. The long range objective of this research was to circumvent apomixis and breed ergot resistant types.

A fertile dallisgrass biotype from Uruguay was crossed with common dallisgrass to produce improved and more fertile types (Bennett et al. 1969). This biotype, known as yellow-anthered dallisgrass, is a sexual tetraploid with 40 chromosomes which pair as 20 bivalents during metaphase I of meiosis (Bashaw and Forbes 1958, Bashaw and Holt 1958). Since it is sexual, it was used as the maternal parent when crossed with common dallisgrass (Bennett et al. 1969). Crossability between the two types was very low (0.04%) and only two hybrids were recovered. Both were sexual and had 45 chromosomes that associated as 20 bivalents and 5 univalents at meiosis (Bennett et al. 1969). In subsequent generations, the five univalents lagged during anaphase and were lost as micronuclei. By the F_3 generation, all progeny had only 40 chromosomes that paired as 20 bivalents. These plants were fertile and reproduced sexually but phenotypically resembled the yellow-anther biotype. None were desirable forage types and all were susceptible to ergot.

The cytogenetic findings from these hybrids were significant because they demonstrated that the 40 chromosomes in the yellow-anthered biotype are homologous with 40 of the 50 chromosomes in the meiotically irregular apomictic common biotype. This information provided the basis for a comprehensive phylogenetic investigation to identify the diploid progenitors of common dallisgrass using the meiotically stable, sexual yellow-anthered biotype as a cytological substitute for the common biotype. The yellow-anthered biotype was crossed with a number of diploid and tetraploid *Paspalum* species and the chromosome pairing in the hybrids revealed that both genomes in yellow-anthered dallisgrass were homologous with chromosomes in several diploid species belonging to the Paniculata (J),

Quadriferia (I), and Virgata (I) taxonomic groups (Burson et al. 1973, Burson and Quarin 1982, Burson 1983, Burson and Quarin 1992). Chromosomes in diploid *Paspalum intermedium* Munro. ex Morong and diploid *Paspalum jurgensii* Hackel had the highest level of homology with the chromosomes of the yellow-anthered biotype, and these two diploid species are considered the closest progenitors of dallisgrass (Burson et al. 1973). Later, hybrids were recovered from crosses between *P. jurgensii* and *P. intermedium,* and the lack of chromosome pairing in these hybrids confirmed that the genomes in these two species are different (Burson, 1981a). The yellow-anthered and common biotypes were subsequently assigned the genomic formulas IIJJ and IIJJX, respectively (Burson 1983). The diploid contributor of the X genome was not identified (Burson 1983, Espinoza and Quarin 2000).

These findings established the genome composition of both yellow-anthered and common dallisgrass but provided little insight how the common biotype originated. Significant progress was not made until hybrids were recovered from crosses between the yellow-anther biotype and an apomictic hexaploid ($2n = 6x = 60$) dallisgrass type from Uruguay. This biotype is known as the Uruguayan dallisgrass and its chromosomes pair as 30 bivalents during metaphase I (Burson et al. 1991). It is a robust, upright type that produces more forage and persists better under grazing than common dallisgrass (Venuto et al. 2003). Because of these superior traits, an apomictic accession was released as the cultivar 'Sabine' (Burson et al. 2009).

The percent crossability between the yellow-anthered and Uruguayan biotypes was only 0.17%; however, 41 morphologically variable F_1 hybrids were recovered. Most had $2n = 5x = 50$ chromosomes that associated as 20 bivalents and 10 univalents at metaphase I (Burson 1991a). The genome formulas IIJJX and IIJJXX were assigned to the F_1 hybrids and the Uruguayan biotype, respectively (Burson 1991a). A few hybrids resembled the yellow-anthered biotype but were more robust. Most of the remaining hybrids resembled the Uruguayan biotype but some were less upright in growth habit. However, two hybrids closely resembled common dallisgrass. All hybrids were fertile and produced F_2 progeny. The progeny from the F_1 hybrids that resembled the yellow-anthered biotype were variable and lacked vigor, indicating sexual reproduction and inbreeding depression. The F_2 progeny from the hybrids that resembled the Uruguayan biotype were uniform, indicating apomictic reproduction, and some were more vigorous and had potential as a forage grass but lacked the persistence of the Uruguayan biotype. Progeny from the F_1 hybrids similar to common dallisgrass were uniform, again indicating apomictic reproduction. These F_2 plants were identical to common dallisgrass. Because these hybrids and common dallisgrass were morphologically, cytologically, and reproductively similar, it was proposed that common dallisgrass originated from natural crosses between the sexual yellow-anthered biotype or a closely related sexual tetraploid and the apomictic, hexaploid Uruguayan biotype (Burson 1991a). These similarities support this hypothesis but the question regarding the relationship between members of the X genome in the common and Uruguayan biotypes remained.

To learn more about the X genome, a vegetative clone of a sexual yellow-anthered × common dallisgrass F_1 hybrid ($2n = 45$) (Bennett et al. 1969) that had been maintained for approximately 30 years was used, because it had five chromosomes of the X genome from common dallisgrass. This hybrid was pollinated with the Uruguayan biotype and only two hybrids were recovered (Burson 1991b). One had 52 chromosomes and the other had 53, and they associated at metaphase I as 22 bivalents and 8 univalents and 23 bivalents and 7 univalents, respectively. Twenty bivalents in both hybrids were paired members of the I and J genomes from both parents, and the two and three extra bivalents were pairing between members of the two X genomes from the common and Uruguayan biotypes (Burson 1991b). The univalents (seven and eight) were members of the X genome from the Uruguayan biotype which did not have homologs to pair with. These extra bivalents demonstrate that at least three and possibly four or five chromosomes of the X genome in these two biotypes are similar. From this, we proposed the X genomes in these two biotypes are similar. This strongly supports the hypothesis that apomictic, pentaploid common dallisgrass (IIJJX) originated from a natural cross between the sexual, tetraploid yellow-anthered dallisgrass (IIJJ) (or a similar type) and an apomictic, hexaploid (IIJJXX) similar to the Uruguayan biotype. Since both biotypes occur naturally in Uruguay and are sympatric with common dallisgrass, this event could have occurred eons ago in what is now western Uruguay (Burson 1991a). Even though the crossability between these two biotypes is extremely low, this probably was not a single event, but occurred multiple times over an extended period. As demonstrated in the man-made crosses between the yellow-anthered and Uruguayan biotypes mentioned above, the resulting hybrids were phenotypically diverse, and those that reproduced sexually would have been meiotically unstable and hence would have eliminated chromosomes of the X genome in one or two generations (Burson 1991a). If common dallisgrass had not reproduced by apomixis, this biotype would not exist today. This is a classic example of how apomixis can provide an escape from sterility and also preserve the genomic composition of a meiotically irregular plant that originated from wide hybridization.

Identifying the original contributor of the X genome is not only important in elucidating the phylogeny of dallisgrass; members of this genome also apparently possess genes that influence several important traits. Apomixis is expressed only in the dallisgrass biotypes and hybrids that have at least 10 chromosomes of the X genome, because when none or fewer than 10 of these chromosomes are present, the plants reproduce sexually (Bennett et al. 1969, Burson 1991a, Burson 1995). This indicates that the genes controlling apomixis are located on some of the X chromosomes. This also may be true for desirable forage traits, because biotypes and hybrids with 10 or more X chromosomes were superior forage types.

Bahiagrass and Other Paspalum Species

Of all the *Paspalum* species, bahiagrass is the most widely planted and grown for forage. Until recently, all tetraploid cultivars ($2n = 4x = 40$) were native apomictic ecotypes (Acuña et al. 2007) that were collected in the wild and eventually released as cultivars (Argentine, Paraguay 22, Wilmington) (Alderson and Sharp 1994). Pensacola bahiagrass is an exception in that it is a sexual, diploid cultivar ($2n = 2x = 20$). It was discovered growing in the port area of Pensacola, Florida, USA and was assumed to have been accidently introduced on a ship from South America (Burton 1967). The grass was eventually released as the cultivar 'Pensacola', and presently it is the most widely grown bahiagrass cultivar in the southeastern USA (Alderson and Sharp 1994). Using germplasm that can be traced back to Pensacola bahiagrass, breeders have used traditional breeding methods to develop improved sexual diploid cultivars ('Tifton 9' and 'TifQuik') (Burton 1989, Anderson et al. 2011). In 2012, an apomictic tetraploid cultivar 'Boyero UNNE' was developed and released by breeders at the Univeridad Nacional del Nordeste, Facultad de Ciencias Agrarias in Corrientes, Argentina (C.L. Quarin, personal communication). This cultivar was selected from hybrids resulting from crosses between an induced sexual tetraploid and an apomictic tetraploid. The induced sexual tetraploid parent originated from germplasm that was produced when the chromosomes in Pensacola bahiagrass were doubled with colchicine (Quarin et al. 2003).

All interspecific hybrids involving bahiagrass have been used to determine genome relationships. It was determined that the bahiagrass chromosomes are not homologous with chromosomes of the I and J genomes in dallisgrass (Burson et al. 1973), and the diploid and tetraploid ecotypes were assigned the genome formulas NN and NNNN, respectively (Burson 1981b, Quarin and Burson 1983, Quarin et al. 1984). Other species of the Notata group also have forms of the N genome (Quarin and Burson 1983, Quarin et al. 1984), but species in other taxonomic groups do not possess the N genome (Burson 1981b, Quarin, 1983).

Seashore paspalum is another economically important *Paspalum* species. Most ecotypes are sexual diploids ($2n = 2x = 20$) (Burson et al. 1973, Gould 1975) but triploid (Schwartz et al. 2013c) and tetraploid types (Gould 1975) have been reported. It is used as a turf grass for athletic fields, golf courses, parks, and home lawns, especially where salinity is an issue. The better turf types were produced by selecting desirable ecotypes growing in nature or selecting desirable types from controlled full-sib matings. Like bahiagrass, all interspecific hybrids with diploid seashore paspalum as a parent were made primarily for phylogenetic studies. Its chromosomes are not homologous with members of the I, J, or N genomes (Burson et al. 1973, Burson 1981b), and the species was assigned the genome designation of DD (Burson 1981a). Its chromosomes also are not homologous with those of other diploid species belonging to other *Paspalum* taxonomic groups (Burson 1981a, b, Quarin and Burson 1983).

Polyploidization

Polyploids can originate from the products of abnormal cytological events that occur during mitosis or meiosis. The unreduced gametes produced from irregular meiosis result in polyploidy when they are involved in either self- or cross-pollination. Fertilization of unreduced gametes is the most common means that ploidy levels are increased in plants (Harlan and de Wet 1975, de Wet 1979). This can occur in three ways: (i) fertilization of an unreduced egg cell by a reduced sperm ($2n + n$), (ii) fertilization of a reduced egg cell by an unreduced sperm nucleus ($n + 2n$), and (iii) union of an unreduced egg cell with an unreduced sperm nucleus ($2n + 2n$). Of these, fertilization of an unreduced egg by a reduced sperm nucleus ($2n + n$) is the most common and widespread (Harlan and de Wet 1975). Unreduced gametes are produced in both sexual and apomictic plants. Most unreduced gametes in sexual plants occur because of failure of chromosome reduction during meiosis I or failure of cytokinesis during meiosis II of mega- and/or microsporogenesis. Both of these irregular behaviors prevent chromosome reduction and the resulting products have an unreduced chromosome number, the same as the somatic tissues. In apomicts, the meiotic process is circumvented during the development of the female gametophyte (embryo sac) and all cells, including the egg cell, have unreduced chromosome numbers. Therefore, in apomicts there is always a source of unreduced eggs. Shortly before anthesis, the unreduced egg cell initiates mitosis and develops into an embryo (via parthenogenesis) that is genetically identical to the maternal plant because fertilization is not involved. However, in rare instances, the unreduced egg cell is fertilized by a reduced sperm cell ($2n + n$) and this increases the ploidy level of the resulting plant. This phenomenon not only increases ploidy level but is a means of producing genetically diverse offspring that sometimes are superior to their apomictic maternal parents (Pepin and Funk 1971, Burson and Hussey 1996).

The frequency of $2n + n$ fertilization in apomicts is usually very low, which negates the possibility of it being a reliable means of improving obligate or highly apomictic species. Pentaploid, apomictic common dallisgrass was pollinated with 15 different diploid *Paspalum* species and only five produced $2n + n$ hybrids (Espinoza and Quarin 2000). Percent crossability ranged from zero to only 0.81%; however, most were in the 0.04% to 0.07% range. Bashaw and Hignight (1990) reported the frequency of $2n + n$ fertilization in crosses between an apomictic pentaploid ($2n = 5x = 45$) buffelgrass, *Pennisetum ciliare* (L.) Link syn. *Cenchrus ciliaris* L., and tetraploid ($2n = 4x = 36$) birdwoodgrass, *Cenchrus setigerus* Vahl, as 1.3%. They concluded this was high enough that $2n + n$ fertilization could be used to circumvent the apomictic barrier and improve obligate apomicts. In another study, 201 interspecific hybrids were recovered from crosses between *Pennisetum flaccidum* Griseb. and *Pennisetum mezianum* Leeke (both facultative apomicts) and 54% (109) of them were $n + n$ hybrids and 46% (92) were $2n + n$ hybrids (Bashaw et al. 1992). Most $2n + n$

hybrids were more fertile than the $n + n$ hybrids. Kindiger and Dewald (1994) reported $2n + n$ fertilization in eastern gamagrass, *Tripsacum dactyloides* (L.) L. intraspecific hybrids as high as 27%.

These frequencies of $2n + n$ fertilization are high enough to merit using this phenomenon as a breeding tool to improve obligate or high apomictic species, but an understanding of what controls this event is necessary before it can be successfully utilized. Martinez et al. (1994) provided some insight into what initiates this event. They determined the frequency of $2n + n$ hybrids increased in apomictic bahiagrass when its stigmas were pollinated 2 or 3 days prior to normal anthesis. Using electron microscopy, Vielle et al. (1995) investigated the ultrastructural development of egg cells in apomictic and sexual embryo sacs of buffelgrass and determined that a cell wall developed and enclosed the unreduced egg cell before anthesis in the apomict, but a cell wall did not enclose the entire reduced egg cell in the sexual genotype until after fertilization. These findings provide some insight into what Martinez et al. (1994) reported in that the cell wall surrounding the unreduced apomictic egg cell may prevent a sperm nucleus from entering the egg at anthesis; however, when pollinated prior to anthesis, the sperm enters the egg cell because the cell wall is absent. Burson et al. (2002) self-pollinated apomictic and sexual buffelgrass accessions and also cross-pollinated them with birdwoodgrass pollen at 3, 2, 1, and 0 days before anthesis. More $2n + n$ selfed and hybrid plants were recovered when pollination took place 2 and 3 days prior to anthesis, and the frequency of $2n + n$ hybridization was as high as 8.2% for some accessions. This supports the ultrastructural findings of Vielle et al. (1995). All *Pennisetum* species and eastern gamagrass mentioned above are protogynous, and the stigmas are exserted from the florets several days prior to anthesis. This flowering behavior promotes early pollination, which obviously increases the chances for $2n + n$ fertilization in apomictic species. Because this phenomenon occurs at a sufficiently high frequency in species (i.e. *Pennisetum* and *Tripsacum* spp.) with the protogynous flowering behavior, it can be used as a tool to introduce genetic variation into uniform apomictic species for the purpose of developing new and improved germplasm.

Zoysia

Zoysia Willdenow are perennial grasses assigned to the family Gramineae (Poaceae), subfamily Chloridoideae and tribe Zoysieae, and are speculated to have originated near southeastern Asia and Indonesia (Engelke and Anderson 2003). Seed and plant samples were first introduced into the United States in the early 20th century by United States Department of Agriculture (USDA) researchers Frank N. Meyer and C.V. Piper (Childers and White 1947, Grau and Radko 1951). In total, over 1000 unique *Zoysia* accessions have been imported to the United States. Genotypes have been collected from as far north as 43°N latitude to as far south as 9°N latitude, indicating the adaptability of zoysiagrass to many climates. Considerable variation has been noted among accessions for color, winter dormancy, shade tolerance, growth habits, seed-set, leaf texture, disease resistance, and growing conditions (Engelke 2000, Engelke and Anderson 2003, Murray and Engelke 1983).

Most cytological studies of *Zoysia* spp. have determined that $2n = 40$ is the most common chromosome number (Arumuganathan et al. 1999, Chen and Hsu 1962, Christopher and Abraham 1974, Forbes 1952, Murray et al. 2005, Schwartz et al. 2010, Tateoka 1955). There is an account of one diploid plant, $2n = 20$, collected from Sri Lanka (Gould and Soderstrom 1974) and one report of a octaploid sample identified in Florida (Harris-Shultz et al. 2014). Forbes (1952) theorized that *Zoysia* spp. were diploid in nature, but did not disregard the possibility that the basic chromosome number could be 5 or 10. Chen and Hsu (1962) suspected that the basic chromosome number was 10 based on cytological studies in other eragrostoid grasses. This was later confirmed by Gould (1968). Zoysiagrasses are currently described as allotetraploids ($2n = 4x = 40$) based on restriction fragment length polymorphism (RFLP) linkage mapping and analysis (Yaneshita et al. 1999). The absence of anaphase bridges, fragments, lagging chromosomes, univalents or multivalents during meiosis I in conjunction with the formation of 95% viable pollen (Forbes 1952) supports the allotetraploid classification.

In 1951, the USDA-ARS and the United States Golf Association (USGA) Green Section released a turf-type *Zoysia japonica* Steudel named 'Meyer' (Grau and Radko 1951). This cultivar is often credited as the first zoysiagrass, although an improved *Z. matrella* (L.) Merrill was released as 'FC13521' in 1930 by the Alabama Agricultural Experiment Station in Auburn. Ian Forbes studied the cytological and morphological variation among many zoysiagrasses and recommended that *Z. japonica, Z. matrella,* and *Z. tenuifolia* Thiele be considered varieties of one species based on their cross compatibility. 'Emerald' was released in 1955 as a result of these efforts and was described as a *Z. japonica* × *Z. tenuifolia* hybrid (Forbes et al. 1955, Forbes 1952).

More recently, Anderson (2000) suggested that the *Zoysia* genus be reclassified to include eight additional species (*Z. macrantha* Desvaux, *Z. macrostachya* Franchet & Savatier, *Z. minima* (Colenso) Zotov, *Z. pacifica* (Goudswaard) Hotta & Kuroki, *Z. pauciflora* Mez, *Z. planifolia* Zotov, *Z. seslerioides* (Balansa) Claton & Richardson, and *Z. sinica* Hance) based on morphological variation and nuclear DNA constitution. Genetic variability has been characterized through random amplification of polymorphic DNA (RAPD) (Choi et al. 1997, Weng et al. 2007), restriction length polymorphism (RFLP) (Anderson 2000, Yaneshita et al. 1999, Yaneshita et al. 1997), amplified fragment length polymorphism (AFLP) (Cai et al. 2004, Kimball et al. 2012), simple sequence repeat (SSR) (Cai et al. 2005, Tsuruta et al. 2005), and resistance gene analog (RGA) (Harris-Shultz et al. 2012) marker analyses to discern relatedness among various zoysiagrass species. Using the current taxonomic classifications, zoysiagrasses are often incorrectly assigned when based on morphology alone, but molecular markers have helped reduce this confusion. Most mistakes could be eliminated if the grouping of species was based on reproductive compatibility (Mayr 1948) rather than by genotypic and morphological variability.

Kim (1983) and Weng et al. (2007) postulated that speciation occurred through geographic isolation rather than by genetic changes associated with an increase or decrease in ploidy level. Cross-compatibility does exist between a number of the *Zoysia* species (Forbes 1952, Hong and Yeam 1985), although Engelke and Anderson (2003) observed low germination percentages between a few interspecific combinations. Schwartz et al. (2010) found the range in variation for nuclear DNA content between *Z. japonica, Z. macrantha, Z. matrella, Z. minima, Z. pacifica, Z. pauciflora* and select interspecific hybrids to be less than reported for other tetraploid grass genotypes within the same species. Seed fertility, germination, and subsequent contamination of vegetatively propagated zoysiagrass fields have affected the purity and performance of many cultivars over time. This problem has not been observed in the sterile interspecific *Cynondon dactylon* × *C. transvaalensis* triploid hybrids during the past 60 years. In an effort to reproduce the success of the bermudagrass breeding model, we began treating tetraploid zoysiagrass seed with colchicine to induce chromosome doubling during 2009 (Schwartz et al. 2013a). Research is currently underway to determine if superior, sterile zoysiagrasses can be created by hybridizing genotypes with higher ploidy levels ($2n > 4x$) in a genus where cross-compatibility and fertility between species is relatively high, inbreeding depression after creating an auto-allooctaploid is a possibility, and the genetic control of chromosome pairing during meiosis is not well understood.

These initial efforts led to the creation of four M_0 octaploid genotypes, in addition to one cytochimera (Schwartz et al. 2013a). Average stomatal length of the four octaploids was 28% larger than observed in tetraploid zoysiagrass. Pollen diameter of these octaploids was significantly larger than found in the tetraploids, but pollen stainability was relatively unchanged by the colchicine treatments. Cross-pollination between the four colchicine-induced octaploids, and two tetraploid genotypes, led to the development of 136 M_1 octaploid and 210 M_1 hexaploid zoysiagrasses, respectively (Schwartz et al. 2013b). Further crossing using tetraploid, hexaploid, and octaploid parent lines resulted in the creation of 261 M_2 pentaploid and 55 M_2 septaploid genotypes. More pollinations were required to produce the pentaploid and septaploid zoysiagrasses due to poor seed set and germination of these crosses.

One of the primary goals of this research was to produce sterile *Zoysia* genotypes. Of the 136 octapoid and 210 hexaploid plants created, most were unexpectedly pollen and seed fertile, relatively easy to cross, and consistently set viable seed. Sixteen of the 55 septaploid plants have yet to produce seed, but seeds from 22 of the fertile 39 plants have germinated. There are currently 40 of the original 261 pentaploid zoysiagrasses that have not set seed, although a much lower percentage of the plants have produced seed that has germinated (22 of the 221) compared to the pentaploid genotypes.

A sterile turfgrass cultivar is only valuable if it is healthy and consistently performs well in many diverse environments. Preliminary observations of vigor and persistence have been made for the novel zoysiagrass germplasm developed in Tifton. Generally, most of the colchicine-derived octaploid genotypes grow slowly, are shallow rooted, and do not overwinter well. Conversely, most of the hexaploid zoysiagrasses have been nearly indistinguishable from tetraploids over the past three years during varying environmental conditions. The adaptation and persistence of the pentaploid and septaploid plants appears to be genotype-specific, as some are vigorous, but many have been adversely affected by disease and cold-stress, while others have died.

Further evaluation of these zoysiagrass polyploids for their long-term turfgrass performance is the next step in determining the value of this breeding procedure for the improvement of *Zoysia* species. It would stand to reason that incomplete sterility in a superior pentaploid or septaploid genotype could significantly reduce the aggregate potential for seed-contamination of a cultivar if grown on hundreds or thousands of acres. Another avenue that should be researched is the induction of a diploid zoysiagrass through pollen culture or *via* intergeneric crossing. If diploid genotypes become available for cultivar development, zoysiagrass breeders could more closely reproduce the methodology used to create triploid bermudagrasses, an effort that resulted in the high industry and consumer expectations that are in place today.

Summary

There are many examples of agronomically or economically important characteristics in wild species, such as disease tolerance or quality, that are difficult to incorporate into related cultivated species due to differences in ploidy levels. Sterility, or at least reduced fertility, is common when wild × cultivated crosses are made, if hybrids can be produced at all. This is unacceptable when seed yield is of primary concern. However, germplasm can be successfully transferred by using colchicine to double chromosome numbers, developing bridging hybrids, taking advantage of genetic mechanisms controlling chromosome/genome segregation, introducing genetic variation in apomictic species by reconstituting the species, and by using the apomictic mechanism to build new ploidy levels from within and between species. When vegetative propagation is a possibility, as in the case of *Cynodon* and *Zoysia,* the chances of successfully using germplasm from polyploid wild species in a breeding program has a much higher potential (Hanna et al. 2013). Numerous interspecific *Paspalum* hybrids have been produced, but none were released as improved cultivars, primarily because there were sterility issues in the F_1 hybrids and vegetative propagation was not practical. However, most hybrids were successfully used in phylogenetic studies to elucidate the genomic relationships between a number of different *Paspalum* species. Also, two of the three diploid progenitors of apomictic, pentaploid common dallisgrass were identified, and a hypothesis of how this important apomictic biotype originated was proposed. In the future, comparative genomics will likely reveal specific shifts in ploidy level, interchromosomal rearrangements, and major genome restructuring that resulted in the evolutionary separation of the wild and cultivated species that we use in our breeding programs today.

References

Acuña C.A., A.R. Blount, K.H. Quesenberry, W.W. Hanna and K.E. Kenworthy. 2007. Reproductive characterization of bahiagrass germplasm. Crop Sci. 47: 1711-1717.

Akiyama, Y., S. Goel, Z. Chen, W. Hanna and P. Ozias-Akins. 2006. *Pennisetum squamulatum*: Is the predominant cytotype hexaploid or octaploid? J. Heredity 97: 521-524.

Alderson J. and C.R. Sharp. 1994. Grass varieties in the United States. USDA-SCS Agric. Handb. 170, U.S. Gov. Print. Office, Washington, DC.

Anderson, W.F., R.N. Gates, and W.W. Hanna. 2011. Registration of 'TifQuik' bahiagrass. J. Plant Reg. 5: 147-150.

Anderson, S.J. 2000. Taxonomy of *Zoysia* (Poaceae): morphological and molecular variation, Ph.D. dissertation. Texas A&M Univ., College Station, TX.

Arumuganathan, K., S.P. Tallury, M.L. Fraser, A.H. Bruneau and R. Qu. 1999. Nuclear DNA content of thirteen turfgrass species by flow cytometry. Crop Sci. 39: 1518-1521.

Bashaw, E.C. and I. Forbes, Jr. 1958. Chromosome numbers and meiosis in dallisgrass *Paspalum dilatatum* Poir. Agron. J. 50: 441-445.

Bashaw, E.C. and E.C. Holt. 1958. Megasporogenesis, embryo sac development and embryogenesis in dallisgrass, *Paspalum dilatatum* Poir. Agron. J. 50: 753-756.

Bashaw, E.C. and K.W. Hignight. 1990. Gene transfer in apomictic buffelgrass through fertilization of a unreduced egg. Crop Sci. 30: 571-575.

Bashaw, E.C., M.A. Hussey and K.W. Hignight. 1992. Hybridization (N + N and 2N + N) of facultative apomictic species in the *Pennisetum* agamic complex. Int. J. Plant Sci. 153: 466-470.

Bennett, H.W., B.L. Burson and E.C. Bashaw. 1969. Intraspecific hybridization in dallisgrass, *Paspalum dilatatum* Poir. Crop Sci. 9: 807-809.

Burson, B.L. 1981a. Cytogenetic relationships between *Paspalum jurgensii* and *P. intermedium*, *P. vaginatum* and *P. setaceum* var. *ciliatifolium*. Crop Sci. 21: 515-519.

Burson, B.L. 1981b. Genome relations among four diploid *Paspalum* species. Bot. Gaz. 142:592-596.

Burson, B.L. 1983. Phylogenetic investigations of *Paspalum dilatatum* and related species. p. 170-173, *In* J.A. Smith and V.W. Hays (eds) Proc. 14th Int. Grassland Congr. Lexington, KY. 15-24 June 1981. Westview Press, Boulder, CO.

Burson, B.L. 1991a. Genome relationships between tetraploid and hexaploid biotypes of dallisgrass, *Paspalum dilatatum*. Bot. Gaz. 152: 219-223.

Burson, B.L. 1991b. Homology of chromosomes of the X genomes in common and Uruguayan dallisgrass, *Paspalum dilatatum*. Genome 34: 950-953.

Burson, B.L. 1995. Genome relationship and reproductive behavior of intraspecific *Paspalum dilatatum* hybrids: Yellow-anthered × Uruguaiana. Int. J. Plant Sci. 156: 326-331.

Burson, B.L. and M.A. Hussey. 1996. Breeding apomictic forage grasses. p. 226-230. *In* M.J. Williams (ed) Proc. Am. Forage Grassl. Counc. Vancouver, BC, Canada. 13-15 June 1996. AFGC, Georgetown, TX.

Burson, B.L. and C.L. Quarin. 1982. Cytology of *Paspalum virgatum* and its relationship with *P. intermedium* and *P. jurgensii*. Can. J. Genet. Cytol. 24: 219-226.

Burson, B.L. and C.L. Quarin. 1992. Cytological relationship between *Paspalum dilatatum* and diploid cytotypes of *P. brunneum* and *P. rufum*. Genome 35: 332-336.

Burson, B.L., H.S. Lee, and H.W. Bennett. 1973. Genome relations between tetraploid *Paspalum dilatatum* and four diploid *Paspalum* species. Crop Sci. 13: 739-743.

Burson, B.L., P.W. Voigt, and G.W. Evers. 1991. Cytology, reproductive behavior and forage potential of hexaploid dallisgrass biotypes. Crop Sci. 31: 636-641.

Burson, B.L., B.C. Venuto and M.A. Hussey. 2009. Registration of 'Sabine' dallisgrass. J. Plant Reg. 3: 132-137.

Burson, B.L., M.A. Hussey, J.M. Actkinson, and G.S. Shafer. 2002. Effect of pollination time on the frequency of $2n + n$ fertilization in apomictic buffelgrass. Crop Sci. 42: 1075-1080.

Burton, G.W. 1943. Coastal bermudagrass. Georgia Coastal Plain Exp. Stn. Circ. No. 30.

Burton, G.W. 1967. A search of the origin of Pensacola bahiagrass. Econ. Bot. 21: 379-382.

Burton, G.W. 1989. Registration of 'Tifton 9' Pensacola bahiagrass. Crop Sci. 29: 1326.

Burton, G.W., R.N. Gates and G.M. Hill. 1993. Registration of Tifton 85 bermudagrass. Crop Sci. 33: 644-645.

Cai, H., M. Inoue, N. Yuyama and S. Nakayama. 2004. An AFLP-based linkage map of zoysiagrass (*Zoysia japonica*). Plant Breed. 123: 543-548.

Cai, H.W., M. Inoue, N. Yuyama, W. Takahashi, M. Hirata and T. Sasaki. 2005. Isolation, characterization and mapping of simple sequence repeat markers in zoysiagrass (*Zoysia* spp.). Theor. Appl. Genet. 112: 158-166.

Chase, A.A. 1929. The North American species of *Paspalum*. Contr. U.S. Natl. Herb. 28: 1-310.

Chen, C.C. and C.C. Hsu. 1962. Cytological studies on Taiwan grasses. 2. Chromosome numbers of some miscellaneous tribes. J. Jap. Bot. 37: 300-313.

Childers, N.F. and D.G. White. 1947. Manila grass for lawns. Puerto Rico Agric. Exp. Stn. Circ. No. 26: 1-16.

Choi, J., B. Ahn and G. Yang. 1997. Classification of zoysiagrasses (*Zoysia* spp.) native to the southwest coastal regions of Korea using RAPDs. J. Korean Soc. Hortic. Sci. 38:789-795.

Christopher, J. and A. Abraham. 1974. Studies on the cytology and phylogeny of South Indian grasses. II. Sub-family *Eragrostoideae*. Cytologia 39: 561-571.

de Wet, J.M.J. 1979. Origins of polyploids. p. 3-15. *In* W.H. Lewis (ed). Polyploidy: Biological Relevance. Plenum Press, New York.

Dujardin, M. and W.W. Hanna. 1984. Cytogenetics of double cross hybrids between *Pennisetum americanum* × *P. purpureum*am phiploids and *P. americanum* × *P. squamulatum* interspecific hybrids. Theor. Appl. Genet. 69: 97-100.

Dujardin, M. and W.W. Hanna. 1987. Inducing male fertility in crosses between pearl millet and *Pennisetum orientale* Rich. Crop Sci. 27: 65-68.

Dujardin, M. and W. W. Hanna. 1989. Developing apomictic pearl millet characterization of a BC_3. J. Genet. Plant Breed. 43: 145-151.

Engelke, M.C. 2000. Widely used for centuries, zoysiagrass is a time-tested reservoir of genetic diversity. Diversity 16: 48-49.

Engelke, M.C. and S. Anderson. 2003. Zoysiagrass. pp. 271-285. *In* M.D. Casler and R.R. Duncan (eds). Turfgrass biology, genetics and breeding. John Wiley & Sons, Hoboken, NJ.

Espinoza, F. and C.L. Quarin. 2000. $2n + n$ hybridization of apomictic *Paspalum dilatatum* with diploid *Paspalum* species. Int. J. Plant Sci. 161: 221-225.

Evers, G.W. and B.L. Burson. 2004. Dallisgrass and other *Paspalum* Species, p. 681-713. *In* Moser, L.E., Burson, B.L. and Sollenberger, L.E. (eds.) Warm-Season (C_4) Grasses, Agron.Monogr. 45, ASA/CSSA/SSSA, Madison, WI. 1171 pp.

Forbes, I., B.P. Robinson and J.M. Latham. 1955. Emerald *Zoysia*-an improved hybrid lawn grass for the South. U.S. Golf. Assn. J. 7: 23-26.

Forbes, I., Jr. 1952. Chromosome numbers and hybrids in *Zoysia*. Agron. J. 44: 194-199.

Gonzalez, B. and W.W. Hanna. 1984. Morphological and fertility responses in isogenic triploid and hexaploid pearl millet × napiergrass hybrids. J. Hered. 75: 317318.

Gould, F.W. 1968. Grass systematics. McGraw-Hill Book Company, New York.
Gould, F.W. and T.R. Soderstrom. 1974. Chromosome numbers of some Ceylon grasses. Can. J. Bot. 52: 1075-1090.
Gould, F.W. 1975. The grasses of Texas. Texas A & M Univ. Press, College Station, TX.
Grau, F.V. and A.M. Radko. 1951. 'Meyer' (Z-52) *Zoysia*. U.S. Golf. Assn. J. 4: 30-31.
Hanna, W.W. 1989. Characteristics and stability of a new cytoplasmic-nuclear male-sterile source in pearl millet. Crop Sci. 29: 1457-1459.
Hanna, W.W. 1990. Transfer of germplasm from the secondary to the primary gene pool in *Pennisetum*. Theor. Appl. Genet. 80: 200-204.
Hanna, W.W. 2000a. Tift 93 Pearl millet. Plant Variety Protection number 9600191. May 8, 2000.
Hanna, W.W. 2000b. Millet, pearl, 'Tift 8593. Plant Variety Protection number 9600192. October 12, 2000.
Hanna, W.W. 2000c. Millet, pearl, 'Tifleaft 3'. Plant Variety Protection number 9600190. August 30, 2000.
Hanna, W.W. and E.C. Bashaw. 1987. Apomixis: its identification and use in plant breeding. Crop Sci. 27: 1136-1139.
Hanna, W.W., K. Braman and B. Schwartz. 2010. 'Tift 17' and Tift '23' Hybrid Ornamental Pennisetums. HortScience 45: 135-138. (Marketed as 'Princess Caroline' and 'Princess Molly', respectively.)
Hanna, W.W., G.W. Burton and A.W. Johnson. 1990. Registration of 'Tifton 10' turf bermudagrass. Crop Sci. 30: 1355-1356.
Hanna, W.W., B.J. Chaparro, B.W. Mathews, J.C. Burns, L.E. Sollenberger and J.R. Carpenter. 2004. Perennial Pennisetums. p. 503-535. In Warm-season (C_4) grasses, Number 45, L.E. Moser, B.L. Burson, and L.E. Sollenberger (eds.), Amer. Soc. Agron., Crop Sci. Amer., and Soil Sci. Amer. Madison, WI.
Hanna, W.W. and W.G. Monson. 1980. Yield, quality, and breeding behavior of pearl millet *x* napiergrass interspecific hybrids. Agron. J. 72: 358-360.
Hanna, W., J. Wilson and P. Timper. 2005. Registration of pearl millet line Tift 454. Crop Sci. 45: 2670-2671.
Hanna, W., P. Raymer and B. Schwartz. 2013. Warm-season grasses: biology and breeding. p. 543-590. *In*: J.C. Stier, B.P. Horgan and S. A. Bonos (eds).Turfgrass: Biology, Use, and Management. Agronomy Monograph No. 56. Madison, WI.
Hanna, W. and W. Anderson. 2008. Development and impact of vegetative propagation in forage and turf bermudagrass. Celebrate the Centennial (A Supplement to Agronomy Journal). S-103 to S-107.
Harlan, J.R. and J.M.J. de Wet. 1971. Toward a rational classification of cultivated plants. Taxon 20: 509-517.
Harlan, J.R. and J.M.J. de Wet. 1975. On Ö Winge and a prayer: The origins of polyploidy. Bot. Rev. 41: 361-391.
Harlan, J.R., J.M.J. de Wet, K.M. Rawal, W.W. Huffine and J.R. Deakin. 1970. A guide to the species of Cynodon (Gramineae). Okla. Agric. Exp. Sta. Bull. B-673.
Harris-Shultz, K.R., S.R. Milla-Lewis and J.A. Brady. 2012. Transferability of SSR and RGA markers developed in *Cynodon* spp. to *Zoysia* spp. Plant Mol. Biol. Rep. 30: 1264-1269.
Harris-Shultz, K.R., S. Milla-Lewis, A.J. Patton, K. Kenworthy, A. Chandra, F.C. Waltz, G.L. Hodnett and D.M. Stelly. 2014. Detection of DNA and ploidy variation within vegetatively propagated zoysiagrass cultivars. J. Am. Soc. Hortic. Sci. 139: 547-552.
Hong, K.H. and D.Y. Yeam. 1985. Studies on interspecific hybridization in Korean lawn grasses (*Zoysia* spp.). J. Korean Soc. Hortic. Sci. 26: 169-178.
Jauhar, P.P. 1981. Cytogenetics and breeding of pearl millet and related species. Alan R. Liss, New York.
Kim, J.H. 1983. A taxonomic study of the genus *Zoysia* Willd. in Korea. Kor. J. Plant Taxon. 13: 41-53.
Kimball, J.A., M.C. Zuleta, K.E. Kenworthy, V.G. Lehman and S. Milla-Lewis. 2012. Assessment of genetic diversity in *Zoysia* species using amplified fragment length polymorphism markers. Crop Sci. 52: 360-370.

Kindiger, B. and C.L. Dewald. 1994. Genome accumulation in eastern gamagrass, *Tripsacum dactyloides* (L.) L. (Poaceae). Genetica 92: 197-201.

Martinez, E.J., F. Espinoza and C.L. Quarin. 1994. B_{III} progeny ($2n + n$) from apomictic *Paspalum notatum* obtained through early pollination. J. Hered. 85: 295-297.

Mayr, E. 1948. The bearing of the new systematics on genetical problems. The nature of species. Adv. Genet. 2: 205-237.

Moraes Fernandes, M.I.B. de, I.L. Barreto and F.M. Salzano. 1968. Cytogenetic, ecologic, and morphologic studies in Brazilian forms of *Paspalum dilatatum*. Can. J. Genet. Cytol. 10: 131-138.

Muldoon, D.K. and C.J. Pearson. 1979. The hybrids between *Pennisetum americanum* and *Pennisetum purpureum*. Herbage Abst. 49: 189-199.

Murray, B.G., P.J. de Lange and A.R. Ferguson. 2005. Nuclear DNA variation, chromosome numbers and polyploidy in the endemic and indigenous grass flora of N.Z. Ann. Bot. 96: 1293-1305.

Murray, J.J. and M.C. Engelke. 1983. Exploration for zoysiagrass in Eastern Asia. USGA Green Section Record 21: 8-12.

Osgood, R.V., W.W. Hanna and T.L. Tew. 1997. Hybrid seed production of pearl millet × napiergrass triploid hybrids. Crop Sci. 37: 998-999.

Quarin, C.L. 1974. Relaciones citotaxonomica entre *Paspalum almum* Chase y *P. hexastachyum* Parodi (gramineae). Bonplandia 3: 115-127.

Quarin, C.L. 1983. Hibridos interespecificos de *Paspalum notatum* × *P. modestum*. Bonplandia 5: 235-242.

Quarin, C.L. and B.L. Burson. 1983. Cytogenetic relations among *Paspalum notatum* var. *saurae*, *P. indecorum*, and *P. vaginatum*. Bot. Gaz. 144: 433-438.

Quarin, C.L., B.L. Burson and G.W. Burton. 1984. Cytology of intra- and interspecific hybrids between two cytotypes of *Paspalum notatum* and *P. cromyorrhizon*. Bot. Gaz. 145: 420-426.

Quarin, C.L., M.H. Urbani, A.R. Blount, E.J. Martinez, C.M. Hack, G.W. Burton and K.H. Quesenberry. 2003. Registration of Q4188 and Q4205 sexual tetraploid germplasm lines of bahiagrass. Crop Sci. 43: 745-746.

Pepin, G.W. and C.R. Funk. 1971. Intraspecific hybridization as a method of breeding Kentucky bluegrass (*Poa pratensis* L.) for turf. Crop Sci. 11: 445-448.

Schwartz, B.M., K.R. Harris-Shultz, R.N. Contreras, C.S. Hans and S.A. Jackson. 2013a. Creation of hexaploid and octaploid zoysiagrass using colchicine and breeding. Crop Sci. 53: 2218-2224.

Schwartz, B.M., R.N. Contreras, W.W. Hanna and S.A. Jackson. 2013b. Manipulating the chromosome number of zoysiagrass. *In* ASA-CSSA-SSSA Abstracts [375-1]. Amer. Soc. Agron., Madison, WI.

Schwartz, B.M., R.N. Contreras, K.R. Harris-Shultz, D.L. Heckart, J.B. Peake and P.L. Raymer. 2013c. Discovery and characterization of turf-type triploid seashore paspalum. Hort Science 48:1424-1427.

Schwartz, B.M., K.E. Kenworthy, M.C. Engelke, A.D. Genovesi, R.M. Odom, and K.H. Quesenberry. 2010. Variation in 2C nuclear DNA content of *Zoysia* spp. as determined by flow cytometry. Crop Sci. 50: 1519-1525.

Singh, M., J.A. Conner, Y. Zeng, W.W. Hanna, V.E. Johnson and P. Ozias-Akins. 2010. Characterization of apomictic BC7 and BC8 pearl millet: Meiotic chromosome behavior and construction of an ASGR-carrier chromosome-specific library. Crop Sci. 50: 892-902.

Taliaferro, C.M. 2003. Bermudagrass (*Cynodon* (L.) Rich). pp. 235-246. *In*: Turfgrass Biology, Genetics, and Breeding, M.D. Casler and R.M. Duncan (eds). John Wiley and Sons, Inc., Hoboken, New Jersey.

Taliaferro, C.M., D.L. Martin, J.A. Anderson and M.P. Anderson. 2006. Patriot turf bermudagrass. U.S. Plant Patent PP16, 801.

Tateoka, T. 1955. Karyotaxonomy in *Poaceae*. III. Further studies of somatic chromosomes. Cytologia 20: 296-306.

Tischler, C.R. and B.L. Burson. 1995. Evaluating different bahiagrass cytotypes for heat tolerance and leaf epicuticular wax content. Euphytica 84: 229-235.

Tsuruta, S.-I., M. Hashiguchi, M. Ebina, T. Matsuo, T. Yamamoto, M. Kobayashi, M. Takahara, H. Nakagawa and R. Akashi. 2005. Development and characterization of simple sequence repeat markers in *Zoysia japonica* Steud. Grassland Sci. 51: 249-257.

Venuto, B.C., B.L. Burson, M.A. Hussey, D.D. Redfearn, W.E. Wyatt and L.P. Brown. 2003. Forage yield, nutritive value, and grazing tolerance of dallisgrass biotypes. Crop Sci. 43: 295-301.

Vielle, J-Ph., B.L. Burson, E.C. Bashaw and M.A. Hussey. 1995. Early fertilization events in the sexual and aposporous egg apparatus of *Pennisetum ciliare* (L.) Link. Plant J. 8: 309-312.

Weng, J.H., M.J. Fan, C.Y. Lin, Y.H. Liu and S.Y. Huang. 2007. Genetic variation of *Zoysia* as revealed by random amplified polymorphic DNA (RAPD) and isozyme pattern. Plant Production Sci. 10: 80-85.

Wu,Y., D.L. Martin, C.M. Taliaferro, J.A. Anderson and J.Q. Martin. 2013. Northbridge turf bermudagrass. Plant Patent PP24, 116.

Yaneshita, M., S. Kaneko and T. Sasakuma. 1999. Allotetraploidy of *Zoysia* species with $2n = 40$ based on a RFLP genetic map. Theor. Appl. Genet. 98: 751-756.

Yaneshita, M., R. Nagasawa, M.C. Engelke and T. Sasakuma. 1997. Genetic variation and interspecific hybridization among natural populations of zoysiagrasses detected by RFLP analyses of chloroplast and nuclear DNA. Genes Genet. Syst. 72: 173-179.

Yukio, A., S. Goel, Z. Chen, W. Hanna and P. Ozias-Akins. 2006. *Pennisetum squamulatum*: Is the predominant cytotype hexaploid or octaploid? J. Hered. 97: 521-524.

Chapter 13

Interploid and Interspecific Hybridization for Kiwifruit Improvement

A.R. Ferguson[1,*] and J.-H. Wu[2]

ABSTRACT

The kiwifruit of commerce are large-fruited selections of *Actinidia chinensis* var. *chinensis* and *A. chinensis* var. *deliciosa*. The main constraints to kiwifruit breeding include dioecy, the long generation time, the difficulties of genetic analysis in polyploids, and the complexity of some key traits. In *Actinidia* there is variation in ploidy both within and between taxa. Most taxa are diploid but there are also tetraploids, hexaploids, octoploids and decaploids in diminishing frequency. The high basic chromosome number, $x = 29$, suggests that diploids of *Actinidia* are palaeopolyploids. Many taxa contain ploidy races, especially the *A. arguta* and *A. chinensis* species complexes. Thus the ploidy of any genotype used in breeding programmes and the ploidy of offspring should be checked: they are not always at the expected ploidy. In wild populations, there appears to have been frequent hybridization and reticulate evolution influenced by geographic patterns of distribution with transitional forms suggesting considerable gene flow between sympatric *Actinidia* taxa. Controlled crossing has also often been successful although differences in ploidy can be expected to complicate interspecific crosses or even crosses within a taxon. Ploidy in *Actinidia* can be manipulated by chromosome doubling, plantlet regeneration from pollen or unfertilized embryos, use of numerically unreduced gametes, or culture of endosperm tissue. Protoplast fusion to produce allopolyploids needs further development. Success in kiwifruit breeding programmes will depend on a good understanding of the reproductive biology of *Actinidia* and an awareness of the commercial requirements of new cultivars.

[1,2] The New Zealand Institute for Plant & Food Research Ltd, Private Bag 92169, Auckland Mail Centre, Auckland 1142, New Zealand, E-mail: jinhu.wu@plantandfood.co.nz

* Corresponding author, E-mail: ross.ferguson@plantandfood.co.nz

Introduction

The kiwifruit of commerce are large-fruited selections of *Actinidia chinensis* Planch. (Actinidiaceae). A recent revision of the genus *Actinidia* (Li et al. 2007a) has reduced what were previously considered as two separate species to varieties of the one species: *A. chinensis* var. *chinensis* and *A. chinensis* var. *deliciosa* (A. Chev.) A. Chev. – the latter previously *A. deliciosa* (A. Chev.) C.F. Liang et A.R. Ferguson. Fruit of *A. chinensis* var. *deliciosa* are the well-known "traditional" kiwifruit with brown hairy skins and emerald green fruit flesh – much of the scientific literature will refer to this variety as *A. deliciosa*. Fruit of *A. chinensis* var. *chinensis* have smoother, less hairy skins and the fruit flesh is usually yellow, although in some genotypes it may be green or partially red. Much smaller quantities of a third species, *A. arguta* (Sieb. et Zucc.) Planch. ex Miq., are also being produced commercially: they are now usually promoted as kiwiberries.

Breeding Aims

The New Zealand kiwifruit marketer Zespri Group Limited, commonly known as Zespri, has summarised (in 2015) what it considers necessary in new kiwifruit cultivars: "Adding new products to Zespri's portfolio attracts new consumers to the kiwifruit category, earns more shelf space with retailers and offers growers the opportunity to expand or diversify their businesses and improve profitability...the attributes we need for a successful Zespri [kiwifruit] cultivar [are] great taste, good yield, size and storage, Psa tolerance, strong health attributes [and] consumer liking...". Psa is the bacterial canker of kiwifruit, a disease caused by virulent strains of *Pseudomonas syringae* pv. *actinidiae*.

These aims are common to most kiwifruit improvement programmes throughout the world: an emphasis on fruit novelty, flavour, fruit size, flesh colour, length of storage life, convenience to consumers, environmental adaptation and vine productivity. The contributions of kiwifruit to human health and enjoyment, such as vitamin C content, also frequently gain attention. Novelty is considered particularly important in expanding the kiwifruit category, complementing rather than cannibalising existing products (Martin and Luxton 2005), and this has encouraged an emphasis on breeding fruit that will occupy new marketing niches (Jaeger et al. 2003, Wismer et al. 2005).

The main constraints to kiwifruit breeding include dioecy, the difficulty and expense of determining the contributions of the male parents by progeny trials, the long generation time, the difficulties of genetic analysis in polyploids, and the complexity of some key traits, with many of those associated with fruit quality being quantitatively inherited. Kiwifruit are vines which require strong and expensive support structures: they grow vigorously and growth must be controlled to ensure fruiting. Furthermore, they have only recently been domesticated, existing cultivars are at best only several generations removed from the wild, and our knowledge of their reproductive biology is narrow and generally based on detailed studies of only a very small number of genotypes. There is great genetic diversity within the genus, but the germplasm available to most kiwifruit breeders is very limited and inadequately represents the diversity in the wild. This chapter concentrates mainly on the diversity in ploidy within the genus *Actinidia,* the opportunities for ploidy manipulation, the extensive interspecific hybridization in the wild and the potential of interspecific hybridization for kiwifruit improvement programmes.

The breeding aims are ambitious. Any new cultivar must appeal to consumers but must also be profitable for growers and marketers. Priorities can vary: exporting countries such as Chile and New Zealand place particular emphasis on fruit storage life to allow shipping around the world and orderly marketing over an extended season. In Italy, the main kiwifruit cultivar 'Hayward' matures late in the season and is sometimes affected by autumn frosts: an Italian priority has therefore been early harvest maturity or at least maturation earlier than that of 'Hayward' (Testolin and Ferguson 2009). In parts of China, there was a preference for cultivars that can be harvested late in the season when temperatures are cooler (Huang and Ferguson 2001) but in areas that are further north, such as Shaanxi, and at higher altitudes, early-maturing cultivars are preferred. In all countries the recent advent of bacterial canker of kiwifruit, Psa, has resulted in much greater importance being placed on disease tolerance or resistance. There is also more interest in unique, proprietary cultivars, protected by plant variety rights, and available to growers only as part of arrangements that include limitations on the areas planted, technical advice on growing and exclusive marketing arrangements. Examples of such cultivars are 'Hort16A' (Zespri® Gold Kiwifruit) and 'Zesy002' (Zespri® SunGold Kiwifruit), marketed by Zespri, and 'Jintao', marketed as Jingold™ by Consorzio Kiwigold®.

The identification of a promising new selection is just the beginning of commercialization (Martin and Luxton 2005). There must also be research into pre- and post-harvest treatments to optimise fruit quality, shelf space in the major retailers must be secured, and potential consumers must be made aware of the new product and its unique qualities. For countries that rely on kiwifruit exports, the requirements of importing countries for food safety and traceability must be satisfied as must the standards of GlobalGAP. It is therefore not surprising that the costs of branding, promotion and marketing of a new cultivar can be 10–12 times the cost of the original work to develop the cultivar (Martin and Luxton 2005).

The Genus *Actinidia*

The genus *Actinidia,* family Actinidiaceae, contains about 50 species of climbing or scrambling, perennial, normally deciduous plants (Huang et al. 2014). The probable centre of current evolution of the genus is between the

Yangzi and Pearl Rivers, China, in a zone between approximately 25 and 30° north (Liang 1983). This area has the greatest abundance and greatest diversity of *Actinidia* species. The area is characterised by warm, moist environments where *Actinidia* species are now largely confined to the higher hills and mountains in relatively damp, shady or semi-shady areas in gullies, especially along streams, under the tree canopy or on the edge of the forest or in clearings (Huang and Ferguson 2007). *Actinidia* plants can scramble along the ground or form thickets but usually climb trees wherever possible, normally fruiting only when exposed to the light (Gao and Xie 1990). Some *Actinidia* species extend much further to higher latitudes as far as 50° north, where they can occur almost at sea level, and one species to just south of the equator. The distribution of *Actinidia* is therefore typical of many Chinese flora, comprising considerable diversity within China itself with outlier species in adjoining countries.

A characteristic feature of all *Actinidia* species is the structure of the gynoecium of the pistillate flower. The ovary is free and superior, formed by the fusion of many carpels but leaving free the radiating styles, often reflexed (for illustration, see McGregor 1976). Each locule contains many ovules.

All members of the genus appear morphologically to be androdioecious (i.e., individual plants look as if either morphologically hermaphrodite or male). The pistillate, apparently hermaphroditic flowers appear perfect with a well-developed ovary and styles and stamens, whereas staminate flowers have only a rudimentary small gynoecium with poorly developed styles and no ovules. However, the anthers of pistillate flowers of some *Actinidia* species have been shown to release non-viable pollen, and these species can therefore be defined functionally as cryptically dioecious (Schmid 1978, Kawagoe and Suzuki 2004, Mizugami et al. 2007). This is probably true of all *Actinidia* species. Dioecism is not absolute and occasional gender-inconstant genotypes have been observed both in commercial orchards (McNeilage 1991) and in the wild (Tang and Jiang 1995, Zhu and Zhang 1995). A self-setting clone of *A. arguta* 'Issai' has also been described (Mizugami et al. 2007). Furthermore, some wild *Actinidia* from southern Japan, probably related to *A. callosa* Lindl. (I. Kataoka pers. comm.), are hermaphroditic and self-setting (Matsumoto et al. 2013). Dioecism and the occurrence of gender-inconstant genotypes have important consequences for kiwifruit breeding programmes.

Fruit of *Actinidia* are defined botanically as berries. Internally, they can be divided into an outer pericarp, an inner pericarp, which includes the locules containing two radial rows of many small seed within a mucilaginous matrix, and a central core or columella (Schmid 1978, Ferguson 1984). *Actinidia* species are cultivated for their fruit. There is tremendous variation both within and between species in fruit characteristics, as well as in vine morphology and behaviour (Ferguson 1990a, Huang et al. 2004, Huang and Ferguson 2007, Nishiyama 2007, Huang et al. 2014). One of the main aims of breeding is to introgress desirable fruit characteristics to obtain new kiwifruit cultivars of commercial potential (Gamble et al. 2010).

Actinidia Species

Actinidia taxa are often difficult to circumscribe unequivocally because they can be very variable and there is an intermingling of both morphological and molecular characteristics (Huang and Ferguson 2007, Hseih et al. 2011). Herbarium collections are often inadequate: some taxa are incompletely described and there is a need for herbarium specimens from the entire distribution range of taxa to determine species variability (Li et al. 2009, 2011). Furthermore, taxonomists vary in the broadness of the concepts that they use. Species that are widespread geographically have often been subdivided on the basis of morphological differences into taxa that occupy discrete regions, but other taxonomists have not found such differences to be consistent or correlated. Thus the most recent treatment of the genus *Actinidia* in China (Li et al. 2007a, b) reduced by about a third the number of species accepted and also reduced drastically the number of infraspecific taxa. Li et al. (2009) discuss difficulties with some of the better known taxa: for example, whether *A. chinensis* var. *chinensis* and *A. chinensis* var. *deliciosa* are better considered as separate species or should be combined as varieties of the one species, and whether the retention of *A. arguta* and *A. melanandra* Franch. as distinct species can be justified. More extensive collections and detailed studies of apparently related taxa are required to resolve such problems. A previously well-known taxon, *A. arguta* var. *purpurea* (Rehder) C.F. Liang (syn. *A. purpurea* Rehder) has been submerged in *A. arguta* (Li et al. 2007a, b) as the colouration of the fruit could not be consistently correlated with vegetative characteristics.

Repeated attempts have been made to group *Actinidia* species into sections (Dunn 1911, Huang and Ferguson 2007, Li et al. 2009) using morphological features such as the degree of pubescence, the nature of the leaf hairs, ovary shape, the presence or absence of lenticels on the fruit and whether the pith in the stem was lamellate or solid. However, the characters used to separate the sections were not always consistent and molecular evidence indicated that many of the sections proposed were polyphyletic. The most recent revision of the genus (Li et al. 2007a) does not attempt to subdivide the genus into sections.

Ploidy and Ploidy Races in Wild *Actinidia* Populations

Basic Chromosome Number

The basic chromosome number in *Actinidia* is $n = 29$ (Lu et al. 1984, Xiong and Huang 1988: Xiong et al. 1985, 1993, McNeilage and Considine 1989, Yan et al. 1994, 1997a). Such a high number could indicate that diploid *Actinidia* species are cryptic polyploids or rediploidized palaeopolyploids (Grant 1963,

Goldblatt 1980, McNeilage and Considine 1989, Yan et al. 1997a, He et al. 1998, 2005). This conclusion is supported by primer pairs for microsatellites apparently amplifying several distinct loci (Huang et al. 1998, Testolin et al. 1999, 2001, Fraser et al. 2005). Further evidence was provided by Shi et al. (2010), who concluded that *Actinidia* had undergone a number of whole genome duplications: the most recent, about 28.3 million years ago, being specific to the genus after it had separated from related members of the Ericales. Gene duplication has probably facilitated the observed high heterozygosity in *Actinidia* (Messina et al. 1991, Huang et al. 1997, Testolin and Ferguson 1997, Huang et al. 2004, Liu et al. 2010).

Ploidy Races in Actinidia

Chromosome counts and flow cytometry have shown that in *Actinidia* taxa there is a regular pattern of diploids (2*x*), tetraploids (4*x*), hexaploids (6*x*), and octoploids (8*x*) in diminishing frequency (Xiong 1992, Yan et al. 1994, Ferguson et al. 1997, Huang et al. 2004, Huang and Ferguson 2007, Huang et al. 2014). Heptaploids (7*x*) (Kataoka et al. 2010) and decaploids (10*x*) (Li et al. 2013b) have also been observed in wild populations of *A. arguta* and occasional dodecaploids (12*x*) and mixoploids have been raised from seed collected from fruit of wild *A. chinensis* var. *deliciosa* plants (A.R. Ferguson unpublished). Dodecaploids have also been detected among somaclones of *A. chinensis* var. *deliciosa* 'Hayward' (Boase and Hopping 1995).

In general, ploidy has been determined for only a small number of genotypes of any particular *Actinidia* taxon. Furthermore, these genotypes usually come from a limited number of accessions. Nevertheless, there is now convincing evidence of intrataxon variation in ploidy, most commonly diploid and tetraploid races, in at least 15 *Actinidia* taxa, although the extent of the variation depends to some extent on where taxonomic boundaries are established (Huang and Ferguson 2007, Huang et al. 2014). Detailed studies of several wild *Actinidia* populations show that there can be even greater complexity in ploidy.

Ploidy Races in the Actinidia arguta Complex

Flow cytometry of genotypes and named selections from the wild as well as some of unknown origin has revealed tetraploid, hexaploid and heptaploid forms of *A. arguta* in Japan, whereas plants of *A. hypoleuca* Nakai, sometimes reduced to *A. arguta* var. *hypoleuca* (Nakai) Kitam., were found to be diploid (Watanabe et al. 1990, Phivnil et al. 2005, Mizugami et al. 2007). Clones of *A. arguta* 'Issai' were reported as being either hexaploid (Watanabe et al. 1990) or heptaploid (Mizugami et al. 2007).

Subsequent studies of samples of *A. arguta* collected from recorded localities in the wild showed that whereas the tetraploid forms were distributed throughout Japan, the diploid forms (*A. arguta* var. *hypoleuca*) were limited to warm, coastal areas in southwestern Kyushu and the hexaploid forms to the much colder, northern regions of Honshu. The diploid and the hexaploid forms could generally be distinguished morphologically from the tetraploid forms. Two heptaploid and one octoploid plant were also found in northern Honshu (Kataoka et al. 2010). Kuroda et al. (2011) likewise found that plants of *A. arguta* var. *hypoleuca* were consistently diploid and although one plant of what appeared morphologically to be *A. arguta* var. *arguta* was diploid, others so identified were all tetraploid. However, chloroplastic DNA polymorphism indicated that the tetraploid plants of *A. arguta* var. *arguta* from the southern island of Kyushu were genetically closer to the one diploid plant of *A. arguta* var. *arguta* observed and the plants of *A. arguta* var. *hypoleuca* from the same region than to the tetraploid plants of *A. arguta* var. *arguta* from northern Japan (Kuroda et al. 2011).

Similar variation in ploidy complexity has been observed in populations of *A. arguta* in China. The most detailed studies were of seven adjacent, but separate, wild populations in the eastern Daba Mountains of Shaanxi Province (Li et al. 2013b, Lai et al. 2015). Tetraploids (36%), hexaploids (11%), octoploids (33%) and decaploids (20%) were present. Two populations were largely tetraploid but three other populations consisted of octoploids and decaploids, and the remaining two populations contained several ploidy levels. All the populations studied had red or purple fruit and would therefore probably have been formerly classified as *A. arguta* var. *purpurea* (syn. *A. purpurea*), but would now be included in *A. arguta* var. *arguta* (Li et al. 2007a). Tetraploid and octoploid forms of *A. arguta* var. *purpurea* had previously been identified (Huang and Ferguson 2007).

Ploidy Races in the Actinidia chinensis Species Complex

Wild populations of the *A. chinensis* species complex exhibit great diversity in ploidy. The two varieties are widely distributed throughout China (for maps, see Ferguson 1990b, Huang and Liu 2014): *A. chinensis* var. *chinensis* is largely found in warmer, lowland areas to the east, *A. chinensis* var. *deliciosa* more inland to the west at higher, colder altitudes. Initial studies on a limited number of accessions from various parts of China suggested that whereas *A. chinensis* var. *deliciosa* was always hexaploid, there were diploid and tetraploid races of *A. chinensis* var. *chinensis*, the tetraploid forms coming from a restricted part of China (Xiong 1991, 1992, Yan et al. 1994). Subsequent, more comprehensive studies have shown that this conclusion is incorrect and that tetraploid forms of *A. chinensis* var. *chinensis* are more widespread. The geographic distributions of *A. chinensis* var. *chinensis* and *A. chinensis* var. *deliciosa* often overlap, particularly in southeastern Shaanxi, southwestern Henan, western Hubei and Hunan and in eastern Guizhou, with *A. chinensis* var. *deliciosa* usually at the higher altitudes. In such areas where the two species coexist, there is gradation in morphological characteristics between "standard" *A. chinensis* var. *chinensis* and "standard" *A. chinensis* var. *deliciosa* accompanied by a transition in ploidy with plants at 2*x*, 4*x*, 5*x* and 6*x* (Zeng et al. 2009, Li et al. 2010a, Liu et al. 2011, 2015, Huang and Liu 2014). The relative proportions of the different cytotypes vary, possibly in response to altitude, niche disturbance and vegetation cover (Liu et al. 2015).

The Need to Check the Ploidy of Individual Genotypes

As more and more plants in *ex situ* germplasm collections or in wild populations are studied it has become obvious that the ploidy of an individual *Actinidia* genotype should never be assumed. Ploidy races have been found in an increasing number of *Actinidia* taxa and are probably yet to be discovered in other taxa, especially those that are morphologically variable or that are geographically widespread. In the *A. chinensis* species complex, diploid *A. chinensis* var. *chinensis* genotypes are readily distinguished from hexaploid *A. chinensis* var. *deliciosa* genotypes, but in sympatric populations there is often a morphological gradation between the extremes. Tetraploid genotypes of *A. chinensis* var. *chinensis* in the Plant & Food Research *Actinidia* germplasm collection usually flower several weeks after diploid accessions and before *A. chinensis* var. *deliciosa* genotypes, but this may simply be a coincidence as the tetraploid genotypes of known provenance came from a limited area in China. Fortunately, ploidy in *Actinidia* can be readily checked by flow cytometry (Ferguson et al. 2009).

Interspecific and Interploidy *Actinidia* Hybrids in Nature

The rather confusing mixtures of *Actinidia* taxa often lacking discrete taxonomic boundaries and the intermingling of characters, both morphological and molecular, are probably due to the frequent occurrence of hybridization and reticulate evolution influenced by geographic patterns of distribution. Transitional forms suggest considerable hybridization between taxa with overlapping distributions. Liang noted (1982a, b) that there was a steady progression of characteristics in the *A. chinensis* species complex from coastal Zhejiang, where *A. chinensis* var. *chinensis* has small leaves and very soft fruit hairs, to the stiff-haired, larger leafed *A. chinensis* var. *deliciosa* found in Sichuan and Shaanxi. He also thought (Liang 1984) that some of the *Actinidia* species that had been described might be natural hybrids, e.g., that *A. chengkouensis* C.Y. Chang might be a hybrid between *A. chinensis* var. *deliciosa* and a species from the *Strigosae,* and that *A. stellatopilosa* C.Y. Chang might be a hybrid between *A. chinensis* var. *deliciosa* and *A. trichogyna* Franch. Likewise, *A. hubeiensis* H.M. Sun et R.H. Huang might be a hybrid between *A. chinensis* var. *chinensis* and a member of the *Maculate* such as *A. callosa* Lindl. (Sun and Huang 1994).

Molecular studies support the conclusion that there has been extensive hybridization within *Actinidia*. Chat et al. (2004) concluded that many *Actinidia* species appear to be polyphyletic, and to have undergone reticulate evolution with frequent hybridization in the wild. Their conclusion, that at least one quarter of the *Actinidia* taxa studied had resulted from at least one episode of hybridization, was based on apparent incongruences between the observed and expected patterns of inheritance of mitochondrial and chloroplast DNA. It was assumed that in *Actinidia*, chloroplasts were paternally inherited and mitochondria maternally inherited (Cipriani et al. 1995, Testolin and Cipriani 1997, Chat et al. 1999, Lee et al. 2003). However, other work has shown that inheritance of chloroplasts in *Actinidia* interspecific crosses is not quite so clear cut and although chloroplasts are in most cases inherited paternally, they can, depending on the taxa and the particular genotypes, also be inherited maternally, or even occasionally biparentally (Jung et al. 2003, Li et al. 2013a). Nevertheless, these rare exceptions do not negate the general conclusion and other results are consistent with frequent interspecific hybridization having occurred in the wild (e.g., Cipriani et al. 1998, Li et al. 2002, 2007c). Yao et al. (2015) concluded from a study of exon-primed intron-crossing markers that interspecific hybridization is more likely than incomplete lineage sorting to be the explanation for the incongruent relationships between plastid and nuclear DNA.

Hybridization is still occurring and there can be considerable gene flow between sympatric *Actinidia* taxa (Zhang et al. 2007, Liu et al. 2008, 2010, 2015, Li et al. 2010a, Huang and Liu 2014). Some species maintain their genetic integrity despite gene flow, and this suggests selection against hybrids (Liu et al. 2010). Introgressive hybridization between *A. eriantha* Benth. and *A. latifolia* (Gardn. et Champ.) Merr. seemed to occur only in certain localities. Gene flow was particularly common between sympatric *A. chinensis* var. *chinensis* and *A. chinensis* var. *deliciosa* with complex patterns of ploidy. Tetraploid plants are thought to be hybrids and are intermediate in morphology between diploid *A. chinensis* var. *chinensis* and hexaploid *A. chinensis* var. *deliciosa*. Introgression breeding based on germplasm collected from natural hybrid zones could be a useful addition to traditional breeding programmes for kiwifruit (Huang and Liu 2014). Such programmes would depend, of course, on ready access to wild populations of germplasm. Existing Chinese cultivars selected from the wild come largely from the Xuefeng mountains, Mufu mountains, Qinling mountains and Daba mountains, which are centres of morphological polymorphism and ploidy variation in the *A. chinensis* species complex (Li et al. 2010b).

Tetraploid plants of *A. chinensis* var. *chinensis* also occur in other parts of China (Xiong 1991, 1992, Yan et al. 1994), well away from the natural distribution of *A. chinensis* var. *deliciosa*, suggesting that tetraploid *A. chinensis* is polyphyletic.

Controlled Crossing and Introgression in *Actinidia*

As nearly all *Actinidia* genotypes are male or female, selfing is not usually an issue and successful crossing implies transfer of pollen from a staminate flower on a male plant to a pistillate flower on a female plant. This helps to maintain the high heterozygosity in *Actinidia*. Differences in ploidy can be expected to complicate interspecific crosses or even crosses within a taxon: even crosses at the same ploidy level within a taxon are not always successful. This could be due to many different reasons, including incongruity within the taxon with development of some reproductive isolation.

Interploidy crosses are useful in enhancing the genetic diversity of F_1 hybrids (Li et al. 2014). However, such crosses in *Actinidia* can be difficult. Pollination may fail or the developing seed may abort early in development, possibly because of endosperm failure. Normal endosperm development involves the fusion of two nuclei from the mother and one nucleus from the pollen parent, a combination of two maternal genomes and one paternal genome. Deviations from the normal 2 maternal genomes: 1 paternal genome ratio if the parents are at different ploidy levels often lead to endosperm degeneration. The success of an interploidy cross can depend on which parent has the higher ploidy: it can also be affected by the production of unreduced gametes (Johnston et al. 1980). Pringle (1986) concluded that in *Actinidia* success was more likely if the female parent has the lower ploidy, but this conclusion has not been supported by other workers (Hirsch et al. 2001).

Properly pollinated *Actinidia* fruit contain many hundreds of seeds. Although many interploidy crosses fail, a few viable seed can often result, and these produce plants that are frequently at an initially bewildering array of ploidies. For example, crossing a hexaploid *A. chinensis* var. *deliciosa* female with 24 diploid *A. chinensis* var. *chinensis* males gave viable seedlings in only 17 of the 24 crosses. Most (75%) of the seedlings were at the expected tetraploid level, but there were also plants at the triploid, pentaploid and hexaploid levels, with three possible aneuploids and one mixoploid (Seal et al. 2012). Crossing hexaploid *A. chinensis* var. *deliciosa* females with diploid *A. eriantha* males produced mainly tetraploid seedlings but also some triploids and some pentaploids. Tetraploid offspring can be readily explained by fertilization of $3x$ eggs by $1x$ sperm nuclei from pollen of the diploid males. Triploids could result from parthenocarpic development of ovules of the hexaploid. Any hexaploids could result by pollen contamination or by parthenocarpic development of a $2n$ egg from the hexaploid female, and the pentaploids are likely to have arisen from fertilization by numerically unreduced ($2n$) gametes from the male diploids (Seal et al. 2012). Most (87%) of these pentaploids were male and this distorted the overall sex ratio. The preponderance of males was ascribed to the failure of homologous chromosomes to separate during First Division Restitution of meiosis, resulting in gametes containing two sets of non-sister chromatids and hence both X and Y chromosomes.

Likewise, crosses between a hexaploid *A. chinensis* var. *deliciosa* cultivar and two diploid *A. chinensis* var. *chinensis* males resulted in triploid, tetraploid, pentaploid, heptaploid and octoploid offspring (Rao et al. 2012). A cross between *A. chinensis* var. *chinensis* ($2x$) and *A. melanandra* ($4x$) produced diploids, triploids, tetraploids and hexaploids (Harvey et al. 1995) when embryos were rescued. Uncontrolled crosses between *A. arguta* (4*x*) females and an *A. chinensis* var. *deliciosa* (6*x*) polliniser produced pentaploids, but also some diploid plants which molecular analyses indicated were parthenocarpic (Chat and Dumoulin 1997).

It is possible that in such experiments there could be selective loss of some ploidy levels or of aneuploids during seedling raising because of their lower viability (Zhang et al. 2010). Some plants of unexpected ploidy could result from trace contamination of the pollen used or from extraneous wind-borne pollen when bags were opened for hand pollination (Harvey et al. 1995).

Many other interspecific *Actinidia* crosses have been described. One of the first was the production of the hybrid *A.* × *fairchildii* by pollinating female flowers of *A. arguta* with pollen from *A. chinensis* var. *deliciosa* (Fairchild 1927, Li 1952). Michurin (1949) produced a number of *Actinidia* cultivars, sometimes by hybridizing *A. arguta* and *A. kolomikta*. One of these may be the cultivar known as 'Ananasnaya' (sometimes shortened to 'Anna'), but it is also possibly a different genotype that was introduced into the United States under this name (Strik and Hummer 2006). Other interspecific crosses that have been listed include those of Pringle (1986), Xiong et al. (1987), Wang et al. (1989, 1994), Mu et al. (1990a, b), Ke et al. (1992), Qian and Yu (1992), White and Beatson (1993), Testolin and Costa (1994), An et al. (1995), Harvey et al. (1995), Liang and Mu (1995), Hirsch et al. (2001), Kataoka et al. (2003), Fan et al. (2004), Lee et al. (2004), Beatson et al. (2007), Cho et al. (2007), Guthrie et al. (2007), Zhang et al. (2010), Seal et al. (2012) and Fraser et al. (2013). Some of the crosses listed were very successful with thousands of hybrid plants being raised, e.g., *A. chinensis* var. *chinensis* × *A. eriantha* (Wang et al. 2000), but success often depended on the genotypes used (e.g., as found by Harvey et al. 1992). Many crosses failed. Fruit were not set or the seed failed at various stages (Harvey et al. 1992, Hirsch et al. 2001) so that the fruit that were set contained only a small number of viable seeds. Embryos might not have developed, or the endosperm might have failed and was inadequate to support continued embryo development (Harvey et al. 1992). Embryo rescue can help, particularly when the parents are at the same ploidy level, but is less successful in interploidy crosses (Mu et al. 1990b, 1992, Hirsch et al. 2001).

Although some F_1 hybrids are considered to have immediate commercial potential, e.g., 'Jinyan' (Zhong et al. 2011), further crossing or backcrossing is often necessary and this requires the F_1 hybrids to be fertile (Sybenga 1992). Furthermore, introgression of characters depends on chromosome pairing between parental genomes at meiosis and generally, chromosome pairing in *Actinidia* hybrids is greater the more closely related the parents are. High rates of chromosome pairing can also occur in hybrids resulting from wide crosses, but in such cases, the pairing may be within (autosyndetic) rather than between (allosyndetic) parental genomes (Datson et al. 2006). Knowing the type of chromosome pairing in hybrids helps decisions as to whether the production of F_2 hybrids or backcrossing to one or other of the parents is the better breeding strategy. In at least some *Actinidia* interspecific hybrids (*A. arguta* × *A. melanandra* and *A. macrosperma* C.F. Liang × *A. valvata* Dunn) there was little or no recombination between the different parental genomes; chromosome pairing occured predominantly between chromosomes of the individual parental genomes (Datson et al. 2006), indicating that production of F_2 hybrids is the more promising option.

Pollenizer Ploidy

In commercial kiwifruit orchards fruit are often set by application to female flowers of pollen collected from male vines. Ideally, it might be expected that such pollen would be collected from male plants of the same ploidy level as the fruiting parent, but the male flowers of diploid and tetraploid *A. chinensis* var. *chinensis* are smaller and produce less pollen than those of hexaploid *A. chinensis* var. *deliciosa*. Most kiwifruit pollen available commercially for pollinating *A. chinensis* var. *chinensis* (both diploid and tetraploid) and *A. chinensis* var. *deliciosa* is therefore from selected *A. chinensis* var. *deliciosa* males. As expected, pollination when the male and the fruiting plants are at different ploidy levels can result in extensive seed abortion (Harvey et al. 1992, Seal et al. 2013a), and this can affect aspects of fruit quality other than simply fruit size (Seal et al. 2013a). However, the effects are genotype-dependent: for example, with the diploid cultivar 'Hort16A', pollination with pollen from a tetraploid and a hexaploid actually increased fruit weight over that resulting from pollination from diploid and tetraploid *A. chinensis* var. *chinensis* males (Seal et al. 2013b). This indicates the need to check the behaviour of individual *Actinidia* genotype combinations rather than generalising.

Manipulation of Ploidy in *Actinidia*

It should be possible (Wu 2012) to manipulate ploidy in *Actinidia* through:

- chromosome doubling, either spontaneous or induced, to produce autopolyploids;
- plantlet regeneration from pollen or unfertilized embryos;
- use of numerically unreduced gametes, particularly in interploidy crosses;
- culture of endosperm tissue to produce triploids;
- protoplast fusion to produce allopolyploids.

Most of these methods involve the ploidy screening of many hundreds of plants, preferably by flow cytometry confirmed by some chromosome counts, followed by the selection of the preferred cytotypes (Ferguson et al. 2009).

Chromosome Doubling

Spontaneous doubling of chromosome numbers has been detected in commercial orchards of *A. chinensis* var. *chinensis* 'Hort16A' (a diploid cultivar) by identifying occasional shoots carrying fruit that are much larger than normal (Martin 2005). Some plants derived by grafting these shoots onto rootstocks have been shown by flow cytometry to be mixoploid ($2x + 4x$): others are stable tetraploids. When flowers on such plants were pollinated using pollen from tetraploid *A. chinensis* var. *chinensis* males, all the offspring produced were tetraploid (Ferguson et al. 2009). Such sports could be used to incorporate the excellent fruit characteristics of 'Hort16A' into tetraploid *A. chinensis* var. *chinensis* breeding populations.

Ploidy in *Actinidia* can also be doubled by use of antimitotic agents such as oryzalin or colchicine to disrupt mitosis. Colchicine treatment of somatic tissues from mature vines of *A. chinensis* var. *chinensis* has been successfully combined with use of flow cytometry to identify the autotetraploid plants produced (Wu et al. 2009, 2011). Doubling the number of chromosomes of four female diploid *A. chinensis* var. *chinensis* genotypes resulted in fruit which were, on average, 50–60% larger than fruit of their respective diploid progenitors, a much bigger increase than would be expected by use of vine management techniques (Wu et al. 2012). Such an increase in fruit size implies that autotetraploids of kiwifruit selections with fruit of otherwise inadequate size might have commercial potential in their own right. There were also changes in fruit shape that were stable on vegetative propagation. However, unexplained variation in fruit characteristics exists among each group of autotetraploid regenerants and selection among the regenerants might be required (Wu et al. 2013). Colchicine treatment has also been used to produce octoploid plants of *A. arguta* from tetraploids (Liu et al. 2011), but the consequences for fruit size have not been described.

As well as increasing fruit size, doubling the chromosome number might be used to facilitate interploidy crossing in *Actinidia*. However, doubling the chromosome numbers of male plants will result in two Y chromosomes, which will distort sex ratios in the offspring.

Unreduced Gametes

Unreduced gametes usually arise through irregularities in meiosis (Ramanna and Jacobsen 2003) and they are thought to have been important in the polyploidization and evolution of many plants. Occurrence of unreduced gametes in both male and female plants of *A. chinensis* var. *chinensis* was first inferred from the ploidy analysis of progenies from interploidy crosses and confirmed by observation of dyad, triad and tetrad formation during microsporogenesis (Yan et al. 1997b). Unreduced gametes seem to be produced by many different *Actinidia* genotypes (Seal et al. 2012) and have great potential in transferring desirable characters from lower to higher ploidy levels. Two of the most promising yellow-fleshed tetraploid cultivars of *A. chinensis* var. *chinensis* released in New Zealand are derived from production of 2*n* eggs by diploids crossed with tetraploid males (R.G. Lowe unpublished data). Sexual polyploidization through the production of unreduced gametes has the advantage of increasing heterozygosity and reducing inbreeding depression (Bretagnolle and Thompson 1995). However, unreduced gametes from male plants will produce male offspring.

Production of Haploids

Production of haploids through culture of pollen or of anthers has not so far proved successful in *Actinidia* (Fraser and Harvey 1986, Wu 2015). However, pollination by lethally-irradiated pollen can induce parthenogenesis in *A. chinensis* var. *deliciosa*, resulting in the production of trihaploid plants (Pandey et al. 1990, Chalak and Legave 1996, 1997). *A. chinensis* var. *deliciosa* is amongst the few plants in which the process can lead to the production of mature seed without embryo rescue being required. Such trihaploid kiwifruit plants are usually weak and sterile. They can double in chromosome number either spontaneously or after treatment with antimitotic agents such as colchicine or oryzalin (Chalak and Legave 1996). Such doubled trihaploids do not seem to have been used further and there are no reports of induced parthenogenesis being attempted in *Actinidia* at other ploidy levels, although parthenogenesis might also result from interploidy crosses as described above.

Endosperm Culture

Endosperm tissue from diploid plants pollinated by pollen from a diploid is triploid (Hoshino et al. 2011), and plants regenerated from the endosperm should therefore be triploid. Large numbers of plants have been successfully raised by tissue culture of *Actinidia* endosperm tissue (Gui et al. 1982, Huang et al. 1983, Mu et al. 1990a). The disadvantage is that endosperm tissue often shows a high degree of polyploidy or aneuploidy (Thomas and Chaturvedi 2008), and this is certainly the case with *Actinidia* where considerable chromosomal variation is observed in regenerants from endosperm culture (Huang and Tan 1988, Mu et al. 1992, Gui et al. 1993, Góralski et al. 2005) or from other immature tissues (Ollitraut-Sammarcelli et al. 1994, Hirsch et al. 2001). If triploid *Actinidia* plants are wanted, it is probably more efficient to cross diploid and tetraploid plants.

Protoplast Fusion

Protoplast fusion could allow ploidy levels to be manipulated or, more usefully, allow the genomes of two female plants to be combined directly. Despite many attempts, there has been limited success in regenerating plants from *Actinidia* protoplasts (Mii and Ohashi 1988, Tsai 1988, Cai et al. 1993, Xiao et al. 1993a, b, Oliveira et al. 1994, Derambure and Hirsch 1995, Zhang et al. 1998, Zhu et al. 2001, Gan et al. 2003, Raquel and Oliveira 1996,

Xiao and Hirsch 1996, Hu et al. 1998, Oliveira and Fraser 2005, Wang and Gleave 2012). The main challenges seem to be the preparation of sterile cultures, the mucilage, common in many *Actinidia* species, which makes it difficult to achieve get consistent protoplast preparations, and the lack of satisfactory regeneration systems (Wu 2010). Furthermore, extensive variation in chromosome number can occur during regeneration (He et al. 1995, Zhang et al. 1997). There has so far been only very limited success in fusing *Actinidia* protoplasts (Lindsay et al. 1995, Wu 2010): somatic hybrids between *A. chinensis* var. *chinensis*, $2n = 2x$, and *A. chinensis* var. *deliciosa*, $2n = 6x$ (Xiao and Han 1997), and between *A. chinensis* var. *chinensis*, $2n = 2x$, and *A. kolomikta* (Maxim. et Rupr.) Maxim., $2n = 2x$, (Xiao et al. 2004) have been reported. Some of the putative somatic hybrids were at the expected ploidy level but others were not, possibly because of the regeneration process.

Cultivars Resulting from Interspecific or Interploidy Crosses

Most kiwifruit cultivars are direct selections from the wild or are only one or two generations removed from the wild (Huang and Ferguson 2007): very few have so far resulted from controlled breeding programmes and even fewer from interspecific or interploidy crosses.

'Jinyan' is a yellow-fleshed, late-maturing kiwifruit cultivar described as having large, attractive fruit of good taste (Zhong et al. 2011, 2012). It originated from pollination of a large-fruited selection of *A. eriantha* by pollen of *A. chinensis* var. *chinensis*. It shows no obvious morphological characters of *A. eriantha*, appearing as if it were simply a large-fruited *A. chinensis* var. *chinensis*, but its hybrid origin was confirmed by mitochondrial DNA sequence analyses (Zhong et al. 2012). The plant is tetraploid whereas *A. eriantha* is normally diploid. About 7500 ha of 'Jinyan' have been planted in China (Ferguson 2015) and this makes 'Jinyan' the first interspecific *Actinidia* cross to be grown on a substantial commercial scale.

Some kiwiberry cultivars have resulted from interspecific crosses, mainly aiming to introduce red skins and red flesh. 'Hortgem Rua' (Horticulture and Food Research Institute of New Zealand 2004) and 'Ken's Red' (Strik 2005) result from crosses between *A. arguta* and *A. melanandra*. A number of cultivars are also described as being hybrids between *A. arguta* and *A. purpurea*, e.g., 'Figurnaja', 'Kijewskaja Gibrydnaja' and 'Kijewskaja Krupnopłodnaja' (Bieniek 2012) and 'Bingo' (Latocha 2012). However, *A. purpurea* is now treated as *A. arguta* (Li et al. 2007a, b) so these should now be considered as intraspecific crosses, although possibly interploidy crosses depending on the ploidy of the parents as although most genotypes of *A. arguta* are tetraploid, there were tetraploid and octoploid cytotypes in what was previously classified as *A. arguta* var. *purpurea*.

'Zesh004' is a hybrid, the product of an intraspecific but interploidy cross between *A. chinensis* var. *chinensis* and *A. chinensis* var. *deliciosa*. 'Zesh004' (commonly known as Green14: the fruit marketed as Zespri® Sweet Green

Kiwifruit), is being grown commercially on a small scale in New Zealand, France and Italy, with about 300 ha planted in New Zealand (KVH 2015). It is being promoted because of the perceived sweetness of the fruit, but whether it has sufficient marketing advantages over the existing green cultivar 'Hayward' has yet to be determined (Zespri 2015). The Chinese cultivar 'Huayou' is also thought to be a hybrid between *A. chinensis* var. *chinensis* and *A. chinensis* var. *deliciosa*, but is closer in appearance to *A. chinensis* (Ferguson et al. 2012).

Several new cultivars from New Zealand have resulted from intraspecific interploidy crosses involving production of unreduced gametes. The most successful of these is *A. chinensis* var. *chinensis* 'Zesy002' (commonly known as Gold3: the fruit marketed as Zespri® SunGold Kiwifruit). Its fruit have a light yellow flesh but its main advantage is its comparative resistance to Psa. By August 2015, nearly 4000 ha had been planted in New Zealand and 7,000 ha worldwide as a replacement for the much more Psa-susceptible 'Hort16A'. Already annual exports of yellow-fleshed kiwifruit from New Zealand have surpassed the pre-Psa exports of 'Hort16A'. The second cultivar whose breeding also involves production of unreduced gametes is *A. chinensis* var. *chinensis* 'Zesy003' (commonly known as Gold9: the fruit marketed as Zespri® Charm Kiwifruit). About 160 ha are currently planted in New Zealand, but its commercial prospects are currently being re-assessed (Zespri 2015).

Future Prospects

The cultivars that have so far emerged from interploidy or interspecific crosses, 'Jinyan', 'Zesh004', 'Zesy002' and 'Zesy003', can be considered as variants of the existing yellow-fleshed and green-fleshed types of kiwifruit. They are readily recognized by consumers as kiwifruit and they can be sold as such, no matter how they originated. They will probably be joined by further useful variants of "standard" kiwifruit produced by interploidy or interspecific crosses.

Truly novel types of kiwifruit, such as 'Kiri', from the cross between *A. arguta* and *A. chinensis* var. *deliciosa* backcrossed to *A. chinensis* var. *deliciosa* (White and Beatson 1993), have so far had little success. It may be that such thin-skinned fruit are too difficult to handle satisfactorily or have too short a storage life for large-scale production and extended marketing. It may be that that the management and handling systems so far developed are suitable for only certain types of fruit with thicker skins.

However, fruit characteristics are not the only features that are selected for. Real progress could be made in using interploidy or interspecific crosses to introgress other vine characteristics. As an example, dioecism is a major constraint to kiwifruit production (Huang and Ferguson 2007). Hermaphroditism is stable and inheritable in the promising hermaphrodite breeding lines of *A. chinensis* var. *deliciosa* derived originally from fruiting males (McNeilage and Steinhagen 1998, McNeilage et al. 2007). The main limitation to introgressing this character into other *Actinidia* species is that *A. chinensis* var. *deliciosa* is usually hexaploid. Endodormancy and inadequate winter chilling restrict yields in many kiwifruit cultivars. *A. rufa* has a lower chilling requirement for breaking endodormancy than *A. chinensis* var. *chinensis* and *A. chinensis* var. *deliciosa* (Phivnil et al. 2004) and had useful drought and waterlogging tolerance when it was used as a rootstock for *A. chinensis* var. *deliciosa* (Nagata et al. 1997). These characteristics might also be usefully introgressed, especially as both diploid and tetraploid cytotypes of *A. rufa* have been reported (Matsumoto et al. 2013). Diploid *A. rufa* has high cross compatibility with diploid *A. chinensis* var. *chinensis* (Matsumoto et al. 2011): tetraploid forms of *A. rufa* may show similar compatibility with tetraploid *A. chinensis* var. *chinensis*.

Breeding kiwifruit is challenging. Results are often unpredictable and, as shown by interploidy crosses, the unexpected should always be expected. Nevertheless there is great genetic diversity in the wild germplasm, diversity that is the starting point for breeding programmes. Success will depend on a good understanding of the reproductive biology of *Actinidia* and an awareness of the commercial requirements of new cultivars.

Acknowledgements

We thank L.G. Fraser, S. Z. Hanley, M.A. McNeilage and A.G. Seal for helpful comments.

References

An, H.-X., D.-R. Cai, X.-J. Mu, B. Zheng and Q.G. Shen. 1995. New germplasm of interspecific hybridisation in *Actinidia*. Acta Hort. Sin. 22: 133-137.

Beatson, R.A., P.M. Datson, P.M. Harris-Virgin and L.T. Graham. 2007. Progress in the breeding of novel interspecific *Actinidia* hybrids. Acta Hort. 753: 147-151.

Bieniek, A. 2012. Yield, morphology and biological value of fruits of *Actinidia arguta* and *Actinidia purpurea* and some of their hybrid cultivars grown in north-eastern Poland. Acta Sci. Pol., Hortorum Cultus 11(3): 117-130.

Boase, M.R. and M.E. Hopping. 1995. DNA dodecaploid plants detected among somaclones of *Actinidia deliciosa* var. *deliciosa* cv. Hayward. Plant Cell Rep. 14: 319-323.

Bretagnolle, F. and J.D. Thompson. 1995. Gametes with the somatic chromosome number: mechanisms of their formation and role in the evolution of autopolyploid plants. New Phytol. 129: 1-22.

Cai, Q.-G. [Tsai, C.-K.], Y.-Q. Qian [Y.-C. Chien], S.-Q. Ke and Z.-C. He. 1993. Regeneration of plants from protoplasts of kiwifruit (*Actinidia deliciosa*). pp. 3-17. *In*: Y.P.S. Bajaj (ed.). Plant Protoplasts and Genetic Engineering IV. Biotechnology in Agriculture and Forestry Vol. 23. Springer-Verlag, Berlin, Heidelberg.

Chalak, L. and J.M. Legave. 1996. Oryzalin combined with adventitious regeneration for an efficient chromosome doubling of trihaploid kiwifruit. Plant Cell Rep. 16: 97-100.

Chalak, L. and J.M. Legave. 1997. Effects of pollination by irradiated pollen in Hayward kiwifruit and spontaneous doubling of induced parthenogenetic trihaploids. Scientia Hort. 68: 83-93.

Chat, J., L. Chalak and R.J. Petit. 1999. Strict paternal inheritance of chloroplast DNA and maternal inheritance of mitochondrial DNA in intraspecific crosses of kiwifruit. Theor. Appl. Genet. 99: 314-322.

Chat, J. and P.Y. Dumoulin. 1997. Flow cytometry and RAPD markers applied to parentage analysis within *Actinidia* genus. Acta Hort. 444: 109-111.

Chat, J., B. Jáuregui, R.J. Petit and S. Nadot. 2004. Reticulate evolution in kiwifruit (*Actinidia*, Actinidiaceae) identified by comparing their maternal and paternal phylogenies. Am. J. Bot. 91: 736-747.

Cho, H.S., Y.S. Jo, I.S. Liu and C.S. Ahn. 2007. Characteristics of *Actinidia deliciosa* × *A. arguta* and *A. arguta* × *A. deliciosa* hybrids. Acta Hort. 753: 205-209.

Cipriani, G., R. Testolin and R. Gardner 1998. Restriction-site variation of PCR-amplified chloroplast DNA regions and its implication for the evolution and taxonomy of *Actinidia*. Theor. Appl. Genet. 96: 389-396.

Cipriani, G., R. Testolin and M. Morgante. 1995. Paternal inheritance of plastids in interspecific hybrids of the genus *Actinidia* revealed by PCR-amplification of chloroplast DNA fragments. Mol. Gen. Genet. 247: 693-697.

Datson, P., R. Beatson and T. Harris-Virgin. 2006. Meiotic chromosome behaviour in interspecific *Actinidia* hybrids. pp. 869-874. *In*: C.F. Mercer (ed.), Breeding for Success: Diversity in Action. Proc. 13th Australasian Plant Breeding Conf., Christchurch, 18-21 April, 2006.

Derambure, A. and A.-M. Hirsch. 1995. Obtention de protoplastes à partir de différents clones *d'Actinidia deliciosa* (Kiwi) et d'espèces botaniques du genre *Actinidia* résistantes au froid. Acta Bot. Gallica: Bot. Lett. 142: 5-21.

Dunn, S.T. 1911. A revision of the genus *Actinidia*, Lindl. J. Linn. Soc. London, Bot. 39: 390-410.

Fairchild, D. 1927. The fascination of making a plant hybrid being a detailed account of the hybridization of *Actinidia arguta* and *Actinidia chinensis*. J. Hered. 18: 49-62.

Fan, P.-G., H.-X. An, D.-R.Cai and X.-J. Mu. 2004. Interspecific hybridization between species of *Actinidia* L. and breeding of superior selection. J. Fruit Sci. 21: 208-211.

Ferguson, A.R. 1984. Kiwifruit: a botanical review. Hort. Rev. 6: 1-64.

Ferguson, A.R. 1990a. Kiwifruit (*Actinidia*). Acta Hort. 290: 603-653.

Ferguson A.R. 1990b. Botanical nomenclature: *Actinidia chinensis*, *Actinidia deliciosa*, and *Actinidia setosa*. pp. 36-57 + 2 plates. *In*: I.J. Warrington and G.C. Weston (eds). Kiwifruit: Science and Management. Ray Richards Publ. and New Zealand Soc. Hort. Sci., Auckland.

Ferguson, A.R. 2015. Kiwifruit in the world - 2014. Acta Hort. 1096: 33-46.

Ferguson, A.R., H.-W. Huang and R. Testolin. 2012. Kiwifruit. *In*: C.E. Finn and J.R. Clark(eds). Register of new fruit and nut cultivars, list 46. HortScience 47: 547-550.

Ferguson, A.R., I.E.W. O'Brien and G.-J. Yan. 1997. Ploidy in *Actinidia*. Acta Hort. 444: 67-71.

Ferguson, A.R.: J.-L. Zhang, A.M. Duffy, R.A. Beatson, C.-H. Cheng, P.M. Datson, R.G. Lowe, M.A. McNeilage, P.M. Harris-Virgin, A.G. Seal and J.-H. Wu. 2009. Ploidia e uso della citometria a flusso nel miglioramento genetico dell'actinidia. *Italus Hortus* **16**(5) (Atti del IX Convegno Nazionale dell'Actinidia, Viterbo-Latina, 6-8 October, 2009): 78-83.

Fraser, L.G. and C.F. Harvey. 1986. Somatic embryogenesis from anther-derived callus in two *Actinidia* species. Scientia Hort. 29: 335-346.

Fraser, L.G., M.A. McNeilage, G.K. Tsang, C.F. Harvey and H.N. De Silva. 2005. Cross-species amplification of microsatellite loci within the dioecious, polyploid genus *Actinidia* (Actinidiaceae). Theor. Appl. Genet. 112: 149-157.

Fraser, L.G., A.G. Seal, M. Montefiori, T.K. McGhie, G.K. Tsang, P.M. Datson, E. Hilario, H.E. Marsh, J.K. Dunn, R.P. Hellens, K.M. Davies, M.A. McNeilage, H.N. De Silva and A.C. Allan. 2013. An R2R3 MYB transcription factor determines red petal colour in an *Actinidia* (kiwifruit) hybrid population. BMC Genom. 14: 28.

Gamble, J., L. Axten, M. Wohlers, I. Hallett and A. Seal. 2010. The sensory profiles of kiwifruit hybrids involving *Actinidia eriantha*. Acta Hort. 880: 47-53.

Gan, L., X. Xiong, R. Wang, J.B. Power and M.R. Davey. 2003. Plant regeneration from cell suspension protoplasts of *Actinidia deliciosa*. Acta Hort. 610: 197-202.

Gao, X.-Z. and M. Xie. 1990. A survey of recent studies on *Actinidia* species in China. Acta Hort. 282: 43-52.

Goldblatt, P. 1980. Polyploidy in Angiosperms: Monocotyledons. pp. 219-239. *In*: W.H. Lewis (ed.). Polyploidy: Biological Relevance. Plenum Press, New York.

Góralski, G., M. Popielarska, H. Ślesak, D. Siwińska and M. Batycka. 2005. Organogenesis in endosperm of *Actinidia deliciosa* cv. Hayward cultured in vitro. Acta Biologica Cracoviensia Ser. Bot. 47(2): 121-128.

Grant, V. 1963. The Origin of Adaptations. Columbia Univ. Press, New York.

Gui, Y., S. Hong, S. Ke and R.M. Skirvin. 1993. Fruit and vegetative characteristics of endosperm-derived kiwifruit (*Actinidia chinensis* F) plants. Euphytica 71: 57-62.

Gui, Y.-L., X.-J. Mu and T.-Y. Xu. 1982. Studies on morphological differentiation of endosperm plantlets of Chinese gooseberry *in vitro*. Acta Bot. Sin. 24: 216-221.

Guthrie, R.S., J.J. Luby, D.S. Bedford and S.T. McNamara. 2007. Partial dominance in *Actinidia kolomikta* interspecific hybrids. Acta Hort. 753: 211-218.

Harvey, C.F., L.G. Fraser and J. Kent. 1992. Actinidia seed development in interspecific crosses. Acta Hort. 297: 71-78.

Harvey, C.F., L.G. Fraser, J. Kent, S. Steinhagen, M.A. McNeilage and G.-J. Yan. 1995. Analysis of plants obtained by embryo rescue from an interspecific *Actinidia* cross. Scientia Hort. 60: 199-212.

He, Z.-C., Q.-G. Cai, S.-Q. Ke, Y.-Q. Qian and L.-M. Xu. 1995. Cytogenetic studies on regenerated plants derived from protoplasts of *Actinidia deliciosa*. I. Variation of chromosome number of somatic cells. J. Wuhan. Bot. Res. 13(2): 97-101.

He, Z.-C., J.-Q. Li, Q. Cai and Q. Wang. 2005. The cytology of *Actinidia, Saurauia* and *Clematoclethra* (Actinidiaceae). Bot. J. Linn. Soc. 147: 369-374.

He, Z.-C., S.-M. Wang, H. Huang and H.-Q. Huang. 1998. Study on chromosome numbers of 6 species and 1 variety in *Actinidia* Lindl. J. Wuhan Bot. Res. 16: 299-301.

Hirsch, A.M., R. Testolin, S. Brown, J. Chat, D. Fortune, J.M. Bureau and D. De Nay. 2001. Embryo rescue from interspecific crosses in the genus *Actinidia* (kiwifruit). Plant. Cell Rep. 20: 508-516.

Horticulture and Food Research Institute of New Zealand, Ltd. 2004. Kiwi plant named 'Hortgem Rua'. US Plant Patent 14625 P3.

Hoshino, Y., T. Miyashita and T.D. Thomas. 2011. In vitro culture of endosperm and its application in plant breeding: Approaches to polyploidy breeding. Scientia Hort. 130: 1-8.

Hsieh T.-Y., S.-M. Ku, C.-T. Chien and Y.-T. Liou. 2011. Classifier modeling and numerical taxonomy of *Actinidia* (Actinidiaceae) in Taiwan. Bot. Studies 52: 337-357.

Hu, J., X. Xiong, Q. Zhang and L. Gan. 1998. Studies on technology of protoplast culture and plant regeneration of *Actinidia deliciosa*. J. Hunan Agri. Univ. 24: 184-190.

Huang, H.-W., F. Dane, Z.-Z. Wang, Z.-W. Jiang, R.-H. Huang and S.-M. Wang. 1997. Isozyme inheritance and variation in *Actinidia*. Heredity 78: 328-336.

Huang, H.[-W.] and A.R. Ferguson. 2001. Kiwifruit in China. New Zealand J. Crop Hort. Sci. 29: 1-14.

Huang, H-W. and A.R. Ferguson. 2007 [incorrectly published as Ferguson, A.R. and H.-W. Huang] Genetic resources of kiwifruit: domestication and breeding. Hort. Rev. 33: 1-121.

Huang, H.-W. and Y.-F. Liu. 2014. Natural hybridization, introgression breeding, and cultivar improvement in the genus *Actinidia*. Tree Genet. Genomes 10: 1113-1122.

Huang, H.-W., Y. Wang, Z.-H. Zhang, Z.-W. Jiang and S.-M. Wang. 2004. *Actinidia* germplasm resources and kiwifruit industry in China. HortScience 39: 1165-1172.

Huang H.-W., C.-H. Zhong, Z.-W. Jiang, X.-W. Li, D.-W. Li, Y.-F. Liu, Z.-Z. Li, X.-H. Yao, S.-M. Wang, J.-J. Gong, T. Shi, L. Zhang, S.-W. Jia and B. Chen. 2014. The Genus *Actinidia*, a World Monograph. Science Press, Beijing.

Huang, W.-G., G. Cipriani, M. Morgante and R. Testolin. 1998. Microsatellite DNA in *Actinidia chinensis*: isolation, characterisation, and homology in related species. Theor. Appl. Genet. 97: 1269-1278.

Huang, Z.-G., Y.-L. Huangfu and L.-Y. Xu. 1983. Triploid plantlets of Chinese gooseberry obtained by tissue culture. Kexue Tongbao 28: 112-117.

Huang, Z.-G. and C.-Y. Tan. 1988. Chinese gooseberry, kiwifruit (*Actinidia* spp.). pp. 166-180. *In*: Y.P.S. Bajaj (ed.). Biotechnology in Agriculture and Forestry, Vol. 6 Crops II. Springer-Verlag, Berlin, Heidelberg.

Jaeger, S.R., K.L. Rossiter, W.V. Wismer and F.R. Harker. 2003. Consumer-driven product development in the kiwifruit industry. Food Qual. Pref. 14: 187-198.
Johnston, S.A., T.P.M. den Nijs, S.J. Peloquin and R.E. Hanneman, Jr. 1980. The significance of genic balance to endosperm development in interspecific crosses. Theor. Appl. Genet. 57: 5-9.
Jung, Y.-H., S.-C. Kim, M. Kim, K.-H. Kim, H.-M. Kwon and M.-Y. Oh. 2003. Chloroplast inheritance patterns in *Actinidia* hybrids determined by single stranded conformation polymorphism analysis. Mol. Cells 15: 277-282.
Kataoka, I., K. Kokudo, K. Beppu, T. Fukuda, S. Mabuchi and K. Suezawa. 2003. Evaluation of characteristics of *Actinidia* inter-specific hybrid 'Kosui'. Acta Hort. 610: 103-108.
Kataoka, I., T. Mizugami, J.G. Kim, K. Beppu, T. Fukuda, S. Sugahara, K. Tanaka, H. Satoh and K. Tozawa. 2010. Ploidy variation of hardy kiwifruit (*Actinidia arguta*) resources and geographic distribution in Japan. Scientia Hort. 124: 409-414.
Kawagoe, T. and N. Suzuki. 2004. Cryptic dioecy in *Actinidia polygama*: a test of the pollinator attraction hypothesis. Can. J. Bot. 82: 214-218.
Ke, S.-Q., R.-H. Huang, S.-M. Wang, Z.-T. Xion[g] and Z.-W. Wu. 1992. Studies on interspecific hybrids of *Actinidia*. Acta Hort. 297: 133-139.
Kuroda, K., K. Sakai, T. Haji and A. Wakana. 2011. Geographical diversity in ploidy level and chloroplast DNA of Japanese *Actinidia*. Acta Hort. 913: 175-179.
KVH. 2015. Kiwifruit Vine Health Inc. http: //www.kvh.org.nz/statistics (accessed 8 September 2015).
Lai, J.-J., Z.-Z. Li,Y.P. Man, R. Lei and Y-.C. Wang. 2015. Genetic diversity of five wild *Actinidia arguta* populations native to China as revealed by SSR markers. Scientia Hort 191: 101–107.
Latocha, P. 2012. Some morphological and biological features of 'Bingo' – a new hardy kiwifruit cultivar from Warsaw University of Life Sciences (WULS) in Poland. Rocznik Polskiego Towarzystwa Dendrologicznego 60: 61-67.
Lee, B.S., M.H. Lee, S.I. Hwang, S.C. Kim, S.S. Han and U. Lee. 2004. Characteristics of F1 seeds of the cross between *A. arguta* and *A. chinensis*. Korean J. Breed. 36: 345-349.
Lee, B.S., E.W. Noh, J.S. Lee, S.I. Hwang, M.H. Lee and J.S. Li. 2003. Paternal inheritance of plastid genome in the interspecific cross between *Actinidia arguta* and *A. deliciosa*. Kor. J. Breed. 35: 209-212.
Li D.-W., Y.-F. Liu, X.-W. Li, J.-Y. Rao, X.-H. Yao and C.-H. Zhong. 2014. Genetic diversity in kiwifruit polyploid complexes: insights into cultivar evaluation, conservation, and utilization. Tree Genet. Genom. 10: 1451-1463.
Li, D.-W., Y.-F. Liu, C.-H. Zhong and H.-W. Huang. 2010a. Morphological and cytotype variation of wild kiwifruit (*Actinidia chinensis* complex) along an altitudinal and longitudinal gradient in central-west China. Bot. J. Linn. Soc. 164: 72-83.
Li, D.-W., X.-Q. Qi, X.-W. Li, L. Li, C.-H. Zhong and H.-W. Huang. 2013a. Maternal inheritance of mitochondrial genomes and complex inheritance of chloroplast genomes in *Actinidia* Lind.: evidences from interspecific crosses. Mol. Genet. Genomics 288: 101-110.
Li, D.-W, C.-H. Zhong, Y.-F. Liu and H.-W. Huang. 2010b. Correlation between ploidy level and fruit characters of the main kiwifruit cultivars in China: implication for selection and improvement. New Zealand J. Crop Hort. Sci. 38: 137-145.
Li, H.-L. 1952. A taxonomic review of the genus *Actinidia*. J. Arnold Arb. 33: 1-61.
Li, J.-Q., H.-W. Huang and T. Sang. 2002. Molecular phylogeny and infrageneric classification of *Actinidia* (Actinidiaceae). Syst. Bot. 27: 408-415.
Li, J.-Q., X.-W. Li and D.D. Soejarto. 2007a. Actinidiaceae. pp. 334-360. *In*: Z.-Y. Wu, P.H. Raven and D.Y. Hong (eds). Flora of China, Vol. 12, Science Press, Beijing, Missouri Botanical Garden, St Louis.
Li, X.-W., J.-Q. Li and D.D. Soejarto. 2009. Advances in the study of the systematics of *Actinidia* Lindley. Front. Biol. China 4: 55-61.
Li, X.-W., J.-Q. Li and D.D. Soejarto. 2007b. New synonyms in Actinidiaceae from China. Acta Phytotaxon. Sin. 45: 633-660.
Li, X.W., X. Li, J.Q. Li and D.D. Soejarto 2011. Progress in the phylogeny and taxonomy of *Actinidia* during the past decade. Acta Hort. 913: 71-76.

Li, Z.-Z., M. Kang, H.-W. Huang, R. Testolin, Z.-W. Jiang, J.-Q. Li, Y. Wang and G. Cipriani. 2007c. Phylogenetic relationships in *Actinidia* as revealed by nuclear DNA genetic markers and cytoplasmic DNA sequence analysis. Acta Hort. 753: 45-58.

Li, Z.-Z., Y.-P. Man, X.-Y. Lan and Y.-C. Wang. 2013b. Ploidy and phenotype variation of a natural *Actinidia arguta* population in the east of Daba Mountain located in a region of Shaanxi. Scientia Hort. 161: 259-265.

Liang, C.-F. 1982a. New taxa of the genus *Actinidia* Lindl. Guihaia 2: 1-6.

Liang, C.-F. 1982b. An addition to the infraspecific taxa of *Actinidia chinensis* Planch. Acta Phytotaxon. Sin. 20: 101-104.

Liang, C.-F. 1983. On the distribution of Actinidias. Guihaia 3: 229-248.

Liang, C.-F. 1984. *Actinidia*. pp. 196-268, 309-324. *In*: K.-M. Feng (ed.). Flora Reipublicae Popularis Sinicae, 49/2. Science Press, Beijing.

Liang, T.-B. and X.-J. Mu. 1995. Observation of pollen tube behaviour and early embryogenesis following interspecies pollination between *Actinidia deliciosa* and *A. arguta*. Acta Bot. Sin. 37: 607-612.

Lindsay, G.C., M.E. Hopping, H. Binding and G.K. Burge. 1995. Graft chimeras and somatic hybrids for new cultivars. New Zealand J. Bot. 33: 79-92.

Liu, C., X. Sun, H. Dai and Z. Zhang. 2011. In vitro induction of octoploid plants from tetraploid *Actinidia arguta*. Acta Hort. 913: 185-190.

Liu, Y.F., D.W. Li and H. Huang. 2011. Cytotype variation in the natural populations of *Actinidia chinensis* complex in China. Acta Hort. 913: 157-162.

Liu, Y.-F., D.-W Li, L. Yan and H.-W. Huang. 2015. The microgeographical patterns of morphological and molecular variation of a mixed ploidy population in the species complex *Actinidia chinensis*. PLoS ONE 10(2): e0117596.

Liu Y.-F., Y.-L. Liu and H.-W. Huang. 2010. Genetic variation and natural hybridization among sympatric *Actinidia* species and the implications for introgression breeding of kiwifruit. Tree Genet. Genomes 6: 801-813.

Liu, Y.-L., Z.-Z. Li, Z.-W. Jiang, Y.-F. Liu and H.W. Huang. 2008. Genetic structure and hybridization introgression in natural populations of two closely related *Actinidia* species, *A. chinensis* and *A. deliciosa*. Chinese J. Plant Ecol. 32: 704-718.

Lu, L.-X., M.-C. Tao and Y.-X. Pan. 1984. Observation on the chromosome and meioses of pollen mother cells in "Maohua-Gooseberry" (*Actinidia eriantha* Benth.). Fujian Nongxueyuan Xuebao (J. Fujian Agr. Coll.) 13: 25-30.

McGregor, S.E. 1976. Insect pollination of cultivated crop plants. Agr. Handb. 496, US Dept. Agriculture.

McNeilage, M.A. 1991. Gender variation in *Actinidia deliciosa,* the kiwifruit. Sex. Plant Reprod. 4: 267-273.

McNeilage, M.A. and J.A. Considine. 1989. Chromosome studies in some *Actinidia* taxa and implications for breeding. New Zealand J. Bot. 27: 71-81.

McNeilage, M.A., A.M. Duffy, L.G. Fraser, H.D. Marsh and B.J. Hofstee. 2007. All together now: the development and use of hermaphrodite breeding lines in *Actinidia deliciosa*. Acta Hort. 753: 191-197.

McNeilage, M.A. and S. Steinhagen. 1998. Flower and fruit characters in a kiwifruit hermaphrodite. Euphytica 101: 69-72.

Martin, P. 2005. Spot the difference. New Zealand Kiwifruit J. May/June 2005: 12-15.

Martin, R.A. and P. Luxton. 2005. The successful commercialisation of ZESPRI™ Gold Kiwifruit. Acta Hort. 694: 35-40.

Matsumoto, H., K. Beppu and I. Kataoka. 2013. Identification of hermaphroditism and self-fruitfulness in the wild *Actinidia* found in warm region of Japan. Hort. Res. (Japan) 12: 361-366.

Matsumoto, H., T. Seino, K. Beppu, K. Suezawa, T. Fukuda and I. Kataoka. 2011. Characteristics of interspecific hybrids between *Actinidia chinensis* kiwifruit and *A. rufa* native to Japan. Acta Hort. 913: 191-196.

Messina, R., R. Testolin and M. Morgante. 1991. Isozymes for cultivar identification in kiwifruit. HortScience 26: 899-902.

Michurin, I.V. 1949. Selected works. Foreign Languages Publishing House, Moscow.

Mii, M. and H. Ohashi. 1988. Plantlet regeneration from protoplasts of kiwifruit. *Actinidia chinensis* Planch. Acta Hort. 230: 167-170.

Mizugami,T., J.G. Kim, K. Beppu, T. Fukuda and I. Kataoka. 2007. Observation of parthenocarpy in *Actinidia arguta* selection 'Issai'. Acta Hort 753: 199-203.

Mu, S.K. [Mu, X.-J.], L.G. Fraser and C.F. Harvey. 1990a. Initiation of callus and regeneration of plantlets from endosperm of *Actinidia* interspecific hybrids. Scientia Hort. 44: 107-117.

Mu, S.K.[Mu, X.-J.], L.G. Fraser and C.F. Harvey. 1990b. Rescue of hybrid embryos of *Actinidia* species. Scientia Hort. 44: 97-106.

Mu, X.-J., D.-R. Tsai, H.-X. An and W.-L. Wang. 1992. Embryology and embryo rescue of interspecific hybrids in *Actinidia*. Acta Hort. 297: 93-97.

Mu, X.-J., W.-L. Wang, D.-R. Cai and H.-X. An. 1990b. Embryology and embryo rescue of an interspecific cross between *Actinidia deliciosa* cv. Hayward and *A. eriantha*. Acta Bot. Sin. 32: 425-431.

Nagata, K., K. Morinaga and F. Ikeda. 1997. Photosynthetic ability of kiwifruit (*Actinidia deliciosa*) and characteristics of Shima-sarunashi (*Actinidia rufa*) rootstock. Bull. Shikoku Natl. Agric. Exp. Stn. No. 61: 159-165.

Nishiyama, I. 2007. Fruits of the *Actinidia* genus. Adv. Food Nutr. Res. 52: 293-324.

Oliveira, M.M., J.G. Barrosa, M. Martins and M.S. Pais. 1994. II.1. Genetic transformation in *Actinidia deliciosa* (kiwifruit). pp. 193-214. *In*: Y.P.S. Bajaj (ed.). Plant Protoplasts and Genetic Engineering V. Biotechnology in Agriculture and Forestry, vol. 29. Springer-Verlag, Berlin, Heidelberg.

Oliveira, M.M. and L.G. Fraser. 2005. 1.1 *Actinidia* spp. Kiwifruit. pp. 2-27. *In*: R.E. Litz (ed.). Biotechnology of Fruit and Nut Crops. (Biotechnology in Agriculture No. 29). CABI Publ., Wallingford, UK.

Ollitraut-Sammarcelli, F., J.M. Legave, N. Michaux-Ferrière and A.M. Hirsch. 1994. Use of flow cytometry for rapid determination of ploidy level in the genus *Actinidia*. Scientia Hort. 57: 303-313.

Pandey, K.K., L. Przywara and P.M. Sanders. 1990. Induced parthenogenesis in kiwifruit (*Actinidia deliciosa*) induced through the use of lethally irradiated pollen. Euphytica 51: 1-9.

Phivnil, K., K. Beppu, R. Mochioka, T. Fukuda and I. Kataoka. 2004. Low-chill trait for endodormancy completion in *Actinidia arguta* Planch. (Sarunashi) and *A. rufa* Planch. (Shima-sarunashi), indigenous *Actinidia* species in Japan and their interspecific hybrids. J. Japan. Soc. Hort. Sci. 73: 244-246.

Phivnil, K., K. Beppu, T. Takamura, T. Fukuda and I. Kataoka. 2005. Flow cytometric assessment of ploidy in native resources of *Actinidia* in Japan. J. Am. Pomol. Soc. 59: 44-49.

Pringle, G.J. 1986. Potential for hybridization in the genus *Actinidia*. Plant Breeding Symposium DSIR. Agron Soc. New Zealand Special Publn. 5: 365-368.

Qian, Y.-Q. and D.-P. Yu. 1992. Advances in *Actinidia* research in China. Acta Hort. 297: 51-55.

Ramanna, M.S. and E. Jacobsen. 2003. Relevance of sexual polyploidization for crop improvement – A review. Euphytica 133: 3-18.

Rao, J.-Y., Y.-F. Liu and H.-W. Huang. 2012. Analysis of ploidy segregation and genetic variation of progenies of different interploidy crosses in *Actinidia chinensis*. Acta Hort. Sin. 39: 1447-1456.

Raquel, M.H. and M.M. Oliveira 1996. Kiwifruit leaf protoplasts competent for plant regeneration and direct DNA transfer. Plant Sci. 121: 107-114.

Schmid, R. 1978. Reproductive anatomy of *Actinidia chinensis* (Actinidiaceae). Bot. Jahrb. Syst. Pflanzengesch. Pflanzengeogr. 100: 149-195.

Seal, A.G., J.K. Dunn, H.N. De Silva, T.K. McGhie and R.C.M. Lunken. 2013a. Choice of pollen parent affects red flesh colour in seedlings of diploid *Actinidia chinensis* (kiwifruit). New Zealand J. Crop Hort. Sci. 41: 207-218.

Seal, A.G., J.K. Dunn and Y.L. Jia. 2013b. Pollen parent effects on fruit attributes of diploid *Actinidia chinensis* 'Hort16A' kiwifruit. New Zealand J. Crop Hort. Sci. 41: 219-229.

Seal, A.G., A.R. Ferguson, H.N. de Silva and J.-L. Zhang 2012. The effect of 2*n* gametes on sex ratios in *Actinidia*. Sex. Plant Reprod. 25: 197-203.

Shi, T., H.-W. Huang and M.S. Barker. 2010. Ancient genome duplications during the evolution of kiwifruit (*Actinidia*) and related Ericales. Ann. Bot. 106: 497-504.

Strik, B. 2005. Growing kiwifruit. PNW 507. Pacific Northwest Extension Publication. Oregon State University, University of Idaho, Washington State University.

Strik, B.C. and K.E. Hummer. 2006. 'Ananasnaya' hardy kiwifruit. J. Amer. Pomol. Soc. 60: 106-112.

Sun, H.-M. and R.-H. Huang. 1994. *Actinidia hubeiensis* – a new species of *Actinidia*. J. Wuhan Bot. Res. 12: 321-323.

Sybenga, J. 1992. Cytogenetics in Plant Breeding. Springer-Verlag, Berlin, Heidelberg.

Tang, S.-X. and S.-F. Jiang. 1995. Detection and observation on the bud mutation of *Actinidia chinensis* Planch. Acta Hort. 403: 71-73.

Testolin, R. and G. Cipriani. 1997. Paternal inheritance of chloroplast DNA and maternal inheritance of mitochondrial DNA in the genus *Actinidia*. Theor. Appl. Genet. 94: 897-903.

Testolin, R. and G. Costa 1994. Il miglioramento genetico dell'actinidia. Riv. Frutticolt. Ortofloricolt. 56(1): 31-42.

Testolin, R. and A.R. Ferguson. 1997. Isozyme polymorphism in the genus *Actinidia* and the origin of the kiwifruit genome. Syst. Bot. 22: 685-700.

Testolin, R. and A.R. Ferguson. 2009. Kiwifruit (*Actinidia* spp.) production and marketing in Italy. New Zealand. J. Crop Hort. Sci. 37: 1-32.

Testolin, R., W.-G. Huang and G. Cipriani. 1999. Towards a linkage map in kiwifruit (*Actinidia chinensis* Planch.) based on microsatellites and saturated with AFLP markers. Acta Hort. 498: 79-84.

Testolin, R., W.G. Huang, O. Lain, R. Messina, A. Vecchione and G. Cipriani. 2001. A kiwifruit (*Actinidia* spp.) linkage map based on microsatellites and integrated with AFLP markers. Theor. Appl. Genet. 103: 30-36.

Thomas, T.D. and R. Chaturvedi. 2008. Endosperm culture: a novel method for triploid plant production. Plant Cell Tiss. Organ Cult. 93: 1-14.

Tsai, C.-K. [Cai Q.-G.]. 1988. Plant regeneration from leaf callus protoplasts of *Actinidia chinensis* Planch. var. *chinensis*. Plant Sci. 54: 231-235.

Wang S.-M., H.-W. Huang, Z.-W. Jiang, Z.-H. Zhang, S.-R. Zhang and H.-Q. Huang. 2000. Studies on *Actinidia* breeding by species hybridization between *A. chinensis* and *A. eriantha* and their hybrids' progenies. pp. 126-132. *In*: H.-W. Huang (ed.). Advances in *Actinidia* Research. Science Press, Beijing.

Wang, S.-M., R.-H. Huang, X.-W. Wu and N. Kang. 1994. Studies on *Actinidia* breeding by species hybridization. J. Fruit Sci. 11: 23-26.

Wang, S.-M., X.-W. Wu, R.-H. Huang, Z.-T. Xiong and S.-Q. Ke. 1989. Preliminary report on fluctuation of interspecific crosses of Chinese gooseberry. J. Wuhan Bot. Res. 7: 399-402.

Wang, T. and A. P. Gleave. 2012. Applications of biotechnology in kiwifruit (*Actinidia*). pp. 3-30. *In*: E.C. Agbo (ed.). Innovations in Biotechnology, ISBN: 978-953-51-0096-6, InTech,: http://www.intechopen.com/books/innovations-in-biotechnology/applications-of-biotechnology-in-kiwifruit-actinidia

Watanabe, K., B. Takahashi and K. Shirato. 1990. Chromosome numbers in kiwifruit (*Actinidia delciosa*) and related species. J. Japan. Soc. Hort. Sci. 58: 835-840.

White, A. and R. Beatson. 1993. Evaluation of a new kiwifruit hybrid. pp. 161-163. *In*: Proc. Australasian Postharvest Conf., Gatton College, University of Queensland, Lawes, Queensland, 20-24 September 1993.

Wismer, W.V., F.R. Harker, F.A. Gunson, K.L. Rossiter, K. Lau, A.G. Seal, R.G. Lowe and R. Beatson. 2005. Identifying flavour targets for fruit breeding: A kiwifruit example. Euphytica 141: 93-104.

Wu, J.-H. 2010. Manipulation of ploidy level of *Actinidia* by *in vitro* tissue culture for kiwifruit breeding. Ph.D. thesis, University of Auckland.

Wu, J.-H. 2012. Manipulation of ploidy for kiwifruit breeding and the study of *Actinidia* genomics. Acta Hort. 961: 539-546.

Wu, J.-H. 2015. Callus and plantlets produced from in vitro culture of anthers of *Actinidia chinensis*. Acta Hort. 1083: 115-121.

Wu, J.-H., A.R. Ferguson and B.G. Murray. 2009. *In vitro* induction of autotetraploid *Actinidia* plants and their field evaluation for crop improvement. Acta Hort. 829: 245-250.

Wu, J.-H., A.R. Ferguson and B.G. Murray. 2011. Manipulation of ploidy for kiwifruit breeding: in vitro chromosome doubling in diploid *Actinidia chinensis* Planch. Plant Cell Tiss. Org. Cult. 106: 503-511.

Wu, J.-H., A.R. Ferguson, B.G. Murray, A.M. Duffy, Y. Jia, C. Cheng and P.J. Martin. 2013. Fruit quality in induced polyploids of *Actinidia chinensis*. HortScience 48: 701-707.

Wu, J.-H., A.R. Ferguson, B.G. Murray, Y. Jia, P.M. Datson and J. Zhang. 2012. Induced polyploidy dramatically increases the size and alters the shape of fruit in *Actinidia chinensis*. Ann. Bot. 109: 169-179.

Xiao, X.-G. and A.-M. Hirsch. 1996. Microcallus formation from leaf mesophyll protoplasts in the genus *Actinidia* Lindl. Plant Cell Rep. 15: 896-899.

Xiao, Z.-A. and B.-W. Han. 1997. Interspecific somatic hybrids in *Actinidia*. Acta Bot. Sin. 39: 1110-1117.

Xiao, Z.-A., D.-X. Shen and B.-N Lin. 1993a. Culture and plant regeneration of *Actinidia deliciosa* protoplasts isolated from cotyledon callus. J. Wuhan Bot. Res. 11: 247-252 + plate.

Xiao, Z.-A., D.-X. Shen and B.-N. Lin. 1993b. Plant regeneration from protoplasts of *Actinidia chinensis* Planch. Acta Bot. Sin. 34: 736-742.

Xiao, Z.-A., L.-C. Wan and B.-W. Han. 2004. An interspecific somatic hybrid between *Actinidia chinensis* and *Actinidia kolomikta* and its chilling tolerance. Plant Cell Tiss. Organ Cult. 79: 299-306.

Xiong, Z.-T. 1991. Studies on the phylogenetic relationship between *Actinidia chinensis* and *A. deliciosa* using cladistic analysis. Guihaia 11: 36-39.

Xiong, Z.-T. 1992. Chromosome studies on cultivated Chinese gooseberry. Guihaia 12: 79-82.

Xiong, Z.-T. and Huang. 1988. Chromosome numbers of 10 species and 3 varieties in *Actinidia* Lindl. Acta Phytotaxon. Sinica 26: 245-247.

Xiong, Z.-T., R.-H. Huang and X.-W. Wu. 1985. Observations on the chromosome numbers of 4 species in *Actinidia*. J. Wuhan. Bot. Res. 3: 219-224.

Xiong, Z.-T., R.-H. Huang, P.-A. Yuan and X.-W. Wu. 1993. Cytological observations on the meiosis of PMC and pollen development of tetraploid in *Actinidia chinensis*. Guihaia 13: 139-142.

Xiong, Z.-T., S.-M. Wang and R.-H. Huang. 1987. Studies on the hybridization between *Actinidia chinensis* var. *chinensis* and *A. eriantha*. J. Wuhan Bot. Res. 5: 321-328.

Yan, G.-J., A.R. Ferguson, and M.A. McNeilage. 1994. Ploidy races in *Actinidia chinensis*. Euphytica 78: 175-183.

Yan, G.-J., A.R. Ferguson, M.A. McNeilage and B.G. Murray. 1997a. Numerically unreduced (2*n*) gametes and sexual polyploidization in *Actinidia*. Euphytica 96: 267-272.

Yan, G.[-J.], J. Yao, A.R. Ferguson, M.A. McNeilage, A.G. Seal and B.G. Murray. 1997b. New reports of chromosome numbers in *Actinidia* (Actinidiaceae). New Zealand J. Bot. 35: 181-186.

Yao X.-H., L. Liu, M.-K. Yan, D.-W. Li, C.-H. Zhong and H.-W. Huang. 2015. Exon primed intron-crossing (EPIC) markers reveal natural hybridization and introgression in *Actinidia* (Actinidiaceae) with sympatric distribution. Biochem. Sys. Ecol. 59: 246-255.

Zeng, H., D.-W. Li and H.-W. Huang. 2009. Distribution pattern of ploidy variation of *Actinidia chinensis* and *A. deliciosa*. J. Wuhan Bot. Res. 27: 312-317.

Zespri. 2015. Annual Review 2014/15. Zespri Group Ltd, Mt Maunganui.

Zhang, L., Z.-Z. Li, Y.-C. Wang, Z.-W. Jiang, S.-M. Wang and H.-W. Huang. 2010. Vitamin C, flower color and ploidy variation of hybrids from a ploidy-unbalanced *Actinidia* interspecific cross and SSR characterization. Euphytica 175: 133-143.

Zhang, T., Z.-Z. Li, Y.-L. Liu, Z.-W. Jiang and H.-W. Huang. 2007. Genetic diversity, gene introgression and homoplasy in sympatric populations of the genus *Actinidia* as revealed by chloroplast microsatellite markers. Biodiv. Sci. 15: 1-22.

Zhang, Y.-J., Y.-Q. Qian, X.-J. Mu, Q.-G. Cai, Y.-L. Zhou and X.-P. Wei. 1998. Plant regeneration from *in vitro*-cultured seedling leaf protoplasts of *Actinidia eriantha* Benth. Plant Cell Rep. 17: 819-821.

Zhang, Y.-J., Y.-Q. Qian, Q.-G. Cai, X.-J. Mu, X.-P. Wei and Y.-L. Zhou. 1997. Somaclonal variation in chromosome number and nuclei number of regenerated plants from protoplasts of *Actinidia eriantha*. Acta Bot. Sin. 39(2): 102-105 + plate.

Zhong, C., S. Wang, H. Huang and Z. Jiang. 2011. 'Jinyan'- a superior yellow-fleshed kiwifruit cultivar with excellent storage quality. Acta Hort. 913: 135-143.

Zhong, C.-H., S.-M. Wang, Z.-W. Jiang and H.-W. Huang. 2012. 'Jinyan', an interspecific hybrid kiwifruit with brilliant yellow flesh and good storage quality. HortScience 47: 1187-1190.

Zhu, D.-Y., Y.-H. Qin, Y.-B. Zhi, M.-L. Yi, Z.-K. Chen and Z. Hu. 2001. Protoplast culture and cell colony regeneration of *Actinidia arguta*. J. Henan Agri. Univ. 35: 221-224.

Zhu, D.-Y. and J.-H. Zhang. 1995. Variation of dioecism in *Actinidia chinensis*. Acta Agric. Uni. Henanensis 1995 (1): 1-5.

Chapter 14

Oat Improvement and Innovation Using Wild Genetic Resources (Poaceae, *Avena* spp.): Elevating "Oats" to a New Level and Stature

Eric N. Jellen[1,*], Eric W. Jackson[1] and Peter J. Maughan[2]

ABSTRACT

Common oat (*Avena sativa*, $2n = 6x = 42$) has been steadily declining in harvested acreage across much of its traditional cultivation range, in spite of the crop's tremendous health benefits. Wild and primitive domesticated genetic resources in *Avena* – including taxa at lower ploidy levels – represent esoteric opportunities to produce novel, innovative oat crops and products to stem this tide. Creative approaches for exploiting these resources will first require deconvolution of cryptic genotypic variation in exotic and elite germplasm. Through access to this knowledge, which is rapidly being accumulated by genome scientists, oat breeders will be able to refine improvement processes involving wide hybridization and introgression for a variety of traits, both single-gene and complex; domestication of wild and improvement of primitive domesticated taxa, especially the highly variable and stress-tolerant A-genome diploids; and creation of novel synthetic polyploids.

Introduction

Common oat (*Avena sativa*, $2n = 6x = 42$, AACCDD genomes) is an allohexaploid of temperate Old World origin. As the sixth or seventh most important world cereal, it is part of an interfertile biological species complex that also includes wild-weedy (*A. sterilis*), feral-weedy (*A. fatua*), subtropical forage (*A. byzantina*), and hulless (*A. sativa* subsp. *nuda*) taxa or races. In light of emerging evidence for Near Eastern pre-domestication of the wild

[1] 4105 LSB, Department of Plant & Wildlife Sciences, Brigham Young University, Provo, UT 84602, USA, Email: jellen@byu.edu, Jeff_maughan@byu.edu

[2] General Mills, Inc., Crop Bioscience, Wheat Innovation Center, 1990 Kimball Ave. Manhattan, KS 66502, USA, Email: eric.jackson@genmills.com

* Corresponding author

animated oat *A. sterilis* approximately 12,000+ years ago, oat was one of the foundational foods responsible for the agricultural revolution and the rise of Eurasian civilization (Weiss et al. 2006).

Avena species relationships were recently covered in an exhaustive review by Loskutov and Rines (2011), so our intent is not to repeat this effort. Basically, *Avena* allotetraploids consist of two species groups. The first is an AABB biological species complex comprised of wild and semi-domesticated Ethiopian endemics along with the common Mediterranean weed *A. barbata* and, possibly, the Moroccan endemic *A. agadiriana*. The second is a group of three large-seeded (Section *Pachycarpa*) wild oats having related but varying genome constitutions of AC (CD) and native to heavy clay soils in the central-western Mediterranean basin and Maghreb. *Avena* diploids include an assortment of Mediterranean and Southwest Asian A- and C-genome groups, the former including a wild-weedy-domesticated complex of subgenome constitution A_sA_s. The genus also includes a single perennial autotetraploid, *A. macrostachya*, from the Algerian Atlas Mountains and having genomes most closely related to the C genome (Leggett 1990).

Common oat improvement since the 19th century has been characterized by small but consistent incremental yield gains. The prospect of jeopardizing these yield gains, combined with often-formidable hybridization barriers, has largely deterred breeders from exploiting exotic wild-weedy oat taxa of differing ploidy levels except as sources of targeted resistance genes (reviewed by Loskutov and Rines 2011). However, *A. sterilis* has been used repeatedly as a source of exotic genes for improving a number of cultivated traits, most notably by the oat breeding program at Iowa State University during the latter three decades of the 20th Century (Frey 1994; Thro and Robertson 2011). Additionally, *A. fatua* has been suggested as a rich gene source for increasing yield and other characters (Burrows 1970; Frey 1991).

When compared with diploid small grained species rice (*Oryza sativa*) and barley (*Hordeum vulgare*), oat breeders are historically limited by a much more constraining set of market-driven breeding objectives. While it is primarily a human food, rice nonetheless has a wide range of diverse uses and market classes based on grain size, starch composition and texture, fragrance, etc. Barley's three principal economic uses – for food, feed, and malting – likewise encouraged selection across a broader range of characters than is the case in oat. In both crops, as well as in the even more diverse cereal maize (*Zea mays*), an intriguing question to a geneticist is to what extent has this diversity of economic/commercial use been driven, or at least facilitated, by the diploid genomic structure of the organism? Logic says that selective breeding, including identification of useful genetic variation, has been more straightforward in these diploid crops than in polyploid oat or wheat.

Even when compared with wheat (*Triticum aestivum*), oat is likewise in an unfavorable competitive position from a breeding standpoint – regardless of the disparity in international research interest in the two crops. While both are hexaploids of similar genome sizes and with comparable obstacles of genetic buffering and extreme epistasis, wheat has benefited by having a relatively stable genome, including subgenomes that have retained their homoeologous identities and can be traced back to diploidspecies or diploid-genome groups. Utilization of these ancestor-diploids for breeding, particularly *Aegilops tauschii* (DD), has been facilitated in this species by the retention of homologous pairing capacity and in other diploids by the availability of the *ph1*-based mechanism promoting homoeologous recombination.

The presence of numerous intergenomic translocations, particularly between the C and D genomes, are now regarded as significant obstacles to recombination in certain areas of the oat genome, even among cultivated genotypes within the primary oat breeding gene pool. Oliver et al. (2013) assembled the first consensus hexaploid oat map using a physical chromosome mapping strategy involving monosomic hybrids; this effort was recently expanded to a consensus map involving almost 10000 markers in 12 biparental oat populations (J. Schlueter and N. Tinker, personal communication). The presence of large and fairly common intergenomic translocations 7C-17A (Jellen et al. 1994; Jellen and Beard 2000) and 3C-14D (Jellen et al. 1997), as well as numerous smaller rearrangements only detectable in genetic maps or whole-genome sequences, are significant obstacles for single-gene isolation via recombination within the cultivated hexaploid gene pool itself (Sanz et al. 2010; Oliver et al. 2013).

As mentioned above, the systematics of *Avena* and history of oat breeding has been comprehensively reviewed in several recent publications, most notably by Loskutov and Rines (2011) and Thro and Robertson (2011) and, more recently, Boczkowska et al. (2015). One significant modification to the phylogenetic context in *Avena* is the emerging recognition that the set of large-seeded Section *Pachycarpa* allotetraploids – *A. insularis, A. magna,* and *A. murphyi* – have genome constitutions that are variants of CD and not AC; this has been supported by recent molecular (Yan et al. 2014) and cytogenetic observations (Sanz et al. 2010; Badaeva et al. 2011). Additionally, now-abundant and mounting DNA sequence data are confirming that the C subgenome is much more divergent than the A and D are to each other (personal observations).

A major focus of discussions at international oat conferences over the course of the past decade or two has been on how to elevate the crop's importance and enhance its value. This is widely viewed as a matter of crop preservation due to three principal factors: 1) the continued dominance, and even expansion, of major commodity crops like maize; 2) steady declines in harvested oat acreage and commodity prices; and 3) the emergence of previously obscure, and increasingly high-value, specialty crops. In the present article, we will address the question of how the oat genetics and breeding community might be able to exploit wild *Avena* resources to enhance the value, versatility, and productivity of oat-related crops and crop products. These strategies have included breeding approaches to produce domesticated polyploids composed of novel subgenome combinations; domestication of the wild, large-seeded Section *Pachycarpa* allotetraploids, especially *A. magna*, using domestication syndrome genes from *A. sativa*; and a proposal to unleash the tremendous range of ecological adaptation, morphological and favorable biochemical variation in the A_sA_s diploid species group by intentional re-domestication of the small-seeded forage oat *A. strigosa* as a seed crop. All three of these strategies, moving forward, would require a commitment on the part of the oat milling and processing industry to fundamentally reexamine existing practices, commercial uses, and marketing strategies of oat products.

Obstacles to Transformational Improvement of the Hexaploid Oat Gene Pool

Hexaploid Oat Production and Climate Change

Oat has, as its primary production assets, the ability to tolerate poor, low-fertility, sandy soils. Primary abiotic constraints to expanded oat production include its relatively poor winter hardiness (Livingston and Elwinger 1995) and seed yield reductions under conditions of extreme heat (Coffman 1939). Principal pathogens of importanceto worldwide oat producers include crown (*Puccinia coronata* f. sp. *avenae*) and stem (*Puccinia graminis* f. sp. *avenae*) rusts; powdery mildew (*Blumeria graminis* f. sp. *avenae*); *Fusarium* head blight; and barley yellow dwarf virus (BYDV). Increasing global temperatures would be expected to increase severity of many of these types of diseases (reviewed by Pautasso et al. 2012).

In a world where climatic conditions are rapidly shifting due to rising atmospheric CO_2 levels, increasing annual variation for temperature and moisture require crops to be more yield-stable (Ray et al. 2015). Interestingly, Castellanos-Frias et al. (2014) demonstrated that favorable habitat for *A. sterilis* – and by extension *A. sativa* cultivars possessing *A. sterilis*-derived adaptation genes – should dramatically expand northward under models projecting current climate-change trajectories. Presumably, such conditions would favor the northward expansion of winter oat cultivation in Europe and North America. Holland et al. (2000) identified recurrent selection to be a potentially promising breeding approach for improving hexaploid oat's environmental adaptation, as measured by yield stability across environmentally diverse test locations in Iowa, Idaho, and Norway. Notably, out of the 20 cultivars or lines included in their germplasm set, seven (28%) had significant *A. sterilis* parentage. Surveys of the pedigrees of 2012 and 2013 Uniform Midseason Oat Performance Nursery (UMOPN) entries revealed at least 24/36 (66%) and 28/36 (78%), respectively, had wild oats in their pedigree – mostly from *A. sterilis* via the cultivars 'Bond' and its descendant 'Ogle', or through the synthetic hexaploid 'Amagalon' (*A. magna* × *A. longiglumis*) and its descendant cultivar 'HiFi' (more on this below).

While these results are encouraging from the standpoint of mere survival of oat as a minor niche grain through an impending era of global warming, the greater question for the oat community is whether this current trajectory is acceptable: namely, might there be a "breakthrough" that can somehow elevate the oat crop to a more competitive position in the international foods market?

Exotic Major Oat Resistance Genes, Durability, and Linkage Drag

The dream of plant breeders and pathologists has been to discover, and successfully introgress, a major source of single-gene resistance that would remain effective for many years. Unfortunately, such "silver bullets" have been very few and far between, and the breakdown of major resistance genes originating in the primary oat genepool (*A. byzantina*, *A. fatua*, *A. sativa*, and *A. sterilis*) has led pathologists to screen *Avena* secondary and even tertiary genepools for major resistance genes (Tan and Carson 2013). An emerging, alternative strategy is to focus on breeding using durable, partial, adult-plant resistance (APR); however, this approach often involves a difficult screening process that must discriminate between the two resistance types (Leonard 2002). Recent mapping indicates that upwards of 75% of the APR in the most-studied source – MN841801 – may be due to the effect of a single QTL (Lin et al. 2014) and is therefore, arguably, another example of single-gene resistance. Regardless of whether the genes confer seedling resistance or APR, if they originate in exotic germplasm then the sexual transfer of these genes is expected to come at the cost of undesirable linkages that could slow down the rate of yield gain in common oats. This may also potentially create gene pools comprised of germplasm harboring common alien segments that would be recalcitrant to recombination in the alien introgression regions with non-members of the gene pool. A third issue with interploidy resistance-gene transfer has been interference or suppression of the alien resistance. A fourth issue is the retention (proliferation?) of intra-varietal heterogeneity – a familiar issue for oat geneticists involved since the late 1980's in mapping efforts with cultivars 'Ogle', 'Sun II', and 'Kanota'.

Carson (2011) noted the rapidity with which *Puccinia coronata* has overcome major-gene seedling resistance in the United States – not uncommonly, within five years of resistant-cultivar release. A similar situation has occurred in Australia with the breakdown of 19 major genes for resistance to the stem rust pathogen (Adhikary and McIntosh 2001). More recently, Sanchez-Martin et al. (2012) noted efforts to anticipate the inevitable spread in Europe and North America of emerging *P. coronata* races resistant to the most successful and durable *A. strigosa*-derived resistance gene, *Pc94*. This has not only been true with genes for rust resistance; Roderick et al. (2000) reviewed the historical record in documenting the breakdown of powdery mildew resistance genes in Britain, including an OMR 4 resistance gene from *A. barbata* that broke down even before it could be deployed in resistant cultivars.

As mentioned previously, the literature reveals that oat has such an extensive degree of chromosomal rearrangement that sexual and parasexual (i.e., use of embryo rescue, anther culture, oat × maize hybridization, etc.) transfer of even single, major resistance genes is impossible without bringing along undesirable wild traits through linkage drag. Aung et al. (2010) concluded that potent crown rust resistance genes from A-genome diploid oats could only be successfully transferred to *A. sativa* through production of disomic additions or in synthetic octoploids. Hsam et al. (1997) noted that deployment of *A. barbata*-derived powdery mildew resistance was hindered by a reduction in tillering in *A. sativa* hybrids carrying this gene source.

Alien resistance-gene suppression was reviewed by Loskutov and Rines (2011). A classic example in oat was the *Pc38*-induced suppression of crown rust from two sources: the *A. sterilis*-derived gene *Pc62* (Wilson and McMullen 1997) and the *A. strigosa Pc94* gene (Chong and Aung 1996). Moreover, the suppressive effect of *Pc38* was dosage-dependent when the two (normally) linked genes *Pc38* and *Pc62* were present together in varying copy numbers, as demonstrated by recovery and manipulation of duplicate-deficient gametes in selfed, selected progenies from translocation heterozygotes.

Partial APRto crown rust in *A. sativa* has been studied most thoroughly in line MN841801, which has retained its resistance since its introduction into uniform oat field nurseries in the 1980's (Leonard 2002). Lin et al. (2014) identified a major QTL on chromosome 14D that is responsible for upwards of 75% of the APR effect, using SNP markers in three mapping populations involving MN841801 as a parent. Previously, Portyanko et al. (2005), Zhu and Kaeppler (2003), Zhu et al. (2003), Barbosa et al. (2006), and Acevedo et al. (2010) had identified eight other QTL affecting this trait using other molecular marker systems. Lin et al. (2014) reasoned that these apparently contradictory results might have been a consequence of methodological differences, such as race-composition of the test pathogen, molecular marker types, confounding with flowering time QTL, and/or heterogeneity in the original MN841801 parent.

Dilkova et al. (2000) analyzed a series of *A. strigosa*-derived crown rust-resistant materials from the Wisconsin oat breeding program and found that years of wide hybridization, thermal neutron irradiation, and introgression with selection for retained resistance had resulted in phenotypically normal oat lines carrying an alien disomic substitution (DCS1789), intergenomic 6C-21D translocation (N770-165-2-1), and, in the case of MAM17-4, a karyotypically "normal" line displaying some meiotic irregularities. These included ~20% of PMCs with univalents, multivalents, or abnormal bivalents in selfed progeny or in the F_1 offspring of crosses to MAM17-4's backcross *A. sativa* parent. Interestingly, the *A. strigosa* chromosome pair transferred to DCS1789 was substituting for a missing pair of 12D chromosomes. Babiker et al. (2015) recently identified chromosome 12D in *A. sativa* cv. 'CDC Boyer' as the location of a significant partial-resistance QTL for crown rust. Based on its published pedigree, CDC Boyer does not share parentage with DCS1789 or any of its known descendants.

The issue of intra-varietal heterogeneity may or may not be related to selection for disease resistance per se. It is interesting that the 3C-14D interchange present in some strains of 'Sun II' (Jellen et al. 1997) has been retained in some other cultivars descended from 'Markische Landsorte' – its presumed source – including 'Flaemingsnova', 'Gelbhafer', and 'Lang' (Jellen, unpublished). Other varieties such as 'Kanota' are heterogeneous for the 7C-17D translocation (Jellen, unpublished) and the original Kanota monosomic series from Japan was heterogeneous for a 7C-14D translocation (Jellen et al. 1993). In the Kanota case, appearance of a unique (to date) 7C-14D translocation suggests that this rearrangement may have arisen in one or more haploid plants from which the Kanotaaneuploids were originally derived. Jellen (unpublished) detected evidence for meiotic abnormalities, including production of aneuploids, from anther culture-derived 'Aslak' × 'Matilda' hybrids (Tanhuanpaa et al. 2008).

Impact of Genomic Buffering on Breeding Progress

In hexaploid crop species like oat and wheat (*Triticum aestivum*), genomic buffering and complex epistatic gene interactions should theoretically be major impediments to rapid breeding progress in the absence of molecular marker and/or gene sequence data for genotypic selection. Using examples from a number of different domestication genes, Dubcovsky and Dvorak (2007) discussed how hexaploid wheat's molecular variation-generating and buffering mechanisms balance each other out in a system that delivers "piecemeal" phenotypic variation for a crop whose adaptational range has gradually expanded under artificial selection. Hexaploid oat, likewise, has a genome rife with proliferating, stress-activated retrotransposons that would be conducive to generating new variation at the molecular level (Kimura et al. 2001).

A negative side of genetic buffering is that it restricts breeders' ability to generate novel, expressed phenotypic variation for common oat improvement. The presence of two or three functional orthologs of any gene of interest renders the likelihood of inducing, or identifying natural, null mutations by phenotypic inspection alone nearly impossible. Even partial loss- or gain-of-function mutations may require expensive quantitative screening to be identified. While DNA sequencing is a powerful tool for identifying such mutants, the costs associated with large-scale genotyping by sequencing may be prohibitive in a low-value crop like oat.

Specialized Interploidy *Avena* Breeding Efforts

Domestication of Avena magna

Ladizinsky (1995) described the transfer of genes for the "domestication syndrome" – reduced awns, light-colored lemma, and glabrous, nonabscising spikelets – from 42-chromosome *A. sativa* to 28-chromosome *A. magna* and *A. murphyi*. The primary purpose of these experiments was to investigate whether the large-seeded Section *Pachycarpa* allotetraploids could be domesticated while retaining their elevated protein content. Oliver et al. (2011) created a molecular marker-based linkage map from a recombinant inbred line (RIL) population derived from one of Ladizinsky's domesticated *A. magna* lines, Ba 13-13, crossed to a wild *A. magna*, and identified 14 segregating linkage groups, two of which carried the domestication genes in distal map locations. This map represented an important step toward development of an *A. magna* breeding program; current efforts are producing advanced breeding lines having acceptable agronomic characteristics, yield, and seed protein contents in excess of 25% (unpublished data). However, as noted by Ladizinsky (1995), *A. magna* is susceptible to barley yellow dwarf virus (BYDV) and so its commercial cultivation may not be practical in traditional oat growing areas where aphids that vector BYDV are common.

Synthetic Polyploids: Amagalon, Strimagdo, Macrosativa, etc.

Another approach for utilizing wild *Avena* germplasm is to create new genome combinations by sexual and parasexual manipulation of amphiploids (i.e., embryo rescue, hormone treatments, colchicine-doubling, mutagenesis, etc.). These synthetics would be new species, like triticales, and would therefore be essentially sexually isolated. However, novel genome combinations might result in desirable phenotypes for oats with special biochemical or agronomic characteristics. A few examples of interesting synthetic polyploid *Avena* genotypes are presented below.

'Amagalon' was a novel synthetic hexaploid (Rothman 1984) derived from a cross between *A. magna* CI 8330 ($2n = 28$, CCDD subgenomes) and *A. longiglumis* CW57 ($2n = 14$, $A_lA_l$subgenomes). Though not a "domesticated" phenotype, is has nonetheless been extremely important as a source of disease resistance internationally. Rooney et al. (1994) and Adhikary and McIntosh (2001) confirmed that Amagalon has *Pc91* crown rust and *Pg-a* stem rust resistances. These resistances have shown reasonable durability, as they have been deployed though Amagalon parentage into a number of important breeding lines and cultivars such as 'Drover' and 'HiFi' (see Table 3 in McCartney et al. 2011).

Ladizinsky (2000) generated relatively stable synthetic hexaploid lines from doubled and selfed *A. strigosa* 'Saia' × domesticated *A. magna* triploid hybrids ('Strimagdo', A_sA_sCCDD). While some of these Strimagdo lines had excellent seed set with large seeds (resembling domesticated *A. magna*) and light-colored hulls, all inherited the dominant susceptibility to BYDV present in their *A. magna* parent and the dominant reduction in protein content (16%) from Saia. On the positive side, the lines did express Saia's resistance to crown rust.

Lapinski (unpublished) has experimented extensively with synthetic octoploids derived from *A. sativa* × *A. macrostachya* ($2n = 4x = 28$, AACCDDMM) crosses. (Note: the MM subgenome designation has not been used before and is introduced here merely for the purpose of convenience.) While the overriding goals were to improve oat winterhardiness and introduce perennial growth habit from *A. macrostachya*, they noted the octoploids possessed substantially larger seed than the hexaploid *A. sativa* parents, evidently due to favorable alleles or gene combinations with the very small-seeded *A. macrostachya*. Cytological examination of these octoploids revealed that they carried a single set of 14 chromosomes from *A. macrostachya* with a full complement of morphologically normal *A. sativa* chromosomes (Jellen, unpublished). Unfortunately, these particular octoploids did not have a sufficient level of winterhardiness to warrant their direct commercialization in the harsh winter environment of Poland.

Improvement of A_sA_s Diploid Oat

Variation and Potential of the A. strigosa Biological Species Group

Apart from the *Avena* hexaploids, the most diverse biological species group in the genus are the diploids having the A_sA_s genome (Rajhathy and Morrison 1959; Rajhathy 1961; Baum and Fedak 1985; Leggett 1987). This group includes a number of different cultivated and wild/weedy taxa, depending on the taxonomic source, that are interfertile and share karyotypic affinity. Domesticated taxa include *A. strigosa, A. brevis, A. hispanica,* and hulless *A. nuda*. Wild and weedy taxa of this group include spikelet-dispersing *A. atlantica* and the two floret-shattering ecotypes, *A. hirtula* and *A. wiestii*. *Avena wiestii* is adapted to sandy soils in arid deserts; *A. hirtula* to stony, calcareous Mediterranean zone soils; and *A. atlantica* to slightly heavier soils on coastward slopes.

Loskutov and Rines (2011) reviewed and summarized potential sources of quality traits, disease resistances, and ecological adaptation of this diploid species complex. As a group, the A_s-genome diploids carry superb resistance to powdery mildew, the rusts, BYDV, smut, and *Septoria* leaf blight. They also contain elevated levels of protein, oil, starch, avenanthramides, and tocotrienols, as well as the highest percentages of beta-glucans in the genus *Avena* (Welch et al. 2000; Redaelli et al. 2016).

Given the tremendous amount of variation, including domestication genes, in the A_s-genome diploid group, perhaps it makes sense to take a closer look at the potential for developing grain oat varieties of *A. strigosa*. Although *A. strigosa* groats are smaller and thinner than *A. sativa*, they are comparable in shape to some strains of rice, and could potentially be attractive as a more nutritious rice substitute. While *A. strigosa* plants are notoriously tall and spindly, there is abundant variation within the species group for shorter stature and panicle density. There may also be variation within the group – and if not, mutagenesis might be relatively straightforward – for novel value-added *A. strigosa* cultivars, for example high-amylose or low-amylose (waxy or sticky), high beta-glucan, and high antioxidant varieties with tremendous nutritional and/or cosmetic-industry appeal. Given the species' tolerance for a wide range of environmental conditions, *A. strigosa* might also provide highly water-efficient cultivars for irrigated desert conditions. Another area of possible interest would be to investigate if the seed size-enhancing effect of *A. macrostachya* would also be present in *A. strigosa* × *A. macrostachya* allotetraploids.

Conclusion

Many in the oat breeding, genetics, and processing community sense that common oat has gotten stuck in a rut that is leading to the side of the road or, worse, to a precipitous cliff. At a time when consumers are demanding new, novel, and highly nutritious food products, it seems obvious that oat breeders and processors should, given emerging knowledge on the structure and sequence of the oat genomes and the exercise of their imaginations, be able to deliver exactly those kinds of products and raw materials by better exploiting a genus that includes the most highly nutritious cereal in the world, oat.

References

Acevedo, M., E.W. Jackson, J. Chong, H.W. Rines, S. Harrison and J.M. Bonman. 2010. Identification and validation of quantitative trait loci for partial resistance to crown rust in oat. Phytopathology 100: 511-521.

Adhikary, K.N. and R.A. McIntosh. 2001. Identification of stem rust and leaf rust resistance genes in Amagalon oats. Aust. J. Agric. Res. 52: 1011-1016.

Aung, T., P. Zwer, R. Park, P. Davies, P. Sidhu and I. Dundas. 2010. Hybrids of *Avena sativa* with two diploid wild oats (CIav6956) and (CIav7233) resistant to crown rust. Euphytica 174: 189-198.

Babiker, E.M., T.C. Gordon, E.W. Jackson, S. Chao, S.A. Harrison, M.L. Carson, D.E. Obert and J.M. Bonman. 2015. Quantitative trait loci from two genotypes of oat (Avena sativa) conditioning resistance to Puccinia coronata. Phytopathology 105: 239-245.

Badaeva, E.D., O.Y. Shelukhina, O.S. Dedkova, I.G. Loskutov and V.A. Pukhalskyi. 2011. Comparative cytogenetic analysis of hexaploid *Avena* L. species. Russ. J. Genet. 47: 691-702.

Barbosa, M.M., L.C. Federizzi, S.C.K. Milach, J.A. Martinelli and G.C. Thorne. 2006. Molecular mapping and identification of QTL's associated to oat crown rust partial resistance. Euphytica 150: 257-269.

Baum, B.R. and G. Fedak. 1985. *Avenaatlantica*, a new diploid species of the oat genus from Morocco. Can. J. Bot. 63: 1057-1060.

Boczkowska, M., W. Podyma and B. Lapinski. 2015. Oat. pp. 159-226. *In*: Singh, M. and H.D. Upadhyaya (eds.). Genetic and Genomic Resources for Gain Cereals Improvement. Academic Press, San Diego.

Burrows, V.D. 1970. Yield and disease-escape potential of fall-sown oats possessing seed dormancy. Can. J. Plant Sci. 50: 371-378.

Carson, M.L. 2011. Virulence in oat crown rust (*Puccinia coronata* f. sp. *avenae*) in the United States from 2006 through 2009. Plant Dis. 95: 1528-1534.

Castellanos-Frias, E., D. Garcia de Leon, A. Pujadas-Salva, J. Dorado and J.L. Gonzalez-Andujar. 2014. Ann. Appl. Biol. 165: 53-61.

Chong, J. and T. Aung. 1996. Interaction of the crown rust resistance gene *Pc94* with several *Pc* genes. In: Kema G.H.J., R.E. Niks and R.A. Daamen (ed.) Proceedings of the 9th European and Mediterranean cereal rusts and powdery mildews conference, 2-6th Sept 1996, Lunteren, The Netherlands. European and Mediterranean Cereal Rust Foundation, Wageningen, The Netherlands, pp. 172-175.

Coffman, F.A. 1939. Heat resistance in oat varieties. Agron. J. 31: 811-817.

Dilkova, M., E.N. Jellen and R.A. Forsberg. 2000. C-banded karyotypes and meiotic abnormalities in germplasm derived from interploidy crosses in *Avena*. Euphytica 111: 174-184.

Dubcovsky, J. and J. Dvorak. 2007. Genome plasticity a key factor in the success of polyploidy wheat under domestication. Science 316: 1862-1866.

Frey, K.J. 1991. Genetic resources of oats, pp. 15-24. *In:* Use of Plant Introductions in Cultivar Development. CSSA Special Publication, Part 1, No. 17, Madison, WI, USA.

Frey, K.J. 1994. Remaking a crop gene pool: the case history of *Avena*. Proc. Natl. Sci. Counc., Repub. China, Part B: Life Sci. 18: 85-93.

Holland, J.B., A. Bjornstad, K.J. Frey, M. Gullord, D.M. Wesenberg and T. Buraas. 2000. Recurrent selection in oat for adaptation to diverse environments. Euphytica 113: 195-205.

Hsam, S.L.K., N. Peters, E.V. Paderina, F. Felsenstein, K. Oppitz and F.J. Zeller. 1997. Genetic studies of powdery mildew resistance in common oat (*Avena sativa* L.) I. Cultivars and breeding lines grown in Western Europe and North America. Euphytica 96: 421-427.

Jellen, E.N. and J. Beard. 2000. Geographical distribution of a chromosome 7C and 17 intergenomic translocation in cultivated oat. Crop Sci. 40: 256-263.

Jellen, E.N., B.S. Gill and T.S. Cox. 1994. Genomic in situ hybridization differentiates between A/D- and C-genome chromatin and detects intergenomic translocations in polyploidy oat species (genus *Avena*). Genome 37: 613-618.

Jellen, E.N., H.W. Rines, S.L. Fox, D.W. Davis, R.L. Phillips and B.S. Gill. 1997. Characterization of 'Sun II' oat monosomics through C-banding and identification of eight new 'Sun II' monosomics. Theor. Appl. Genet. 95: 1190-1195.

Jellen, E.N., W.L. Rooney, R.L. Phillips and H.W. Rines. 1993. Characterization of the hexaploid oat *Avena byzantina* cv. Kanotamonosomic series using C-banding and RFLPs. Genome 36: 962-970.

Kimura, Y., Y. Tosa, S. Shimada, R. Sogo, M. Kusaba, T. Sunaga, S. Betsuyaku, E. Yukiko, H. Nakayashiki and S. Mayama. 2001. *OARE-1*, a *Ty1-copia* retrotransposon in oat activated by abiotic and biotic stresses. Plant Cell Physiol. 42: 1345-1354.

Ladizinsky, G. 1995. Domestication via hybridization of the wild tetraploid oats *Avena magna* and *A. murphyi*. Theor. Appl. Genet. 91: 639-646.

Ladizinsky, G. 2000. A synthetic hexaploid (2n = 42) oat from the cross of *Avena strigosa* (2n = 14) and domesticated *A. magna* (2n = 28). Euphytica 116: 231-235.

Leggett, J.M. 1987. Interspecific hybrids involving the recently described diploid taxon *Avena atlantica*. Genome 29: 361-364.

Leggett, J.M. 1990. A new triploid hybrid between *Avena eriantha* and *A. macrostachya*. Cereal Res. Commun. 18: 97-110.

Leonard, K.J. 2002. Oat lines with effective adult plant resistance to crown rust. Plant Dis. 86: 593-598.

Lin, Y., B.N. Gnanesh, J. Chong, G. Chen, A.D. Beattie, J.W. Mitchell Fetch, H.R. Kutcher, P.E. Eckstein, J.G. Menzies, E.W. Jackson and C.A. McCartney. 2014. A major quantitative trait locus conferring adult plant partial resistance to crown rust in oat. BMC Plant Biol. 14: 250.

Livingston, D.P. and G.F. Elwinger. 1995. Improvement of winter hardiness in oat from 1935 to 1992. Crop Sci. 35: 749-755.

Loskutov, I.G. and H.W. Rines. 2011. Avena. pp. 109-183. *In:* C. Kole (ed.). Wild Crop Relatives: Genomic and Breeding Resources: Cereals. Springer, Heidelberg.

McCartney, C.A., R.G. Stonehouse, B.G. Rossnagel, P.E. Eckstein, G.J. Scoles, T. Zatorski, A.D. Beattie and J. Chong. 2011. Mapping of the oat crown rust resistance gene *Pc91*. Theor. Appl. Genet. 122: 317-325.

Oliver, R.E., E.N. Jellen, G. Ladizinsky, A.B. Korol, A. Kilian, J.L. Beard, Z. Dumlupinar, N.H. Wisniewski-Morehead, E. Svedin, M. Coon, R.R. Redman, P.J. Maughan, D.E. Obert and E.W. Jackson. 2011. New Diversity Arrays Technology (DArT) markers for tetraploid oat (*Avena magna* Murphy et Terrell) provide the first complete oat linkage map and markers linked to domestication genes from hexaploid *A. sativa* L. Theor. Appl. Genet. 123: 1159-1171.

Oliver, R.E., N.A. Tinker, G.R. Lazo, S. Chao, E.N. Jellen, M.L. Carson, H.W. Rines, D.E. Obert, J.D. Lutz, I. Shackelford, A.B. Korol, C.P. Wight, K.M. Gardner, J. Hattori, A.D. Beattie, A. Bjornstad, J.M. Bonman, J.-L. Jannink, M.E. Sorrells, G.L. Brown-Guedira, J.W. Mitchell Fetch, S.A. Harrison, C.J. Howarth, A. Ibrahim, F.L. Kolb, M.S. McMullen, J.P. Murphy, H.W. Ohm, B.G. Rossnagel, W. Yan, K.J. Miclaus, J. Hiller, P.J. Maughan, R.R. Redman Hulse, J.M. Anderson, E. Islamovic and E.W. Jackson. 2013. SNP discovery and chromosome anchoring provide the first physically-anchored hexaploid oat map and reveal synteny with model species. PLoS ONE 8: e58068.

Pautasso, M., T.F. Doring, M. Garbelotto, L. Pellis and M.J. Jeger. 2012. Impacts of climate change on plant diseases – opinions and trends. European Journal of Plant Pathology 133: 295-313.

Portyanko, V.A., G. Chen, H.W. Rines, R.L. Phillips, K.J. Leonard, G.E. Ochocki and D.D. Stuthman. 2005. Quantitative trait loci for partial resistance to crown rust, *Puccinia coronata*, in cultivated oat, *Avena sativa* L. Theor. Appl. Genet. 111: 313-324.

Rajhathy, T. 1961. Chromosome differentiation and speciation in diploid *Avena*. Can. J. Genet. Cytol. 3: 372-377.

Rajhathy, T. and J.W. Morrison. 1959. Chromosome morphology in the genus *Avena*. Can. J. Bot. 37: 331-337.

Ray, D.K., J.S. Gerber, G.K. MacDonald and P.C. West. 2015. Climate variation explains a third of global crop yield variability. Nature Comm. 6: 5989.

Redaelli R., L. Dimberg, C.U. Germeier, N. Berardo, S. Locatelli and L. Guerrini. 2016. Variability of tocopherols, tocotrienols and avenanthramide contents in European oat germplasm. Euphytica 207: 273-292.

Roderick, H.W., E.R.L. Jones and J. Sebesta. 2000. Resistance to oat powdery mildew in Britain and Europe: a review. Ann. Appl. Bot. 136: 85-91.

Rooney, W.L., H.R. Rines and R.L. Phillips. 1994. Identification of RFLP markers linked to crown rust resistance genes *Pc91* and *Pc92* in oat. Crop Sci. 34: 940-944.

Rothman, P.G. 1984. Registration of four stem rust and crown rust resistant oat germplasm lines. Crop Sci. 24: 1217-1218.

Sanchez-Martin, J., D. Rubiales, J.C. Sillero and E. Prats. 2012. Identification and characterization of sources of resistance in *Avena sativa, A. byzantina* and *A. strigosa* germplasm against a pathotype of *Puccinia coronata* f. sp. *avenae* with virulence against the *Pc94* resistance gene. Plant Path. 61: 315-322.

Sanz, M.J., E.N. Jellen, Y. Loarce, M.L. Irigoyen, E. Ferrer and A. Fominaya. 2010. A new chromosome nomenclature system for oat (*Avena sativa* L. and *A. byzantina* C. Koch) based on FISH analysis of monosomic lines. Theor. Appl. Genet. 121: 1541-1552.

Tan, M.Y.A. and M.L. Carson. 2013. Screening wild oat accessions from Morocco for resistance to *Puccinia coronata.* Plant Disease 97: 1544-1548.

Tanhuanpaa, P., R. Kalendar, A.H. Schulman and E. Kiviharju. 2008. The first doubled haploid linkage map for cultivated oat. Genome 51: 560-569.

Thro, A.M. and L. Robertson. 2011. Dedication: Kenneth J. Frey Oat breeder, educator, and champion of plant breeding. Plant Breeding Reviews 34: 1-36.

Weiss, E., M.E. Kislev and A. Hartmann. 2006. Autonomous cultivation before domestication. Science 312: 1608-1610.

Welch, R.W., J.C.W. Brown and J.M. Leggett. 2000. Interspecific and intraspecific variation in grain and groat characteristics of wild oat (*Avena*) species: very high groat (1→3),(1→4)-beta-D-glucan in an *Avena atlantica* genotype. J. Cereal Sci. 31: 273-279.

Wilson, W.A. and M.S. McMullen. 1997. Dosage dependent genetic suppression of oat crown rust resistance gene *Pc-62*. Crop Sci. 37: 1699-1705.

Yan, H.-H., B.R. Baum, P.-P. Zhou, J. Zhao, Y.-M. Wei, C.-Z. Ren, F.-Q. Xiong, G. Liu, L. Zhong, G. Zhao and Y.-Y. Peng. 2014. Phylogenetic analysis of the genus *Avena* based on chloroplast intergenic spacer *psb*A-*trn*H and single-copy nuclear gene *Acc1*. Genome 57: 267-277.

Zhu, S. and H.F. Kaeppler. 2003. Identification of quantitative trait loci for resistance to crown rust in oat line MAM 17-5. Crop Sci. 43: 358-366.

Zhu, S., K.J. Leonard and H.F. Kaeppler. 2003. Quantitative trait loci associated with seedling resistance to isolates of *Puccinia coronata* in oat. Phytopathology 93: 860-866.

Chapter 15

Interspecific Hybridization of Chestnut

S. Pereira-Lorenzo[1,*], R. Costa[2], S. Anagnostakis[3], U. Serdar[4], T. Yamamoto[5], T. Saito[5], A.M. Ramos-Cabrer[1], Q. Ling[6], T. Barreneche[7], C. Robin[8], R. Botta[9], C. Contessa[9], M. Conedera[10], L.M. Martín[11], A. Martín[12], J. Gomes-Laranjo[13], F. Villani[14] and J.E. Carlson[15]

[1] Universidade de Santiago de Compostela, Departamento de Producción Vegetal, EPS, Campus de Lugo, 27002 Lugo, Spain, E-mail: santiago.pereira.lorenzo@usc.es

[2] Instituto Nacional de Investigação Agrária e Veterinária I.P., Quinta do Marquês, Av. da República 2780-159 Oeiras, Portugal, E-mail: rita.lcosta@iniav.pt

[3] The Connecticut Agricultural Experiment Station, Box 1106, New Haven, CT 06504, USA, E-mail: Sandra.Anagnostakis@ct.gov

[4] Ondokuz Mayis University, Agriculture Faculty, Horticulture Department, Samsun, Turkey, E-mail: userdar@omu.edu.tr

[5] NARO Institute of Fruit Tree Science, National Agriculture and Food Research Organization, Fujimoto 2-1, Tsukuba, Ibaraki 305-8605, Japan, E-mail: toshiya@affrc.go.jp, saito@affrc.go.jp

[6] Department of Horticulture, Beijing Agricultural College, Beijing University, Beijing 102206, China, E- mail: qinlingbac@126.com

[7] INRA-Université de Bordeaux, UMR 1332 de Biologie du Fruit et Pathologie, F-33140 Villenave d'Ornon, France, E-mail: Teresa.Barreneche@bordeaux.inra.fr

[8] INRA-Université de Bordeaux, UMR 1202 Biogeco, 33612 Cestas, France, E-mail: cecile.robin@bordeaux.inra.fr

[9] Università degli Studi di Torino, Dipartimento di Scienze Agrarie, Forestali e Alimentari – DISAFA, Largo Paolo Braccini 2, 10095 Grugliasco (TO), Italy, E-mail: roberto.botta@unito.it

[10] Swiss Federal Institute for Forest, Snow and Landscape Research (WSL), A Ramèl 18, CH-6594 Cadenazzo, Switzerland, E-mail: marco.conedera@wsl.ch

[11] Universidad de Córdoba, Escuela Técnica Superior de Ingenieros Agrónomos y de Montes, Edificio Gregor Mendel (C-5), Campus de Rabanales, 14071-Córdoba, Spain, E-mail: lm.martin@uco.es

[12] Universidad de Extremadura, Departamento de Ingeniería del Medio Agronómico y Forestal, Centro Universitario de Plasencia, Avenida Virgen del Puerto, 2, 10600 Plasencia (Cáceres), Spain, E-mail: angelamartin@unex.es

[13] University of Tras-Os-Montes (UTAD) CITAB, Vila Real, Portugal, E-mail: jlaranjo@utad.pt

[14] National Research Council (CNR-IBAF), Viale Marconi, 2, 05010 Porano (TR), Italy fiorella. E-mail: villani@ibaf.cnr.it

[15] Department of Ecosystem Science and Management, 323 Forest Resources Building, Pennsylvania State University, University Park, PA, USA 16802, E-mail: jec16@psu.edu

* Corresponding author

ABSTRACT

Chestnut (*Castanea*) is a tree genus distributed throughout the northern hemisphere in natural stands, orchards, and coppices. As a multipurpose tree, chestnut is used to produce timber, nuts, tannins, and other related products. Interspecific hybridization was first done in 1894 in the USA, in the 1910s in Japan, and in the 1920s in Europe. In the USA, blight resistant Chinese (*C. mollissima*) and Japanese (*C. crenata*) species have been used since 1930 as the pollen parents in a backcrossing program with American chestnuts (*C. dentata*) as the recurrent parent, in order to produced blight resistant timber chestnuts. In Europe, Japanese chestnut trees were introduced because of their resistance to ink disease (caused by *Phytophthora* spp.). A clonal collection of hybrids of Japanese chestnut crossed with European chestnut (*C. sativa*) is used to produce nuts and timber, and for rootstocks for local cultivars. The pellicle of the Japanese chestnut cultivars is not easily removed, but this has been overcome with the newly released Japanese chestnut hybrid 'Porotan', with the easy-peel pellicle trait controlled by a single recessive gene (P/p) from native Japanese chestnuts. After more than 100 years of chestnut hybridisation, new technologies and the recently reported genome sequence will allow us to face new threats and to maintain the rich genetic and cultural heritage of chestnut.

Introduction

Chestnut trees (*Castanea* spp.), chestnuts and chinquapins, belong to the family *Fagaceae*. Species are found all over the world in temperate zones: in China *C. mollissima* Blume (Chinese chestnut), *C. henryi* (Skan) Rehd. & E.H. Wils. (willow leaf or pearl chestnut), *C. seguinii* Dode, and *C. davidii* Dode; in the Korean Peninsula and in Japan *C. crenata* Sieb. et Zucc. (Japanese chestnut); in Europe and Turkey, *C. sativa* (European or sweet chestnut) and, less frequently, interspecific hybrids; and in North America, *C. dentata* (Marsh.) Borkh. (American chestnut) and *C. pumila* Mill. (Allegheny chinquapin) in the Appalachian Mountain Range; *C. ozarkensis* (Ashe) (Ozark chinquapin) on the Ozark plateau, and *C. floridana* (Sarg.) Ashe (Florida chinquapin) and *C. alnifolia* Nutt. (trailing chinquapin) in small areas of southern Florida, USA.

C. dentata (Marsh.) Borkh. and *C. sativa* Mill. were confirmed to be homoploid (share the same chromosome number) by Warschefsky et al. (2014) based on the previous studies by Li and Dane (2013) for the American chestnut (*C. dentata*), and by Villani et al. (1999) and Mattioni et al. (2013) for the European chestnut (*C. sativa*). Hybridization was first done in 1894 in the USA (Van Fleet 1914). *C. sativa* hybrids were made in the 1920s in Europe (Gallastegui 1926, Vieira Natividade 1947, Schad et al. 1952), in 1926 in Spain (Gallastegui 1926), and *C. crenata* hybrids in Japan in 1917 (Kajiura 1978). These crosses were made to introgress resistance to diseases into the species, and to improve the agronomic and forestry characteristics. Hybrid clones have been released as named cultivars in Japan, the USA, and in Europe, but they still have minor importance among natural stands (France, Spain and Portugal; Pereira-Lorenzo et al. 2010). Some hybrid cultivars and seedlings are used to produce nuts (France, the USA, and Japan) and timber (mainly in Spain), and as rootstocks for resistance to *Phytophthora* spp. (France, Portugal, and Spain). Nuts of hybrids are often larger than those of wild type *C. sativa* trees, and their seedlings grow more rapidly to produce timber while maintaining the wood quality. The main limitation of growing hybrids of Asian chestnuts with *C. sativa* in Europe is their susceptibility to spring frosts due to earlier blooming. Chestnut hybrids will have a more important role in the restoration of threatened chestnut timber stands in the near future, as is happening in the USA with the American chestnut (Anagnostakis 2012, Jacobs et al. 2013).

Chestnut Germplasm

The genetic resources of chestnut include cultivars, seedlings, and species trees from Europe, Asia, and North America (Avanzato and Bounous 2009). A total of 27 countries reported native chestnut trees, mainly in the Northern Hemisphere, but chestnuts are also grown in Argentina, Brazil, and Chile in South America, and Australia and New Zealand in Oceania. Three main chestnut species (*C. sativa, C. mollissima,* and *C. crenata*) and their interspecific hybrids are cultivated in Europe, and two in Asia (*C. mollissima* and *C. crenata*). Cultivation involves grafting of selected cultivars onto rootstocks (seedlings or selected interspecific hybrids), planting at regular distances, and pruning. Conservation of chestnut cultivars is a priority in Europe (France, Italy, Portugal, Switzerland, and Spain) and in Asia (China and Japan), with national repositories for cultivars, hybrids, and natural populations.

Hybridization was first done in Europe in 1920 to combat ink disease (*Phytophthora* spp.), which was introduced in the eighteenth century (Gallastegui 1926, Vieira Natividade 1947, Schad et al. 1952). Introduced Asian species, which were resistant to ink disease, did not adapt well to European climatic conditions and their different agronomic and forestry characteristics made them unsuitable for replacing the disease-susceptible *C. sativa* trees (Lafitte 1946). However, in some valleys in the French Pyrenees (Basque country), *C. crenata* was introduced in 1909 and adapted well. Cultivars were then selected and registered in the French Official Catalogue of Fruit Species, which are now commercialized under the names 'Ederra', 'Ipharra', and 'Marki'.

Genomic Resources for *Castanea*

A number of molecular marker systems have been used in *Castanea* for applications such as cultivar identification, population genetics, linkage analysis, and marker-assisted selection. These marker systems consisted initially of isoenzymes, followed by random amplified polymorphic DNAs

(RAPDs), inter-simple sequence repeats (ISSRs), and amplification fragment length polymorphisms (AFLPs), and more recently by simple sequence repeats (SSRs) and single nucleotide polymorphisms (SNPs). Given that SSRs and SNPs are rapidly becoming the markers of choice due to their usefulness, high transferability across related taxa, prevalence in the genome, and amenability to automated, high-throughput, genome-wide analyses, we have chosen to focus primarily on the development and use of these newer marker systems.

Development of Molecular Markers

There are currently four main sources from which SSR or SNP markers are being developed: genomic DNA libraries enriched for repeat-containing sequences; expressed sequence tags (ESTs); bacterial artificial chromosome (BAC) sequencing, and whole genome sequencing efforts. Due to a lack of DNA sequence data available for the Fagaceae, and in particular for *Castanea*, most SSRs currently available have been developed from repeat-enriched genomic DNA libraries. A small number of SSR markers developed from *Quercus* sequences are proving useful in *Castanea* (Aldrich et al. 2003, Barreneche et al. 2004). Of those developed from *Fagus* (Tanaka et al. 1999, Pastorelli et al. 2003), comparatively fewer are proving to be useful within *Castanea* (Kremer et al. 2007, Kubisiak, unpublished data). The transferability of SSRs from other genera within the Fagaceae to *Castanea* will largely depend upon their evolutionary distance, with higher levels of transferability expected for more closely related genera such as *Castanopsis* (Manos et al. 2001).

Sequence data for at least 83 SSR primer pairs developed specifically from *Castanea* are currently publicly available. Forty-six SSR primer pairs developed from *C. sativa* have been characterized (Buck et al. 2003, Marinoni et al. 2003). Yamamoto et al. (2003) characterized 15 SSR primer pairs developed from *C. crenata*. More recently, 22 SSR primer pairs were developed from *C. mollissima* and characterized in both *C. mollissima* and *C. crenata* (Inoue et al. 2009). In general, these markers appear to be highly transferable across *Castanea*. SSR markers are already proving useful for cultivar identification and typing (Boccacci et al. 2004, Beccaro et al. 2004, Pereira-Lorenzo et al. 2011), linkage analysis (Sisco et al. 2005) and QTL analyses (Kubisiak, unpublished data; Casasoli et al. 2004, 2006). Expressed sequence tags (ESTs) from *Quercus robur* and *Quercus petraea* expressed during drought stress and bud burst were used to develop 283 SSR markers (Durand et al. 2010), and their transferability between three *Fagaceae* genera (*Quercus, Castanea,* and *Fagus*) was tested. Fifty SSR markers were suitable for chestnut linkage analysis in *C. sativa* (Durand 2009) and nine EST-SSRs were used to assess diversity in Spanish, Greek and Italian chestnut populations (Martin et al. 2010).

Utilizing recently released transcriptome data (Serrazina et al. 2015), 43 SSR markers were developed from ESTs differentially expressed in

European chestnut (*C. sativa*) and Japanese chestnut (*C. crenata*) in response to inoculation with the most severe chestnut pathogen in Europe, *Phytophthora cinnamomi* (Pc). (Santos et al. 2015a). The novel EST-SSR markers developed were validated for the four chestnut species involved in American (TACF) and Portuguese breeding programs: *C. sativa, C. dentata, C. crenata,* and *C. mollissima*. Furthermore, these SSRs are potentially functional genetic markers that may contribute to: (**1**) marker-assisted selection for the facilitation of chestnut breeding; (**2**) establishing, more accurately and rapidly, marker-trait associations for identification of QTLs for Pc resistance; and (**3**) developing universal gene-specific markers for genome synteny studies within Fagaceae and related taxa. As Pc has also devastated important food crops and native forest (Hardham 2005), these results may be useful for other woody species such as *Eucalyptus* spp., *Quercus* spp., *Fagus* spp., *Juglans* spp. and many ornamental trees and shrubs that are attacked by Pc.

Previously, the availability of DNA sequence data for *Castanea* was extremely limited (Connors et al. 2001). A genomic tool development project for various members of the Fagaceae was initiated in 2006 (Sederoff et al. 2008), with *C. mollissima* being a key model species. A large component of this project was focused on the transcriptomes of *C. mollissima* and *C. dentata* (Carlson et al. 2007, 2008). Large EST databases were created with significant numbers of sequence contigs showing similarity to predicted proteins in *Populus trichocarpa*. For *C. dentata*, 28 890 unique transcript contigs were assembled (version 1), and a total of 6 605 SSRs and 11 924 SNPs were identified. For *C. mollissima*, 48 335 unique transcript contigs were assembled (version 2), and a total of 24 655 SSRs and 41 584 SNPs were identified. The transcript contigs and predicted DNA markers are available to download as a resource for the community interested in the genetics, breeding, and biotechnology of *Castanea*. An integrated web-based resource for the *Castanea* genetics/genomics community [Fagaceae Genomic Database (FGD): http://www.fagaceae.org] was developed (Ficklin et al. 2007), and relevant sequence information, homology results, genetic/physical map information, predicted SSRs, SNPs, and other genomic data are posted, including publically available data for *C. sativa, C. crenata,* and *Quercus* spp.

More than 400 SSR markers have now been developed in Japanese chestnut (*C. crenata*) (Yamamoto et al. 2003, Nishio et al. 2011a, b). Nishio et al. (2011b) developed a total of 366 SSR markers derived from Japanese chestnut, comprising 220 genomic SSRs from enriched genomic libraries and 146 EST–SSRs from large-scale EST sequencing analysis. The EST–SSRs showed less polymorphism than the genomic SSRs and were more transferable to other species (*C. mollissima, C. sativa,* and *C. dentata*).

Another large component of the genomic tool development project is the production of genetic and physical mapping resources for *C. mollissima*. A BAC library has been constructed which consists of ~20 × coverage (http://www.fagaceae.org/progress/NE1015/Tomkins_FingerprintingAndcDNAUpdate.ppt). The entire library was fingerprinted by high information content fingerprinting, leading to the identification of a subset of clones consisting of the minimum tiling path representing the physical map for *C. mollissima* (Fang et al. 2013). An improved physical map was recently published (Staton et al. 2015) consisting of 1 300 contigs and 12 941 singlets from 126 449 fingerprinted BACs. This physical map was integrated with the genetic map using 858 shared sequence-based markers, 1390 overgo (cDNA) probes and 42 970 BAC-end sequences (Fang et al. 2013) (available at http://www.hardwoodgenomics.org/content/chinese-chestnut-physical-map and at GenBank: HN270092-HN275251, JY172573-187037). The integrated genetic/physical map for *C. mollissima* provides a platform for targeted genome sequencing of regions harbouring resistance genes and for gene cloning studies in *Castanea* (Staton et al. 2015). The BACs spanning the three major blight-resistance QTL were sequenced, in which 782 genes were annotated, showing a diversity of putative molecular and biological functions (Staton et al. 2015). Fifteen of the genes were annotated with the "defense response" GO term, and are being further studied for roles in blight resistance.

Three partial gene encoding proteins described as pathogenesis-related were isolated and cloned from infected, resistant chestnuts: a cystatin, a beta 1,3 glucanase isoform and a thaumatin-like protein gene using the RT-PCR technology (Serrazina 2004). The nucleotide sequences and amino acid-deduced sequences have high homology with resistance gene sequences from other plant species in the GenBank database. A partial gene encoding AOC (allene oxide cyclase) was also cloned, and found to be similar to the *Lycopersicum esculentum* gene. Gene expression analysis by Northern Blotting of *aoc, cist, gluc* and *pttum* suggested a relationship of these genes with *C. sativa* genes conferring resistance to *P. cinnamomi* (Serrazina 2004). Tobacco explants were transformed by particle bombardment in order to study the effect of over-expression of the isolated genes on tobacco plant resistance to *P. cinnamomi*. The observation of inoculated transformed and non-transformed plantlets suggests that the constitutive expression of *aoc, cist, gluc,* and *pttaum* genes (separately) attenuates the pathogenic effects of *P. cinnamomi* in transformed tobacco plants (Serrazina 2004).

To gain insight into the genetic expression which leads to the formation of ectomycorrhiza, a cDNA microarray was constructed and used to study the interaction of *C. sativa* roots and *Pisolithus tinctorius* during the first hours of contact (Sebastiana 2006). Statistical analysis of microarray results identified a set of 32 *C. sativa* genes and eight *Pisolithus tinctorius* genes with altered expression in response to the interaction between the two organisms. Differentially expressed genes identified in *C. sativa* roots had significant sequence similarities to proteins involved in cellular processes such as defence response, protein maturation/degradation, cell wall modification, primary metabolism, signal transduction, and cytoskeletal organization. Fungal genes regulated by the interaction with *C. sativa* roots had significant sequence similarities to proteins involved in cell wall structure, protein maturation/degradation, and cellular organization (Sebastiana 2006).

To elucidate chestnut defence mechanisms to *Phytophthora cinnamomi*, root transcriptomes of the susceptible species *C. sativa* and the resistant species *C. crenata* were compared after *P. cinnamomi* inoculation (Serrazina et al. 2015). Four cDNA libraries were constructed: two from *C. sativa* root samples (inoculated and non-inoculated with *P. cinnamomi*) and two from *C. crenata* root samples (inoculated and non-inoculated). Pyrosequencing produced 771 030 reads; 15 683 contigs were assembled for *C. sativa* and 16 828 for *C. crenata*. GO annotation revealed terms related to stress as "response to stimulus, "transcription factor activity," or "signaling" for both transcriptomes.

The reports of Barakat et al. (2009, 2012) provided the first insights into chestnut resistance to *C. parasitica* using high-throughput RNA-seq. The response of chestnut to *C. parasitica* and *P. cinnamomi* may be comparable, as fungi and oomycetes share similar infection mechanisms (Latijnhouwers et al. 2003). When comparing Chinese and American chestnut responses to *C. parasitica* with the Japanese and European chestnut responses to *P. cinnamomi*, similar differentially expressed genes (DEG) were found that fell into the following functional categories: (1) regulation of biotic stress response (ATPase transporter, Pyridine nucleotide-disulphide oxidoreductase); (2) hypersensitive response (HR) and cell wall lignification (Peroxidase); (3) HR recovery (Arginine decarboxylase, Manganese superoxide dismutase); (4) anti-fungal enzymes (Thaumatin-like protein, β-1,3-glucanase, Chitinase); (5) anti-fungal metabolite synthesis (family 1 Cytochrome P450 glycosyltransferase, Abscisic acid 8′-hydroxylase, Squalene monooxygenase, UDP-glucosyltransferase); (6) cell wall synthesis (β-expansin); and (7) stress recovery (ABC transporter family, Glyceraldehyde 3-phosphate dehydrogenase). Other shared *Castanea* responses to both pathogens include DEGs such as kinase genes involved in pathogen recognition and jasmonic acid pathway activation, gene regulation by Myb TF andethylene-responsive TF, and genes of the 26S proteasome regulatory unit. The response of all four species to both pathogens further includes genes from the flavonoid pathway that promote phytoalexin synthesis. Differential gene expression analysis revealed that in *C. crenata* there were more genes expressed that were related to biotic stress after inoculation with pathogens than were expressed in *C. sativa*. Those genes for both species were those involved in regulation of plant immune response and stress adaptation and recovery.

Genome-Wide Selection for Advancing Introgression Breeding

To provide a foundation for genome-wide marker-assisted selection for advancing breeding programs and for discovery of genes for resistance to the chestnut blight fungus (*C. parasitica*), a genome sequence was produced for Chinese chestnut (*C. mollissima*). The first draft of the *C. mollissima* cultivar 'Vanuxem' genome was announced at The Plant & Animal Genome Conference XXII, on January 12, 2014 (Carlson et al. 2014).

The first draft genome assembly covered 724.4 Mbp of the estimated 800 Mb chestnut genome in 41 270 genome sequence scaffolds. A consensus of 38 146 genes was predicted and annotated, confirmed with gene expression (EST) data. The scaffold genome sequences are publicly available at the National Center for Biotechnology Information website (accession number GCA_000763605.1) and can be downloaded and searched at the Hardwood Genomics website (http://hardwoodgenomics.org/chinese-chestnut-genome). The chestnut genome project was supported by The Forest Health Initiative (http://www.foresthealthinitiative.org). A new version, covering 98% (784 Mbases) of the cv. 'Vanuxem' genome assembled into 12 pseudo-chromosome sequences using the integrated genetic-physical map, will soon be submitted for publication. This version will also be available for querying and for download at the above hardwood genomics website. The reference genome pseudo-chromosomes should serve well for whole genome analyses such genotyping-by-sequencing for following the progress of introgression of disease-resistance into American and European species of *Castanea*. Preliminary data, showing over 80% micro-synteny to other Fagaceae genomes, indicates that the reference genome should also prove valuable for translation of genetic information into other Fagaceae species.

Genetic Diversity Within the Genus *Castanea*

Chestnut morphological characteristics were recorded by Breviglieri's (1955) "Scheda Castanografica", after the UPOV chestnut guideline (1988) for cultivar identification and, more recently, were applied to the European cultivars (Pereira-Lorenzo et al. 1996a, 2006, Gomes-Laranjo et al. 2007, Costa et al. 2008) and different chestnut species (Oraguzie et al. 1998). The first diversity studies in chestnut used isoenzymes. Sawano et al. (1984) studied 16 clones (ten Japanese, three Chinese, and two hybrids), and later Wen and Norton (1992) studied 22 Chinese cultivars. The European cultivars showed great genetic variability, with multiple genotypes obtained per cultivar indicating their polyclonal origin (Costa et al. 2008, Pereira-Lorenzo et al. 1996a, b, 2011), probably from cross pollination between lines, but possibly also from mutations. The genetic diversity of 216 chestnut accessions was assessed by Nishio et al. using 12 SSR markers, including 142 Japanese chestnuts (*C. crenata*), 38 Chinese chestnuts (*C. mollissima*), 2 European chestnuts (*C. sativa*), 9 American chestnuts (*C. dentata*), and 23 Japanese–Chinese hybrids (Nishio et al. 2011a). The mean values of H_o and H_e in the Japanese chestnuts were 0.65 and 0.65, respectively. Three major groups were identified, corresponding to Japanese, Chinese and American chestnuts. Japanese–Chinese hybrid cultivars were most closely related to the Chinese chestnut group. Genetic relationships were identified using 22 SSR markers, including 23 Chinese chestnuts (*C. mollissima*), 42 Japanese chestnuts (*C. crenata*), and 5 Japanese–Chinese hybrids (Inoue et al. 2009). The H_o and H_e values of Chinese chestnuts (H_o: 0.68 and H_e: 0.68) were higher than those of Japanese chestnuts (H_o: 0.59 and H_e: 0.58).

Genetic Diversity Within Chinese Chestnut: C. mollissima

For the cultivated population in China, allozyme genetic variability in nine populations of *C. mollissima* was investigated using starch gel electrophoresis at 12 loci coding for 11 enzyme systems by Zhang and Liu in 1998. The level of genetic polymorphism and population differentiation was greater in *C. mollissima* than was usual for an outbreeding species, with the proportion of polymorphic loci (P) 71.3%, average number of alleles (A) 2.27, and average expected heterozygosity (He) 0.346. About 89.3% of the total genetic variation was partitioned within populations. The Nei's genetic distance ranked from 0.010~0.177. A principal component analysis of the data based on allelic frequencies revealed a spatial pattern of genetic variability in Chinese chestnut, and provided some data facilitating selection of enzyme markers. Based on previous research, the effect of artificial selection on the genetic diversity of C. *mollissima* was also studied, and it could be inferred that the southwest area of China was the possible centre of genetic diversity in this species (Zhang and Liu 1998). When eighty-nine traditional cultivars of Chinese chestnut were investigated for genetic diversity, cultivar identification and genetic relationships using 15 isozyme alleles from 9 markers, high genetic diversity was observed in major chestnut production provinces, such as Zhejiang, Shandong, Hubei and Jiangsu, and eighty-four out of 89 cultivars could be differentiated. The UPGMA dendrogram showed that most of the cultivars of Shandong, Hubei, Jiangsu and Henan provinces were clustered together (Bao and Huang 2002).

The genetic diversity and population structure of Chinese chestnut cultivars was later assessed by AFLP analysis by Xu Hong-mei in 2004. The percentage polymorphism was 60.77%, and the average genetic diversity was h=0.1895. Genetic diversity within populations ranged from 0.1631 in the Beijing population to 0.1803 in the Hubei population, and genetic variation within populations was higher than between populations. Gene flow for four populations was over two, which suggested that sufficient genetic exchanges had taken place to prevent genetic differentiation made by genetic drift. Spatial autocorrelation analysis of 45 cultivars showed a randomly distributed spatial pattern, but some loci displayed a partial spatial structure (Xu Hong-Mei 2004).

The genetic diversity of wild Chinese chestnut (*C. mollissima*) was further evaluated using ten SSR markers, and 84 alleles were successfully amplified from 69 accessions of the four wild populations (Tian et al. 2009, Cheng et al. 2012, Ai et al. 2011). The numbers of alleles per locus ranged from 4 to 13, with an average of 8.4. The values of average number of alleles (Na) and effective number of alleles (Ne) were 8.4 and 4.998, respectively. Average values of H_e and PIC (polymorphism information content) were 0.777 and 0.739, respectively. However, the highest genetic diversity was found in the Shaanxi population. The low genetic differentiation (G_{st} = 0.141) showed that genetic variation mainly occurred within the wild populations (Huang et al. 2010). The diversity of cultivated chestnuts and closely related species was also studied using chloroplast SSRs: Na, Ne, H_e, and Nei's index values obtained were 3.25, 2.554, 0.606 and 0.320, respectively. The diversity of cpSSR loci was much lower than that revealed by nuclear SSRs. Furthermore, the intra-specific genetic diversity was high but varied greatly between different groups, and the wild chestnuts in China showed the highest genetic diversity. The results have important implications for understanding the population genetics and evolutionary patterns in the genus *Castanea* and provide baseline data for formulating conservation and management strategies (Cheng 2015). Wild chestnut diversity in different local areas was also studied by SSR, cpSSR and AFLP markers (Tian et al. 2009, Cheng et al. 2012, Ai et al. 2011).

Genetic Diversity Within North American Species C. dentata, C. pumila and C. ozarkensis

Studies with isoenzymes suggested that *C. dentata* was the least variable of the North American species, with an H_e of ~0.18 (mean expected heterozygosity) (Huang et al. 1994b, Huang et al. 1998), followed by *C. pumila* with an H_e of ~0.30 (Fu and Dane 2003) and *C. ozarkensis* with an H_e of ~0.27 (Dane et al. 1999), similar to that found in other woody plant species (Hamrick and Godt 1989). Huang et al. (1998) also showed evidence for possible glacial refugium in southern populations of *C. dentata* after evaluating the variation in ~1000 trees from 18 sample sites using RAPD and SSR markers. This is consistent with the hypothesis of a single metapopulation that continued to evolve due to genetic drift (Kubisiak and Roberds 2006). Asian and European *Castanea* (Villani et al. 1991a, b, Huang et al. 1994a, Lang et al. 2007) levels of genetic diversity based on isoenzymes appear to be similar to those reported for *C. seguinii* (H_e ~0.20), and levels reported for *C. pumila* appear similar to those reported for other *Castanea*; *C. mollissima* (H_e ~0.31), *C. henryi* (H_e ~0.26), and *C. sativa* (H_e ~0.24).

Genetic Diversity Within European Chestnut C. sativa

In natural chestnut populations in Europe, isozyme studies showed that genetic variability increased from the East (Turkey) to the West (Italy) (Pigliucci et al. 1990a, b, Villani et al. 1991a, b, 1993, Aravanopoulos et al. 2002). Two main centers of variability in European cultivated chestnut were identified in the Iberian Peninsula using SSR markers, one in the north and a second in the center (Pereira-Lorenzo et al. 2010, 2011). Ten cultivar groups were differentiated: four in northern Spain, five in the centre of the Iberian Peninsula, and one in southern Spain which was related to those in the center of the Iberian Peninsula. In the same study (Pereira-Lorenzo et al. 2010), it was demonstrated that cultivar origin and the diversification process was a combination of clonal propagation of selected seedlings, hybridization, and mutations, which led to high levels of diversity. In Portugal and Spain, clones accounted for 33% of the trees because of grafting.

In Italy, 33 SSR loci were isolated in chestnut (Marinoni et al. 2003), and several oak loci (Steinkellner et al. 1997, Kampfer et al. 1998) were found to be polymorphic in *Castanea sativa* (Boccacci et al. 2004). Microsatellites are preferred for the DNA genotyping of cultivars aimed at identification, and have been used in many studies of characterization of *C. sativa* Mill. cultivated germplasm (Marinoni et al. 2003, Gobbin et al. 2007, Martín et al. 2009, Pereira-Lorenzo et al. 2011). Twenty cultivars from the North West Italian germplasm were characterized at 14 SSR polymorphic loci (Marinoni et al. 2003) andthe total number of alleles was 90, ranging from 4 to 10 per locus, with an average of 6.4. The mean expected heterozygosity was 0.72 (range: 0.65-0.83). The average observed heterozygosity (H_o) was 0.793 (range: 0.35-0.95). Martin et al. (2010) published the genetic characterisation of 94 accessions, corresponding to 26 different cultivar names from different Italian Regions, with 7 SSR loci. Three different genotypes were found within the Marrone type cultivars 'Marrone di Chiusa Pesio', 'Marrone Combai' and a large group including predominantly 'Marrone' cultivars from Central Italy. Cluster analysis showed a separation between South Italian cultivars and cultivars from Central and North Italy. Mellano et al. (2012) studied 105 cultivars, 58 of them from Italy. These cultivars are planted in the Centro Regionale di Castanicoltura (Chiusa Pesio- CN, Piemonte), which is a collection field of chestnut germplasm including regional, national, and international accessions. These authors presented SSR data for 31 Italian accessions showing the presence of 22 unique genotypes and confirming synonymy among a group of Marrone cultivars: 'Marrone di Gemonio' (Lombardia), 'Marrone di Roccamonfina' (Campania), 'Marrone di Castel del Rio', Marrone di Zocca' (Emilia Romagna), 'Marrone Caprese Michelangelo' (Toscana), 'Marrone di San Mauro Saline' (Veneto), 'Marrone di Segni' (Lazio), and 'Marrone Val Susa' (Piemonte). Further work was carried out within the EU project MANCHEST, and 121 North Italian accessions (Piemonte Region), including 39 Marrone individuals, were characterized (Botta et al. 2006). Fifty-two genotypes were identified by the markers and were described by chemical and morphological traits. Sixty-eight grafted chestnut trees were evaluated using 10 SSRs (simple sequence repeats) loci and 20 morphological descriptors (Torello Marinoni et al. 2013). The H_o ranged from 0.64 to 0.89, with an average of 0.75; and H_e varied from 0.59 to 0.83, with an average of 0.72. Thirty-six different genotypes were identified and the analysis of the genetic structure revealed that four gene pools contributed to the formation of this germplasm. Cultivars tended to group into a main gene pool based on their growing area and prevalent use. Among morphological traits, nut and leaf shape, nut hairiness and male flower type were the most discriminant characters associated with the genetic structure.

The variability of the Spanish chestnut cultivars evaluated by isoenzymes (Pereira-Lorenzo et al. 1996b, 2006) demonstrated that, for the most part, a main clone was predominant in orchards (over 60% of the samples), but intra-cultivar variability was important, probably due to the use of seedlings of those main cultivars by the growers. The H_o and H_e values obtained with isoenzymes in the Spanish chestnut cultivars were, on average, 0.398 and 0.333 respectively (Pereira-Lorenzo et al. 2006). The excess of heterozygotes found in Galician chestnuts that were at least 300 years old was similar to that found in natural populations of *C. dentata* that were over 70 years old in Virginia, USA (Stilwell et al. 2003). Work in Portugal and Spain with microsatellites confirmed the variability found with isoenzymes in Spanish cultivars (Pereira-Lorenzo et al. 2006, 2010, Ramos-Cabrer et al. 2006) and heterozygosity was significantly higher than in other species. In the Portuguese cultivar 'Judia' Dinis et al. (2011) found only slight genotypic polymorphisms between selections from different regions of Trás-os-Montes.

Genetic Diversity in Japanese Chestnut *C. crenata*

In Japan, thirty Japanese chestnut accessions (*C. crenata*) were evaluated by 14 polymorphic SSR markers. These included 12 cultivars and 6 wild trees which originated in Japan, and 6 cultivars and 6 wild trees which originated in the Korean Peninsula (Yamamoto et al. 2003). The average H_o and H_e values were 0.50 and 0.54, respectively. No differences in allele composition were observed between cultivated and wild chestnuts or between trees with Japanese and Korean origins. The results could indicate that the Japanese chestnuts originating from Japan and the Korean Peninsula have similar genetic backgrounds, and that cultivated chestnuts could have been selected from wild chestnuts. The genetic diversity of wild chestnut (*C. crenata*) populations in northern Japan showed a high level of heterozygosity in wild populations (Tanaka et al. 2005). The H_O and H_e values in the chestnut (*C. crenata*) populations (H_O: 0.727 and H_e: 0.780) were similar to other *Fagaceae*, such as *Fagus sylvatica* (0.727 and 0.753) (Pastorelli et al. 2003), *F. orientalis* (0.697 and 0.740) (Pastorelli et al. 2003), and *Quercus rubra* (0.679 and 0.737) (Aldrich et al. 2002).

In order to clarify the breeding history, the genealogy, and spreading pattern of Japanese chestnut cultivars in Japan, Nishio et al. (2014) evaluated the population structure of 60 native Japanese chestnut cultivars by hierarchical clustering and Bayesian model-based clustering (Nishio et al. 2014). Both analyses resulted in two main clusters: one with cultivars from the Tanba region (the central part of modern Kyoto Prefecture and the east-central part of Hyogo Prefecture), the other with cultivars from other areas of Japan. Additionally, parent–offspring relationships obtained by 175 SSR markers suggested that native cultivars were originated in the Tanba region, and then were spread throughout Japan. It is assumed that local cultivars were established after crossing local chestnuts with Tanba cultivars, followed by selection, and introgression of favourable traits from the Tanba cultivars.

Mating System in *Castanea*

Chestnut is a monoecious species with male flowers in catkins and female flowers that develop at the base of bisexual catkins. Many interspecific hybrids are male sterile with astaminate catkins that do not produce pollen. Brachystaminate catkins can produce very limited quantities of pollen, while longistaminate catkins are those that produce more pollen. Astaminate catkins are more frequent in some of the best "marron" type cultivars. Male-sterility has been found in up to 21% of Spanish cultivars and 64% of Portuguese cultivars. Up to 8% of the cultivars in Spain and Portugal had astaminate catkins, and 13% had brachystaminate catkins. Those cultivars require pollinizers with longistaminate catkins (Pereira-Lorenzo et al. 2006, Pimentel-Pereira et al. 2007).

The first crossing studies of male sterility were made by Soylu in Turkey (1990). Seedlings obtained from controlled crosses among chestnut cultivars (*C. sativa* Mill.) were used. A total of 428 seedlings were observed during their 3rd and 4th years of growth, when most of them produced first catkins. As the result of that study it was suggested that male sterility is governed by two genes which are thought to be nearly equal in their expression. A genetic model based on two genes and five morphotypes was proposed: astaminate (xxzz), brachystaminate (xxZz), mesostaminate (Xxzz), and longistaminate long/short (XXZZ/XxZz). Some dominance relationships between the two genes and modifier genes could also be involved.

Pistillate flowers have six to eight styles whose tips are hollow at full bloom. The female flowers have seven (rarely six or eight) carpels. Each flower has 10 to 16 anatropous ovules. The mono-embryonic seeds (marron type) have been correlated with a high occurrence of anomalies, such as delayed embryo sac differentiation and the presence of supernumerary nuclei in the embryo sac (Botta et al. 1995). Very little is known about the genetic system controlling mating and the self-incompatibility system in chestnut, although it is considered to be a gametophytic type (Breviglieri 1951, Brewbaker 1957, Jaynes 1975). Cross pollination is necessary for nut production.

Inoue et al. (2012) investigated the self-compatibility of 51 genotypes of chestnut: 43 *C. crenata,* three *C. mollissima,* four *C. crenata* × *C. mollissima* and one *C. seguinii.* Thirteen putative self-pollinated seedlings obtained from five cultivars were analyzed for parentage by using SSR markers. As a result of SSR analysis, eight offspring including six from 'Toyotamawase', one from 'Ibuki' and one from 'Oomine' were identified as derived from self-pollination, while the other five offspring from two cultivars were shown to have originated from cross-pollination. 'Toyotamawase' showed the highest rate of bur set (62.5%) and seed set (33.3%), and all the analyzed seedlings from this cultivar were shown to have originated from self-pollination, suggesting that 'Toyotamawase' has some self-fertility, whereas the other chestnut cultivars have strong self-sterility.

Hybridisation Between Species

Castanea spp. all have $2n = 24$ chromosomes (Jaynes 1972), the number characteristic of most of the Fagaceae studied to date (Mehra et al. 1972, Ohri and Ahuja 1991, D'Emerico et al. 1995). A European collection of 71 *Castanea* hybrids (from France, Portugal, and Spain) was evaluated with SSRs, obtaining a H_e value of 0.722 (Pereira-Lorenzo et al. 2011). Hybrids showed the highest allelic richness with an average of 4.0 and were followed by *C. crenata* (3.8), *C. sativa* (2.9), *C. mollissima* (2.8) and *C. henryi* (2.7). In general, a rather high degree of fertility has been observed among interspecific hybrids (Jaynes 1964, 1972). When hybrids were made between *C. mollissima* and *C. henryi*, 18 (4%) of the 450 sampled offspring had intermediate leaf phenotypes between the two species (Liu et al. 2009). Three cultivars of *C. mollissima* ('Yanhong', 'Huaihuang' and 'Huaijiu') were used as female parents, with *C. henryi* as the male parent. Average seed setting rate over three years was only 23.6%, the lowest was 6.5%, and the empty bur rate was high (average 50.1%); the highest was 77.5%. Pollination and seed setting characteristics among different parental combinations in different years were not the same: significant differences in seed setting rate were observed between different years. However, the choice of female parent played an important role in deternining the seed setting rate. 'Huaihuang' × *C. henryi* and 'Huaijiu' × *C. henryi* produced more nuts than 'Yanhong' × *C. henryi* (Liu et al. 2013).

Although chromosome pairing appears to be normal in many interspecific hybrids, the presence of segregation distortion in some mapping populations (Gillet and Gregorius 1992, Kubisiak et al. 1997, Casasoli et al. 2001, Doucleff et al. 2004) and abnormal pairing in F_1 hybrid pollen mother cells (Faridi et al. 2008), suggest that significant chromosomal differences such as translocations and/or inversions are likely to exist among *Castanea* species. An attempt to generate chestnut triploids from colchicine-treated chestnut seedlings was reported by Genys (1963). A better understanding of genomic differences between species will be important for effective breeding using interspecific hybrids. A tractable genome size and abundant genetic and genomic resources make *Castanea* a good candidate for future targeted- or whole-genome sequencing (Kremer et al. 2007).

Cross-Pollination to Produce Interspecific Hybrids

Meiosis in pollen mother cells occurs 10 to 15 days before anthesis, in the first week of June in Italy (Botta et al. 1995). Pollen viability was observed to vary from 81.3 ± 6.1 % based on fluorochromatic reaction, to 58.2 ± 7.0 % in hanging drops and 50.1 ± 4.5 % germination on agar media (Botta et al. 1995).

Pollen is easily collected from the longistaminate pollinizers and desiccated to be stored in a refrigerator for short-term storage, or in a freezer for long-term storage. Emasculation and female flower isolation is needed to avoid unknown pollination. Male flowers from the bisexual catkins must be also removed. Bags cover the female flowers and hand pollination with fresh catkins brushed over pistils was detailed by Van Fleet in 1914 and Nienstaedt in 1956. Bags are removed at harvest time and nuts are collected when burs begin to crack and nuts begin to turn brown.

Interspecific Hybridization for Chestnut Improvement

Hybridization was the strategy used for many years to overcome pest and diseases: in the USA for blight disease on *C. dentata*, in Japan for gall wasp (*Dryocosmus kuriphilus*) infestation of *C. crenata*, and in Europe for *Phytopthora* spp. (ink disease). Hybridization was also used in Japan to incorporate the easy-peeling trait into nuts of Japanese cultivars. The first controlled crosses were made in 1894 in the USA (Van Fleet 1914), in 1926 in Spain (Gallastegui 1926) and in 1917 in Japan (Kajiura 1978). Hybrid clones have been released in Japan, the USA, and Europe (France, Spain and Portugal; Pereira-Lorenzo et al. 2010) and some of these are commercialized for nut production, timber production, and rootstocks. The International Society for Horticultural Science list of chestnut cultivars and their characteristics can be found at http://www.ct.gov/caes/chestnutcultivars.

Interspecific Hybridization to Introgress Disease and Insect Resistance into European Cultivars

Japanese chestnut trees were introduced in south-western France because of their resistance to ink disease (*Phytophthora* spp.), to try to maintain nut production (Lafitte 1946). They were also used as rootstocks for grafting the susceptible cultivars. However, they proved to be unsuitable for the environmental conditions, because they were prone to early frost damage due to earlier blooming. Grafting European chestnut cultivars on Japanese rootstocks was also a problem because of possible delayed graft-incompatibility reactions between rootstock and scion. Moreover, Asian species showed phenotypic traits that were not well accepted by growers and consumers, such as poor quality of the nuts and poor timber stature. For that reason, hybridization of Japanese chestnuts with European chestnuts was one of the first breeding programs begun in forest/fruit species in France, Portugal and Spain. Systematic hybridization and selection programmes started in the late 1940s in many European countries in order to select good fruit cultivars resistant to both ink disease and chestnut blight disease.

In France, Schad et al. (1952) began a breeding program to select interspecific hybrids between *C. sativa* and *C. crenata* resistant to ink disease. This program initially exploited natural hybrids (*C. crenata* × *C. sativa*), giving rise to cultivars resistant to ink disease and suitable for fruit production ('Bournette' and 'Précoce Migoule') or as rootstocks ('Marsol' and 'Maraval'). 'Marigoule', another natural hybrid, was released first for reforestation, then it was used as rootstock, and it is currently used for fruit production. The breeding program then turned towards controlled hybridizations, allowing the release of cultivars such as 'Bouche de Betizac', which is resistant to ink disease, tolerant to chestnut blight, and was recently found to be resistant to gall wasp infestation (*Dryocosmus kuriphilus*). In parallel, new root-stocks resistant to ink disease were also created ('Maridonne' and 'Marlhac'). Since 1971, the breeding work was carried out at INRA Bordeaux (Chapa et al. 1990, Salesses et al. 1993a, b), which provided hybrids more similar to the European species, and which are now the base of the French chestnut production (both rootstocks and nut producers). The traits sought in the progenies from *C. sativa* and *C. crenata* for rootstocks were: improved tolerance to ink disease (compared to the European chestnut), medium to high vigour, adaptation to French climates (cold, in particular), and ease of propagation by mound layering. For chestnut selections for nut production, aims were to improve nut size (compared to the European chestnut), produce nuts that were one per bur nuts without pellicle intrusions, and improve nut quality (compared to the Japanese parent). The Japanese chestnuts are more resistant to *Phytophthora* but are less suitable as rootstock, being susceptible to cold injuries and having a smaller trunk diameter than the more vigorous *C. sativa*. They have large nuts, but these are of lower quality of aroma and sweetness. Several hybrid selections suitable for rootstocks cultivars were obtained from the French breeding program and some of them, such as 'Marigoule' and 'Maraval', are used also for fruit production. The graft affinity of rootstocks with the European cultivars is variable and not always very good; the level of tolerance to *Phytophthora* is variable but higher than in *C. sativa*.

A breeding program started in 1986 aimed at obtaining cultivars with higher quality that contained the resistance to *Cryphonectria* present in *C. mollissima*. The genepool used for breeding included genotypes of *C. mollissima*, *C. crenata*, hybrids of *C. sativa* and *C. crenata* and cultivars of *C. sativa*. From 1980 to the present, the breeding strategy has consisted of enlarging the genitors' genetic base, increasing not only the number and the origin of *C. sativa* genitors, but also the exotic ones (*C. crenata* and *C. mollissima*). The breeding program was oriented in two directions, for nut production and root-stock selection. For nut varieties, the breeding objectives were to develop cultivars tolerant to chestnut blight and resistant to abiotic stresses (frost and water logging). In addition, selections were made for nuts of good organoleptic qualities and "marron" type development for the fresh market, and nuts with easy peeling for the processing industry. For root-stocks, the breeding criteria were tolerance to ink disease, resistance to abiotic stress, and wide grafting compatibility. More than 20 000 first generation interspecific hybrids were obtained by controlled crosses and evaluated. At the end of the breeding process six nut cultivars and 30 root-stocks were retained. One of the new nut cultivars, 'Bellefer', has been submitted to the Community Plant Variety Office (CPVO) to obtain plant breeders' rights on a European level.

A similar breeding program was initiated in Spain between *C. crenata* and *C. sativa* by Gallastegui (1926), and later on by Urquijo (1944), obtaining some vigorous hybrids used for timber production, and others for rootstocks, with good compatibility to cultivars (Urquijo 1944, 1957, Pereira-Lorenzo and Fernández-López 1997). In Portugal, the first interspecific hybridizations were initiated in 1947 by Bernardino Barros Gomes. The male parents were the *C. crenata* cultivar 'Tamba' (Gomes Guerreiro 1948, 1957) and in some cases, *C. crenata* pollen was brought from Spain (Vieira Natividade 1947). Those hybrids were studied with SSRs and compared to the main chestnut species, and it was found that a total of 18 out of 61 identified interspecific hybrids were Japanese-European (30%), confirming that *C. crenata* was the main source of resistance to ink disease used in the European breeding program, as stated by Urquijo (1944, 1957).

The first introduction and evaluation of Asian chestnut germplasm in Italy (seedlings of *C. crenata* and *C. mollissima*) dates back to 1920-1930, and was aimed to select material for nut production, use in forestry, and for improving resistance to diseases (Bagnaresi 1956). In the early 1970s, The Institute of Arboriculture (now DISAFA) of the University of Turin introduced *C. crenata* fruit cultivars from Japan and started a breeding program aimed at obtaining selections more tolerant to chestnut blight with high yield and good nut quality. Thousands of seedlings were obtained from open pollinated cultivars 'Tsukuba' and 'Tanzawa' and were selected for the desired traits. Two new cultivars ('Primato' and 'Lusenta') were selected within the 'Tsukuba' progeny. 'Primato' was released in 1986; and is resistant to gall wasp and ripens early (Paglietta 1986, 1992, Sartor et al. 2015). More recently (2008-2010) interspecific hybridization was carried out at DISAFA of University of Turin between *C. sativa* ('Madonna') and the Euro-Japanese hybrid 'Bouche de Bétizac' (Dini et al. 2012). The aim of the program was to obtain progeny to map the trait of resistance to gall wasp found in 'Bouche de Bétizac'. Further aims are the study of nut traits and the selection of genotypes with improved tolerance to diseases and early ripening, while bearing nuts of quality comparable with that of the European parent. Selection and map construction is currently in progress.

In Portugal interspecific hybridization has been on-going since 2006, and two full-sib progenies were obtained from artificial controlled crosses *C. sativa* × *C. crenata* and *C. sativa* × *C. mollissima* to introgress *P. cinnamomi* (ink disease) and *C. parasitica* (chestnut blight) resistance genes from Asian species into the susceptible European chestnut species (Costa et al 2011). A primary goal is to use these chestnut hybrids to perform DNA marker-trait association for resistance QTL identification. As a first step, the Chinese chestnut transcriptome-derived molecular markers were used for linkage map construction and QTL detection. Additionally, a reliable phenotyping method was established to evaluate the ink disease resistance of each individual progeny plant. This method uses two types of inoculation tests to determine the response of individual chestnut seedlings from two mapping populations to inoculation with *P. cinnamomi*: root inoculation using replicates of each genotype obtained by micropropagation and excised shoot inoculation test (Santos et al. 2015b). The variable days of survival was the most important indicator of resistance to ink disease, because differences in response between genotypes were maximized in the root inoculation test. The lesion caused by the inoculation of excised shoots is considered to be an indirect measure of resistance. The results showed that, similar to root inoculated plants, the resistance to *P. cinnamomi* in the shoots is related to the confinement of the lesion to point of inoculation. For the most resistant genotypes, the surrounding tissues dried, limiting the progression of the lesion. In this study, shoot internal lesions were evaluated for the first time as a parameter to assess chestnut resistance to *P. cinnamomi*. This parameter was chosen because it indicates the spread of the pathogen from the roots and collar to the aerial vascular system. This is important for determining the degree of plant resistance, as the rapid invasion of the pathogen into the phloem and xylem may affect water and nutrient movement through the shoots, causing death. The strong favourable genetic correlation observed between the two variables suggests that indirect selection could be made by lesion progression rate, which is easily measured. In this study, the resistance phenotype of each progeny was accurately determined by measuring lesion progression rate. Lesion progression rate was the variable selected to perform DNA marker–variable association, for future QTL identification. Thirty five percent of genotypes were selected as the most resistant so far, and they are being propagated by micropropagation to be used as rootstocks.

Complex Interspecific Hybridization for Multiple Trait Improvement in Turkey

Hybrid chestnut seeds were imported by Ondokuz Mayis University in Turkey from The Connecticut Agricultural Experiment Station, USA in 2005. These were 30 seeds of *C. crenata* from a hand pollinated cross of two trees planted in Connecticut in 1876, both are blight resistant and cold hardy, and 50 seeds of a complex hybrid from a hand pollinated cross of 'King Arthur' × 'Lockwood'. The former is a *mollissima/seguine* hybrid and the latter is a *crenata/sativa/dentata* hybrid. First they were planted in pots. After one year (in 2006), surviving plants were planted into a genetic sources collection orchard at The Black Sea Agricultural Research Institution with 7 × 5 m spaces. Beginning in 2006, plant growth, yield and some pomological characteristics of the hybrid genotypes were examined (Serdar and Macit 2010).

Some complex hybrid chestnut seedlings were grafted on 3- to 8-year-old wild Turkish rootstock (*C. sativa* Mill) in May 2012. Clones of tree A-100 had the best survival (Serdar et al. 2014). The gall wasp was observed for the first time in Yalova province in Turkey in 2014. To determine the resistance levels of the genotypes to gall wasp, they were grafted on 3- to 6-year-old wild rootstock (*C. sativa* Mill.) in May 2015 in this province. Amongst complex hybrids, A-25, A-100 and A-14 may be good enough to release as cultivars. In addition, A-25 and A-41 may be useful as dwarfing rootstocks.

Dwarfing culture and close planting can improve the yield and quality of *C. mollissima,* and would be important for growers, but there were no suitable cultivars available. Therefore, rapid selection for new varieties for dwarf culture and close planting became the new breeding objective in Turkey. Chestnuts types known as "weeping chestnuts" grow shorter. The conventional breeding cycle of chestnuts is long, and it would be an advantage to shorten breeding time, and speed up the breeding process. Work is underway to lay the foundation for marker-assisted breeding to do this in the future. Identification of molecular markers linked to the weeping branch traits of Chinese chestnut and molecular marker assisted breeding are being studied. Molecular markers linked to weeping branch trait were detected with AFLP and SRAP markers coupled with the bulked segregant analysis method. In weeping branch varieties and vertical branch varieties, five SRAP markers (me4Me1-160, me4Me1-12, me4Me2-24, me4Me2-130, me5Me2-200) were screened that may have been linked to the weeping branch traits of Chinese chestnut. Using the backcross hybrids, with the supporting nursery technology and Marker Assisted Selection, two SRAP markers (me4Me1-160 and me4Me1-120), were identified as linked to the weeping branch traits of *C. mollissima* Blume. The result validation showed that the weeping branch traits of *C. mollissima* might be controlled by independent recessive genes (Liu Ting 2011).

Introgressing Resistance to Chestnut Blight in the USA

In the USA, a backcrossing program using blight resistant Chinese and Japanese species as the donor and American chestnuts (*C. dentata*) as the recurrent parent has successfully produced blight resistant timber chestnuts (Diskin et al. 2006, Anagnostakis, 2012). The first hybrid reported in the USA (in 1894) was made by Walter Van Fleet using pollen from an American chestnut on a European (or European-American) cultivar 'Paragon' (Van Fleet 1914, 1920). The first Japanese-American hybrid cultivar was named 'Daniel Boone' and was made in 1899 by amateur nut grower George W. Endicott (Taylor and Gould 1914, Detlefsen and Ruth 1922). Between 1900 and 1921, crosses including native chinquapin (*C. pumila*) and Japanese chestnut were made by Van Fleet and one of his best hybrids (S8) was planted at the Bell, Maryland research farm (Nienstaedt 1948, Anagnostakis 2012). A seedling of Van Fleet's S-8 (its row and tree location in the nursery) was a cross made by Arthur Graves in Connecticut in 1937 with a forest-type Japanese to produce the cultivar 'Essate Jap'. From 1937 until 1960, hybrids were obtained and tested in the U.S.A. for their resistance to chestnut blight and their fitness throughout North America. The longest-continuing chestnut breeding program in the United States is that in Connecticut. From 1930 onwards, breeders focused on making hybrids that were combinations of species, looking for single ideal progeny that could be propagated clonally. The Connecticut Agricultural Experiment Station trees are also being used by the American Chestnut Foundation. Two first-generation-backcross trees

[(Chinese × American) × American], are now over sixty years old. Diller planted the hybrid 'Clapper' in 1946 (Diller and Clapper 1969), and Graves and Nienstaedt the other old BC_1 tree, named 'Graves', in 1953. Both of these have timber form, good blight resistance, and acceptable nuts. BC_3-F_2 generation progeny are now being planted and tested, with 93% of seedlings showing morphological characteristics of American chestnut (Diskin et al. 2006). New hybrids are now being developed to include resistance to the Asian chestnut gall wasp, using Ozark chinquapins (*C. ozarkensis*), which are resistant to infestation (Anagnostakis 2012).

Interspecific Hybridization for Chestnut Improvement in Japan: Easy-Peeling Pellicles

Previously, all Japanese chestnut cultivars were thought to have difficult-to-peel pellicles, which increase the labour and cost for removing the pellicle from the nut during processing. Therefore, a pellicle that is easier to peel has been an important objective of Japanese chestnut breeding programs. 'Porotan' is a newly released Japanese chestnut cultivar characterized by the easy-peel pellicle trait. 'Porotan' was obtained from the cross of 550-40 × 'Tanzawa', and both parents are Japanese chestnuts with difficult-peeling pellicles. Out of 59 offspring of the cross, 12 hybrids had an easy-peeling pellicle and 47 had a difficult-to-peel pellicle (1:3 for monogenic inheritance) (Takada et al. 2012). Thirty-nine offspring from 'Tanzawa' × 'Porotan' segregated in a ratio of 19 difficult-to-peel pellicle to 20 easy-peeling pellicle (1:1 ratio for monogenic segregation). The easy-peeling pellicle trait of 'Porotan' is controlled by a single recessive gene, designated as *P*/*p*. The genotypes of 'Tanzawa' and 550-40 are heterozygous (*Pp*), and the 'Porotan' genotype is homozygous recessive (*pp*). A total of 11 SSR markers were established showing significant linkages to the *p* gene from genetic analysis using two F_1 populations, 'Tanzawa' (*P*/*p*) × 'Porotan' (*p*/*p*) and 550-40 (*P*/*p*) × 'Tanzawa' (*P*/*p*) (Nishio et al. 2013). Molecular markers linked to the easy-peeling-pellicle trait have been used for marker-assisted selection, which will greatly improve Japanese chestnut breeding.

C. sativa and *C. mollissima* were introduced into Japan from the Meiji period (1868-1911) onward, as breeding resources. It was unfortunate that *C. sativa* grew very poorly in Japan due to its high susceptibility to chestnut blight disease, and that *C. mollissima* was not adaptable to the Japanese climate (Shimura 1984). Beginning in 1907, Japanese private breeders brought nuts of *C. mollissima* from China, and selected individuals showing adaptability to the Japanese climate from *C. mollissima* seedlings (Isaki 1978). They succeeded in selecting some cultivars of *C. mollissima* with easy pellicle removal. However, these cultivars did not flourish due to serious damage by the chestnut gall wasp, which susceptibility was discovered in 1941. Gall wasp spread rapidly throughout the country, limiting the use of these lines, but the selected cultivars were used as parents for subsequent breeding programs. Since both *C. sativa* and *C. dentata* grew poorly in Japan due to their high susceptibility to chestnut blight disease, almost all hybridizations in Japan have been made between *C. mollissima* and *C. crenata*. The first hybridization of chestnut in Japan was done in 1917 by Tanikawa at the Horticultural Research Station, presently known as the Institute of Fruit Tree Science, National Agriculture and Food Research Organization (hereafter NIFTS) (Kajiura 1978). Beginning around 1930, chestnut breeding programs were also initiated in Kanagawa, Niigata, and Hyogo Prefectures (Tano 1954, Shimura et al. 1971, Kanato 1973). The major purpose of these programs was to introgress easy pellicle peel-ability into Japanese chestnut from *C. mollissima*. However, no easy pellicle peeling offspring were obtained in F_1 seedlings between *C. mollissima* and *C. crenata*, and easy pellicle selections obtained from *C. mollissima* × *C. mollissima* were not released due to their susceptibility to gall wasp (Kanato 1973). However, some F_1 cultivars of *C. mollissima* and *C. crenata* were released by the private breeders. These hybrids do not all have easy pellicle-peeling abilities and almost all of them did not flourish due to susceptibility to gall wasp. However, 'Riheiguri', which is a chance seedling discovered from a mixed planting orchard of *C. crenata* and *C. mollissima* in Gifu Prefecture, was released in 1950 by Tsuchida. In 2012, 'Riheiguri' accounted for about 8% of the total planting area due to its large nut size and superior nut quality. Morphological traits and genetic structure (Nishio et al. 2011a) showed that 'Riheiguri' was probably a hybrid between *C. crenata* and *C. mollissima*.

An organized chestnut breeding program supported by Japanese governmental funds was initiated in 1947 by NIFTS (Shimura et al. 1971, Kotobuki et al. 1999). The initial goal was to develop cultivars from the intraspecific crosses of *C. crenata* which ripen early or slightly later, and which excel in yield, nut size, and nut quality. Crosses of *C. crenata* and C *mollissima* were made to develop cultivars with easy pellicle peeling. At the beginning of the program, 34 interspecific crosses were made using 18 maternal and 24 paternal parents (Machida 1984). No interspecific hybrid progeny were selected because nut size or nut quality did not meet the selection criterion. However, 13 lines were selected from intraspecific crosses of *C. crenata*. Resistance to gall wasp was added as a breeding goal in 1952, and therefore interspecific hybridization crosses were fewer, since *C. mollissima* was susceptible to gall wasp. Three *C. crenata* cultivars ('Tanzawa', 'Ibuki', 'Tsukuba') which have early to medium ripening, were released as resistant to gall wasp in 1959 and were spread widely. The hybridization program was restarted in 1966 (Machida 1984). Seventy crosses were made and 1390 offspring were obtained. Tanaka and Kotobuki (1992) investigated pellicle removability of up to 200 interspecific hybrid selections. The pellicle peelability of F_1 hybrids was found to be influenced by the pollen parent. Nuts from crosses with *C. mollissima* pollen showed easy pellicle peelability, whereas nuts from crosses with *C. crenata* pollen had difficult pellicle peelability. Selections with easy pellicle peelability were found among progeny from cross combinations of siblings among the F_1s and from back crossing with *C. mollissima* cultivars. However, those easy pellicle removal selections obtained from the program were not released as cultivars due to inferior agronomic traits. From the 1990s, F_1 hybrids (interspecific hybrids between *C. crenata* and *C. mollissima*) have been used as parents, especially for improving the eating quality of nuts in the breeding program of NIFTS. Recently, 'Shuuhou' was released, which is a descendent of 'Riheiguri' (a putative hybrid of *C. crenata* × *C. mollissima*) (Kotobuki et al. 2005). 'Mikuri' which is an offspring of 'Shuuhou', was released as a cultivar with good eating quality although its pellicle is as difficult to peel as that of *C. crenata* (Saito et al. 2015). However, another released cultivar 'Porotan', which is an intraspecific cross of *C. crenata*, has a pellicle which is as easily removed as that of Chinese chestnut (Saito et al. 2009).

Interspecific Hybrid Mapping Populations

Linkage relationships between isoenzymes and morphological traits in some interspecific crosses have been found (Huang et al. 1996), and molecular maps have been developed (Kubisiak et al. 1997, Casasoli et al. 2001, Kubisiak et al. 2013). Mapping progenies have been obtained in Europe by hybridisation based on European chestnut (*C. sativa*) in order to generate genetic maps of key phenotypic traits. The National Research Council (CNR-IBAF) conserves a full-sib mapping family from which the first genetic map of *C. sativa* was constructed (Casasoli et al. 2001). From this population, QTLs were identified for adaptive traits such bud phenology, growth and carbon isotopic discrimination (which provides an indirect measure of plant water use efficiency) (Casasoli et al. 2004). INRA-Bordeaux maintains two mapping families suitable for mapping resistance to the key diseases *Phytophthora* and *Cryphonectria* based on interspecific hybrids (*C. sativa* × *C. mollissima*, and *C. sativa* × *C. crenata*). DISAFA of University of Torino (UNITO) conserves 250 individual progeny obtained by crossing between the gall wasp-resistant hybrid 'Bouche de Bétizac' (*C. sativa* × *C. crenata*; Dini et al. 2012) and European chestnut (cultivar 'Madonna') showing segregation for gall wasp resistance and for phenological and nut traits. INIAV maintains 60 progenies of two hybrid crosses (*C. sativa* × *C. crenata* and *C. sativa* × *C. mollissima*) showing segregation for resistance to *P. cinnamomi*.

Conclusions

After more than 100 years of chestnut hybridisation, chestnut breeding programs will face new threats such as gall wasp, and other pests and diseases not even known yet. In order to meet these challenges, new technologies will be incorporated in combination with previously existing methods and resources. The use of the recently reported genome sequence will be extended to the other chestnut species and hybrids which, in combination with phenotypic traits, will allow us to maintain these trees, as others did before us, with a promising future.

Acknowledgements

To the Spanish Government for the project "An integrated approach to sustainable management of chestnut in Spain" (AGL2013-48017). A. Martín is grateful to the Secretaría General de Ciencia y Tecnología de la Consejería de Economía, Competitividad e Innovación from the Regional Government of Extremadura (Spain) for the financial support.

References

Ai C-X., G-N. Shen, K. Zhang, S-L. Tian and L. Xu. 2011. Genetic diversity on wild populations of chestnut in Qinba mountain area of west China with AFLP markers. Journal of Plant Genetic Resources 3: 408-412.

Aldrich, P.R., C.H. Michler, W. Sun and J. Romero-Severson. 2002. Microsatellite markers for northern red oak (Fagaceae: *Quercus rubra*). Mol. Ecol. Notes. 2(4): 472-474.

Aldrich, P.R., G.R. Parker, C.H. Michler and J. Romero-Severson. 2003. Whole-tree silvic identifications and the microsatellite genetic structure of a red oak species complex in an Indiana old-growth forest. Can. J. Forest. Res. 33: 2228-2237.

Anagnostakis, S.L. 2012. Chestnut breeding in the United States for disease and insect resistance. Plant Dis. 96: 1392-1403.

Aravanopoulos, F.A., A.D. Drouzas and P.G. Alizoti. 2002. Electrophoretic and quantitative variation in chestnut (*Castanea sativa* Mill.) in Hellenic populations in old-growth natural and coppice stands. For. Snow Landsc. Res., 76(3): 429-434.

Avanzato and G. Bounous. 2009. Following chestnut footprints (*Castanea sativa* L.). Damiato, Bounous (eds). Scripta Horticulturae Scripta Hort. 9. International Society for the Horticultural Science. ISBN: 978-90-6605-632-9.

Bagnaresi, U. 1956, Osservazioni morfo-biologiche sulle provenienze di castagno giapponese coltivate in Italia. Pubblicazione n. 3, Centro Studio sul Castagno (C.N.R.), Firenze. La Ricerca Scientifica (Suppl.) 26: 7-48.

Bao Z. and H. Huang. 2002. Analysis of genetic diversity and genetic relationships of Chinese chestnut. Acta Hortic. Sin. 1: 13-19.

Barakat, A., D.S. DiLoreto, Y. Zhang, C. Smith, K. Baier, W.A. Powell, N. Wheeler, R. Sederoff and J.E. Carlson. 2009. Comparison of the transcriptomes of American chestnut (*Castanea dentata*) and Chinese chestnut (*Castanea mollissima*) in response to the chestnut blight infection. BMC Plant. Biol. 9: 51. doi: 10.1186/1471-2229-9-51.

Barakat, A., M. Staton, C-H. Cheng, J. Park, N.B.M. Yassin, S. Ficklin, C-C. Yeh, F. Hebard, K. Baier, W. Powell, S.C. Schuster, N. Wheeler, A. Abbot, J.E. Carlson and R. Sederoff. 2012. Chestnut resistance to the blight disease: insights from transcriptome analysis. BMC Plant. Biol. 12: 38. doi: 10.1186/1471-2229-12-38.

Barreneche, T., M. Casasoli, K. Russell, A. Akkak, H. Meddour, C. Plomion, F. Villani and A. Kremer. 2004. Comparative mapping between *Quercus* and *Castanea* using simple-sequence repeats (SSRs) Theor. Appl. Genet. 108: 558-566.

Beccaro, G.L., R. Botta, D. Torello Marinoni, A. Akkak and G. Bounous. 2004. Application and evaluation of morphological, phoenological and molecular techniques for the characterization of *Castanea sativa* Mill. cultivars. Acta Hort. 693: 453-457.

Boccacci, P., A. Akkak, D. Torello Marinoni, G. Bounous and R. Botta. 2004. Typing European chestnut (*Castanea sativa* Mill.) cultivars using oak simple sequence repeat markers. Hort. Sci. 39: 1212-1216.

Botta, R., G. Vergano, G. Me and R. Vallania. 1995. Floral biology and embryo development in chestnut (*Castanea sativa* Mill.). Hort.Sci. 30(6): 1283-1286.

Botta, R., P. Guaraldo, M.G. Mellano and G. Bounous. 2006. DNA typing and quality evaluation of chestnut (*Castanea sativa* Mill.) cultivars. International Symposium "Optimization, Productivity and Sustainability of Chestnut Ecosystems in Mediterranean Europe" Catania (Italy), 23-26 February 2005. Advances Hort. Sci. 1: 96-100.

Breviglieri, N. 1951. Research on the flower and fruit biology in *Castanea sativa* and *Castanea crenata* in Vallombrosa territory. Publication n.1, Centro di Studio sul Castagno (C.N.R.) La Ricerca Scientifica (Suppl.) 21: 15-49.

Breviglieri, N. 1955. Indagini ed osservazioni sulle migliori varietà italiane di Castagno. Centro di Studio Sul Castagno, Pubblicazione N.2, Supplemento a la Ricerca Scientifica, pp. 27-164.

Brewbaker, J.L. 1957. Pollen cytology and incompatibility systems in plants. J. Hered. 48: 217-277.

Buck, E.J., M. Hadonou, C.J. James, D. Blakesley and K. Russell. 2003. Isolation and characterization of polymorphic microsatellites in European chestnut (*Castanea sativa* Mill.) Mol. Ecol. Notes, 3 (2), pp. 239-241.

Carlson, J., W. Powell, J. Tomkins, S. Ficklin and C. Smith. 2007. GS20 454 sequencing of the chestnut transcriptome: results of a pilot study. Plant & Animal Genomes XV Conference, January 13-17, 2007, San Diego, CA. W127.

Carlson, J., A. Barakat, D.S. DiLoreto and C. Smith. 2008. The chestnut transcriptome. Plant & Animal Genomes XVI Conference, January 12-16, 2008, San Diego, CA. P25.

Carlson, J.E., M.E. Staton, C. Addo-Quaye, N. Cannon, L.P. Tomsho, S. Ficklin, C. Saski, R. Burhans, D. Drautz, T.K. Wagner, N. Zembower, S.C. Schuster, A.G. Abbott, C.D. Nelson and F.V. Hebard. 2014. The Chestnut Genome Project. Plant & Animal Genome Conference XXII, San Diego, CA, January 12, 2014, Abstract W307.

Casasoli, M., C. Mattioni, M. Cherubini and F. Villani. 2001. A genetic linkage map of European chestnut (*Castanea sativa* Mill.) based on RAPD, ISSR and isozyme markers. Theor. Appl. Genet. 102: 1190-1199.

Casasoli, M., D. Pot, C. Plomion, M.C. Monteverdi, T. Barreneche, M. Lauteri and F. Villani. 2004. Identification of QTLs affecting adaptive traits in *Castanea sativa* Mill. Plant, Cell Environ. 27: 1088-1101.

Casasoli, M., J. Derory, C. Morera-Dutrey, O. Brendel, I. Porth, J-M. Guehl, F. Villani and A. Kremer. 2006. Comparison of QTLs for adaptive traits between oak and chestnut based on an EST consensus map. Genetics 172: 533-546.

Connors, B.J., C.A. Maynard and W.A. Powell. 2001. Expressed sequence tags from stem tissue of the American chestnut, *Castanea dentata*. Biotech. Letters 23:1407-1411.

Costa, R., C. Ribeiro, T. Valdiviesso, S. Afonso, O. Borges, J. Soeiro, H. Costa, L. Fonseca, C. Augusta, M.H. Cruz, M. Salazar, F. Matos Soares, J. Sequeira, A. Assunção, P. Correia and M.J. Lima. 2008. Variedades de Castanha das Regiões Centro e Norte de Portugal. INRB.I.P

Costa, R., C. Santos, F. Tavares, H. Machado, J. Gomes-Laranjo, T. Kubisiak and C.D. Nelson. 2011. Mapping and transcriptomic approaches implemented for understanding disease resistance to *Phytophthora cinammomi* in *Castanea* sp. BMC Proc. 5: O18. doi: 10.1186/1753-6561-5-S7-O18.

Chapa, J., P. Chazerans and J. Coulie. 1990. Multiplication vegetative du chataignier. Amelioration par greffage de printemps et bouturage semi-ligneux. L'Arboriculture Fruitiere 431: 41-48.

Cheng, L-L., H-D. Feng, Q. Rao, W. Wu, M. Zhou, G-L. Hu and W-G. Huang. 2012. Diversity of wild chestnut chloroplast DNA SSRs in Shiyan. Journal of Fruit Science 3: 382-386.

Cheng, L-L., G-L. Hu, S-C Su and W-G. Huang. 2015. Diversity of cultivated chestnuts and its close relative species using chloroplast SSRs. Acta Agricultural Boreali-sinica, 2: 145-149.

D'Emerico, S., P. Bianco, P. Medagli and B. Schirone. 1995. Karyotype analysis in *Quercus* spp. (Fagaceae). Silvae Genet. 44: 66-70.

Dane, F., L.K. Hawkins and H. Huang. 1999. Genetic variation and population structure of *Castanea pumila* var. *ozarkensis*. J. Amer. Soc. Hort. Sci. 124: 666-670.

Detlefsen, J.A. and W.A. Ruth. 1922. An orchard of chestnut hybrids. J. Hered. 13: 305-314.

Diller, J.D. and R.B. Clapper. 1969. Asiatic and hybrid chestnut trees in the eastern United States. J. Forest. 67: 328-331.

Dini, F., C. Sartor and R. Botta. 2012. Detection of a hypersensitive reaction in the chestnut hybrid 'Bouche de Bétizac' infested by *Dryocosmus kuriphilus* Yasumatsu. Plant Physiol. Bioch. ISSN 0981-9428, 60: 67-73.

Dinis, L.T., F. Peixoto, T. Pinto, R. Costa, R.N. Bennett and J. Gomes-Laranjo. 2011. Study of morphological and phenological diversity in chestnut trees ("Judia" variety) as a function of temperature sum, Environ. Exp. Bot., 70: 110-120.

Diskin, M., K.C. Steiner and F.V. Hebard. 2006. Recovery of American chestnut characteristics following hybridization and backcross breeding to restore blight-ravaged *Castanea dentata*. For. Ecol. Manage. 223: 439-447.

Doucleff, M., Y. Jin, F. Gao, S. Riaz, A.F. Krivanek and M.A. Walker. 2004. A genetic linkage map of grape, utilizing *Vitis rupestris* and *Vitis arizonica*.Theor. Appl. Genet. 109(6), 1178-1187.

Durand, J. 2009. PhD Thesis « Cartographie comparée chez les Fagacées ». Université de Bordeaux pp178. (in French).

Durand, J., C. Bodénès, E. Chancerel, J.M. Frigerio, G. Vendramin, F. Sebastiani, A. Buonamici, O. Gailing, H-P. Koelewijn, F. Villani, C. Mattioni, M. Cherubini, P.G. Goicoechea, A. Herrán, Z. Ikaran, C. Cabané, S. Ueno, S. Alberto, P.Y. Dumoulin, E. Guichoux, de A. Daruvar, A. Antoine Kremer and C. Plomion. 2010. A fast and cost-effective approach to develop and map EST-SSR markers: oak as a case study. BMC Genomics 11: 570.

Fang, G.C., B.P. Blackmon, M.E. Staton, C.D. Nelson, T.L. Kubisiak, B.A. Olukolu, D. Henry, T. Zhebentyayeva, C.A. Saski, C.H. Cheng, M. Monsanto, S. Ficklin, M. Atkins, L.L. Georgi, A. Barakat, N. Wheeler, J.E. Carlson, R. Sederoff and A.G. Abbott. 2013. A physical map of the Chinese chestnut (*Castanea mollissima*) genome and its integration with the genetic map. Tree Genet. Genomes 9(2): 525-537.

Faridi, N., C.D.Nelson, H. Banda, M. Abdul Majid, T.L. Kubisiak, F.V. Hebard, P.H. Sisco and R. Phillips 2008. Cytogenetic analysis of a reciprocal translocation in F1 hybrid between American and Chinese chestnuts. Plant & Animal Genomes XVI Conference, January 12-16, 2008, San Diego, CA. W346.

Ficklin, S., C. Smith, P. Sisco, N. Wheeler, J. Carlson, R. Sederoff and J. Tomkins. 2007. *Fagaceae* genomic database (FGD): an integrated web-based resource for tree genomics. Plant & Animal Genomes XV Conference, January 13-17, 2007, San Diego, CA. P848.

Fu, Y. and F. Dane. 2003. Allozyme variation in endangered *Castanea pumila* var. *pumila*. Ann. Bot. 92: 223-230.

Gallastegui, C. 1926. Técnica de la hibridación artificial del castaño. Boletín de la Real Sociedad Española de Historia Natural. Tomo XXVI, pp. 94-88.

Genys, J.B. 1963. One-Year data on colchicine-treated chestnut seedlings. Chesapeake Science. 4: 57-59.

Gillet, E. and H.R. Gregorius. 1992. What Can Be Inferred from Open-Pollination Progenies about the Source of Observed Segregation Distortion?—A Case Study in *Castanea sativa* Mill. Silvae Genet, 41, 82-87.

Gobbin, D., L. Hohl, L. Conza, M. Jermini, C. Gessler and M. Conedera. 2007. Microsatellite-based characterization of the *Castanea sativa* cultivar heritage of southern Switzerland. Genome, 50(12), 1089-1103.

Gomes Guerreiro, M. 1948. Acerca do uso da análise discriminatória: comparação entre duas castas de castanhas. – Sep. Das Publicações da Direcção Geral dos Serviços Florestais e Aquícolas, Vol. XV, Tomo I e II, p. 137-151.

Gomes Guerreiro, M. 1957. Castanheiros: alguns estudos sobre a sua ecologia e o seu melhoramento genético. Instituto Superior de Agronomía, Lisboa.

Gomes-Laranjo, J., J.P. Coutinho, F. Peixoto and J. Araujo-Alves. 2007. Ecologia do castanheiro. In J Gomes-Laranjo, J Ferreira-Cardoso, E Portela, CG Abreu, eds, Castanheiros. UTAD, Vila Real, pp. 109-149.

Hamrick, J.L. and M.J.W. Godt. 1989. Allozyme diversity in plants. *In:* Brown, A.H.D., Clegg, M.T., Kahler, A.L., and Weir, B.S., eds. Plant population genetics, breeding and genetic resources. Sunderland, MA: Sinauer Associates Inc.

Hardham, A.R. 2005. *Phytophthora cinnamomi*. Mol. Plant. Pathol. 6: 589-604.

Huang, H., F. Dane and J.D. Norton. 1994a. Genetic analysis of 11 polymorphic isozyme loci in chestnut species and characterization of chestnut cultivars by multi-locus allozyme genotypes. J. Amer. Soc. Hort. Sci., 119(4), 840-849.

Huang, H., F. Dane and J.D. Norton. 1994b. Allozyme diversity in Chinese, Seguin and American chestnut (*Castanea* spp.). Theor. Appl. Genet., 88: 981-985.

Huang, H., F. Dane and T.L. Kubisiak, 1998. Allozyme and RAPD analysis of the genetic diversity and geographic variation in wild populations of the American chestnut (*Fagaceae*). Amer. J. of Bot. 85: 1013-1021.

Huang, H., W.A. Carey, F. Dane and J.D. Norton. 1996. Evaluation of Chinese chestnut cultivars for resistance to Cryphonectria parasitica. Plant Dis 80: 45-47.

Huang, W-G., L-L. Cheng, Z-J. Zhou and J-L. Liu. 2010 SSR analysis on genetic diversity of wild Chinese chestnut populations[J]. Journal of Fruit Science, 2010, 02: 227-232.

Inoue, E., L. Ning, H. Hara, S. Ruan and H. Anzai. 2009. Development of simple sequence repeat markers in Chinese chestnut and their characterization in diverse chestnut cultivars. J. Amer. Soc. Hort. Sci. 134: 610-617.

Inoue, E., T. Homma, M. Sasaki, T. Gonai and M. Kasumi. 2012. Evaluation of self-fertility among chestnut varieties on shading treatment of the flower and genetic analysis of the SSR loci in self-pollinated seedlings. Hort. Res. (Japan) 11 (2): 199-203. (in Japanese with English summary)

Isaki, H. 1978. Principles and practice on chestnut culture. Hakuyusya, Tokyo. (in Japanese)

Jacobs, D.F., H.J. Dalgleish and C.D. Nelson. 2013. A conceptual framework for restoration of threatened plants: the effective model of American chestnut (*Castanea dentata*). New Phytol. 197: 378-393.

Jaynes, R.A. 1964. Interspecific crosses in the genus *Castanea*. Silvae Genet. 13: 146-154.

Jaynes, R.A. 1972. Genetics of chestnut. USDA Forest Serv. Res. Pap. WO-17, 13 p.

Jaynes, R.A. 1975. Chestnuts. Pages 490-503 in: Janick, J and Moore, J.N. (eds.) Advances in Fruit Breeding. Purdue Univ. Press, West Lafayette, IN.

Kajiura, I. 1978. New information for fruit tree breeding during the Meiji and Taisho periods in Japan. News Fruit Tree Res. Stn. 9: 8-9. (in Japanese).

Kampfer, S., C. Lexer, J. Glossl and H. Steinkellner. 1998. Characterization of (GA)n microsatellite loci from Quercus robur. Hereditas 129: 183-186.

Kanato, K. 1973. Cultivars and breeding. Other fruit trees. In: Japan. Soc. Hort. Sci. (eds.), Whole book of horticulture. Yokendo. Tokyo, pp. 35-42. (in Japanese)

Kotobuki, K., T. Saito, Y. Kashimura and M. Shoda. 1999. Chestnut breeding program in National Institute of Fruit Tree Science, Japan. Acta Hort. 494: 323-326.

Kotobuki, K, T. Saito, Y. Sawamura, Y. Machida, I. Kajiura, Y. Sato, R. Masuda, K. Abe, A. Kurihara, T. Ogata, O. Terai, T. Nishibata, M. Shoda, Y. Kashimura, T. Kozono, H. Fukuda, T. Kihara and K. Suzuki. 2005. New Japanese chestnut cultivar 'Shuuhou' Bull. Natl. Inst. Fruit Tree Sci. 4: 29-36. (in Japanese with English summary)

Kremer, A., M. Casasoli, T. Barreneche, C. Bodénès, P. Sisco, T. Kubisiak, M. Scalfi, S. Leonardi, E. Bakker, J. Buiteveld, J. Romero-Severson, K. Arumuganathan, K. Derory, C. Scotti-Saintagne, G. Roussel, M.E. Bertocchi, C. Lexer, I. Porth, F. Hebard, C. Clark, J. Carlson, C. Plomion, H.P. Koelewijn and F. Villani. 2007. Chapter 5 Fagaceae Trees. In: Genome Mapping and Molecular Breeding in Plants Vol. 7; Forest Trees. (Kole, C. Ed.), Springer Publ, Leipzig, Germany.

Kubisiak, T.L., F.V. Hebard, C.D. Nelson, J. Zhang, R. Bernatzky, H. Huang, S.L. Anagnostakis and R.L. Doudrick. 1997. Molecular mapping of resistance to blight in an interspecific cross in the genus *Castanea*. Phytopathology, 87(7), 751-759.

Kubisiak, T.L. and J.H. Roberds. 2006. Genetic structure of American chestnut populations based on neutral DNA markers. *In:* Steiner, K.C. and Carlson, J.E., eds. Restoration of American

Chestnut to Forest Lands – Proceedings of a Conference and Workshop. May4-6, 2004, The North Carolina Arboretum, Natural Resources Report NPS/NCR/CUE/NRR – 2006/001, National Park Service, Washington, DC.

Kubisiak, T.L., C.D. Nelson, M.E. Staton, T. Zhebentyayeva, C. Smith, B.A. Olukolu, G-C. Fang, F.V. Hebard, S. Anagnostakis, N. Wheeler, P.H. Sisco, A.G. Abbott and R.R. Sederoff. 2013. A transcriptome-based genetic map of Chinese chestnut (*Castanea mollissima*) and identification of regions of segmental homology with peach (*Prunus persica*). Tree Genet. Genomes 9(2), 557-571.

Latijnhouwers, M., P.J.G.M. deWit and F. Govers. 2003. Oomycetes and fungi: similar weaponry to attack plants. Trends Microbiol 11: 462-469. doi: 10.1016/j.tim.2003.08.002.

Lafitte, G. 1946. Le Châtaignier Japonais en Pays Basque. Mendionde 69.

Lang, P., F. Dane, T.L. Kubisiak and H. Huang. 2007. Molecular evidence for an asian origin and a unique westward migration of species in the genus *Castanea* via Europe and North America. Mo.l Phylogenet. Evol. 43: 49-59.

Li, X. and F. Dane. 2013. Comparative chloroplast and nuclear DNA analysis of *Castanea* species in the southern region of the USA. Tree Genet. Genomes 9: 107-116.

Liu, G-B., Y-P. Lan, J. Cao, W-I. Lan, L. Zhou, J-B. Wang and J-L Liu. 2013. Study on the distant hybridization compatibility between *Castanea mollissima* and *C. henry* [J]. Northern Horticulture (15): 1-4.

Liu Ting. 2011. Identification of molecular marker linked to the weeping branch traits of *Castanea mollissima* and Molecular Assisted Breeding [D]. Beijing Forestry University.

Liu, Y., Z.L. Ning, J. Wang, M. Kang and H.W. Huang. 2009. Study on leaf phenotypic variation of sympatric population of *Castanea mollissima* and *C. henry*[J]. J. of Wuhang Botany Research 2009, 27(5): 480-488.

Machida, Y. 1984. Progress and achievement in fruit tree science of Horticultural Research Station. Chestnut breeding. In: Fruit Tree and Vegetable Research Stations in Ministry of Agriculture, Forestry and Fisheries. (eds.), Chronicles of Horticultural Research Station. Katsumi publishing Co., Ltd. Tokyo, pp. 15-17. (in Japanese).

Manos, P.S., Z-K. Zhe-Kun Zhou and C.H. Cannon. 2001. Systematics of *Fagaceae*: phylogenetic tests of reproductive trait evolution. Int. J. Plant Sci., 162(6): 1361-1379.

Marinoni, D., A. Akkak, G. Bounous, K.J. Edwards and R. Botta. 2003. Development and characterization of microsatellite markers in *Castanea sativa* (Mill.). Mol. Breeding 11: 127-136.

Martín, M.A., J.B. Alvarez, C. Mattioni, M. Cherubini, F. Villani and L.M. Martín. 2009. Identification and characterisation of traditional chestnut varieties of southern Spain using morphological and simple sequence repeat (SSRs) markers. Ann. Appl. Biol. 154: 389-398.

Martin, M.A., C. Mattioni, M. Cherubini, D. Taurchini and F. Villani. 2010. Genetic diversity in European chestnut populations by means of genomic and genic microsatellite markers. Tree Genet. Genomes 6: 735-744.

Mattioni, C., M.A. Martin, P. Pollegioni, M. Cherubini and F. Villani. 2013. Microsatellite markers reveal a strong geographical structure in European populations of *Castanea sativa* (*Fagaceae*): evidence for multiple glacial refugia. Am. J. Bot. 100(5): 951-961.

Mehra, P.N., A.S. Hans and T.S. Sareen. 1972. Cytomorphology of Himalayan *Fagaceae*. Silvae Genet. 21: 102-109.

Mellano, M.G., G.L. Beccaro, D. Donno, D.T. Marinoni, P. Boccacci, S. Canterino, A.K. Cerutti, and G. Bounous. 2012. *Castanea* spp. biodiversity conservation: Collection and characterization of the genetic diversity of an endangered species. Genet. Resour. Crop. Ev. 59 (8): 1727-1741.

Nienstaedt, H. 1948. Notes on the Chestnut: Breeding, Culture, and Botanical Characters of Species and Hybrids. Master's Thesis, Yale School of Forestry, 104 p + XIII.

Nienstaedt, H. 1956. Receptivity of the pistillate flowers and pollen germination test in genus *Castanea*. Zeitschrift Forstgenetik Forstplanzenz 5: 40-45.

Nishio, S., T. Yamamoto, S. Terakami, Y. Sawamura, N. Takada and T. Saito. 2011a. Genetic diversity of Japanese chestnut cultivars assessed by SSR markers. Breed. Sci. 61: 109-120.

Nishio, S., T. Yamamoto, S. Terakami, Y. Sawamura, N. Takada, C. Nishitani and T. Saito. 2011b. Novel genomic and EST-derived SSR markers in Japanese chestnuts. Sci. Hortic. 130: 838-846.

Nishio, S., N. Takada, T. Yamamoto, S. Terakami, T. Hayashi, Y. Sawamura and T. Saito. 2013. Mapping and pedigree analysis of the gene that controls the easy peel pellicle trait in Japanese chestnut (*Castanea crenata* Sieb. et Zucc.). Tree Genet. Genomes 9: 723-730.

Nishio, S., H. Iketani, H. Fujii, T. Yamamoto, S. Terakami, N. Takada and T. Saito. 2014. Use of population structure and parentage analyses to elucidate the spread of native cultivars of Japanese chestnut. Tree Genet. Genomes 10: 1171-1180.

Ohri, D. and M.R. Ahuja. 1991. Giesma C-banding in *Fagus sylvatica* L., *Betula pendula* Roth and *Populus tremula* L. Silvae Genet. 40: 72-75.

Oraguzie, N.C., D.L. McNeil, A.M. Paterson and H.M. Chapman. 1998. Comparison of RAPD and MorphoNut markers for revealing genetic relationships between chestnut species (*Castanea* spp) and New Zealand chestnut selections. N.Z.J. Crop & Hort. Sci, 26, 2, 109-15.

Paglietta, R. 1986. 'Primato' Nuova cultivar precoce di castagno euro-giapponese. In: Giornate di studio sul castagno. Caprarola (Viterbo) 6 7 November 1986. Comunità Montana dei Cimini, Caprarola (VT): 65-70.

Paglietta, R. 1992. 'Lusenta' a new early ripening Euro Japanese chestnut. In: Wallace, R.D. and Spinella, L.G. (eds.). Proc. World Chestnut Industry Conference, July 8 10, 1992, Morgantown, West Virginia. Chestnut Marketing Assoc., Alachua, Florida, USA: 10-11.

Pastorelli, R., M.J. Smulders, W.P.C. Van't Westende, B. Vosman, R. Gianinini, C. Vettori and G.G. Vendramin. 2003. Characterization of microsatellites markers in *Fagus sylvatica* L. and *Fagus orientalis* Lipsky Mol. Ecol. Notes 3 (1): 76-78.

Pereira-Lorenzo, S. and J. Fernández-López. 1997. Propagation of chestnut cultivars by grafting: methods, rootstocks and plant quality. J. Hortic. Sci. 72(5), 731-739.

Pereira-Lorenzo, S., J. Fernández-López and J. Moreno-González. 1996a. Variability and grouping of Northwestern Spanish Chestnut Cultivars (*Castanea sativa*). I. Morphological traits. J. Am. Soc. Hort. Sci. 121(2): 183-189.

Pereira-Lorenzo, S., J. Fernández-López and J. Moreno-González. 1996b. Variability and grouping of Northwestern Spanish Chestnut Cultivars (*Castanea sativa*). II. Isoenzyme traits. J. Am. Soc. Hort. Sci. 121(2): 190-197.

Pereira-Lorenzo, S., M.B. Díaz-Hernández and A.M. Ramos-Cabrer. 2006. Use of highly discriminating morphological characters and isoenzymes in the study of Spanish chestnut cultivars. J. Am. Soc. Hort. Sci. 131(6): 770-779.

Pereira-Lorenzo, S., R. Costa, A.M. Ramos-Cabrer, C. Ribeiro, C. Serra da Silva, G. Manzano and T. Barreneche. 2010. Variation in grafted European chestnut and hybrids by microsatellites reveals two main origins in the Iberian Peninsula. Tree Genet. Genomes. 6: 701-715.

Pereira-Lorenzo, S., R.M. Lourenço Costa, A.M. Ramos-Cabrer, M. Ciordia-Ara, C.A. Marques Ribeiro, O. Borges and T. Barreneche. 2011. Chestnut cultivar diversification process in the Iberian Peninsula, Canary Islands, and Azores. Genome 54: 301-315.

Pigliucci, M., S. Benedettelli and F. Villani. 1990a. Spatial patterns of genetic variability in Italian chestnut (*Castanea sativa*). Journal Canadien de Botanique, 68: 1962-1967.

Pigliucci, M., F. Villani and S. Benedettelli. 1990b. Geographic and climatic factors associated with the spatial structure of gene frequencies in *Castanea sativa* Mill. from Turkey. J. Genet., 69 (3), 141-149.

Pimentel-Pereira, M., J. Gómes-Laranjo and S. Pereira-Lorenzo. 2007. Análise dos caracteres morfométricos de variedades portuguesas. En: Gomes-Laranjo J., Ferreira-Cardoso J., Portela E., Abreu C.G. (eds) Castanheiros. Programa Agro, Código 499, 95-108. ISBN: 978-972-669-844-9.

Ramos-Cabrer, A.M., M.B. Díaz-Hernández, M. Ciordia-Ara, D. Rios-Mesa, J. Gonzalez-Díaz and S. Pereira-Lorenzo. 2006. Study of Spanish chestnut cultivars using SSR markers. Advances in Horticultural Science, 20(1): 113-116.

Saito, T., K. Kotobuki, Y. Sawamura, K. Abe, O. Terai, M. Shoda, N. Takada, Y. Sato, T. Hirabayashi, A. Sato, T. Nishibata, Y. Kahimura, T. Kozono, H. Fukuda, T. Kihara, K. Suzuki and M.

Uchida. 2009. New Japanese chestnut cultivar 'Porotan'. Bull. Natl. Inst. Fruit Tree Sci. 9: 1-9. (in Japanese with English summary).
Saito, T., K. Kotobuki, Y. Sawamura, N. Takada, T. Hirabayashi, A. Sato, M. Shoda, O. Terai, T. Nishibata, Y. Kashimura, K. Abe, S. Nishio, T. Kihara, K. Suzuki and M. Uchida. 2015. New Japanese chestnut cultivar 'Mikuri' (*Castanea crenata* Sieb. et Zucc). Bull. NARO Inst. Fruit Tree Sci. 16: 1-9. (in Japanese with English summary).
Salesses, G., J. Chapa and P. Chazerans. 1993a. The chestnut in France – Cultivars – Breeding programs. Proceedings of the International Congress on Chestnut, Spoleto, Italy, October 20-23, 331-337.
Salesses, G., L. Ronco, J.E. Chauvin and J. Chapa. 1993b. Amélioration génétique du châtaignier. Mise au point de tests d'évaluation du comportement vis-à-vis de la maladie de l'encre. L'Arboriculture Fruitière. 458: 23-31.
Santos, C., T. Zhebentyayeva, S. Serrazina, C. Dana Nelson and R. Costa. 2015a. Development and characterization of EST-SSR markers for mapping reaction to *Phytophthora cinnamomi* in *Castanea* spp. Sci. Hortic. 194 (2015) 181-187 doi: 10.1016/j.scienta.2015.07.04.
Santos, C., H. Machado, I. Correia, F. Gomes, J. Gomes-Laranjo and R. Costa. 2015b. Phenotyping *Castanea* hybrids for *Phytophthora cinnamomi* resistance. Plant Pathol. doi: 10.1111/ppa.12313.
Sartor, C., F. Dini, D. Torello Marinoni, M.G. Mellano, G.L. Beccaro, A. Alma, A. Quacchia and R. Botta. 2015. Impact of the Asian wasp *Dryocosmus kuriphilus* (Yasumatsu) on cultivated chestnut: Yield loss and cultivar susceptibility. Sci. Hortic. http://dx.doi.org/10.1016/j.scienta.2015.10.004.
Sawano, M., T. Ichii, T. Nakanishi and Z. Kotera. 1984. Studies on identification of chestnut species and varieties by isozyme analysis. Science Reports of Faculty of Agriculture, Kobe University, 16: 67-71.
Schad, C., G. Solignat, J. Grente and P. Venot. 1952. Recherches sur le châtaignier à la Station de Brive. Annales de l'Amélioration des Plantes, 3, 376-458.
Sebastiana, M. 2006. Identificação de genes envolvidos no reconhecimento hóspede/hospedeiro em ectomicorrizas. Ph.D. Dissertation. 127 pp.
Sederoff, R., D. Nielson, C. Smith, J. Tomkins, M. Atkins, B. Blackmon, M. Staton, S. Ficklin, F. Hebard, P. Sisco, J. Carlson, S. Diloreto, A. Barakat, W. Powell, K. Baier, S. Anagnostakis, T. Kubisiak and N. Wheeler. 2008. Genomic tool development for the *Fagaceae*. Plant & Animal Genomes XVI Conference, January 12-16, 2008, San Diego, CA. P494.
Serdar, U. and I. Macit. 2010. New advances in chestnut growing in the Black Sea Region, Turkey. Proc. 1st European Congress on Chestnut. Acta Horticulturae 866: 303-308.
Serdar, U., B. Akyuz and D.W. Fulbright. 2014. Graft success of hybrids on European chestnut rootstock and development of chestnut blight disease. 2nd Symposium of Turkey Forest Entomology and Pathology, 7-9 April, Antalya, Turkey. Symposium Proceeding. 127-131.
Serrazina, S. 2004. Isolamento e Caracterização de Genes de Resistência à Doença da Tinta em *Castanea sativa* Mill. Ph.D. Dissertation. 301 pp.
Serrazina, S., C. Santos, H. Machado, C. Pesquita, R. Viventini, V. Pais, M. Sebastina and R. Costa. 2015. *Castanea* root transcriptome in response to *Phytophthora cinnamomi* challenge. Tree Genet. Genomes. doi: 10.1007/s11295-014-0829-7.
Sisco, P.H., T.L. Kubisiak, M. Casasoli, T. Barreneche, A. Kremer, C. Clark, R.R. Sederoff, F.V. Hebard and F. Villani. 2005. An improved genetic map for *Castanea mollissima/Castanea dentata* and its relationship to the genetic map of *Castanea sativa*. Acta. Hort. 693: 491-495.
Shimura, I. 1984. Origin and distribution. In: Rural Culture Association (ed.) Nogyo-gijyutsu-taikei: Kajyu-hen Vol. 5. Rural Culture Association, Tokyo, pp. 3-6 (in Japanese).
Shimura, I., K. Kanato, M. Yasuno and H. Matsunaga. 1971. Studies on the breeding on the chestnut, *Castanea* Spp. I. Percentage of selected hybrids in progenies obtained from each varieties used as cross parents in chestnut breeding projects. Bull. Hort. Res. Stn. A10: 1-9. (in Japanese with English summary).
Soylu, A. 1990. Heredity of male sterility in some chestnut cultivars (*Castanea sativa* Mill.) XIII. International Horticultural Congress August 27-September 1, 1990, Firenze, Italy, Acta Hortic. 317: 181-185.

Staton, M., T. Zhebentyayeva, B. Olukolu, G.C. Fang, D. Nelson, J.E. Carlson and A.G. Abbott. 2015. Substantial genome synteny preservation among woody angiosperm species: comparative genomics of Chinese chestnut (*Castanea mollissima*) and plant reference genomes. BMC Genomics. 16: 744.

Steinkellner, H., S. Fluch, E. Turetschek, C. Lexer, R. Streiff, A. Kremer, K. Burg and J. Glössl. 1997. Identification and characterization of (GA/CT)(n)-microsatellite loci from *Quercus petraea*. Plant Mol. Biol. 33 (6), pp. 1093-1096.

Stilwell, K.L., H.M. Wilbur, C.R. Werth and D.R. Taylor. 2003. Heterozygote advantage in the American chestnut, *Castanea dentata* (Fagaceae) Am. J. Bot. 90: 207-213.

Takada, N., S. Nishio, M. Yamada, Y. Sawamura, A. Sato, T. Hirabayashi and T. Saito. 2012. Inheritance of the easy-peeling pellicle trait of Japanese chestnut cultivar Porotan. HortScience 47: 845-847.

Tanaka, K. and K. Kotobuki. 1992. Comparative ease of pellicle removal among Japanese chestnut (*Castanea crenata* Sieb. et Zucc.) and Chinese chestnut (*C. mollissima* Blume) and their hybrids. J. Japan. Soc. Hort. Sci. 60: 811-819.

Tanaka, K., Y. Tsumura and T. Nakamura. 1999. Development and polymorphism of microsatellite markers for *Fagus crenata* and the closely related species, F. japonica. Theor. Appl. Genet. 99: 11-15.

Tanaka, T., T. Yamamoto and M. Suzuki. 2005. Genetic diversity of *Castanea crenata* in northern Japan assessed by SSR markers. Breed. Sci. 55: 271-277.

Tano. 1954. Studies on fruit breeding part 1. An exposition on new varieties. J. Niigata Agri. Exp. Stn. 6: 33-40. (in Japanese)

Taylor, W.A. and H.P. Gould. 1914. Promising new fruits. pp. 122-124 IN: Yearbook of Agriculture for 1913. USDA, Washington, D.C.

Tian, H., M. Kang, L. Li, X. Yao and H. Huang. 2009. Genetic diversity in natural populations of *Castanea mollissima* inferred from nuclear SSR markers [J]. Biodiversity Science, 03: 296-302.

Torello Marinoni, D., A. Akkak, P. Guaraldo, P. Boccacci, A. Ebone, E. Viotto, G. Bounous, A.M. Ferrara and R. Botta. 2013. Genetic and morphological characterization of chestnut (*Castanea sativa* Mill.) germplasm in Piedmont (North-western Italy). Tree Genet. Genomes. 9 (4): 1017-1030.Van Fleet, Walter. 1914. Chestnut breeding experience. J. Hered. 5: 19-24.

UPOV. 1988. Draft guidelines for the conduct of tests for distinctness, homogeneity and stability (CHESTNUT). TG/124/1(proj.), 23 p.

Urquijo, P. 1944. Aspectos de la obtención de híbridos resistentes a la enfermedad del castaño. Bol. Veg. Ent. Agr. XIII: 447-462.

Urquijo, P. 1957. La regeneración del castaño. Bol. de Pat. Veg. y Entomología Agrícola, XXII: 217-232.

Van Fleet, W. 1914. Chestnut breeding experience. J. Hered. 5: 19-24.

Van Fleet, W. 1920. Chestnut work at Bell Experiment Plot. Annual report of the Northern Nut Growers Association, 11: 16-21.

Vieira Natividade, J. 1947. Quatro anos na defesa da campanha e Reconstituição dos Soutos. Edição da Junta Nacional das frutas. Lisboa.

Villani, F.S., M. Pigliucci, S. Benedettelli and M. Cherubini. 1991a. Genetic differentiation among Turkish chestnut (*Castanea sativa* Mill.) populations. Heredity 66: 131-136.

Villani, F.S., S. Benedettelli, M. Paciucci, M. Cherubini and M. Pigliucci. 1991b. Genetic variation and differentiation between natural populations of chestnut (*Castanea sativa* Mill.) from Italy. In: S. Fineschi, M.E. Malvolti, F. Cannata, H.H. Hattemer eds. Biochemical markers in the population genetics of forest trees. The Netherlands: SPB Academic Publishing BV Press.

Villani, F., M. Pigliucci, M. Cherubini, O. Sun and L. Parducci. 1993. Genetic diversity of *Castanea sativa* Mill. in Europe: Theorical aspects and applied perspectives. Abstracts, International Congress on Chestnut, Spoleto, Italia.

Villani, F., A. Sansotta, M. Cherubini, D. Cesaroni and V. Sbordoni. 1999. Genetic structure of *Castanea sativa* in Turkey: evidence of a hybrid zone. J. Evol. Biol. 12: 233-244.

Warschefsky, E., R. Varma Penmetsa, D.R. Cook and E.J.B. Von Wettberg. 2014. Am. J. Bot. 101 (10): 1791-1800, 2014. doi: 10.3732/ajb.1400116.

Wen, H. and J. Norton. 1992. Enzyme Variation in Chinese Chestnut Cultivars. Abstract, International Chestnut Conference, Morgantown, USA.

Xu Hong-mei. 2004. Genetic diversity and genetic structure about Chinese chestnut by AFLP analysis and the establishment of database about varietal resource in chestnut[D]. Xinjiang Agricultural University.

Yamamoto, T., T. Tanaka, K. Kotobuki, N. Matsuta, M. Suzuki and T. Hayashi. 2003. Characterization of simple sequence repeats in Japanese chestnut. J. Hort. Sci. Biotech. 78: 197-203.

Zhang H. and L. Liu. 1998. The genetic diversity of *Castanea mollissima* the effect of artificial selection[J]. Acta Botanica Yunnanica 01: 81-88.

Chapter 16

Use of Polyploids, Interspecific, and Intergeneric Wide Hybrids in Sugar Beet Improvement

J. Mitchell McGrath[1,*] and Christian Jung[2]

ABSTRACT

Wide hybrids for sugar beet (*Beta vulgaris*) improvement have been sought for introgression of high levels of resistance and even immunity to diseases and pests, notably to sugar beet cyst nematode. Unfortunately, many such hybrids are difficult to make, and once made, regular chromosome pairing and recombination between donor and recipient chromosomes is limited or non-existent. Taxonomic revisions over the past decade better reflect the biological relationships between species, and in the instance of sugar beet cyst nematode, transfer of resistance was accomplished through translocation of a chromosome segment from what is now recognized as species within a different genus. Other more closely related species also have potential for sugar beet improvement but their use has been more limited. Recent trends to access underexploited germplasm have targeted the wild beet forms of *Beta vulgaris*, and thus few breeding programs are using interspecific hybrids. However, genomic approaches applied to the wild relatives may uncover the genetic basis of high levels of disease and pest resistance and thus allow directed improvement of sugar beet by other means than direct inter-specific and inter-generic hybridization.

Introduction

Sugar beet (*Beta vulgaris* L.) is a recent historical crop derived from other crop types within the same species, i.e. from fodder beet, table beet, and leaf beet (chard). Likely, crop evolution in this genus was one of progressive

[1] J. Mitchell McGrath, USDA-ARS, Sugarbeet and Bean Research, 1066 Bogue Street, 494 PSSB, Michigan State University, East Lansing, MI, USA 48824-1325, E-mail: mitchmcg@msu.edu

[2] Christian Jung, Institute of Crop Science and Plant Breeding, Kiel University, Am Botanischen Garten 1-9, 24118 Kiel, Germany, E-mail: c.jung@plantbreeding.uni-kiel.de

* Corresponding author

improvement, initially from leafy types, through selection for root traits first for human food and then animal fodder, and finally for the ability of the fodder types to accumulate sucrose to high concentrations within a single growing season (Biancardi et al. 2005). The most closely related wild taxon is *Beta vulgaris* spp. *maritima*, whose ecological distribution follows the Mediterranean and European coastlines, where the crop types are thought to have arisen from one or more *maritima* ancestors. There are no reported fertility barriers between these subspecies, and diversity within the *maritima* types has contributed many economically important characters to the cultivated types. As one might expect, there is a great deal of admixture between the freely intercrossing types (Andrello et al. 2016). Greater attention is being placed on 'wide hybrids' between the crop types and subspecies *maritima* today, rather than following earlier attempts to access similar traits in more distantly related species. Contributions of *Beta vulgaris* spp. *maritima* to beet crop improvement have been summarized recently (Biancardi et al. 2012).

Sugar beet is perhaps the most economically important species within the family *Caryophyllales* (*Chenopodiaceae*, Subfamily *Betoideae*, formerly *Amaranthaceae*; Hernández-Ledesma et al. 2015). There have been many revisions of *Beta* and allied taxa. The most salient for this review is that the section *Beta* was recently split into two genera: *Beta* section *Beta* (two species including *Beta vulgaris*) and *Beta* section *Corollinae* (five species), while the former *Beta* section *Procumbentes* has been placed within its own genus *Patellifolia*, which includes the former *Beta procumbens*, *B. webbiana*, and *B. patellaris* (Kadereit et al. 2006, Hernández-Ledesma et al. 2015) (Table 1).

Species from outside *Beta* section *Beta* do not readily hybridize with cultivated beet (Coons 1954, Van Geyt et al. 1990). Also, species from these sections often have hard, dormant seeds, which makes germination difficult (De Bock 1986). Nonetheless, many interspecific hybridization attempts have been made for the purpose of introgressing disease resistance traits into the cultivated germplasm. For example, cultivated Swiss chard was used as a bridging species for crosses with *Beta webbiana* and *B. procumbens*; colchicine-doubled tetraploids of *B. macrocarpa*, *B. v.* subsp. *maritima*, and *B. atriplicifolia* were used as parents from section *Beta* to cross with *B. webbiana* and *B. procumbens*; and red table beet and sea beet were used as a bridge to cross with *B. webbiana* and *B. procumbens* and also with species of the section *Corollinae* (reviewed in Coons 1975, De Bock 1986, Van Geyt et al. 1990). Most F_1 hybrids were sterile and had degenerated root systems, and were grown to maturity only after grafting onto healthy sugar beet stems (Coe 1954, Johnson 1956). Unfortunately, chromosome pairing did not occur between donor and recipient chromosomes, so no recombination was effected.

Molecular markers are useful for characterizing *Beta* germplasm. Between *Beta* and *Patellifolia* sections, 95% of the markers tested were able to discriminate at the species level (Jung et al. 1993). Diversity within crop types has been investigated more intensively within sugar beet, where diversity is reduced relative to subspecies *maritima,* and it is clear that much genetic and allelic diversity remains to be exploited in crop beet improvement (Andrello et al. 2016). Species-specific repeated nucleotide sequences are abundant in *Beta* and allied taxa, as >60% of the genome is composed of repeated sequences in *Beta vulgaris* (Flavell et al. 1974, Dohm et al. 2013). These have been useful in characterizing interspecific hybrids and cytogenetic stocks, as an adjunct to cytological characterization as well as a tool for cytological localization of sequences to chromosomes via fluorescent *in situ* hybridization, such as in *B. vulgaris* interspecific hybrids with *B. procumbens* and *B. patellaris,* and also *B. corolliflora* (reviewed in McGrath et al. 2011).

Table 1. Taxonomy of *Beta* species and allies from Kadereit et al. (2006). Reported chromosome numbers of natural species are according to Goldblatt and Johnson (1979) and Reamon-Büttner et al. (1996). Asterisk indicates the predominant ploidy within the species.

Taxon	Ploidy (x =9)
Beta* sect. *Beta	
B. vulgaris L.	$2x$
B. vulgaris L. subsp. *maritima* (L.) Arcang.	$2x$
B. vulgaris L. subsp. *adanensis* (Pamuk.) Ford-Lloyd & Williams	$2x$
B. macrocarpa Guss.	$2x^*$, $4x$
Beta* sect. *Corollinae	
B. corolliflora Zos. ex Buttler	$4x^*$, $6x$
B. macrorhiza Stev.	$2x$
B. lomatogona Fisch. et May.	$2x^*$, $3x$
B. trigyna Waldst. et Kit.	$5x$, $6x$
B. nana Boiss. et Heidr.	$2x$
***Patellifolia* A. J. Scott et al.**	
P. patellaris (Moq.) A. J. Scott et al.	$2x$, $4x^*$
P. procumbens (Sm.) A.J. Scott et al.	$2x$

Beta vulgaris, including spp. *maritima,* is diploid in nature with $2n = 2x = 18$. Artificial tetraploids can be created, and indeed, triploid and anisoploid commercial sugar beet hybrids were common for much of the 1980s and 1990s. Tetraploid pollinators were favored for triploid induction in crosses with diploid seed parents, since the frequency of unbalanced gametes in tetraploid pollen was observed to be less relative to using a diploid pollinator and tetraploid seed parent (Bosemark 1993). Breeding at the tetraploid level has since been largely abandoned, in large part due to equivalent performance and the relative ease of breeding at the diploid level. All *Beta* species are based on $x = 9$, including the *Patellifolia* species (formerly listed under *Beta*), and have small chromosomes with few cytological landmarks, but high differentiation of repeated sequence elements which has allowed specific chromosome identification in some cases (Paesold et al. 2012).

Polyploidy is present among these taxa, sometimes within species (Lange and De Bock 1989, Castro et al. 2013, and Table 1), and taxonomic revisions coupled with sparse chromosome number data make it difficult to generalize. Each of these species, with the exception of *B. nana*, has been the subject of interspecific hybridization with sugar beet and other crop types. *Beta nana* has only recently been intensively collected from its narrow endemic habitat at higher elevations in Greece (Frese et al. 2009).

Breeding has narrowed the level of heterozygosity in cultivated beet germplasm (McGrath et al. 1999), although genetic diversity within *B. vulgaris* and among all section *Beta* species is relatively high (reviewed in McGrath et al. 2011). The majority of sugar beets grown are hybrids, utilizing the Owens cytoplasmic male sterility system (Owen 1945). Seed parents are also monogerm, a recessive character controlling a single flower, and hence a single seed, per axil (Savitsky 1952), which reduces labor required to thin plants to a uniform stand in growers' fields. Pollinators are often multigerm and often contribute adaption to local disease pressures. Successive improvements in sucrose yield, the driving quality trait of sugar beet, have been made from within the crop types of *Beta vulgaris*, and generally from within the sugar beet gene pool, although the potential contributions of the wild species to agronomic traits remains to be investigated. Sucrose yield is governed by the yield of beets per unit area times the proportion of sucrose in the harvested roots, minus losses during storage and processing. Wild *Beta* species generally have been accessed for disease resistance traits. There is a fairly lengthy list of pathogens and stressors that affect beet growth (Harveson et al. 2009), and undoubtedly the wild species have potential in meeting these challenges. With the exception of *Beta nana*, wide hybrids between *Beta vulgaris* and other *Beta* species have been attempted and have been successful to varying degrees, but while interesting academically, they have not generally made a lasting contribution to the cultivated sugar beet germplasm. Traits for which new sources of resistance would be a priority include Sugar Beet Cyst Nematode (SBCN), rhizomania (*Beet necrotic yellow vein virus*, BYNVV) (McGrann et al. 2009), and Cercospora leaf spot (caused by *Cercospora beticola* Sacc.): in each case current commercial resistances can be traced back to subspecies *maritima* (reviewed in Biancardi et al. 2012).

Hybrids Using *Patellifolia* Species (Formerly *Beta* Section *Procumbentes*)

The greatest effort in using wide hybrids as a genetic donor resource has been placed towards transferring immunity to SBCN from *Patellifolia* species into sugar beet. This effort was fraught with problems and sources of resistance have since been found within *Beta vulgaris* spp. *maritima* (Biancardi et al. 2012): although these types are not immune, they limit the infection and are used in breeding tolerant varieties (Stevanato et al. 2015). The two species *P. procumbens* and *P. patellaris* are immune to the Sugar Beet Cyst Nematode (*Heterodera schachtii*), and are also highly resistant to Cercospora leaf spot, the curly top virus, and rhizomania (Coons et al. 1955, Paul et al. 1992, Mesbah et al. 1997, Reamon-Ramos and Wricke 1992). Formerly, a third species *P. webbiana* belonged to the former *Beta* section *Procumbentes*. However, due to its similarity to *P. procumbens*, *P. webbiana* is no longer regarded as a separate species.

The *Patellifolia* species are the only crossable species which conferred complete resistance to SBCN infection (Jung 1987). Even under extreme infection pressure, between 93% and 98% of the plants did not show any cysts (Yu 1984) although the roots were invaded by *H. schachtii* larvae. This renders the *Patellifolia* resistance even more attractive because the number of nematodes in the soil is reduced during cultivation typical for catch crops. Unfortunately, resistance transfer to *B. vulgaris* had two major obstacles. First, most hybrids were not viable because they did not develop a root system (Löptien 1984). This particular problem was solved by grafting the hybrid seedling scion on to a seedling sugar beet root stock. Tetraploid hybrids using 4*x* *B. vulgaris* by 4*x* *P. patellaris* did produce viable root systems and continued to grow without grafting (Löptien 1984). The second problem was that chromosomes failed to pair during meiosis in the diploid hybrid (Savitsky 1960). This problem was overcome by using tetraploid *B. vulgaris* as a crossing parent (Savitsky and Gaskill 1957). Triploid hybrids had 9 bivalents and 9 univalents at Meiosis I (Filutowicz and Kuzdowicz 1959). Some combinations could be found where the hybrids had a fully developed root system. They turned out to be fully resistant to SBCN, indicating nematode resistance was inherited in a dominant manner.

The next step was to reduce the number of wild beet chromosomes by backcrossing with *B. vulgaris*. For each crossing combination, 8 out of 9 wild beet chromosomes were eliminated. A monosomic addition line in sugar beet with only one wild beet chromosome added ($2n = 18 + 1$) was as resistant as the wild beet itself (Savitsky 1975). During meiosis, the wild beet chromosome formed a univalent, and the lack of recombination prevented any SBCN resistance transfer to a beet chromosome. Monosomic addition lines were only used as transition materials, because their agronomic performance was poor and the transmission rates for the nematode resistance genes were low, ranging between 4 and 40% (Jung 1987). Altogether, four wild beet chromosomes were found to carry genes for nematode resistance (*Hs* genes), three from *P. procumbens* and one from *P. patellaris* (Table 2).

Table 2. Wild beet chromosomes with Sugar Beet Cyst Nematode resistance and their gene nomenclature.

Chromosome	1	7	8
P. procumbens	$Hs1^{pro\text{-}1}$ $Hs1\text{-}2^{pro\text{-}1}$	$Hs2^{pro\text{-}7}$	$Hs3^{pro\text{-}8}$
P. patellaris	$Hs1^{pat\text{-}1}$	–	–

The final step was to transfer *Hs* genes into the sugar beet genome and to breed diploid beets with stable inheritance of the nematode resistance genes from *Patellifolia* species. Due to the lack of recombination between chromosomes of both species, the only way to transfer the genes was to rely on spontaneous translocations. Diploid translocation lines, carrying the genes *Hs1*, *Hs1-2*, and *Hs2*, were selected independently on three occasions (Savitsky 1978, Heijbroek et al. 1988, Brandes et al. 1987).

Work to clone the *Hs* genes focused on the line A906001, which turned out to carry the smallest translocation, while the other translocations suffered from pleiotropic effects such as low sucrose content, poor quality parameters, and low yield, caused by genes on the translocation that were completely linked to the *Hs* genes (linkage drag). The A906001 translocation was enriched with molecular markers. Even repetitive markers could be used because there is a low degree of similarity between repetitive sequences from *B. vulgaris* and *P. procumbens*. Satellite DNA sequences from the *P. procumbens* genome were useful for mapping the translocation, and they were also used as molecular markers for distinguishing resistant plants from susceptible ones (Kleine et al. 1998). Using a YAC (yeast artificial chromosome) based physical map, the size of the translocation was estimated to be ca. 2 Mbp. Complete YAC sequences were used as probes for *in situ* hybridization, which confirmed the physical position of the translocation at the end of sugar beet Chromosome 9 (Desel et al. 2001). The $Hs1^{pro\text{-}1}$ gene was cloned as the first gene conferring resistance to plant parasitic nematodes (Cai et al. 1997). Later, it became clear that, apart from $Hs1^{pro\text{-}1}$, one more resistance gene named *Hs1-2* is also located on the translocation. Complete resistance can only be achieved in the presence of both genes. Schulte et al. (2006) established a complete BAC-based map of the A906001 translocation using translocation-specific molecular markers. By comparing the physical maps of two different translocations, both derived from *P. procumbens* Chromosome 1, they concluded that the overlapping region between both translocations is 350 kb in size, which is the maximum size of the region housing the *Hs1-2* gene.

Today, there are sugar beet varieties on the market that are completely resistant to SBCN due to the *P. procumbens* translocation. However, they are of limited use because they only out-compete susceptible cultivars by more than 20% on severely infested soils: under non-infected growth conditions they suffer from a high yield penalty of 10% or more. Tumorous growths on the shoulders of the resistant beets are another clear disadvantage of these genotypes. There are two options to avoid these problems. First, the *Hs* resistance genes alone can be transferred into susceptible cultivars. Unfortunately, all $Hs1^{pro\text{-}1}$ transgenic lines produced so far do not show the same immunity reaction as the translocation lines (C.J., unpublished data), which is due to the fact that a second resistance gene is present on the translocation. Moreover, although the transgene stems from a close relative, these lines are classified as genetically modified which limits their practical use, at least outside the US. New translocation lines have been produced with shorter translocations with the goal of removing genes causing negative agronomic effects. Shorter translocations were obtained after gamma irradiation of seeds from the translocation line A906001. These lines are a valuable resource for breeding nematode resistant varieties with better quality and yielding potential. Moreover, these lines are extremely helpful to further narrow down the region containing the second nematode resistance gene *Hs1-2*. Work is in progress to clone this gene with the help of chromosome breakage mutants. Presently, the complete sequences of four translocation lines are available (2 resistant, 2 susceptible). By a subtractive genome alignment approach, the critical region was further delimited to ca. 90 kb which localizes the *Hs1-2* gene to a small genomic interval.

Hybrids Using *Beta* Sect. *Corollinae*

Species within *Beta* section *Corollinae* form a coherent group well supported with molecular phylogenetic evidence (Kadereit et al. 2006). However, species relationships suggest close affinity between *B. corolliflora* and *B. lomatogona*, and between *B. macrorhiza* and *B. trigyna*, with *B. nana* forming a distinct branch within this section and not a single member of its own section as previously considered. Polyploidy is common in this group (Table 1), which has facilitated generation of wide hybrids, with the exception of *B. nana* where attempts have not yet been made. *Beta corolliflora* shows tetrasomic inheritance and thus is considered an autotetraploid species (Reamon-Büttner and Wricke 1993). Many potentially useful characters are present in these species, such as disease resistance (curly top, virus yellows, Cercospora leaf spot), monogermity, tolerance to cold, drought, and salinity, and, importantly, apomixis (Savitsky 1969, Van Geyt et al. 1990, Gao et al. 2001). Interspecific hybrids and advanced generations between *B. vulgaris* and species of Section *Corollinae* have been reported (Savitsky 1969, Filutowicz et al. 1971, Cleij et al. 1976, Reamon-Büttner et al. 1996, Gao et al. 2001, Gao and Jung 2002). As with *Patellifolia* species, it is not clear that chromosome recombination between Section *Corollinae* and *B. vulgaris* occurs, and no inter-genomic chromosome translocations have yet been reported. The hybrids and their derivatives from Section *Corollinae* have not yet been used in commercial variety development.

Recent efforts have focused on *Beta corolliflora* with an eye towards transferring apomixis into the sugar beet germplasm (Gao et al. 2001). Fertility was reduced in hybrids but an almost complete set of monosomic alien addition lines were made, minus one attributed to lethal traits carried by that chromosome (Gao et al. 2001). Traits such as cold tolerance, curly top resistance, and Cercospora leaf spot resistance were also targets for these interspecific hybrids (Gao and Jung 2002). One apomictic monosome from *B. corolliflora* (designated M-14) showed a >95% transmission through seed and has been examined extensively at the genomic and proteomic levels for a number of potentially useful traits such as salt tolerance and altered transcriptional regulation (Li et al. 2007, Li et al. 2009, Zhu et al. 2009, Ma et al. 2011, Yang et al. 2013). This work is exciting and promises to identify some of the genes responsible for the underlying traits useful for beet improvement, although their direct utility in sugar beet improvement has not yet been realized.

Hybrids Using *Beta* Sect. *Beta*

The literature regarding intra- and inter-specific hybridization within *Beta* section *Beta* has been summarized recently (McGrath et al. 2011, Biancardi et al. 2012). With the exception of hybrids between cultivated beets and *Beta vulgaris* spp. *maritima*, for which there are few if any barriers to hybridization (De Bock 1986, Ford-Lloyd and Hawkes 1986), there are relatively few reported attempts with other taxa in this section (i.e. *B. v.* spp. *adanensis* and *B. macrocarpa* according to Table 1). The tetraploid forms of *B. macrocarpa* are geographically restricted to the Canary Islands, and appear to be allotetraploids due to their dipoidized meiosis (Lange and de Bock 1989). The perceived genetic variation (or lack thereof in these self-fertile taxa) does not yet appear to harbor genes useful towards sugar beet improvement, unlike *Beta vulgaris* spp. *maritima*. Beets are generally allogamous, governed by a complex gametophytic self-incompatibility system which prevents self-pollination but allows almost any two plants to cross-pollinate (Bruun et al. 1995).

Wild beets were introduced into California, and were subsequently identified as *B. macrocarpa* and *B. vulgaris* spp. *maritima*. Additional types with an intermediate morphology were also reported, and their hybrid nature is suspected (McFarlane 1975, Bartsch and Ellstrand 1999). Abe and Tsuda (1988) reported that backcross offspring from hybrids of *Beta vulgaris* and *Beta macrocarpa* showed deviations from expected segregation ratios, consistent with genetic differentiation of the parental genomes independent of large chromosome segmentations (e.g. inversions and translocations).

At least one germplasm type, F1022, has been released through a population enhancement program aimed at increasing the genetic diversity in sugar beet using *Beta macrocarpa* (as well as other *Beta* section *Beta* taxa with no current recognition such as *B. atriplicifolia* and *B. patula*) as a germplasm donor, along with a suggestion for enhanced Rhizoctonia root disease tolerance derived from these taxa (Campbell 2010).

With regard to hybrids between cultivated beets and *Beta vulgaris* spp. *maritima* (sea beet), an intra-specific cross, it is clear that these materials have been invaluable for sugar beet improvement. In the first and earliest instance, Cercospora leaf spot resistance was introgressed into sugar beet from sea beet collected in the Po Valley of Italy (Munerati et al. 1913), and produced cultivars that still contribute much of the Cercospora leaf spot resistant germplasm used worldwide today (Munerati 1932, Biancardi and Biaggi 1979). Cercospora leaf spot is a fungal disease caused by *Cercospora beticola* that can defoliate the sugar beet crop mid-season if uncontrolled. In the second instance, the discovery of rhizomania ('crazy root'), a major yield reduction agent caused by Beet necrotic yellow vein virus, led to extensive efforts to find resistant germplasm, which was found and deployed from numerous sea beet sources (Biancardi et al. 2002, Panella and Lewellen 2007).

For the sake of completeness, trisomic series have also been produced in beets (Butterfass 1964, Romagosa et al. 1987). Each trisomic line has a unique phenotype, and these lines have not been fixed in a homozygous state since the trisomic does not readily transmit through pollen (Nakamura et al. 1991). Molecular markers applied to the Butterfass series allowed standardization of linkage group nomenclature (Schondelmaier and Jung 1997). Intolerance to mono- and nulli-somy, linkage analyses of molecular markers, and whole genome sequencing supports the diploid nature of *Beta vulgaris*, although ancient polyploidy is inferred from the whole genome sequence (Halldén et al. 1998, Dohm et al. 2013).

Future Considerations

The use of wide hybrids in sugar beet improvement has proven difficult, largely because recombination between genomes is rare or non-existent. This may be because the group as a whole is rather old (diversification of *Patellifolia* is suggested at 30.9 to 15.3 million years ago, Kadereit et al. 2006) leading to a high level of highly repetitive DNA sequence divergence (Schmidt and Heslop-Harrison 1993). The potential for these species to contribute to sugar beet improvement is widely recognized. Thus, newer technologies that will identify the genetic basis of such desirable traits are being deployed where possible, and the transfer of these traits once identified must be accomplished through the use of more precise methods of gene transfer. It is likely that beneficial alleles and genes for most of the major pest and disease pressures affecting sugar beet can be found among the wild relatives, and the use of wide hybrids and in particular monosomic addition lines is likely to remain useful to locate traits to chromosomes, from where they can be more easily accessed and deployed. Most pests and diseases of sugar beet have been problematic for many years, thus the attractiveness of immunity has been a significant driver of wide hybrid investigations historically. Perhaps novel technologies based on the identification of novel resistances in the *Betoideae* and specific gene transfer may finally put these problems to rest.

References

Abe, J. and C. Tsuda. 1988. Distorted segregation in the backcrossed progeny between *Beta vulgaris* and *B. macrocarpa* Guss. Japan J. Breed. 38: 309-318.

Andrello, M., K. Henry, P. Devaux, B. Desprez and S. Manel. 2016. Taxonomic, spatial and adaptive genetic variation of *Beta* section *Beta*. Theor. Appl. Genetics 129: 257-271

Bartsch, D. and N. Ellstrand. 1999. Genetic evidence for the origin of Californian wild beets (genus *Beta*). Theor. Appl. Genet. 99: 1120-1130.

Biancardi, E. and M. De Biaggi. 1979. *Beta maritima* L. in the Po delta. pp. 183-185. *In:* ISCI (ed). Proc Convegno Tenico Internazionale in Commerorazione di Ottavio Munerati. Rovigo, Italy.

Biancardi, E., R. Lewellen, M. De Biaggi, A. Erichsen and P. Stevanato. 2002. The origin of rhizomania resistance in sugar beet. Euphytica 127: 383-397.

Biancardi, E., L.G. Campbell, G.N. Skaracis and M. De Biaggi. 2005. Genetics and Breeding of Sugar Beet. Science Publishers, Enfield, New Hampshire.

Biancardi, E., L.W. Panella and R.T. Lewellen. 2012. *Beta maritima*: the Origin of Beets. Springer Verlag, New York, New York.

Bosemark, N.O. 1993. Genetics and breeding. *In:* D.A. Cooke and R.K. Scott (eds.). The Sugar Beet Crop: Science into Practice, Chapman and Hall, London.

Brandes, A., C. Jung and G. Wricke. 1987. Nematode resistance derived from wild beet and its meiotic stability in sugar beet. Plant Breeding 99: 56-64.

Bruun, L., A. Haldrup, S. Petersen, L. Frese, T. Debock and W. Lange. 1995. Self-incompatibility reactions in wild species of the genus *Beta* and their relation to taxonomical classification and geographical origin. Genetic Resources and Crop Evolution 42: 293-301.

Butterfass, T. 1964. Die chloroplastenzahlen in verschiedenartigen zellen trisomer zuckerruben (*Beta vulgaris* L.). Z. Bot. 52: 46-77.

Cai, D., M. Kleine, S. Kifle, H. Harloff, N.N. Sandal, K.A. Marcker, R.M. Klein-Lankhorst, E.M.J. Salentijn, W. Lange, W.J. Stiekema, U. Wyss, F.M.W. Grundler and C. Jung. 1997. Positional cloning of a gene for nematode resistance in sugar beet. Science 275: 832-834.

Campbell, L.G. 2010. Registration of seven sugarbeet germplasms selected from crosses between cultivated sugarbeet and wild *Beta* species. J. Plant Registrations 4: 149-154.

Castro, S., M.M. Romeiras, M. Castro, M.C. Duarte and J. Loureiro. 2013. Hidden diversity in wild *Beta* taxa from Portugal: Insights from genome size and ploidy level estimations using flow cytometry. Plant Science 207: 72-78.

Cleij, G., T. Debock and B. Lekkerkerker. 1976. Crosses between *Beta vulgaris* L. and *Beta lomatogona* F et M. Euphytica 25: 539-547.

Coe, G.E. 1954. A grafting technique enabling an unthrifty interspecific hybrid of *Beta* to survive. Proc. Am. Society Sugar Beet Technologists 8: 157-160.

Coons, G.H. 1954. The wild species of *Beta*. Proc. Am. Society Sugar Beet Technologists 8: 142-147.

Coons, G.H. 1975. Interspecific hybrids between *Beta vulgaris* L. and the wild species of *Beta*. J. Am. Society Sugar Beet Technologists 18: 281-306.

Coons, G.H., F.V. Owen and D. Stewart. 1955. Improvement of the sugar beet in the United States. Advances in Agronomy 7: 89-139.

De Bock, T.S.M. 1986. The genus *Beta*: Domestication, taxonomy and interspecific hybridization for plant breeding. Acta Horticulturae 182: 335-343.

Desel, C., C. Jung, D. Cai, M. Kleine and T. Schmidt. 2001. High-resolution mapping of YACs and the single-copy gene *Hs1*$^{pro-1}$ on *Beta vulgaris* chromosomes by multi-colour fluorescence *in situ* hybridization. Plant Mol. Biol. 45: 113-122.

Dohm, J.C., A.E. Minoche, D. Holtgräwe, S. Capella-Gutiérrez, F. Zakrzewski, H. Tafer, O. Rupp, T. Rosleff Sörensen, R. Stracke, R. Reinhardt, A. Goesmann, B. Schulz, P.F. Stadler, T. Schmidt, T. Gabaldón, H. Lehrach, B. Weisshaar and H. Himmelbauer. 2013. The genome of the recently domesticated crop plant sugar beet (*Beta vulgaris*). Nature 505: 546-549.

Filutowicz, A., A. Kuzdowicz, J. Trzebinski, K. Pawelski-Kozinski, L. Dalke and B. Jassem. 1971. Breeding and cytological investigations on sugarbeet improvement through interspecific hybridization. Final Report, Project E21-CR-42, Institute of Plant Breeding and Acclimatization, Bydgoszcz, Poland.

Filutowicz, A. and A. Kuzdowicz. 1959. Artbastarde zwischen Zuckerrüben und *Beta patellaris* Moq. Züchter 29: 1979-183.

Flavell, R.B., M.D. Bennet and J.B. Smith. 1974. Genome size and the proportion of repeated nucleotide sequence DNA in plants. Biochem. Genet. 12: 257-269.

Ford-Lloyd, B.V. and J.G. Hawkes. 1986. Weed beets, their origin and classification. Acta Horticulturae 82: 399-404.

Frese, L., R. Hannan, B. Hellier, S. Samaras and L. Panella. 2009. Survey of *Beta nana* in Greece. *In:* L. Frese, C.U. Germeier, E. Lipman and L. Maggioni (eds.). Report of the ECP/GR *Beta* Working Group and World *Beta* Network. Third joint meeting 8-10 March 2006. Tenerife, Spain. Bioversity International, Rome, Italy.

Gao, D., D. Guo and C. Jung. 2001. Monosomic addition lines of *Beta corolliflora* Zoss in sugar beet: cytological and molecular-marker analysis. Theor. Appl. Genetics 103: 240-247.

Gao, D. and C. Jung. 2002. Monosomic addition lines of *Beta corolliflora* in sugar beet: plant morphology and leaf spot resistance. Plant Breeding 121: 81-86.

Goldblatt, P. and D.E. Johnson. 1979 and onwards. Index to plant chromosome numbers. IPCN Chromosome Reports, Missouri Botanical Garden, St. Louis. (http://www.tropicos.org/Project/IPCN).

Halldén, C., D. Ahrén, A. Hjerdin, T. Säll and N.O. Nilsson. 1998. No conserved homoeologous regions found in the sugar beet genome. J. Sugar Beet Res. 35: 1-13.

Harveson, R.M., L.E. Hanson and G.L. Hein. 2009. Compendium of Beet Diseases and Pests (2nd edition). The American Phytopathological Society. St. Paul, Minnesota.

Heijbroek, W., A.J. Roelands, J.H. de Jong, C. van Hulst, A.H.L. Schoone and R.G. Munning. 1988. Sugar beets homozygous for resistance to beet cyst nematode (*Heterodera schachtii* Schm.) developed from monosomic additions of *Beta procumbens* to *B. vulgaris*. Euphytica 38: 121-131.

Hernández-Ledesma, P., W.G. Berendsohn, T. Borsch, S. von Mering, H. Akhani H, S. Arias, I. Castañeda-Noa, U. Eggli, R. Eriksson, H. Flores-Olvera, S. Fuentes-Bazán, G. Kadereit, C. Klak, N. Korotkova, R. Nyffeler, G. Ocampo, H. Ochoterena, B. Oxelman, R.K. Rabeler, A. Sanchez, B.O. Schlumpberger and P. Uotila. 2015. A taxonomic backbone for the global synthesis of species diversity in the angiosperm order *Caryophyllales*. Willdenowia 45: 281-383.

Johnson, R.T. 1956. A grafting method to increase survival of seedlings of interspecific hybrids within the genus, *Beta*. Proc. Am. Society Sugar Beet Technologists 9: 25-31.

Jung, C. 1987. Breeding for nematode resistance in sugar beet. Annal. Biol. 3: 15-25.

Jung, C., K. Pillen, L. Frese, S. Fahr and A. Melchinger. 1993. Phylogenetic-relationships between cultivated and wild species of the genus *Beta* revealed by DNA fingerprinting. Theor. Appl. Genet. 86: 449-457.

Kadereit, G., S. Hohmann and J.W. Kadereit. 2006. A synopsis of *Chenopodiaceae* subfam. *Betoideae* and notes on the taxonomy of *Beta*. Willdenowia 36: 9-19.

Kleine, M., H. Voss, D. Cai and C. Jung. 1998. Evaluation of nematode resistant sugar beet (*Beta vulgaris* L.) lines by molecular analysis. Theor. Appl. Genetics 97: 896-904.

Lange, W., T.S.M. de Bock. 1989. The diploidised meiosis of tetraploid *Beta macrocarpa* and its possible application in breeding sugar beet. Plant Breeding 103: 196-206.

Li, H-Y., C-Q. Ma, B. Yu, C-J. Gao, S-J Zhang, Y. Zhang and D-D. Guo. 2007. Extraction cDNA fragments specially expressed in lines M-14 in sugar beet by mRNA differential display. Bulletin of Botanical Research 27: 465-468.

Li, H., H. Cao, Y. Wang, Q. Pang, C. Ma and S. Chen. 2009. Proteomic analysis of sugar beet apomictic monosomic addition line M14. J Proteomics 73: 297-308.

Löptien, H. 1984. Breeding nematode-resistant beets. I. Development of resistant alien additions by crosses between *Beta vulgaris* L. and wild species of the section *Patellares*. Zeitschrift für Pflanzenzüchtung 92: 208-220.

Ma, C., Y. Wang, Y. Wang, L. Wang, S. Chen and H. Li. 2011. Identification of a sugar beet *BvM14-MADS* box gene through differential gene expression analysis of monosomic addition line M14. J. Plant Physiology 168: 1980-1986.

McFarlane, J.S. 1975. Naturally occurring hybrids between sugarbeet and *Beta macrocarpa* in the Imperial Valley of California. J. Am. Society Sugar Beet Technologists 18: 245-251.

McGrann, G.R.D., M.K. Grimmer, E.S. Mutasa-Göttgens and M. Stevens. 2009. Progress towards the understanding and control of sugar beet rhizomania disease. Mol. Plant. Pathol. 10: 129-141.

McGrath, J.M., C. Derrico and Y. Yu. 1999. Genetic diversity in selected, historical US sugarbeet germplasm and *Beta vulgaris* ssp. *maritima*. Theor. Appl. Genetics 98: 968-976.

McGrath, J.M., L.W. Panella and L. Frese. 2011. *Beta*. pp. 1-28. *In:* C. Kole. (ed.) Wild Crop Relatives: Genomic & Breeding Resources, Industrial Crops. Springer, Heidelberg.

Mesbah, M., O. Scholten, T. De Bock T and W. Lange. 1997. Chromosome localisation of genes for resistance to *Heterodera schachtii*, *Cercospora beticola* and *Polymyxa betae* using sets of *Beta procumbens* and *B. patellaris* derived monosomic additions in *B. vulgaris*. Euphytica 97: 117-127.

Munerati, O. 1932. Sull'incrocio della barbabietola coltivata con la beta selvaggia della costa adriatica. L'Industria Saccarifera Italiana 25: 303-304.

Munerati, O., G. Mezzadroli and T.V. Zapparoli. 1913. Osservazioni sulla *Beta maritima* L. nel triennio 1910-1912. Staz. Sper. Agric. Ital. 46: 415-445.

Nakamura, C., G.N. Skaracis and I. Romagosa. 1991. Cytogenetics and breeding in sugarbeet. pp. 295-314. *In:* T. Tsuchiya and P.K. Gupta (eds.). Chromosome Engineering in Plants: Genetics, Breeding and Evolution. Elsevier, New York, New York.

Owen, F.V. 1945. Cytoplasmically inherited male-sterility in sugar beets. J. Ag. Res. 71: 423-440.

Paesold, S., D. Borchardt, T. Schmidt and D. Dechyeva. 2012. A sugar beet (*Beta vulgaris* L.) reference FISH karyotype for chromosome and chromosome-arm identification, integration of genetic linkage groups and analysis of major repeat family distribution. Plant J. 72: 600-611.

Panella, L. and R. Lewellen. 2007. Broadening the genetic base of sugar beet: Introgression from wild relatives. Euphytica 154: 383-400.

Paul, H., B. Henken, T. De Bock and W. Lange. 1992. Resistance to *Polymyxa betae* in *Beta* species of the section *Procumbentes*, in hybrids with *Beta vulgaris* and in monosomic chromosome additions of *Beta procumbens* in *Beta vulgaris*. Plant Breeding 109: 265-273.

Reamon-Büttner, S., G. Wricke and L. Frese. 1996. Interspecific relationship and genetic diversity in wild beets in section *Corollinae* genus *Beta*: Isozyme and RAPD analyses. Genetic Resources and Crop Evolution 43: 261-274.

Reamon-Büttner, S.M. and G. Wricke. 1993. Evidence of tetrasomic inheritance in *Beta corolliflora*. J. Sugar Beet Res. 30: 321-327.

Reamon-Ramos, S.M. and G. Wricke. 1992. A full set of monosomic addition lines in *Beta vulgaris* from *Beta webbiana* - morphology and isozyme markers. Theor. Appl. Genetics 84: 411-418.

Romagosa, I., L. Cistue, T. Tsuchiya, J. Lasa J and R. Hecker. 1987. Primary trisomics in sugar beet. 2. Cytological identification. Crop Science 27: 435-439.

Savitsky, V.F. 1952. Monogerm sugar beets in the United States. Proc. Am. Society Sugar Beet Technologists 7: 156-159.

Savitsky, H. 1960. Meiosis in an F_1 hybrid between a Turkish wild beet (*Beta vulgaris* ssp. *maritima*) and *B. procumbens*. J. Am. Society Sugar Beet Technologists 11: 49-67.

Savitsky, H. 1969. Meiosis in hybrids between *Beta vulgaris* and *Beta corolliflora* and transmission of resistance to curly top virus. Can. J. Genet. Cytol. 11: 514-521.

Savitsky, H. 1975. hybridization between *Beta vulgaris* and *Beta procumbens* and transmission of nematode (*Heterodera schachtii*) resistance to sugarbeet. Can. J. Genet. Cytol. 17: 197-209.

Savitsky, H. 1978. Nematode (*Heterodera schachtii*) resistance and meiosis in diploid plants from interspecific *Beta vulgaris* × *B. procumbens* hybrids. Can. J. Genet. Cytol. 20: 177-186.

Savitsky, H. and J.O. Gaskill. 1957. A cytological study of F_1 hybrids between Swiss chard and *B. webbiana*. J. Am. Society Sugar Beet Technologists 9: 433-449.

Schmidt, T. and J. Heslop-Harrison. 1993. Variability and evolution of highly repeated DNA sequences in the genus *Beta*. Genome 36: 1074-1079.

Schondelmaier, J. and C. Jung. 1997. Chromosomal assignment of the nine linkage groups of sugar beet (*Beta vulgaris* L.) using primary trisomics. Theor. Appl. Genetics 95: 590-596.

Schulte, D., D. Cai, M. Kleine, L. Fan, S. Wang and C. Jung. 2006. A complete physical map of a wild beet (*Beta procumbens*) translocation in sugar beet. Mol. Genet. Genomics 275: 504-511.

Speckmann, G.J. and T.S.M. de Bock. 1982. The production of alien monosomic additions in *Beta vulgaris* as a source for the introgression of resistance to beet root nematode (*Heterodera schachtii*) from *Beta* species of the section *Patellares*. Euphytica 31: 313-323.

Stevanato, P., D. Trebbi, L. Panella, K. Richardson, C. Broccanello, L. Pakish, A.L. Fenwick and M. Saccomani. 2015. Identification and validation of a SNP marker linked to the gene *HsBvm-1* for nematode resistance in sugar beet. Plant Mol. Biol. Rep. 33: 474-479.

Van Geyt, J.P.C., W. Lange, M. Oleo and T.S.M. De Bock. 1990. Natural variation within the genus *Beta* and its possible use for breeding sugar beet - a review. Euphytica 49: 57-76.

Yang, L., Y. Zhang, N. Zhu, J. Koh, C. Ma, Y. Pan, B. Yu, S. Chen and H. Li. 2013. Proteomic analysis of salt tolerance in sugar beet monosomic addition line M14. J. Proteome Res. 12: 4931-4950.

Yu, M.H. 1984. Resistance to *Heterodera schachtii* in the *Patellares* section of the genus *Beta*. Euphytica 33: 633-640.

Zhu, H., Y-D. Bi, L-J. Yu, D-D Guo and B-C. Wang. 2009. Comparative proteomic analysis of apomictic monosomic addition line of *Beta corolliflora* and *Beta vulgaris* L. in sugar beet. Mol. Biol. Rep. 36: 2093-2098.

Chapter 17

Polyploidy in Watermelon

Wenge Liu[1,*] and Hongju Zhu[2]

ABSTRACT

Watermelon (*Citrullus lanatus*) is an important cucurbit crop, accounting for 7% of the worldwide area devoted to vegetable production. Commercial varieties of watermelon include diploid seed watermelon and triploid seedless watermelon: triploid seedless watermelon is one of the most successful examples of an artificially induced polyploid crop. The triploid seedless watermelons have many advantages relative to homozygous diploid watermelon, such as high yield, strong tolerance and good quality traits. Tetraploid watermelons are used as parental lines in triploid breeding, and are always obtained by treating newly emerged diploid seedlings with colchicine. The early identification of tetraploid plants may require morphological, cytological and even molecular techniques in the process of induced tetraploidy in watermelon. Tetraploids should be evaluated directly for green rind pattern, high seed yield, and other traits such as abundant fertility and high combining ability. Useful tetraploid inbred lines should produce triploid hybrids with excellent yield and quality for the market type and production area of interest. Commercial production of elite triploid hybrids is done by hand in locations where labor is inexpensive, or by bee pollination in isolation blocks. Cultivation of triploid seedless watermelons is similar to seeded watermelon, but there are three major differences: germination difficulties, transplanting requirement, and necessity of planting pollenizer varieties. Due to the advantages of polyploid watermelon, triploid seedless watermelon breeding and cultivation is becoming more and more popular, and has great future potential.

[1,2] Zhengzhou Fruit Research Institute, Chinese Academy of Agricultural Sciences, Zhengzhou, Henan, P.R.China, Zip: 450009, E-mail: liuwenge@caas.cn

* Corresponding author

Introduction

Origin and Distribution of Watermelon

Watermelon is a warm, long-season crop, now widespread in all tropical and subtropical regions of the world and mostly cultivated for its sweet dessert fruit, which are also an important source of water, sucrose, and other phytochemicals. Nutritionists have found that watermelon is a valuable dietary component due to its health-promoting citrulline, vitamin C and antioxidants, including β-carotene and lycopene. Watermelon is an important specialty crop, accounting for 7% of the agricultural area devoted to vegetable crops. The watermelon yield worldwide was 95 211 432 tonnes in 2012 (http://faostat.fao.org/), and the top five watermelon producers are China (70 000 000 tonnes), Turkey (4 044 184 tonnes), Iran (3 800 000 tonnes), Brazil (2 079 547 tonnes) and Egypt (1 874 710 tonnes).

Cultivated watermelon (*Citrullus lanatus*) is a member of the Cucurbitaceae, and is thought to have originated in southern Africa, where it is found growing wild. It reaches maximum genetic diversity there, with sweet, bland and bitter forms (Todd 2008). The culture of watermelons goes back to prehistoric times. The watermelon was cultivated in ancient Egypt, as verified by David Livingstone in the 1850s when he found great tracts of watermelon growing wild in the Kalahari Desert and in semitropical regions of Africa(Rubatzky 2001). Even today, in semi-desert districts of Africa watermelons are cultivated as an important source of water during dry periods. Watermelon was widely distributed throughout the remainder of the world by African slaves and European colonists. It was carried to Brazil, the West Indies, Eastern North America, the islands of the Pacific, New Zealand and Australia. Written records indicate that watermelons were cultivated in Massachusetts as early as 1629, before 1664 by Florida Indians, in 1673 in the Midwest, in 1747 in Connecticut, in 1799 by Indian tribes along the Colorado River and in 1822 in Illinois (Department of Agriculture, Forestry and Fisheries, Republic of South Africa 2011).

Classification of Watermelon

A new taxonomy of watermelon was proposed in 2014, comprising one genus (*Citrullus*), four species (*C. lanatus*, *C. colocynthis*, *C. eccirhosus* and *C. rehmii*), three subspecies (ssp. *lanatus*, ssp. *vulgaris* and ssp. *mucosospermus*) and four varieties (var. *lanatus*, var. *citroides*, var. *vulgaris* and var. *megalospermus*) (Lin 2015). Cultivated watermelon is *Citrullus lanatus*.

Currently, three watermelon ploidy levels are used in production: diploid ($2n = 2x = 22$), triploid ($2n = 3x = 33$) and tetraploid ($2n = 4x = 44$). Commercially, production is mainly of diploid seed watermelon and triploid seedless watermelon.

Diploid watermelons ($2n = 2x = 22$) consist of wild species, open-pollinated varieties, F_1 hybrids and inbred lines. The wild species are not common in watermelon cultivation production, but are often be used to study gene function and evolutionary origins of watermelon and to improve existing watermelon varieties (Yokota et al. 2002). Open-pollinated varieties are developed through several generations of selection, which is generally based upon yield, quality characteristics and disease resistance. Open-pollinated varieties were widespread during the early history of watermelon cultivation, selected mostly by farmers. F_1 hybrids are developed from two inbred lines that have been self-pollinated for several generations and then crossed, now widely cultivated in seed watermelon production (Todd 2008).

Commercial varieties of watermelon also include seedless triploids ($2n = 3x = 33$) produced by crossing female synthetic autotetraploids with a male diploid. The triploid watermelons are referred to as seedless, although they are not truly seedless, but rather have undeveloped seeds that are soft and edible. These rudimentary small seeds are consumed along with the flesh, just as immature seeds are eaten in cucumber.

The tetraploid watermelons ($2n = 4x = 44$) serve two purposes. Tetraploid lines are maintained year after year as parental lines in triploid breeding, but they can also be used as cultivars. However, the area under tetraploid watermelon cultivation is very small.

Triploid Seedless Watermelon

Commercial varieties of watermelon include diploid hybrids and seedless triploids. Triploid seedless watermelon is popular on a commercial scale and fetches high prices in the world market. The fruit of standard, seeded watermelon varieties may contain as many as 1 000 seeds in each fruit. The presence of seeds throughout the flesh makes the removal of seeds while eating difficult. One reason that seedless watermelons are more popular with consumers than seeded varieties is that the consumer does not have to be concerned with and inconvenienced by the seeds while the fruit is being eaten.

The Cultivated Area of Triploid Watermelon

The popularity of triploid seedless watermelon has increased over the last four decades (Todd 2008). During peak watermelon production in the U.S. market, seeded watermelons only comprised around 10% of the market and averaged four to five cents less per pound (U.S. Department of Agriculture 2013). Currently the cultivated area of triploid seedless watermelon in the world is increasing (from about 20% of the total area of watermelon), and the vast majority is concentrated in China, Thailand, South Korea, the United States, Southern America, southeast Asian and southern European countries (Liu et al. 2014). China is the largest country in terms of cultivated watermelon area in the world, accounting for more than 60% of cultivated area and more than 70% of world production. The total area of watermelon production worldwide is basically stable at 1.83 million hm^2, while triploid seedless watermelon cultivation area comprises more than 270 000 hm^2 and is still expanding (Liu et al. 2014).

The Cultivars of Triploid Watermelon

Triploid seedless watermelon variety development is underway by a number of seed companies and research institutes, and new varieties are being developed every year. Depending on the cultivar, triploid watermelon fruit are produced in different sizes: icebox, small, medium, large, or giant; different shapes: round, oval or elongate; different rind patterns: green, narrow stripe, medium stripe, wide stripe, light solid color, or dark solid color; and different flesh colors: white, yellow, orange, or red. Commercially, the most popular seedless watermelons in the USA are red fleshed, oval, and medium-sized (5-8 kg), like the cultivar Tri-X-313 (Todd 2008). The most popular cultivars in China are globe-shaped with red flesh and dark green skin, such as Zhengkangwuzi No. 5 and Dongting No. 1 (Liu et al. 2014). Cultivars with narrow stripes and wide stripes in green skin are also popular.

The Advantages of Triploid Watermelon

Triploid watermelon is unique in that it takes advantage of synthetic polyploidy in breeding commercially important seedless triploids coupled with heterosis due to the dosage effects resulting from various combinations of diploid and tetraploid genomes. Triploid watermelon plants always show vigorous growth, higher fruit number per plant in the field and higher yield than diploid watermelon (NeSmith and Duval 2001). Triploids are resistant to flooding during cultivation, and are also considered to be more tolerant of salt, cold, and fusarium wilt than their homozygous diploid ancestors (Liu et al. 2002, 2003, 2004, 2006; Zhu et al. 2014). Triploid secondary metabolites such as vitamin C, lycopene and citrulline are revealed be present in higher concentrations than in the diploids (Wan et al. 2011), and the flavor is considered to be better than that of the diploids (Leskovar et al. 2004). Triploid watermelons also show other distinct characters such as flesh firmness, thin rind, and possibly longer shelf life (Liu et al. 2014). Seedless watermelon can be sold at a higher price because of these distinct characters.

Tetraploid Watermelon Production

Use of triploid hybrids has provided a method for production of seedless fruit. Kihara began working on seedless watermelons in 1939, and had commercial triploid hybrids available 12 years later (Crow 1994). The development of triploid cultivars adds several problems to the process of watermelon breeding: extra time for the development of tetraploids; additional selection against sterility and fruit abnormalities; choice of parents for reduced seed coat production; the reduction in seed yield per acre obtained by seed companies; reduced seed vigor for the grower; and the necessity of planting diploid pollenizers, which take up one-third of the grower's production field.

Triploid seedless cultivars are produced by crossing a tetraploid ($4x = 44$) inbred line as the female parent with a diploid ($2x = 22$) inbred line as the male parent of the hybrid. The reciprocal cross (diploid female parent) does not produce seeds. The hybrid is a triploid ($3x = 33$), and is female and male sterile. Triploid plants have three sets of chromosomes, and three sets cannot be divided evenly when they go into two daughter cells during meiosis. Since the triploid hybrid is female sterile, the fruit are seedless. Breeders interested in the production of seedless triploid hybrids need to develop tetraploid inbred lines to be used as the female parent in a cross with a diploid male parent. One of the major limiting steps in breeding seedless watermelons is the small number of tetraploid inbred lines available. Development of seedless hybrids will be discussed in the following stages: choice of diploid lines, induction of tetraploid plants, identification of tetraploid plants, tetraploid line development, and hybrid production and testing.

Choice of Diploid Watermelon Lines

Most of the tetraploid lines being used by the seed industry have green rind so that, when crossed with a diploid line with striped rind, it will be easy to separate self-pollinated progeny (which will be seeded fruit from the female parent line) from cross-pollinated progeny (which will be seedless fruit from the triploid hybrid). The grower will want to discard the green fruit so they are not marketed as seedless watermelons by mistake.

Induction of Tetraploid Plants

Many methods have been used effectively in other crops to produce polyploids, including tissue culture regeneration, temperature shock, and X-rays. Traditionally, tetraploid plants have been obtained by treating newly emerged diploid seedlings with colchicine ($C_{22}H_{25}O_6N$), an antimitotic alkaloid which is well known for inhibiting the formation of spindle fibers, and effectively arresting mitosis at the metaphase stage. Since chromosomes have already multiplied but cell division is arrested, polyploid cells are created.

For the seedling treatment method, the diploid line of interest is planted in the greenhouse in flats (8 × 16 cells is a popular size) on heating pads set to keep the soil medium at 29°C for rapid and uniform germination. When the cotyledons first emerge from the soil, the growing point is treated with colchicine to stop chromosome division, resulting in a shoot with four sets of chromosomes rather than two. The colchicine solution is used at a concentration of 1% for small-seeded cultivars, 1.5% for cultivars with medium-sized seeds, and 2% for large-seeded cultivars. Colchicine is applied in the morning and evening for three consecutive days to each seedling, using 1 drop on small- or medium-size-seeded cultivars and 2 drops on large-seeded cultivars. The treatment produces plants that are diploid, tetraploid, or aneuploid, so it is necessary to identify and select the tetraploids in later stages. Treatment of the T0 diploids with colchicine results in about 1-3% tetraploid seedlings (referred to as T1 generation tetraploids). Some diploid cultivars and breeding lines produce a higher percentage of tetraploids than others (Todd 2008).

However, colchicine is carcinogenic and generally less effective than herbicides with a similar mode of action, such as the dinitroanilines oryzalin and trifluralin. The production of polyploid regenerants from tissue culture has been reported for many plant species and has potential application to establish a large number of new watermelon tetraploid breeding lines. As described in other species, polyploidy amongst watermelon plants obtained by adventitious shoot regeneration is common. The advantage of in vitro techniques is that tissue culture can be used to produce non-chimerictrue breeding tetraploids efficiently from a wide range of diploid cultivars.

Identification of Tetraploid Plants

The early identification of tetraploid plants may require morphological, cytological and even molecular techniques in the process of induced tetraploidy in watermelon. Chromosome counting is the usual method to determine ploidy, but this is difficult in watermelon due to the small chromosome size. Chromosome counting is not practical for non-dividing cells in differentiated tissues such as leaves, and can only be carried out in the T1 generation. Flow cytometry is a rapid and exact method for estimating nuclear DNA content. It can be efficiently used for ploidy determination in plants growing in the field and in the greenhouse, and has already been well established in tetraploid watermelon. Induced tetraploid plants of watermelon are often identified by counting the number of chloroplasts per guard cell pair of fully expanded leaves, as diploid and tetraploid watermelon plants possess variable number of chloroplasts in guard cells of stomata (Liu et al. 2005). Ploidy can also be estimated by examining plant morphological traits such as leaf and flower size, and by comparing the size of the pollen grains (about 1.44 times larger than diploid pollen) and the number of colpi (four versus three) (Liu et al. 2003). Tetraploid watermelon can also be effectively identified early in vitro tissue culture because of the difference in salt tolerance between diploid and tetraploid watermelon (Zhao et al. 2015).

Tetraploid Line Development

Tetraploid plants are selected in the T1 generation from the greenhouse flats where they were treated with colchicine. It is then necessary to plant the T2 generation in flats to verify that the plants are tetraploids in that next generation, and transplant the selections to greenhouse pots for self-pollination. Seeds from those selections can then be increased in larger plantings such as field isolation blocks to get sufficient numbers of seeds per tetraploid line to use in triploid hybrid production.

The fertility and seed yield of tetraploid lines will increase over generations of self- or sib-pollination (Todd 2008), probably because plants with chromosome anomalies are eliminated, resulting in a tetraploid line with balanced chromosomes and regular formation of 11 quadrivalents. Seed yield of tetraploid lines in early generations is often only 10-100 seeds per fruit. Another problem with raw tetraploids is poor seed germination, making it difficult to establish uniform field plantings. It may require as much as 10 years of self-pollination before sufficient seeds of tetraploid lines can be produced for commercial production of triploid hybrids. Advanced generations of tetraploid lines usually have improved fertility, seed yield, and germination rate compared to the original lines.

Triploid Hybrid Production and Testing

Tetraploids should be evaluated directly for green rind pattern, high seed yield, and other traits such as male sterility for reduced hand labor in hybrid seed production. However, the major test for tetraploids is as female parents in triploid hybrid seed production after controlled crosses using diploid male parents. The resulting hybrids are tested in yield trials with two rows of triploid plots alternating with one row of diploid plots to assure adequate pollen for fruit set in the triploid hybrids. Useful tetraploid inbred lines should produce triploid hybrids with excellent yield and quality for the market type and production area of interest.

Triploid Watermelon Evaluation

Evaluation of Triploid Hybrids

Evaluation of triploid hybrids is similar to the evaluation of diploid cultivars already discussed previously. There are a few special considerations, however. Triploids are not inherently superior to diploids, so triploid hybrids can be better or worse than their diploid parental lines. Therefore, as in the case of diploid hybrids, many combinations of parental lines should be evaluated in triploid yield trials to identify the lines producing hybrids with the best performance. In general, diploid inbred parents that have poor horticultural performance will produce triploid hybrids having poor performance (Todd 2008).

One problem affecting triploid hybrids is empty seed coats (colored or white) in the fruit. Under some environmental conditions, fruit are produced with large obvious seed coats that are objectionable to consumers. Triploids should be tested for seed coat problems in the fruit during trialing. Seed coats will be large in the hybrids if the parents have large seeds. Seed size is genetically controlled, with at least three genes involved: *l* (long seed gene), *s* (short seed gene), and *tss* (tiny seed). Besides genetic effects, certain environmental conditions seem to increase the number of hard seed coats in triploid hybrids (Todd 2008).

Commercial production of elite triploid hybrids is done by hand in locations where labor is inexpensive, or by bee pollination in isolation blocks. The tetraploid and diploid inbreds are planted together in alternating rows, or in alternating hills within each row. Where labor is cheap, the staminate flowers can be collected from the male (diploid) parent and used to pollinate the pistillate flowers on the female (tetraploid) parent. Pollinated flowers should be capped the previous day to keep bees out, and then covered after pollination to prevent self- or sib-pollination after the cross has been made. The flowers should be tagged with the date so that the fruit can be harvested later.

A method that requires less hand labor is to plant the male and female parents in alternating rows, and to remove all staminate flowers from the female parent rows during peak flowering time, usually a period lasting several weeks. Pistillate flowers on the female parent are tagged on the day they open with the date to assure that the fruit are mature when harvested, and so that only fruit that were pollinated during the time staminate flowers were removed from the female parent are harvested. Seeds that are harvested can also be sorted mechanically for size, weight or density to separate triploid seeds from tetraploid seeds.

Cultivation of Triploid Seedless Watermelon

Although production of triploid seedless watermelons is similar to production of seeded (diploid) watermelons, there are three major differences. Firstly, triploid watermelon seed has more difficulty germinating in the field. The thick seed coat and a large airspace between the underdeveloped embryo and seed coat tissues appear to have a major role in limiting seed germination in triploid watermelon. However, seed nicking or de-coating results indicate that polyploid seed germination is not inhibited by the seed coat alone, but also by high sensitivity to increased moisture content. The improved germination in the presence of 1% or 2% H_2O_2 may result from weakening of the seed coat as H_2O_2 reacts with the seed coat (Duval and NeSmith 2000). Secondly, seedless triploid watermelon crops are established by transplanting, due to the lower seedling production rate. Transplanting watermelons offers several advantages, and allows plants to be produced under greenhouse conditions when outdoor conditions are not conducive to plant growth. Seed-use efficiency also increases, which is especially important with costly hybrid and triploid seed. Soil crusting and damping off, detrimental to seedling growth, can also be eliminated or reduced, and planting depth is more uniform. Transplanting usually results in earlier harvests. Thirdly, since the triploid hybrid is female sterile, triploid hybrid watermelons fail to produce enough pollen to ensure normal pollination. Therefore, pollenizer varieties (diploid, seeded) should be planted nearby to ensure the production of the pollen needed. To avoid confusion during the harvesting and to be able to distinguish between triploid seedless watermelons and seeded watermelons, some experienced growers prefer to use as pollinating varieties lines which markedly differ in external color from the triploid hybrid, as mentioned previously. The proposed set of pollinating varieties has been made up with due regard for plant vigor, pollen producing potential, etc. The recommended triploid hybrid-pollinator ratio is 3:1.

Future Possibilities for Polyploid Watermelon Improvement

With the increasing area of triploid seedless watermelon cultivation in the world, future directions are discussed and several viewpoints are suggested. Innovations are greatly needed in the development of tetraploid watermelon germplasm to promote the quality of triploid seedless watermelon. The utilization of heterosis in tetraploid watermelon should be researched. The breeding of small fruited, seedless watermelon varieties should be accelerated in anticipation of consumer demand. Research related to disease resistance in polyploid watermelon is considered to be of great importance because of the continuous cropping system used for watermelon production. Postharvest storage, transportation and comprehensive processing should also be further optimized. Triploid and tetraploid watermelon have higher yield, higher quality, disease resistance, stress tolerance and higher secondary metabolite contents compared to isogenic diploids, as well as enhanced heterosis. Hence, polyploid watermelon breeding and cultivation is becoming more and more popular, and has great future potential.

References

Crow, J.F. 1994. Hitoshi Kihara, Japan's pioneer geneticist. Genetics 137(4): 891-894.

Department of Agriculture, Forestry and Fisheries, Republic of South Africa. 2011. Watermelon (*Citrullus lanatus*). Directorate Plant Production, Pretoria.

Duval, J.R. and D.S. NeSmith. 2000. Treatment with hydrogen peroxide and seed coat removal or clipping improve germination of `Genesis' triploid watermelon. HortScience. 35(1): 85-86.

Leskovar, D.I., H. Bang, K.M. Crosby, N. Maness, J.A. Franco and P. Perkins-Veazie. 2004. Lycopene, carbohydrates, ascorbic acid and yield components of diploid and triploid watermelon cultivars are affected by deficit irrigation. J. Hortic. Sci. Biotechnol. 79 (1): 75-81.

Lin, D. 2015. A study of systematics for *Citrullus* Schrad. China Cucurbits and Vegetables. 28(5): 1-4.

Liu, W., N. He, S. Zhao and X. Lu. 2014. Annual production and cultivation of triploid seedless watermelon in China. Journal of Changjiang Vegetables. 14: 1-6.

Liu, W., M. Wang and Z. Yan. 2003. Observation and comparison on pollen morphology of different ploidy watermelon. Acta Hortculturae Sinica 30(3): 328-330.

Liu, W., M. Wang and Z. Yan. 2003. Studies on physiological and biochemical characteristics of seedling of different ploidy watermelons under cold-stress. Journal of Fruit Science 20(1): 44-48.

Liu, W., M. Wang, Z. Yan. and H. Zhao. 2004. Effect of cold hardening on SOD, POD activities and on contents of MDA in different ploidy watermelon seedling. Acta. Bot. Boreal.-Occident. Sin. 24(4): 578-582.

Liu, W., Z. Yan and X. Rao. 2005. Comparison of the leaf epidermal ultra-structure morphology of different ploidy watermelon. Journal of Fruit Science. 22(1): 31-34.

Liu, W., Z. Yan, C. Wang and H. Zhang. 2006. Response of antioxidant defense system in watermelon seedling subjected to waterlogged stress, Journal of Fruit Science. 23(6): 860-864.

Liu, W., Z. Yan, H. Zhang and M. Wang. 2002. The study of Salt-Tolerance in Germinating Seeds and Seedling of different ploidy watermelon. China watermelon and melon. 3: 1-2.

NeSmith, D.S. and J.R. Duval. 2001. Fruit set of triploid watermelons as a function of distance from a diploid pollinizer. HortScience 36(1): 60-61.

Rubatzky, V.E. 2001. Origin, distribution, and uses. pp. 21-26. *In*: Donald N. Maynard (eds.). Watermelons characteristics, production, and marketing. ASHS press, Alexandria.

Todd, C.W. 2008. Watermelon. pp. 381-418. *In*: Jaime Prohens and Fernando Nuez (eds.). Vegetables I. Springer, New York.

U.S. Department of Agriculture. 2013. National watermelon report. U.S. Dept. Agr. Agricultural Marketing Service. (http://www.ams.usda.gov) Thomasville, GA.

Wan, X., W. Liu, Z. Yan, S. Zhao, N. He, P. Liu and J. Dai. 2011. Changes of the Contents of Functional Substances Including Lycopene, Citrulline and Ascorbic Acid During Watermelon Fruits Development. Scientia Agricultura Sinica, 44(13): 2738-2747.

Yokota, A., S. Kawasaki, M. Iwano, C. Nakamura, C. Miyake and K. Akashi. 2002. Citrulline and DRIP-1 protein (ArgE homologue) in drought tolerance of wild watermelon. Ann. Bot. 89: 825-832.

Zhao, L., W. Liu, Z. YAN and H. Zhu. 2015. The *In Vitro* tissue induction and salt screening of diploid and tetraploid watermelon. Acta Agiculturae Boreali-occidentalis Sinica. In press.

Zhu, H., W. Liu, S. Zhao, X. Lu, N. He, J. Dou and L. Gao. 2014. Comparison between tetraploid watermelon (*Citrullus lanatus*) and its diploid progenitor of DNA methylation under NaCl stress. Scientia Agricultura Sinica. 47(20): 4045-4055.

Chapter 18

Optimization of Recombination in Interspecific Hybrids to Introduce New Genetic Diversity into Oilseed Rape (*Brassica napus* L.)

Annaliese S. Mason[1] and Anne-Marie Chèvre[2,*]

ABSTRACT

Brassica napus (oilseed rape, rapeseed, canola) is an agriculturally important allotetraploid (2*n* = AACC) species in the *Brassica* genus. However, due to its recent origin from only a few hybridization events between progenitor diploid species *B. rapa* (2*n* = AA) and *B. oleracea* (2*n* = CC), and due to stringent breeding selection pressure for oil-quality traits, *B. napus* has very narrow genetic diversity. This creates a problem for breeders, who need genetic diversity for continual improvement of agronomic traits. Innovative strategies can be proposed using homologous recombination to generate more stable material from crosses with progenitor species *B. rapa* and *B. oleracea* as well as related allopolyploids *B. juncea* (2*n* = AABB) and *B. carinata* (2*n* = BBCC), as all share a genome or subgenome (set of chromosomes) in common with *B. napus*. These methods are useful for trait introgression into *B. napus*, whatever their genetic control, as higher recombination frequencies allow a smaller genomic region to be introgressed with less chance of linkage drag. However, useful traits can be also transferred via homeologous pairing (between chromosomes from different genomes) in interspecific hybrids between *B. napus* and close relatives. This method is difficult and generally results in large genomic introgression regions, with an increased chance of linkage drag (co-introgression of genes with a negative effect on yield traits), but is relevant for traits under monogenic control. The advantages

[1] Plant Breeding Department, IFZ Research Centre for Biosystems, Land Use and Nutrition, Justus Liebig University Giessen, Heinrich-Buff-Ring 26-32, 35392 Giessen, Germany, E-mail: annaliese.mason@agrar.uni-giessen.de

[2] Institut de Génétique, Environnement et Protection des Plantes, INRA, Agrocampus Ouest, UR1, BP35327, 35653 Le Rheu, France, E-mail: anne-marie.chevre@rennes.inra.fr

* Corresponding author

and limitations of each strategy according to the ploidy level of the hybrids are presented.

Introduction

Oilseed rape (*Brassica napus*) is the third most important crop for oil production worldwide. Historically, this crop underwent a burst of agricultural selection in the middle of the 20th century, followed by widespread cultivation in Europe, China, Canada and Australia. Two challenges were rapidly overcome in the improvement of seed quality: firstly, the creation of varieties with low erucic acid content for human consumption, and secondly, decreased glucosinolate content for use of seed meal in animal feed. Oilseed rape varieties which had low glucosinolate and erucic acid content were re-marketed as "canola" in Canada or 00 varieties in Europe, a high-quality oil and meal product that was subsequently in high demand. However, these breeding constraints induced a genetic diversity bottleneck in oilseed rape. *Brassica napus* is also a recent allopolyploid, representing only a fraction of the genetic diversity present in its progenitors (ancestors of extant species *B. rapa* and *B. oleracea*). Fortunately, large, genetically diverse germplasm resources are available for oilseed rape improvement: many related species can be used to introgress novel genetic diversity and useful agricultural traits into this crop through interspecific hybridization (reviewed by Prakash et al. (2009)). Different interspecific hybridization strategies can be developed depending on the species carrying the trait of interest, on the mode of genetic inheritance of the trait and on the genome structure of the donor and recipient species.

Among the Brassiceae tribe, which contains 46 genera, and of the 47 species belonging to the *Brassica* genus (Al-Shehbaz 2012), three diploid species, *Brassica rapa* (genome complement AA, $2n = 20$), *B. oleracea* (genome complement CC, $2n = 18$) and *B. nigra* (genome complement BB, $2n = 16$) hybridized in pairwise combinations to give rise to three natural allopolyploid species: Indian mustard (*B. juncea*, AABB, $2n = 36$), Ethiopian mustard (*B. carinata*, BBCC, $2n = 34$) and oilseed rape (*B. napus*, AACC, $2n = 38$) (U 1935). For the latter, the origin of the natural form is still speculative, with historical and botanical data suggesting a recent occurrence 500 to 600 years ago (Gomez-Campo 1980), but recent sequencing data and phylogenic analyses suggesting an older formation ~7500 years ago (Chalhoub et al. 2014). Genetic mapping (reviewed by Parkin (2011)) and sequencing data (Parkin et al. 2014; Chalhoub et al. 2014; Liu et al. 2014; Wang et al. 2011) revealed that the constitutive genomes of oilseed rape are well conserved compared to the modern genomes of the diploid parent species. The structure of *Brassiceae* diploid species genomes has also been established using the related model plant *Arabidopsis thaliana* as a reference: these data showed that all *Brassiceae* were derived from a common hexaploid ancestor (Parkin et al. 2005; Schranz et al. 2006), with three homeologous copies of each genomic region detected per haploid chromosome complement in Brassiceae species (Ziolkowski et al. 2006; Mandakova and Lysak 2008). From the sequencing of *B. rapa* (Wang et al. 2011), triplicated genomic regions were shown to have differential fractionation (Tang et al. 2012; Cheng et al. 2014; Cheng et al. 2013; Murat et al. 2015) due to the origin of the hexaploid ancestor via crosses between a tetraploid (in which some fractionation had already occurred) and a diploid. As a consequence of this paleopolyploidy, the diploid species contain large genomic regions of homeology within each genome as well as in relation to each other. Despite this high level of homeology, the *Brassica* allopolyploid species have regular meiotic behavior and disomic inheritance. However, these similarities in genome structure can also be efficiently used to introduce new variability into oilseed rape varieties.

Recombination is the main mechanism by which genetic variability is generated in progeny. Homologous recombination is strictly regulated, with one obligate crossover between homologous chromosomes and rarely more than three per homologous chromosome pair in a single meiosis (Mezard et al. 2007). The location of these crossovers depends on different factors. A major factor is chromosome structure, with generally higher frequencies of crossovers in telomeric regions but suppression of recombination in the centromeres. In the case of an allopolyploid species such as oilseed rape, it is possible to use homologous recombination in genomes shared in common between the extant diploid progenitor species (*B. rapa*, $2n$ = AA and *B. oleracea*, $2n$ = CC) and the allopolyploid (*B. napus*, $2n$ = AACC) to introduce genetic variation from the diploids to the allopolyploid. This strategy allows a high level of genome stability to be maintained. By contrast, homeologous recombination between less closely-related genomes is less frequent, although homoeologous recombination has been observed between each of the *Brassica* A, B and C genomes in AB, AC and BC allohaploids (Mizushima (1980) for review).

In oilseed rape haploids ($2n$ = AC), only two meiotic behaviors (high and low pairing between A and C genomes) have been described; these are probably related to two different phylogenic origin events to produce allopolyploid *B. napus* (Cifuentes et al. 2010a, 2010b). A genetic locus regulating homeologous recombination in oilseed rape AC haploid has been identified and analyzed (Jenczewski et al. 2003; Liu et al. 2006; Nicolas et al. 2009; Nicolas et al. 2012), but its role in promoting or preventing homeologous recombination with genomes from related species has still to be demonstrated. It has also been established that homeologous recombination between the A and C genomes can occur naturally, as oilseed rape varieties can carry reciprocal or non-reciprocal translocations (Lombard and Delourme 2001; Osborn et al. 2003; Piquemal et al. 2005; Udall et al. 2005). In the present chapter, we will describe the different strategies that can be undertaken using either homologous or homeologous (between ancestrally related genomes) recombination to introgress genetic variation and traits of interest into *Brassica napus*, showing the interest value and the limits of the material generated, and describing optimal use of these strategies in breeding programs.

Optimal Use of Homologous Recombination in Interspecific Hybrids

Homologous recombination offers the possibility to introgress genomic regions carried by species which share genomes with *B. napus* into the crop via pairing between homologous chromosomes. This strategy is particularly relevant for traits under polygenic control, as it allows multiple genomic regions to be targeted and introgressed simultaneously across the genome. However, homologous recombination can only be used for A and C genomes which have the same structure as those in *B. napus*, whatever their origin. Several different strategies can be applied to introgress characters via homologous recombination. For the sake of simplicity, we will indicate the origin of each A and C genome or subgenome using the first letter of each species as annotation: A^r for *B. rapa*, C^o for *B. oleracea*, B^n for *B. nigra*, A^n and C^n for the *B. napus* A and C genomes respectively, A^j and B^j for the *B. juncea* subgenomes and B^c and C^c for the *B. carinata* subgenomes. Diploid progenitors of *B. napus* are highly diverse for several traits of interest, such as disease resistance, flowering period and seed quality (reviewed by Prakash et al. (2009)). A strategy for accessing this diversity consists of direct crosses between *B. napus* and one of its progenitors (Figure 1).

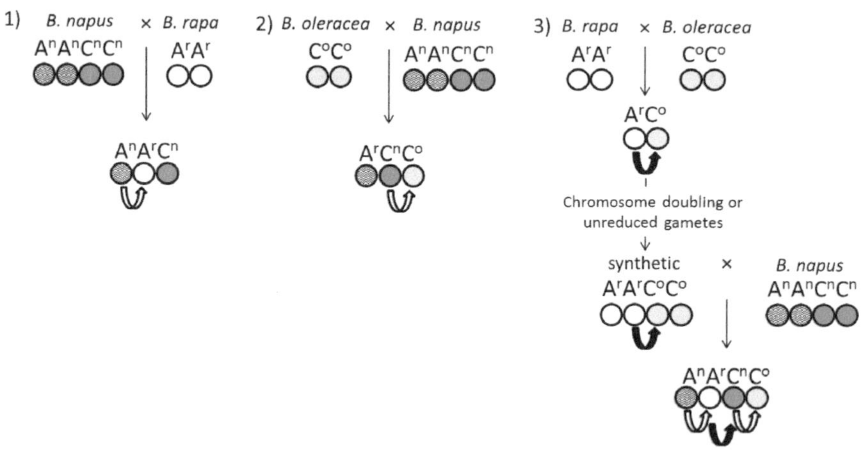

Figure 1. Possible scheme for intercrossing between *Brassica napus* and its two diploid progenitors, *B. rapa* and *B. oleracea*. The A genome is in white for *B. rapa* and white with black lines for *B. napus*. The C genome is in grey for *B. oleracea* and dark grey for *B. napus*. Empty arrows indicate homologous recombination between the A genomes or between the C genomes and black arrow indicates a high probability of homeologous recombination.

The frequency of homologous recombination between A chromosomes in $A^rA^nC^n$ hybrids is higher than the frequency in A^rA^r hybrids, with an additional 4 to 6 crossovers per homologous chromosome pair (Leflon et al. 2010; Nicolas et al. 2009). This high rate of recombination allows reduction of introgression size from the diploid species, an important factor in incorporating useful traits without affecting the agronomic value of the resulting oilseed rape via linkage drag, where many undesirable traits may also be carried on large introgressions in addition to the trait/locus of interest. These AAC hybrids produce gametes with a chromosome number ranging from 10 chromosomes (n = A) to 19 chromosomes (n = AC), following a binomial distribution for inheritance of the (mostly univalent) C genome chromosomes (Leflon et al. 2006). So, in the first backcross generation of the AAC interspecific hybrid to *B. napus*, it is possible to obtain 5 to 10% of plants with an AACC genome structure. This strategy has been previously used to produce Chinese varieties with greater genetic diversity (Qian et al. 2006). It is more difficult to obtain hybrids from crosses between *B. napus* and *B. oleracea* relative to crosses between *B. napus* and *B. rapa* (FitzJohn et al. 2007). However, the gametes produced from $A^nC^nC^o$ hybrids are mainly of two types: unreduced gametes (n = $A^nC^{n/o}C^{n/o}$) or gametes with a reduced (haploid) C genome but inheritance of all A genome chromosomes (n = $A^nC^{n/o}$) (Namai 1987); hence, 5 to 10% of plants also have 38 chromosomes after backcrossing to *B. napus*. Li et al. (2013) proposed to combine genetic diversity from both diploid species by colchicine doubling of $A^nC^nC^o$ hybrids to produce $A^nA^nC^nC^nC^oC^o$ types (which were hypothesized to produce $A^nC^{o/n}C^{o/n}$ gametes) before crosses to *B. rapa* (A^rA^r) in order to directly obtain new types of *B. napus* ($2n$ = $A^{r/n}A^{r/n}C^{o/n}C^{o/n}$).

In order to simultaneously incorporate diversity from both diploid species and based on the knowledge that *B. rapa* and *B. oleracea* are the progenitors of *B. napus* (U 1935), several authors have created synthetic oilseed rape by direct crosses between *B. rapa* and *B. oleracea* followed by colchicine doubling of the A^rC^o F_1 hybrids (Fig. 1). The $A^rA^rC^oC^o$ plants generally show poor seed set. One example has been reported of direct use of a synthetic form, but as a newly bred vegetable crop (Fujii and Ohmido 2011). As oilseed rape is of interest for seed production, the best strategy is generally to cross the synthetic directly with natural *B. napus*. The $A^nA^rC^nC^o$ F_1 hybrids, called semi-synthetics, are then used to produce segregating populations. Several generations of backcrosses has allowed easy introgression of single genes of interest into elite varieties, as has been demonstrated for blackleg resistance (Crouch et al. 1994). With respect to quantitative traits, heterosis (Abel et al. 2005; Seyis et al. 2006) has also been detected in these backcross populations, and QTL for seed yield were also identified using a doubled haploid segregating population between a natural and a synthetic line (Radoev et al. 2008).

Another strategy to explore incorporation of genetic diversity via homologous recombination is to cross the diploid species with the two allopolyploid species carrying the complementary genomes to *B. napus*, i.e. *B. rapa* with *B. carinata* or *B. oleracea* with *B. juncea* (Fig. 2). The ABC hybrids thus produced can then be backcrossed either directly with *B. napus*, relying on the formation of unreduced gametes (n = ABC), or after colchicine doubling as proposed by Li et al. (2004) and Zou et al. (2010). The B genome can then be eliminated by successive backcrosses with *B. napus*, as was carried out by Meng et al. (1998) in order to produce yellow-seeded *B. napus* types. "Intersubgenomic heterosis", or heterosis resulting from interactions between fixed alleles at homeologous loci in the A and C genomes, has been detected in these new oilseed rape types (reviewed by Chen et al. (2011)).

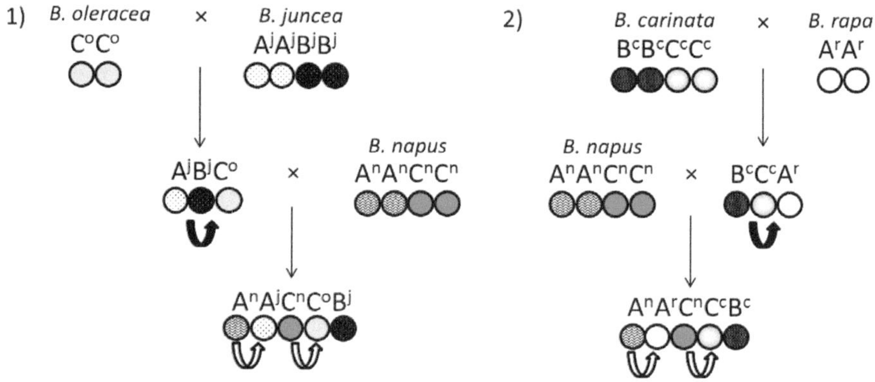

Figure 2. Possible scheme for intercrossing between the two allopolyploid species *B. juncea* and *B. carinata* and the two diploid progenitors of *B. napus*, *B. rapa* and *B. oleracea*. The F_1 hybrids are backcrossed with *B. napus*. The A genome is in white for *B. rapa*, white with black lines for *B. napus* and in white with black points for *B. juncea*. The C genome is in grey for *B. oleracea*, dark grey for *B. napus* and degraded grey for *B. carinata*. The B genome is in black for *B. juncea* and in black with white points for *B. carinata*. Empty arrows indicate homologous recombination between the A genomes or between the C genomes, and black arrows indicate a high probability of homeologous recombination.

The last strategy consists of utilizing the genetic diversity available in the related allopolyploid species which share either the A or the C genome with *B. napus*. Oilseed rape can be crossed with either *B. juncea* or *B. carinata* to generate $A^nA^jC^nB^j$ or $C^nC^cA^nB^c$ hybrids respectively (Figure 3). The analysis of meiotic behavior in such hybrids revealed that the diploid genomes paired preferentially through homologous recombination, although with higher frequencies of multivalent formation (Mason et al. 2010). Similarly to the previous strategy, these F_1 hybrids can subsequently be backcrossed to *B. napus* to eliminate B genome chromosomes.

Optimal Use of Homeologous Recombination in Interspecific Hybrids

Homeologous recombination in oilseed rape improvement requires pairing between related genomes, and relies on the introduction of one or more foreign genomic introgressions. Although this method can and has been used successfully to incorporate novel trait variation for oilseed rape improvement, there are several considerations to be overcome. Firstly, homeologous recombination frequently generates genomic instability in

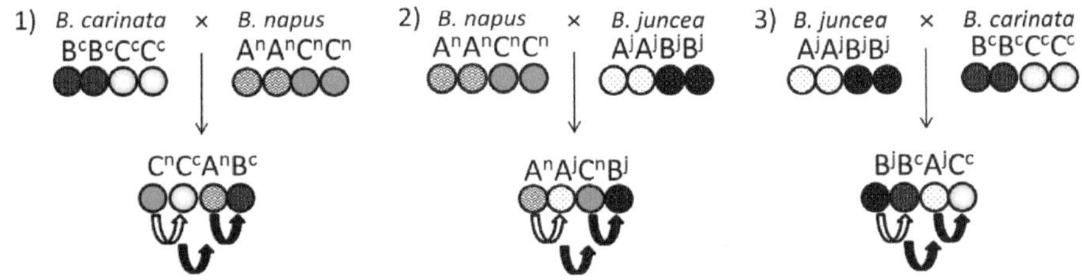

Figure 3. Possible schemes for intercrossing between the three allopolyploid species *B. napus*, *B. juncea* and *B. carinata*. The F_1 hybrids are backcrossed to *B. napus*. The A genome is in white with black lines for *B. napus* and in white with black points for *B. juncea*. The C genome is in dark grey for *B. napus* and degraded grey for *B. carinata*. The B genome is in black for *B. juncea* and in black with white points for *B. carinata*. Empty arrows indicate homologous recombination between the A genomes or the C genomes or the B genomes and black arrows indicate a high probability of homeologous recombination.

the progeny, as chromosome translocations can affect subsequent ability of homologous chromosomes to pair and segregate at meiosis. Generally after a foreign genomic introgression into *B. napus* has occurred with subsequent recovery of a 38 chromosome ($2n$ = AACC) plant, it is difficult to reduce the size of the introgression through recombination, because of the low to null frequency of crossovers between the introgressed region and the crop genome in the corresponding region. Homoelogous recombination can also be used to change the dosage of homeologous genomic regions between the two constitutive A and C genomes of *B. napus* as well as to introgress genomic regions from a related species into *B. napus*. In the latter case, this strategy is applied mainly to traits under monogenic control.

Chromosome pairing is known to be frequent in A^rC^o F_1 hybrids (reviewed by Prakash and Hinata (1980)): on average, 89% of the 19 chromosomes of this hybrid are involved in homeologous pairing between the A^r and C^o genomes per pollen mother cell; this frequency is higher than in A^nC^n haploids of oilseed rape which have a frequency ranging from 72% to 81% (Cifuentes et al. 2010b). These F_1 hybrids can be used either directly to cross with *B. napus* or after colchicine doubling (Figure 1). In the first case, unreduced gametes produced by first division restitution-like mechanisms are commonly observed to transmit the majority of the A^rC^o genome, but show frequent losses of parental loci due to homoeologous recombination (~82% for the A01/C01 homeologous chromosomes), although the genomic regions concerned are usually small (Szadkowski et al. 2011). In the second situation in the colchicine-doubled $A^rA^rC^oC^o$ genome hybrid, 30 to 47% of the chromosomes are involved in homeologous pairing (Szadkowski et al. 2010): loss of loci from the parental genomes is less frequent but the genomic regions involved are large (Szadkowski et al. 2011). In both cases, the first meiosis in resynthesized oilseed rape acts as a genome blender and, in advanced self-pollinated generations of synthetics, whole chromosome losses and substitutions (Xiong et al. 2011) as well as major translocations between the A and C genome can occur (Gaeta et al. 2007; Gaeta and Pires 2010). However, these rearrangements can also lead *de novo* variation in traits such as seed yield (Radoev et al. 2008), flowering time (Pires et al. 2004), seed acid detergent lignin content (Liu et al. 2012) and disease resistance (Zhao et al. 2006). These results demonstrate that recovery of some of these homeologous A-C translocations in a stable genomic background could have potential for oilseed rape improvement.

The *Brassica* B genome has been the focus of several research efforts due to the identification of drought tolerance and resistance to *Leptosphaeria maculans* (blackleg) in this genome. ABC hybrids can be produced by intercrosses between each allopolyploid species and the diploid carrying the complementary genome (i.e. *B. rapa* × *B. carinata, B. napus* × *B. nigra* and *B. juncea* × *B. oleracea*); in most cross combinations homeologous recombination has previously been detected between the A, B and C genomes in the triploid hybrid (Ge and Li 2007). Previously, an $A^nC^nB^n$ triploid hybrid has been crossed directly with *B. napus* to allow the production of *B. napus-B. nigra* addition lines (Jahier et al. 1989; Chevre et al. 1996); from these lines it was shown that the presence of chromosome B4 conferred resistance to blackleg. From one B4 line which showed multivalent formation, indicating that homeologous introgressions may have occurred, an introgressed resistant line ($2n$ = AACC = 38) was obtained (A.M. Chevre unpublished data). Intercrosses between allopolyploid species (Fig. 3) have also been performed to introduce resistance to the same disease; Mason et al. (2010) showed that recombination occurs relatively often (0.3 – 4.0 events per PMC) between homeologous chromosomes from each of the A, B and C genomes in the hybrid types $A^jA^nB^jC^n$, $B^jB^cA^jC^c$ and $C^nC^cA^nB^c$. Roy (1984) crossed *B. napus* with *B. juncea* and, after several backcrosses of the $A^jA^nB^jC^n$ hybrids to *B. napus*, obtained blackleg resistant plants. From these materials, selection for stable meiotic behavior and blackleg resistance produced resistant plants carrying the *Rlm6* gene for resistance to blackleg (Chèvre et al. 1997) introgressed through homeologous recombination on chromosome C3 (Barret et al. 1998). Similarly, Navabi et al. (2010) and Fredua-Agyeman et al. (2014) selected *Sclerotinia*-resistant and blackleg-resistant lines respectively from the progeny of *B. napus* crossed with resistant *B. carinata* plants.

Less related species have also been used to introgress monogenic traits of interest into *B. napus* (reviewed by Prakash et al. (2009)). Generally crosses are performed directly between the related species and *B. napus*. A famous example showing the difficulty of such interspecific introgressions is the introduction of the restorer *Rfo* gene of the cytoplasmic male sterility system "OGURA" from *Raphanus sativus* (R^rR^r, $2n$ = 18) into *B. napus*. Heyn (1976) obtained an introgression of the radish restorer gene into the *B. napus* genome. However, several studies were required to obtain a line with stable meiotic behavior (Pellan-Delourme and Renard 1988; Delourme et al. 1994). Introgression was shown to have occurred through homeologous recombination with the substitution of a *B. napus* region with a large genomic region of radish (Delourme et al. 1998). The recombination rate was zero between the radish region and the corresponding homoeologous region in *B. napus* in the introgression lines: only deletion of the radish region was observed in large segregating populations (Delourme et al. 1998). Only gamma irradiation of pollen and subsequent artificial induction of chromosome breakage allowed elimination of undesirable traits linked to the restorer gene (Primard-Brisset et al. 2005). This result highlights that the initial size of the introgression is the key determinant in rapid selection of plants with high agronomic value. A possible strategy to reduce the size of the initial introgressions is to increase the number of crossovers between related genomes; creating a "burst" of homeologous recombination. One method of inducing higher numbers of crossovers is to manipulate the genomic structure of the initial interspecific hybrids: homeologous recombination involving the B genome was detected at high frequencies in A.B, B.C, BB.A and CC.B interspecific hybrid types where unbalanced genome complements are present (Cui et al. 2012). Homeologous recombination may also be favored in the presence of another diploid genome, as was observed for A-C, B-C and A-B interactions in BBAC, AABC and CCAB interspecific hybrids (Figure 4) (Mason et al. 2010). Nagpal et al. (1996) provided evidence that homeologous recombination is more frequent between the *B. tournefortii* (TT) genome and the A, C or B genomes if an additional diploid genome is present in the hybrid; the authors compared chromosome pairing in TA, TC and TB hybrids produced from crosses between *B. tournefortii* and diploid *Brassica* species to TACC, TBAA and TCAA hybrids produced from crosses between *B. tournefortii* and the three allotetraploid species. The molecular mechanisms which promote homeologous recombination in these hybrids remain unknown. However, these data suggest that the chromosome composition of the hybrids is a key determinant in induction of homeologous recombination.

Conclusions

In the Brassiceae tribe, the conservation of the genome structure allows enlargement of genetic diversity and manipulation of traits in the allopolyploid *B. napus* through homologous and/or homeologous recombination. When traits targeted for introgression into oilseed rape from related species are under polygenic control, many QTL need to be introgressed simultaneously. Hence, a high frequency of recombination between the donor genome and the *B. napus* recipient genomes is required. Therefore, homologous recombination between the highly conserved A and C genomes present in the diploid (*B. rapa* or *B. oleracea*) and allopolyploid (*B. juncea* or *B. carinata*) species is advocated for transfer of polygenic traits into oilseed rape. Additionally, breeders prefer that introgressions remain limited to the genomic region of interest in order to maintain the agronomic value of the crop for other traits. As the size of the introgression region depends on the location of the flanking crossovers that occur, the higher the rate of homologous recombination, the higher the chance to introgress small genomic regions. Leflon et al. (2010) showed that the formation of AAC allotriploids is the most relevant strategy to increase homologous recombination rates. This result has still to be confirmed in CCA hybrids, although greatly elevated homologous recombination rates were detected in the C genome of progeny from a CCAB hybrid (Mason et al. 2015). However, QTL for traits of interest are often carried by homeologous regions in the A and C genomes in *B. napus*, as has been shown for quantitative resistance to blackleg (Fomeju et al. 2014) and for flowering time (Schiessl et al. 2014). Through homeologous recombination between the A and C genomes to generate non-reciprocal translocations, it is possible to change the gene dosage of a specific region, with a possible impact of this new variability on traits.

When a trait is under monogenic control, both homologous and homeologous recombination can be used to introgress the trait into *B. napus*. However, when the donor species shares a homologous genome with *B. napus*, it is easy to limit the introgression to the gene of interest using the strategy of homologous recombination. When the donor species is less related, it has been shown that generally the size of the introgression is larger than desirable for optimal use in breeding programs. Different studies have indicated that the initial genome structure of the F_1 hybrids can play a role in the level of homeologous recombination that occurs. Additionally, when F_1 hybrids are poorly fertile, they generally produce unreduced gametes which are FDR-like, in that chromosome pairing occurs during metaphase I but chromosome are not separated by the first division. Chromatids are separated by the second division such that gametes have the same chromosome number as the mother plant. These unreduced gametes can also be a source of new homeologous recombination events. Similarly, the effect or lack thereof of the genetic control of homeologous pairing described between the A and C genomes of *B. napus* on the control of homeologous pairing with less related species has still to be assessed. Introgressions generally occur between homeologous genomic regions, with the substitution of the corresponding recipient genomic region by the donor one. Production of such introgressed plants requires simultaneous high selection pressure for meiotic stability and for the trait of interest. Decreasing the size of the genomic introgression in advanced generations is difficult, due to the very low rate of recombination that occurs between the introgressed and recipient genomic regions in plants heterozygous for the translocation. Promising results related to homologous recombination show that it is possible to induce bursts in the rate of homologous recombination in some hybrids; this method should be assessed in future in relation to reduction of introgression sizes.

Acknowledgements

ASM is funded by Emmy Noether DFG award MA6473/1-1. This work was partly funded by ANR KBEE project Gewidis.

References

Abel, S., C. Möllers, and H. Becker. 2005. Development of synthetic *Brassica napus* lines for the analysis of "fixed heterosis" in allopolyploid plants. Euphytica 146 (1-2): 157-163.

Al-Shehbaz, I.A. 2012. A generic and tribal synopsis of the Brassicaceae (Cruciferae). Taxon 61 (5): 931-954.

Barret, P., J. Guerif, J.P. Reynoird, et al. 1998. Selection of stable *Brassica napus - Brassica juncea* recombinant lines resistant to blackleg (*Leptosphaeria maculans*). 2. A 'to and fro' strategy to localise and characterise interspecific introgressions on the *B. napus* genome. Theor. Appl. Genet. 96 (8): 1097-1103.

Chalhoub, B., F. Denoeud, S.Y. Liu, et al. 2014. Early allopolyploid evolution in the post-Neolithic *Brassica napus* oilseed genome. Science 345 (6199): 950-953.

Chen, S., M.N. Nelson, A.-M. Chèvre, et al. 2011. Trigenomic bridges for *Brassica* improvement. Crit. Rev. Plant. Sci. 30 (6): 524-547.

Cheng, F., T. Mandakova, J. Wu, Q. Xie, M.A. Lysak and X.W. Wang. 2013. Deciphering the diploid ancestral genome of the mesohexaploid *Brassica rapa*. Plant Cell 25 (5): 1541-1554.

Cheng, F., J. Wu, and X. Wang. 2014. Genome triplication drove the diversification of *Brassica* plants. Hort. Res.1: 14024.

Chèvre, A.M., P. Barret, F. Eber, et al. 1997. Selection of stable *Brassica napus - B. juncea* recombinant lines resistant to blackleg (*Leptosphaeria maculans*). 1. Identification of molecular markers, chromosomal and genomic origin of the introgression. Theor. Appl. Genet. 95 (7): 1104-1111.

Chevre, A.M., F. Eber, P. This, et al. 1996. Characterization of *Brassica nigra* chromosomes and of blackleg resistance in *B. napus - B. nigra* addition lines. Plant Breed. 115 (2): 113-118.

Cifuentes, M., F. Eber, M.O. Lucas, M. Lode, A.M. Chevre and E. Jenczewski. 2010a. Repeated polyploidy drove different levels of crossover suppression between homoeologous chromosomes in *Brassica napus* allohaploids. Plant Cell 22 (7): 2265-2276.

Cifuentes, M., L. Grandont, G. Moore, A.M. Chèvre and E. Jenczewski. 2010b. Genetic regulation of meiosis in polyploid species: new insights into an old question. New Phyt. 186: 29-36.

Crouch, J.H., B.G. Lewis and R.F. Mithen. 1994. The effect of A-genome substitution on the resistance of *Brassica napus* to infection by *Leptosphaeria maculans*. Plant Breed. 112 (4): 265-278.

Cui, C., X.H. Ge, M. Gautam, L. Kang and Z.Y. Li. 2012. Cytoplasmic and genomic effects on meiotic pairing in *Brassica* hybrids and allotetraploids from pair crosses of three cultivated diploids. Genetics 191 (3): 725-U123.

Delourme, R., A. Bouchereau, N. Hubert, M. Renard and B.S. Landry. 1994. Identification of RAPD markers linked to a fertility restorer gene for the Ogura radish cytoplasmic male sterility of rapeseed (*Brassica napus* L). Theor. Appl. Genet. 88 (6-7): 741-748.

Delourme, R., N. Foisset, R. Horvais, et al. 1998. Characterisation of the radish introgression carrying the *Rfo* restorer gene for the Ogu-INRA cytoplasmic male sterility in rapeseed (*Brassica napus* L.). Theor. Appl. Genet. 97 (1-2): 129-134.

FitzJohn, R.G., T.T. Armstrong, L.E. Newstrom-Lloyd, A.D. Wilton and M. Cochrane. 2007. Hybridisation within *Brassica* and allied genera: evaluation of potential for transgene escape. Euphytica 158 (1-2): 209-230.

Fomeju, B.F., C. Falentin, G. Lassalle, M.J. Manzanares-Dauleux and R. Delourme. 2014. Homoeologous duplicated regions are involved in quantitative resistance of *Brassica napus* to stem canker. BMC Genomics 15.

Fredua-Agyeman, R., O. Coriton, V. Huteau, I.A.P. Parkin, A.M. Chevre and H. Rahman. 2014. Molecular cytogenetic identification of B genome chromosomes linked to blackleg disease resistance in *Brassica napus* × *B. carinata* interspecific hybrids. Theor. Appl. Genet. 127 (6): 1305-1318.

Fujii, K. and N. Ohmido. 2011. Stable progeny production of the amphidiploid resynthesized *Brassica napus* cv. Hanakkori, a newly bred vegetable. Theor. Appl. Genet. 123 (8): 1433-1443.

Gaeta, R.T., J.C. Pires, F. Iniguez-Luy, E. Leon and T.C. Osborn. 2007. Genomic changes in resynthesized *Brassica napus* and their effect on gene expression and phenotype. Plant Cell 19 (11): 3403-3417.

Gaeta, Robert T. and J. Chris Pires. 2010. Homoelogous recombination in allopolyploids: the polyploid ratchet. New Phyt. 186: 18-28.

Ge, Xian-Hong, and Zai-Yun Li. 2007. Intra- and intergenomic homology of B-genome chromosomes in trigenomic combinations of the cultivated *Brassica* species revealed by GISH analysis. Chromosome Res. 15: 849-861.

Gomez-Campo, C. 1980. Morphology and morphotaxonomy of the tribe Brassiceae. In *Brassica crops and wild allies: Biology and breeding*, edited by S. Tsunoda, K. Hinata, and C. Gomez-Campo. Tokyo: Japan Scientific Soc. Press.

Heyn, F. 1976. Transfer of restorer genes from *Raphanus* to cytoplasmic male sterile *Brassica napus*. Cruciferae Newsletter 1: 15-16.

Jahier, J., A.M. Chevre, A.M. Tanguy and F. Eber. 1989. Extraction of disomic addition lines of *Brassica napus* - *Brassica nigra*. Genome 32 (3): 408-413.

Jenczewski, E., F. Eber, A. Grimaud, S. Huet, M.O. Lucas, H. Monod and A.M. Chèvre. 2003. *PrBn*, a major gene controlling homeologous pairing in oilseed rape (*Brassica napus*) haploids. Genetics 164 (2): 645-653.

Leflon, M. , F. Eber, J.C. Letanneur, et al. 2006. Pairing and recombination at meiosis of *Brassica rapa* (AA) × *Brassica napus* (AACC) hybrids. Theor. Appl. Genet. 113: 1467-1480.

Leflon, M., L. Grandont, F. Eber, et al. 2010. Crossovers get a boost in *Brassica* allotriploid and allotetraploid hybrids. Plant Cell 22 (7): 2253-64.

Li, M., W. Qian, J. Meng and Z. Li. 2004. Construction of novel *Brassica napus* genotypes through chromosomal substitution and elimination using interploid species hybridization. Chromosome Res. 12 (5): 417-426.

Li, Q.F., J.Q. Mei, Y.J. Zhang, et al. 2013. A large-scale introgression of genomic components of *Brassica rapa* into *B. napus* by the bridge of hexaploid derived from hybridization between *B. napus* and *B. oleracea*. Theor. Appl. Genet. 126 (8): 2073-2080.

Liu, L.Z., A. Stein, B. Wittkop, et al. 2012. A knockout mutation in the lignin biosynthesis gene CCR1 explains a major QTL for acid detergent lignin content in *Brassica napus* seeds. Theor. Appl. Genet. 124 (8): 1573-1586.

Liu, S.Y., Y.M. Liu, X.H. Yang, et al. 2014. The *Brassica oleracea* genome reveals the asymmetrical evolution of polyploid genomes. Nature Comm. 5.

Liu, Zhiqian, Katarzyna Adamczyk, Maria Manzanares-Dauleux, et al. 2006. Mapping *PrBn* and other quantitative trait loci responsible for the control of homeologous chromosome pairing in oilseed rape (*Brassica napus* L.) haploids. Genetics 174 (3): 1583-1596.

Lombard, V. and R. Delourme. 2001. A consensus linkage map for rapeseed (*Brassica napus* L.): construction and integration of three individual maps from DH populations. Theor. Appl. Genet. 103 (4): 491-507.

Mandakova, T. and M.A. Lysak. 2008. Chromosomal phylogeny and karyotype evolution in x = 7 crucifer species (Brassicaceae). Plant Cell 20 (10): 2559-2570.

Mason, A.S., J. Takahira, C. Atri, et al. 2015. Microspore culture reveals complex meiotic behaviour in a trigenomic *Brassica* hybrid. BMC Plant Biol. 15: 173.

Mason, Annaliese S., Virginie Huteau, Frédérique Eber, et al. 2010. Genome structure affects the rate of autosyndesis and allosyndesis in AABC, BBAC and CCAB *Brassica* interspecific hybrids. Chromosome Res. 18 (6): 655-666.

Meng, J., S. Shi, L. Gan, Z. Li and X. Qu. 1998. The production of yellow-seeded *Brassica napus* (AACC) through crossing interspecific hybrids of *B. campestris* (AA) and *B. carinata* (BBCC) with *B. napus*. Euphytica 103 (3): 329-333.

Mezard, C., J. Vignard, J. Drouaud and R. Mercier. 2007. The road to crossovers: plants have their say. Trends Genet. 23 (2): 91-99.

Mizushima, U. 1980. Genome analysis in *Brassica* and allied genera. In *Brassica crops and wild allies: biology and breeding*, edited by S. Tsunoda, Hinata, K., Gomez-Campo, C. Tokyo Japan Scientific Societies Press.

Murat, F., A. Louis, F. Maumus, et al. 2015. Understanding Brassicaceae evolution through ancestral genome reconstruction. Genome Biol. 16: 262.

Nagpal, R., S.N. Raina, Y.S. Sodhi, et al. 1996. Transfer of *Brassica tournefortii* (TT) genes to allotetraploid oilseed *Brassica* species (*B. juncea* AABB, *B. napus* AACC, *B. carinata* BBCC): homoeologous pairing is more pronounced in the three-genome hybrids (TACC, TBAA, TCAA, TCBB) as compared to allodiploids (TA, TB, TC). Theor. Appl. Genet. 92 (5): 566-571.

Namai, H. 1987. Inducing cytogenetical alterations by means of interspecific and intergeneric hybridization in *Brassica* crops. Gamma Field Symp. 26: 41-89.

Navabi, Z.K., S.E. Strelkov, A.G. Good, M.R. Thiagarajah and M.H. Rahman. 2010. *Brassica* B-genome resistance to stem rot (*Sclerotinia sclerotiorum*) in a doubled haploid population of *Brassica napus* × *Brassica carinata*. Can. J. Plant Path. 32 (2): 237-246.

Nicolas, S.D., H. Monod, F. Eber, A.M. Chèvre and E. Jenczewski. 2012. Non-random distribution of extensive chromosome rearrangements in *Brassica napus* depends on genome organization. Plant J.70 (4): 691-703.

Nicolas, Stephane D., M. Leflon, H. Monod, et al. 2009. Genetic regulation of meiotic cross-overs between related genomes in *Brassica napus* haploids and hybrids. Plant Cell 21 (2): 373-385.

Osborn, T.C., D.V. Butrulle, A.G. Sharpe, et al. 2003. Detection and effects of a homeologous reciprocal transposition in *Brassica napus*. Genetics 165 (3): 1569-77.

Parkin, I. 2011. Chasing Ghosts: Comparative Mapping in the Brassicaceae. In *Genetics and Genomics of the Brassicaceae*, edited by R. Schmidt, and I. Bancroft: Springer New York.

Parkin, I.A.P., S.M. Gulden, Sharpe, A. et al. 2005. Segmental structure of the *Brassica napus* genome based on comparative analysis with *Arabidopsis thaliana*. Genetics 171: 765-781.

Parkin, I.A.P., C. Koh, H. Tang et al. 2014. Transcriptome and methylome profiling reveals relics of genome dominance in the mesopolyploid *Brassica oleracea*. Genome Biol. 15 (6): R77.

Pellan-Delourme, R. and M. Renard. 1988. Cytoplasmic male sterility in rapeseed (*Brassica napus* L) - female fertility of restored rapeseed with Ogura and cybrids cytoplasms. Genome 30 (2): 234-238.

Piquemal, J., E. Cinquin, F. Couton, et al. 2005. Construction of an oilseed rape (*Brassica napus* L.) genetic map with SSR markers. Theor. Appl. Genet. 111 (8): 1514-1523.

Pires, J.C., J. Zhao, M.E. Schranz et al. 2004. Flowering time divergence and genomic rearrangements in resynthesized *Brassica* polyploids (*Brassicaceae*). Biological Journal of the Linnean Society 82 (4): 675-688.

Prakash, S. and K. Hinata. 1980. Taxonomy, cytogenetics and origin of crop Brassicas, a review. Opera Botanica 55: 1-57.

Prakash, S.R., S.R. Bhat, C.F. Quiros, P.B. Kirti and V.L. Chopra. 2009. *Brassica* and its close allies: cytogenetics and evolution. Plant Breed. Rev. 31: 21-187.

Primard-Brisset, C., J. P. Poupard, R. Horvais, et al. 2005. A new recombined double low restorer line for the Ogu-INRA cms in rapeseed (*Brassica napus* L.). Theor. Appl. Genet. 111 (4): 736-746.

Qian, W., J. Meng, M. Li, et al. 2006. Introgression of genomic components from Chinese *Brassica rapa* contributes to widening the genetic diversity in rapeseed (*B. napus* L.), with emphasis on the evolution of Chinese rapeseed. Theor. Appl. Genet. 113 (1): 49-54.

Radoev, M., H.C. Becker and W. Ecke. 2008. Genetic analysis of heterosis for yield and yield components in rapeseed (*Brassica napus* L.) by quantitative trait locus mapping. Genetics 179: 1547-1558.

Roy, N.N. 1984. Interspecific transfer of *Brassica juncea*-type blackleg resistance to *Brassica napus*. Euphytica 33: 295-303.

Schiessl, S., B. Samans, B. Huttel, R. Reinhard and R.J. Snowdon. 2014. Capturing sequence variation among flowering-time regulatory gene homologs in the allopolyploid crop species *Brassica napus*. Frontiers Plant Sci. 5 (404): 1-14.

Schranz, M.E., M.A. Lysak and T. Mitchell-Olds. 2006. The ABC's of comparative genomics in the Brassicaceae: building blocks of crucifer genomes. Trends Plant Sci. 11 (11): 535-542.

Seyis, F., W. Friedt and W. Luhs. 2006. Yield of *Brassica napus* L. hybrids developed using resynthesized rapeseed material sown at different locations. Field Crops Res. 96 (1): 176-180.

Szadkowski, E., F. Eber, V. Huteau, et al. 2010. The first meiosis of resynthesized *Brassica napus*, a genome blender. New Phyt. 186 (1): 102-112.

Szadkowski, E., F. Eber, V. Huteau, et al. 2011. Polyploid formation pathways have an impact on genetic rearrangements in resynthesized *Brassica napus*. New Phyt. 191 (3): 884-894.

Tang, H.B., M.R. Woodhouse, F. Cheng, et al. 2012. Altered patterns of fractionation and exon deletions in *Brassica rapa* support a two-step model of paleohexaploidy. Genetics 190 (4): 1563-1574.

U, N. 1935. Genome-analysis in *Brassica* with special reference to the experimental formation of *B. napus* and peculiar mode of fertilization. Jap. J. Bot. 7: 389-452.

Udall, J.A., P.A. Quijada and T.C. Osborn. 2005. Detection of chromosomal rearrangements derived from homeologous recombination in four mapping populations of *Brassica napus* L. Genetics 169 (2): 967-979.

Wang, X.W., H.Z. Wang, J. Wang, et al. 2011. The genome of the mesopolyploid crop species *Brassica rapa*. Nature Genet. 43 (10): 1035-1039.

Xiong, Z.Y., R.T. Gaeta, and J.C. Pires. 2011. Homoeologous shuffling and chromosome compensation maintain genome balance in resynthesized allopolyploid *Brassica napus*. Proc. Natl. Acad. Sci. U.S.A. 108 (19): 7908-7913.

Zhao, J.W., J.A. Udall, P.A. Quijada, C.R. Grau, J.L. Meng and T.C. Osborn. 2006. Quantitative trait loci for resistance to *Sclerotinia sclerotiorum* and its association with a homeologous non-reciprocal transposition in *Brassica napus* L. Theor. Appl. Genet. 112 (3): 509-516.

Ziolkowski, P.A., M. Kaczmarek, D. Babula, and J. Sadowski. 2006. Genome evolution in *Arabidopsis/Brassica*: conservation and divergence of ancient rearranged segments and their breakpoints. Plant J. 47 (1): 63-74.

Zou, J., J. Zhu, S. Huang, et al. 2010. Broadening the avenue of intersubgenomic heterosis in oilseed *Brassica*. Theor. Appl. Genet. 120: 283-290.

Chapter 19

Interspecific Hybridization for Chickpea (*Cicer arietinum* L.) Improvement

Shivali Sharma[1,*], Hari D. Upadhyaya[2], Manish Roorkiwal[3] and Rajeev K. Varshney[4]

ABSTRACT

The narrow genetic base of crop cultivars such as chickpea is one of the major constraints limiting their genetic improvement. Sufficient genetic variability is available in the genus *Cicer*, which comprises eight annual and 35 perennial wild *Cicer* species, for chickpea improvement. These wild species carry many useful genes/alleles for resistance/tolerance to different biotic/abiotic stresses as well as for agronomic and nutrition-related traits, thus offering great potential for the genetic enhancement of cultivated chickpea. In spite of their high potential and importance as new and diverse sources of variation, these wild *Cicer* species have not been used adequately in the chickpea improvement programs. In this chapter, efforts have been made to present a review of the status of different genebanks conserving wild *Cicer* species, and our current understanding of the progress made so far in introgressing the genes/alleles from wild *Cicer* species by various researchers. The need for systematic pre-breeding activities to bridge the gap between the wild *Cicer* species conserved in the genebanks and utilization in breeding programs is discussed. At ICRISAT, Patancheru, India, systematic pre-breeding activities are in progress to enrich the genetic variability in the primary genepool by developing new genepools with a high frequency of useful genes, wider adaptability, and a broad genetic base by using wild *Cicer*

[1] International Crops Research Institute for the Semi-Arid Tropics (ICRISAT), Patancheru 502324, Telangana, India, E-mail: shivali.sharma@cgiar.org
[2] International Crops Research Institute for the Semi-Arid Tropics (ICRISAT), Patancheru 502324, Telangana, India, E-mail: h.upadhyaya@cgiar.org
[3] International Crops Research Institute for the Semi-Arid Tropics (ICRISAT), Patancheru 502324, Telangana, India, Email: m.roorkiwal@cgiar.org
[4] International Crops Research Institute for the Semi-Arid Tropics (ICRISAT), Patancheru 502324, Telangana, India, Email: r.k.varshney@cgiar.org
* Corresponding author

species and chickpea cultivars. The present status and future strategies for the efficient and effective utilization of wild *Cicer* species for generating new variability for chickpea improvement, and the potential of newly developed genomic tools to enhance the efficiency of such pre-breeding activities, is also discussed.

Introduction

Chickpea (*Cicer arietinum* L.) is a self-pollinated, cool season leguminous crop believed to have been domesticated in the Old World about 7000 years ago; it is cultivated primarily for its protein rich seeds. Besides protein, chickpea seeds are also rich sources of fiber, minerals (calcium, potassium, phosphorus, magnesium, iron, and zinc), β-carotene, and unsaturated fatty acids (Jukanti et al. 2012). Chickpea plants have the ability to fix atmospheric nitrogen due to a symbiotic relationship with *Rhizobium* bacteria, meeting up to 80% of its own nitrogen requirements and helping to improve soil fertility.

Chickpea ranks second in global production among food grain legumes (after common beans) with a total production of 13.1 Mt from an area of 13.54 m ha (FAOSTAT 2013). Chickpea is grown in a wide range of environments in over 55 countries in sub-tropical and temperate regions of the world, predominantly in Asia (~89%), followed by Oceania (4.2%), Africa (3.7%), the Americas (2.4%), and Europe (0.8%) (Fig. 1).

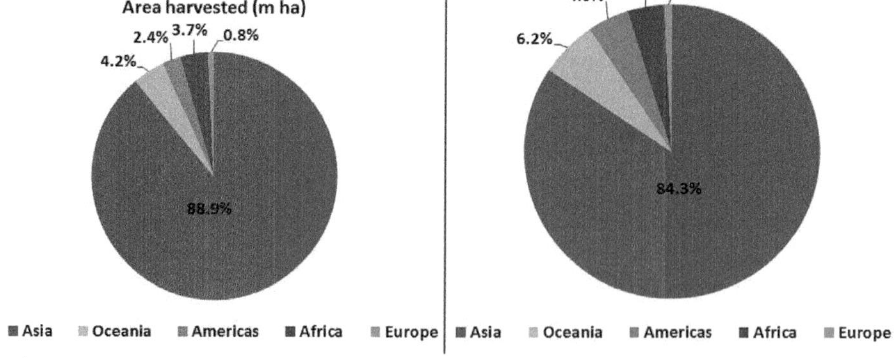

Figure 1. Percentage contribution of different continents towards global chickpea cultivation area and production in 2013.

In terms of production, Asia and Africa together contribute about 88% of world chickpea production whereas the Americas and Oceania together contribute about 11% of world chickpea production. Major chickpea producers are India, Pakistan, Turkey and Myanmar in Asia; Ethiopia in Africa; and Mexico in the Americas. Chickpea area and production statistics in major chickpea growing countries over the last three decades (1985-1994, 1995-2004, and 2005-2014) show a consistent increase in area and production in Australia, Ethiopia, India, and Pakistan but a decline in Turkey (FAOSTAT 2013).

On the basis of size and color of seeds and flowers, seed shape, and pigmentation on plants, two distinct forms of cultivated chickpea are known (Cubero 1975), namely desi (also known as microsperma) and kabuli (also known as macrosperma). Desi types are characterized mostly by pink flowers with pigments on the plants and angular shaped and brown coloured small seeds with a high percentage of fibre, and are primarily grown in South Asia and Africa. Kabuli types have white flowers without pigments on their plants, owl's head shaped, beige coloured large seeds with a low percentage of fibre, and are grown in Mediterranean countries. A third type, intermediate or pea-shaped, is also known to exist, and is characterized by round/pea-shaped medium- to small-sized seeds. Kabuli are grown in about two-thirds of chickpea-growing countries, but desi types dominate production and account for about 85% of total production, while kabuli types account for about 15% of the world chickpea production.

Chickpea production has been constant for several years with low and unstable productivity, because over 80% of chickpeas are cultivated in marginal lands under rainfed conditions (Kumar and van Rheenen 2000). Further, the crop is exposed to a number of biotic (fusarium wilt, ascochyta blight, botrytis gray mold, dry root rot, powdery mildew, pod borer, cutworms) and abiotic (drought, heat, cold, and salinity) stresses and the estimated yield losses by individual pests or diseases vary from 5 – 10% in temperate, and 50 – 100% in tropical regions (van Emden et al. 1988). Development of high yielding, early maturing chickpea cultivars that fit well into a short cropping season and have high levels of resistance/tolerance against biotic/abiotic stresses are the major objectives of chickpea improvement programs. However, the narrow genetic base of cultivated chickpea is one of the major obstacles to sustainable improvement of its productivity, and renders the crop vulnerable to new biotic/abiotic stresses. The narrow genetic base of chickpea is due to a combination of factors: the restricted distribution of the wild progenitor species *C. reticulatum* in southeastern Turkey, the founder effect associated with species domestication, the agricultural selection involved in the shift from a winter to a summer crop and the displacement of landraces with local adaption by modern, genetically uniform varieties (Abbo et al. 2003). Plant genetic resources consisting of landraces, obsolete varieties, breeding lines and crop wild relatives provide ample natural genetic variations for use in chickpea improvement programs. Of these, crop wild relatives are of immense importance as they are genetically far more diverse and have natural defense mechanisms to withstand climate extremities. In this chapter, an effort has been made to review the available information and resources, and to suggest future breeding strategies to ensure a continuous supply of new genetic variation into breeding pipelines by utilizing wild *Cicer* species to develop new cultivars with a broad genetic base.

Taxonomical Relationships and Genepools in the Genus *Cicer*

The genus *Cicer* comprises 44 species (nine annual and 35 perennial) and belongs to the family *Fabaceae,* subfamily Papilionoideae, and tribe Cicereae Alef. (van der Maesen 1972, Muehlbauer 1993). These species are further classified into two subgenera, *Pseudononis* and *Viciastrum,* and four sections, Monocicer, Chamaecicer, Polycicer and Acanthocicer (Table 1) as well as 14 series (van der Maesen 1987) based on geographical distribution, life cycle and morphological characters (see Sharma et al. 2013a). Of the 44 species, *Cicer arietinum,* commonly known as chickpea, is the only domesticated species believed to have originated in the Fertile Crescent region of southeastern Turkey and adjoining Syria (van der Maesen 1987), and is one of the earliest grain crops domesticated in the Old World. Among wild species, *C. reticulatum* and *C. anatolicum* have been identified as the annual and perennial progenitors of chickpea, respectively (Ladizinsky and Adler 1976a, Tayyar and Waines 1996).

Table 1. List of wild *Cicer* species in genus *Cicer* and their distribution.

S. No.	Species	Section	Cycle	Natural occurrence
1.	*C. acanthophyllum* Boriss.	A	P	Afghanistan, Pakistan, Tajikistan
2.	*C. anatolicum* Alef.	P	P	Turkey, Islamic Republic of Iran, Iraq, Armenia
3.	*C. arietinum* L.	M	A	Worldwide
4.	*C. atlanticum* Coss. ex. Maire	P	P	Morocco
5.	*C. balcaricum* Galushko	P	P	Armenia, Azerbaijan, Georgea, Russian Federation
6.	*C baldshuanicum* (Popov) Lincz.	P	P	Tajikistan
7.	*C. bijugum.* Rech. f.	M	A	Turkey, Syrian Arab Republic, Iran, Iraq
8.	*C. canariense* A. Santos & G.P. Lewis	P	P	Spain – Canary Islands
9.	*C. chorassanicum* (Bunge) Popov	C	A	Afghanistan, Islamic Republic of Iran
10.	*C. cuneatum* Hochst. ex A. Rich.	M	A	Ethiopia, Egypt, Eritrea, Saudi Arabia
11.	*C. echinospermum* P.H. Davis	M	A	Turkey, Iraq
12.	*C. fedtschenkoi* Lincz.	P	P	Afghanistan, Iran, Kyrgyzstan, Tajikistan

cont.

13.	*C. flexuosum* Lipsky	P	P	Kyrgyzstan, Tajikistan, Uzbekistan
14.	*C. floribundum* Fenzl	P	P	Turkey
15.	*C. graecum* Orph.	P	P	Greece
16.	*C. grande* (Popov) Korotkova	P	P	Uzbekistan
17.	*C. heterophyllum* Contandr. et al.	P	P	Turkey
18.	*C. incanum* Korotkova	A	P	Tajikistan
19.	*C. incisum* (Willd.) K. Maly	C	P	Armenia, Georgia, Greece, Iran, Lebanon, Syria, Turkey
20.	*C. isauricum* P.H. Davis	P	P	Turkey
21.	*C. judaicum* Boiss.	M	A	Lebanon, Israel,
22.	*C. kermanense* Bornm.	P	P	Islamic Republic of Iran
23.	*C. korshinskyi* Lincz.	P	P	Iran, Tajikistan
24.	*C. laetum* Rassulova & Sharipova	?	?	Tajikistan
25.	*C. luteum* Rassulova & Sharipova	?	?	Tajikistan
26.	*C. macracanthum* Popov	A	P	Afghanistan, India, Pakistan, Tajikistan, Uzbekistan
27.	*C. microphyllum* Benth.	P	P	Afghanistan, China, India, Nepal, Pakistan, Tajikistan
28.	*C. mogoltavicum* (Popov) Korol.	P	P	Tajikistan
29.	*C. montbretti* Jaub. Spach	P	P	Turkey, Bulgaria, Albania, Greece
30.	*C. mutijugum* Maesen	P	P	Afghanistan
31.	*C. nuristanicum* Kitam.	P	P	Afghanistan, India, Pakistan
32.	*C. oxydon* Boiss. & Hohen.	P	P	Afghanistan, Islamic Republic of Iran, Iraq
33.	*C. paucijugum* Nevski	P	P	Kazakhstan; Tajikistan
34.	*C. pinnatifidum* Jaub. & Spach	M	A	Armenia, Cyprus, Turkey, Syrian Arab Republic, Iraq
35.	*C. pungens* Boiss.	A	P	Afghanistan, Tajikistan
36.	*C. rassuloviae* Lincz.	P	P	Tajikistan
37.	*C. rechingeri* Podlech	A	P	Afghanistan
38.	*C. reticulatum* Ladiz.	M	A	Turkey
39.	*C. songaricum* Stephan ex DC.	P	P	Kazakhstan; Kyrgyzstan; Tajikistan; Uzbekistan
40.	*C. spiroceras* Jaub. & Spach	P	P	Islamic Republic of Iran

cont.

41.	*C. stapfianum* Rech. f.	A	P	Islamic Republic of Iran
42.	*C. subaphyllum* Boiss.	P	P	Islamic Republic of Iran
43.	*C. tragacanthoides* Jaub. &. Spach	A	P	Afghanistan, Islamic Republic of Iran, Turkmenistan
44.	*C. yamashitae* Kitam.	M	A	Afghanistan

Table 2. Major holdings of chickpea germplasm around the world.

Country	Institutes/ Genebanks	No. of wild species	No. of wild accessions	No. of cultivated accessions	Total	Source
Australia	Australian Temperate Field Crops Collection (ATFCC)	18	246 (16)	8409 (62)	8655 (62)	http:// apps3.fao. org/wiews/ germplasm_ query. htm?i_l=EN
Bangladesh	Plant Genetic Resources Centre, BARI (PGR-BARI)		–	752 (3)	752 (3)	
Brazil	Embrapa Hortaliças		–	775	775	
Canada	Plant Gene Resources of Canada, Saskatoon Research Centre (PGRC), Agriculture and Agri-Food Canada		–	509 (4)	509 (4)	
Ethiopia	Ethiopian Institute of Biodiversity (EIB)		–	1173 (3)	1173 (3)	
Germany	Genebank, Leibniz Institute of Plant Genetics and Crop Plant Research	3	15 (2)	518 (25)	533 (25)	
Greece	Fodder Crops and Pastures Institute (FCPI)		–	445 (1)	445 (1)	
Hungary	Institute for Agrobotany	5	9	1161 (31)	1170 (31)	

cont.

India	Indian Agricultural Research Institute (IARI)		–	2000	2000	http://apps3.fao.org/wiews/germplasm_query.htm?i_l=EN
	International Crops Research Institute for the Semi-Arid Tropics (ICRISAT)	18	308 (13)	20294 (59)	20602 (60)	http://www.icrisat.org/what-we-do/crops/ChickPea/Project1/pfirst.asp
	National Bureau of Plant Genetic Resources (NBPGR)	10	69 (1)	14635 (23)	14704 (23)	http://apps3.fao.org/wiews/germplasm_query.htm?i_l=EN
	Regional Station Akola, NBPGR		–	813 (1)	813 (1)	http://apps3.fao.org/wiews/germplasm_query.htm?i_l=EN
Iran	College of Agriculture, Tehran University		–	1200	1200	http://apps3.fao.org/wiews/germplasm_query.htm?i_l=EN
	National Plant Gene Bank of Iran, Seed and Plant Improvement Institute (NPGBI-SPII)		–	5700	5700	
Japan	Department of Genetic Resources I, National Institute of Agrobiological Sciences (NIAS)	1	–	681	681	http://apps3.fao.org/wiews/germplasm_query.htm?i_l=EN

cont.

Mexico	Estación de Iguala, Instituto Nacional de Investigaciones Agrícolas		–	1600	1600	http://apps3.fao.org/wiews/germplasm_query.htm?i_l=EN
Pakistan	Nuclear Institute of Agricultural & Biology (NIAB)		–	500	500	http://apps3.fao.org/wiews/germplasm_query.htm?i_l=EN
	Plant Genetic Resources Program (PGRP)	3	89	2057	2146	
	Pulses Research Institute (PRI), Faisalabad		–	520 (1)	520 (1)	
Philippines	Institute of Plant Breeding, College of Agriculture, University of the Philippines, Los Baños College (IPB, UPLB-CA)		–	407	407	http://apps3.fao.org/wiews/germplasm_query.htm?i_l=EN
Russian Federation	N.I. Vavilov All-Russian Scientific Research Institute of Plant Industry (VIR)		–	2091	2091	http://apps3.fao.org/wiews/germplasm_query.htm?i_l=EN
Spain	Instituto Nacional de Investigación y Tecnología Agraria y Alimentaria. Centro de Recursos Fitogenéticos (INIA-CRF)		–	644 (15)	644 (15)	http://apps3.fao.org/wiews/germplasm_query.htm?i_l=EN

cont.

Spain	Junta de Andalucía. Consejería de Agricultura y Pesca. Instituto Andaluz de Investigación y Formación Agraria, Pesquera, Alimentaria y de la Producción Ecológica. Centro Alameda del Obispo (IFAPACOR)		–	608 (23)	608 (23)	http://apps3.fao.org/wiews/germplasm_query.htm?i_l=EN
Syria	International Centre for Agricultural Research in Dry Areas (ICARDA)	11	270 (9)	14636 (60)	14906 (60)	https://www.genesys-pgr.org/wiews/SYR002
Turkey	Plant Genetic Resources Department	4	21 (1)	2054 (1)	2075 (1)	http://apps3.fao.org/wiews/germplasm_query.htm?i_l=EN
Ukraine	Institute of Plant Production n.a. V.Y. Yurjev of UAAS	1	11 (1)	1010 (50)	1021 (50)	
USA	Western Regional Plant Introduction Station, USDA-ARS, Washington State University	21	186 (18)	6009 (55)	6195 (55)	
Uzbekistan	Uzbek Research Institute of Plant Industry (UzRIPI)		–	1055	1055	
Total			1,224	92,256	93,480	

*Numbers in parentheses indicate the number of countries of origin.

Wild *Cicer* Species as a Source of New Variability

Limited genetic variation present in the cultivated type of chickpea germplasm necessitates the exploitation of wild *Cicer* species for genetic improvement. Wild *Cicer* species have been extensively screened by various workers, and several species found to exhibit a very high level of resistance/tolerance to important biotic and abiotic stresses. Based on these studies, promising wild *Cicer* accessions with high levels of resistance/tolerance to *Ascochyta* blight (Singh et al. 1981, Singh and Reddy 1993, Stamigna et al., 2000, Collard et al. 2001, Rao et al. 2003, Croser et al. 2003, Shah et al. 2005, Pande et al. 2006), *Botrytis* grey mold (Stevenson and Haware 1999, Rao et al. 2003, Pande et al. 2006), *Fusarium* wilt (Infantino et al. 1996, Croser et al. 2003, Rao et al. 2003), dry root rot, *Helicoverpa* pod borer (Sharma et al. 2005), drought (Croser et al. 2003, Kashiwagi et al. 2005, Toker et al. 2007), cold (Singh et al. 1990, 1995, Croser et al. 2003, Toker 2005, Berger et al. 2012) and drought and heat (Canci and Toker 2009) have been identified. Besides resistant/tolerant sources, wild *Cicer* species also harbor beneficial alleles/genes for high seed protein (Singh and Pundir 1991, Rao et al. 2003) and improvement of other agronomic traits in cultivated chickpea.

Limitations of Utilizing Wild *Cicer* Species for Chickpea Improvement

Although high levels of various stress resistances are available in wild *Cicer* species, these are not being adequately utilized in chickpea breeding programs. The major limitations are due to different incompatibility barriers (Ahmad et al. 1988) between cultivated and wild species, as well as poor viability and sterility of F_1 hybrids and progenies, and linkage drag. Of the eight annual wild *Cicer* species, *C. reticulatum* and *C. echinospermum* are crossable with the cultivated chickpea, although only *C. reticulatum* results in fertile interspecific hybrids, whereas sterility of F_1 hybrids and progenies is the major limitation associated with the utilization of *C. echinospermum* in crossing programs. For the utilization of the remaining six cross-incompatible annual wild *Cicer* species for chickpea improvement, specialized tissue culture techniques such as *in vitro* application of growth hormones, ovule culture, and embryo culture are required (Badami et al. 1997, Mallikarjuna 1999, Mallikarjuna and Jadhav 2008, Lulsdorf et al. 2005).

In India, especially in southern India, it is difficult to involve wild *Cicer* species in crossing programs with cultivated chickpea because of inappropriate phenology, or want of vernalization (Berger et al. 2005) and appropriate photoperiod requirements. In Mediterranean environments also, wild *Cicer* tend to be a little later than cultivated chickpea, particularly when grown under non-vernalizing conditions (Berger et al. 2005). Due to the long duration of the wild *Cicer* species life cycle, it is difficult to characterize/evaluate these species for important traits, and without this information, it is difficult to select promising donors. Further, the long growth duration of the wild *Cicer* species and the non-synchronization of flowering with cultivated chickpea is yet another problem hindering their frequent utilization in chickpea improvement programs. As well, limited knowledge is available on crossability and patterns of genetic diversity between cultivated and wild relatives.

Owing to these difficulties, wild *Cicer* species are not being utilized frequently in the breeding programs. Under such situations, pre-breeding provides a unique opportunity to expand the genetic variability in cultivated genepool by exploiting wild species. These efforts will ensure a continuous supply of new and useful genetic variability into the breeding pipelines to develop new cultivars with high levels of resistance and a broad genetic base.

Pre-breeding for Broadening the Genetic Base of Chickpea

Pre-breeding involves identification of useful traits and/or genes from unadapted germplasm such as exotic landraces/wild species, creation of new populations/genepools having good agronomic performance and high frequency of useful genes/alleles introgressed from new and diverse unadapted germplasm into a cultivated background, and evaluation of such populations to identify desirable introgression lines (ILs) for ready use by the breeders in breeding programs (Sharma et al. 2013b). Overall, pre-breeding is a time consuming and resource demanding effort (Fig. 2).

Global Success Stories of Utilizing Wild Cicer Species for Chickpea Improvement

Successful interspecific crosses between cultivated chickpea and wild *Cicer* species, *C. reticulatum*, and *C. echinospermum* have been reported by various workers (Ladizinsky and Adler 1976a, b, Jaiswal et al. 1986, Singh and Ocampo 1993, 1997, Pundir and Mengesha 1995, Simon and Muehlbauer 1997, Winter et al. 1999, 2000, Santra et al. 2000, Rajesh et al. 2002, Tekeoglu et al. 2002, Collard et al. 2003a, b, Pfaff and Kahl 2003, Ahmad and Slinkard 2004, Singh et al. 2005, Abbo et al. 2005, Upadhyaya 2008, Madrid et al. 2008). Utilization of wild *Cicer* species has not only enhanced the levels of resistance/tolerance to biotic/abiotic stresses but has also improved the agronomic performance of cultivated chickpea. The most significant contribution of the wild *Cicer* species *C. reticulatum* for chickpea improvement is the development of cyst nematode-resistant chickpea germplasm lines 'ILC 10765' and 'ILC 10766' by involving *C. reticulatum* accession 'ILWC 119' in the crossing program (Malhotra et al. 2002). In similar studies, interspecific derivatives having a high degree of resistance to wilt, foot rot and root rot diseases and high yield derived from *C. arietinum* × *C. reticulatum* crosses (Singh et al., 2005), and derivatives showing resistance to phytophthora root rot derived from *C. arietinum* × *C. echinospermum* (Knights et al. 2008) have been identified.

Derivative lines showing high levels of genetic resistance to botrytis gray mold have also been identified from *C. arietinum* × *C. pinnatifidum* crosses (Kaur et al. 2013). High yielding cold tolerant lines have also been obtained from *C. arietinum* × *C. echinospermum* crosses (ICARDA 1995). In other studies, utilization of *C. reticulatum, C. echinospermum*, and *C. pinnatifidum* in crossing programs with cultivated chickpea has also led to the development of promising high yielding lines with good agronomic performance and seed traits, such as early flowering and high 100-seed weight (Singh et al. 1984, Jaiswal et al. 1986, Singh and Ocampo 1997, Malhotra et al. 2003, Singh et al. 2005, Sandhu et al. 2006, Upadhyaya 2008, Singh et al. 2012a, b).

Regarding utilization of tertiary genepool species for chickpea improvement, success depends upon a range of techniques, including the *in vitro* application of growth regulators after pollination, and ovule culture for saving the aborting embryos. Following various tissue culture techniques, interspecific F_1 hybrids have been produced by using cross-incompatible wild *Cicer* species such as *C. judaicum* (Verma et al. 1990, 1995, Singh et al. 1999), *C. pinnatifidum* (Verma et al. 1990, Badami et al. 1997, Mallikarjuna 1999, Mallikarjuna and Jadhav 2008), *C. cuneatum* (Singh and Singh 1989), and *C. bijugum* (Verma et al. 1990, Singh et al. 1999). Although these studies have not contributed significantly to the development of introgression lines, these experiments have provided insights into the barriers and techniques involved in generating F_1 hybrids. Mallikarjuna and Muehlbauer (2011) proposed that the chances of survival of hybrid shoots after transferring to soil are increased when the hybrid shoots are grafted to chickpea stocks. Further, these interspecific hybrids together with genomic resources have provided tools for chickpea improvement. Using *C. judaicum*, a pre-breeding line 'IPC 71' exhibiting high number of primary branches, more pods per plant and green seeds has been identified for use in chickpea improvement programs (Chaturvedi and Nadarajan 2010).

Achievements and Present Status at ICRISAT

As mentioned earlier, besides cross-incompatibility barriers, the frequent utilization of wild *Cicer* species for chickpea improvement is also hindered due to their different phenology. The annual wild *Cicer* species are predominantly found in western and central Asia, eastern Mediterranean, and in isolated populations adjacent to the African Red Sea coast (Berger et al. 2003), whereas the cultivated chickpea is found in tropical, subtropical, and warm temperate zones. The phenological differences between annual wild and cultivated *Cicer* species have been reported by various workers (Abbo et al. 2002, Summerfield et al. 1989, Robertson et al. 1997, Berger et al. 2005). Under natural field conditions in subtropical regions such as southern India, the wild *Cicer* species are generally late in phenology compared to cultivated chickpea and, therefore, cannot be used frequently in crossing programs for chickpea improvement. At ICRISAT, a comprehensive attempt was made to study the effects of vernalization and extended photoperiod on flowering

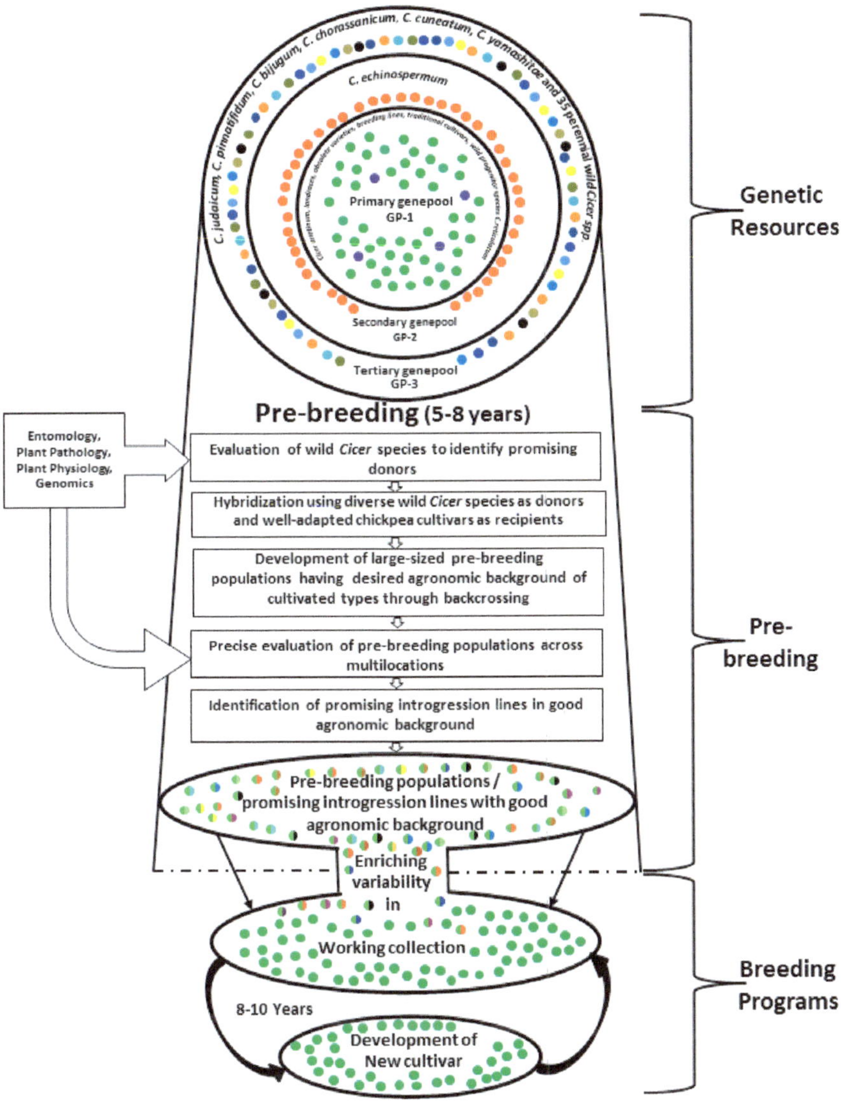

Figure 2. Flowchart of pre-breeding activities for chickpea improvement.

in annual *Cicer* species. All the wild *Cicer* accessions responded to these treatments by decreasing the number of days to first flowering to a greater extent compared to the control condition (Sharma and Upadhyaya 2015). This result showed that vernalization- and photoperiod-responsive genes or alleles control flowering in wild *Cicer* species, and that use of vernalization and extended photoperiod treatments, alone or in combination, can quicken flowering and facilitate the use of wild *Cicer* species for cultivar improvement by synchronizing flowering times between cultivated and wild *Cicer* species (Sharma and Upadhyaya 2015). In *C. reticulatum* and *C. echinospermum*, which are both cross-compatible with cultivated chickpea, response to photoperiod was more significant than response to vernalization. Using extended photoperiod treatments, accessions of *C. reticulatum* and *C. echinospermum* are being utilized in crossing programs to create novel genetic variability for use in chickpea improvement programs.

Extensive screening of wild *Cicer* accessions belonging to eight annual wild *Cicer* species has identified accessions with high levels of resistance to ascochyta blight, botrytis gray mold, dry root rot, and pod borer. Using these promising donors and popular chickpea cultivars, populations are being developed to combine heat tolerance and dry root rot, and short duration and botrytis gray mold following simple (*C. arietinum* × *C. reticulatum* or *C. echinospermum*) and/or complex (*C. arietinum* × (*C. reticulatum* × *C. echinospermum*) crosses. Populations are being developed to introgress pod borer (*Helicoverpa armigera* L.) resistance from *C. reticulatum* and *C. echinospermum* into popular chickpea cultivars. In complex crosses, the objective is to combine the genes from the two different wild species *C. reticulatum* and *C. echinospermum* into a common genetic background. Large-sized advanced backcross populations are being generated using cultivated chickpea as recipient parents, with the objective to recover the maximum genetic background from the cultivated parents with small segments introgressed from the wild species. Following complex crosses, two populations have been developed; one using desi landrace 'ICC 4958', *C. reticulatum* accession 'ICC 17264', and *C. echinospermum* accession 'IG 69978'; and another using kabuli chickpea cultivar 'ICCV 95311', *C. reticulatum* accession 'IG 72933' and *C. echinospermum* accession 'ICC 20192' at ICRISAT, Patancheru, India. In segregating populations (BC_2F_1, BC_1F_2 and 3-way F_2 populations) derived from ICC 4958 × (ICC 17264 × IG 69978) and ICGV 95311 × (IG 72933 × ICC 20192) crosses, considerable variability was observed for morpho-agronomic traits (Fig. 3).

Preliminary evaluation of these populations has resulted in the identification of introgression lines with early flowering (24 days), a high number of pods per plant (up to 740 pods), high po d weight per plant (up to 298.6 g), high numbers of seeds per plant (up to 800 seeds), high seed weight per plant (up to 235.0 g), and high 100-seed weight (up to 52.0 g). Further preliminary screening of a few plants in different populations in each cross following the cut-twig method (Sharma et al. 1995) under controlled environment conditions has identified ILs with high levels of resistance to ascochyta blight and botrytis gray mold. These promising ILs with good agronomic performance in a cultivated genetic background and showing ascochyta blight and botrytis gray mold resistance will be advanced further to develop stable introgression lines for further use in chickpea improvement programs.

Genomic Resources for Chickpea Improvement

Conventional breeding was able to enhance chickpea productivity but could not achieve the desired target of meeting the food requirements of the ever-growing world population. Breeding approaches coupled with modern genomics tools can help to increase chickpea productivity by using marker assisted selection (MAS) approaches in breeding programs (Varshney et al. 2005). Recent advancement in next generation sequencing (NGS) technology has enabled the use of markers to understand mechanisms of complex traits in a much more precise and cost effective manner (Varshney et al. 2014). The last decade has witnessed a tremendous revolution in this area and huge genomic resources have been developed. ICRISAT along with NARS partners has developed large scale genomic and genetic resources, including molecular markers, comprehensive genetic maps, marker-trait associations and QTLs, and has also initiated molecular breeding for various disease resistance and drought tolerance traits in chickpea (Varshney et al. 2013a, Varshney 2015).

Availability of a draft genome sequence has accelerated efforts for chickpea improvement. ICRISAT led the International Chickpea Genome Sequencing Consortium (ICGSC), who used an NGS approach for sequencing the chickpea genome of 'CDC Frontier' (a kabuli chickpea variety) and assembled 544.73 Mb representing 74% of the chickpea genome (Varshney et al. 2013b). Along with the draft chickpea genome, ICGSC also undertook the re-sequencing of 90 cultivated and wild genotypes for the identification of marker(s)/genomic region(s) as a resource for developing superior lines for disease resistance and drought tolerance (Varshney et al. 2013b). In a parallel effort to sequence the chickpea genome, the 'ICC 4958' desi chickpea genotype was sequenced using NGS platforms along with bacterial artificial chromosome (BAC) end sequences and a genetic map (Jain et al. 2013). The chromosomal genomics approach was used to validate these desi and kabuli chickpea genome assemblies, and identified small misassembled regions in the kabuli and large misassembled regions in the desi draft genomes (Ruperao et al. 2014). Recent efforts have increased the length of pseudomolecules up to 2.7 fold and could reduce the gaps in the desi chickpea assembly (Parween et al. 2015).

Availability of the draft genome sequence provides base for accelerated crop improvement efforts, and helps in undertaking large scale genome re-sequencing projects. An NGS-based whole genome re-sequencing (WGRS) approach was used for re-sequencing the chickpea reference set and elite

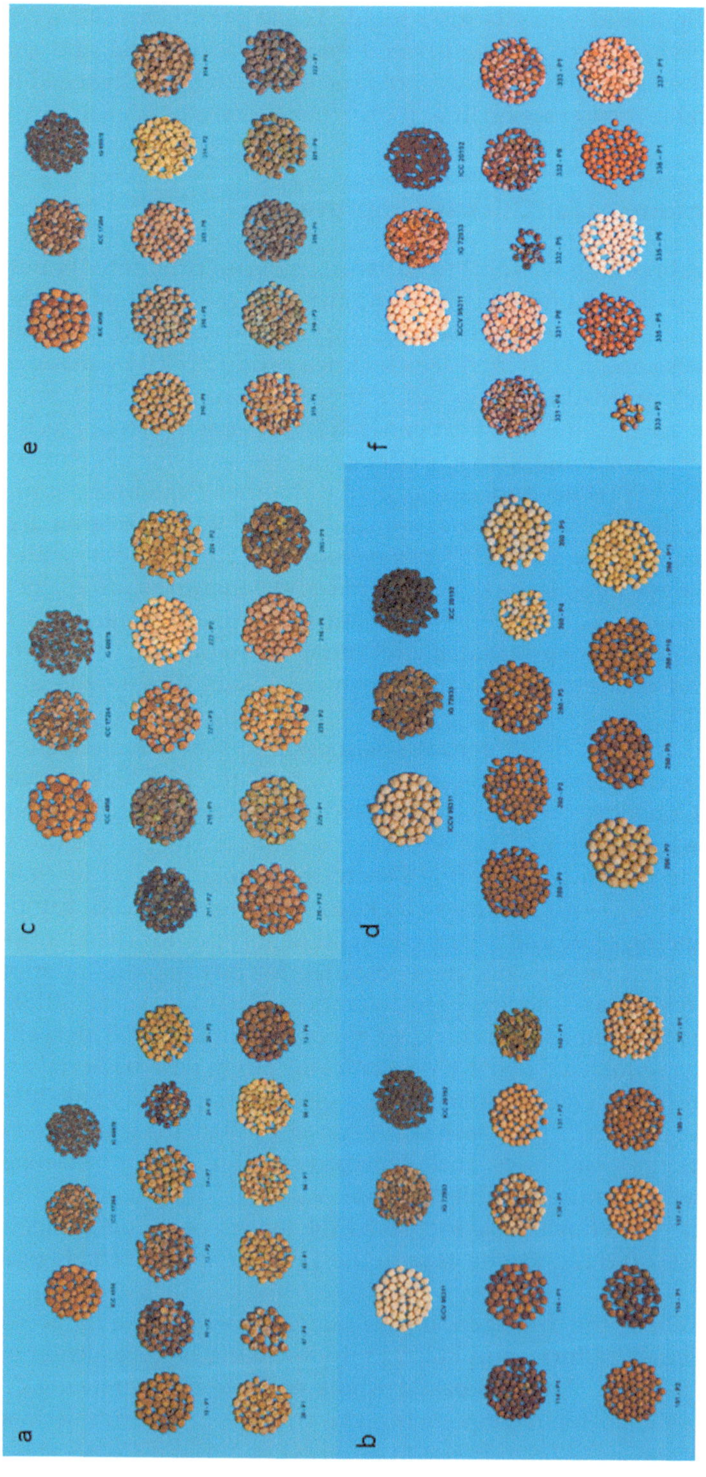

Figure. 3. Variability for seed traits in BC_2F_1 (left: a & b), BC_1F_2 (middle: c & d), and 3-F_2 (right: e & f) populations derived from 4958 × (ICC 17264 × IG 69978) (top) and ICCV 95311 × (IG 72933 × ICC 20192) (bottom) crosses at ICRISAT, Patancheru, India.

chickpea varieties using an Illumina HiSeq 2500 at 5× to 13× coverage. Structural variation analysis revealed ~4.7 million SNPs, and >500 000 Indels and CNVs respectively (unpublished results). Very recently, ICRISAT has also launched "The 3000 Chickpea Genome Sequencing Initiative" where 3000 lines from the global composite collection of chickpea from genebank of ICRISAT and ICARDA will be re-sequenced for identification of novel alleles (see Varshney 2015). These lines include 20 accessions from two wild species (*C. reticulatum* and *C. echinospermum*) and they will provide unique variations in the secondary genepool in addition to the primary genepool.

In addition to the draft genome sequence and re-sequencing efforts, a large number of molecular markers and genotyping platforms have become available in the last decade. Until the last decade chickpea was considered to be an "orphan crop" because very limited genomic resources were available (Varshney et al. 2012). However, a wide range of molecular markers, including simple sequence repeat (SSR), single nucleotide polymorphism (SNP) and DArT markers have become available in the last few years. In addition to markers, large scale transcriptomic resources have also been developed in chickpea. Details about these markers and transcriptomic resources have been reviewed earlier (see Varshney et al. 2013a, Varshney 2015). In addition, transcriptome assemblies have also been developed for the wild species *C. reticulatum* (PI 489777) (Agrawal et al. 2012). Deployment of markers in breeding is largely dependent on the possibility of automation and cost. A vast range of SNP genotyping platforms are available, providing options for the best possible methods depending on the application and cost. In the case of molecular breeding where very few markers are required for selecting the lines, VeraCode assays seem to suit well. Therefore, VeraCode 96-plex SNP assays have been developed for chickpea at ICRISAT, to be used on Illumina's BeadXpress system (Roorkiwal et al. 2013). Another SNP genotyping platform, KASP developed by KBiosciences (known as LGC limited; www.lgcgenomics.com/) also provides a flexible and cost effective assay for SNP genotyping. ICRISAT has developed 2 068 KASP assays in chickpea (Hiremath et al. 2012). These assays are a perfect source of markers for use in the populations developed by wild species.

As mentioned earlier, during the process of domestication and breeding for selective traits, most of the adaptive alleles get ignored and/or lost and selective lethal alleles get fixed, narrowing down the diversity available in the cultivated gene pool. This loss/fixation of alleles restricts efforts for crop improvement beyond a certain boundary, and limits the productivity. Wild relatives are seemingly the key to solve the issue of low genetic diversity, able to provide favorable alleles to the cultivated crop.

In order to harness the potential of wild relatives for crop improvement, it is essential to understand the genetics and diversity of wild relatives of chickpea. Effective use of wild germplasm requires a thorough knowledge of the population genetics and source habitats of the wild material. Efforts using genomics to understand the diversity and domestication of chickpea have identified different possible sources that can be used for introgression of desired alleles into cultivated chickpea. Hybridization-based DArT arrays and SNP-based KASP assays were used to genotype a diverse panel of chickpea representing genotypes from 10 different *Cicer* species from the primary, secondary and tertiary gene pools (Roorkiwal et al. 2014). Genotyping data analysis among species from the different gene pools suggested possible gene flow among these species, and provided indications about the pattern of divergence of wild chickpea. Similarly, high density genotyping using GBS could track the movement of trait-related genomic regions/alleles from wild to cultivated chickpea, and also suggested the geographical origins of the wild *Cicer* accessions (Bajaj et al. 2015).

In addition to genetics and genomics analysis to understand the genetic relationships of wild species and the potential of wild species for enhancing productivity, specialized advanced backcross quantitative trait locus (AB-QTL) populations are being developed. AB-QTL populations involve transfer of favorable allele(s)/gene(s) from wild donor lines to elite cultivated lines in order to develop superior varieties. ICRISAT has developed several AB-QTL populations with the idea of transferring superior alleles from wild germplasm to enhance yield and resistance to biotic diseases in cultivated chickpea. For instance, 'ICC 17160' was crossed with 'ICC 8261' to develop an AB-QTL population. The backcross progenies (BC_4F_1) were genotyped using KASP markers for identification of QTLs linked with traits of interest. A total of 295 polymorphic markers were used for genotyping. Similarly, another population ('JG 11' × 'PI 489777') was developed to bring superior alleles from a donor wild accession ('PI 489777') to a cultivated elite line ('JG 11').

In order to deploy MAS to enhance yield, identification of markers associated with the trait of interest is the first step; this can be done using either bi-parental mapping population-based linkage mapping, or via a germplasm-based association mapping approach. Linkage mapping-based trait mapping studies have already been summarized recently (see Varshney et al. 2015). Recently, interest in trait mapping is shifting from conventional mapping strategies to development of multi-parent mapping populations. These populations include multi-parent advanced generation inter-cross (MAGIC) and nested association mapping (NAM) populations, which provide high resolution genetic mapping and QTL identification (Varshney and Dubey 2009). ICRISAT developed a MAGIC population with 1200 lines using eight parents, including cultivars and elite breeding lines ('ICC 4958', 'ICCV 10', 'JAKI 9218', 'JG 11', 'JG 130', 'JG 16', 'ICCV 97105', and 'ICCV 00108') from India and Africa. These eight diverse parents were crossed in 28 two-way, 14 four-way and 7 eight-way crosses for accumulation of recombination events to allow genome reshuffling to bring about greater genetic diversity (see Varshney 2015).

In order to access allelic diversity in the MAGIC population, the population was re-sequenced using a whole genome re-sequencing approach, which generated a total of 4.67 Tb of clean sequence data. Alignment of re-sequencing data to the reference genome led to the identification of 1 million SNP variants. Similarly, ICRISAT along with its partner is developing NAM population in chickpea with 'ICC 4958' as a common female parent. In total, 14 different crosses have been initiated to generate F_1s. These F_1s are being advanced to generate a minimum of 200 lines from each cross. In addition to these specialized populations, NGS-based mapping technologies were also used for high resolution mapping. For instance, genotyping-by-sequencing (GBS) has been extensively used for constructing high density genetic maps and for QTL analysis (Deokar et al. 2014, Jaganathan et al. 2015, Kujur et al. 2015). In addition, a skim-sequencing based approach was also used for high resolution genetic mapping and QTL analysis in chickpea (Kale et al. 2015).

In order to deploy markers in the chickpea improvement program using genomics assisted breeding (GAB), two approaches are currently been deployed for developing improved lines: marker-assisted backcrossing (MABC), and GS. For addressing the simple traits that are controlled by one or two gene(s)/QTL(s), MABC has been widely accepted and proven successful. In the case of chickpea, MABC was successfully used for intogressing a "*QTL-Hotspot*" harboring several QTLs controlling several drought tolerance-related root traits in the elite chickpea variety 'JG 11' (a drought tolerant variety) (Varshney et al. 2013c). Introgression lines have shown improved performance under rainfed as well irrigated environments as compared to the recipient parent. Similarly, MABC has also been successfully used for introgressing resistance to fusarium wilt (FW) and ascochyta blight (AB) into elite chickpea cultivar 'C 214' (Varshney et al. 2013d). Efforts to pyramid the resistance to FW and AB by crossing introgression lines using MABC are underway. MABC for introgession of *Helicoverpa* resistance is also possible, once markers associated with *Helicoverpa* resistance have become available from the wild species.

Future Strategies

Though the importance of wild *Cicer* species as new and diverse sources of variation is well known, their actual utilization for the genetic improvement of chickpea is still in its infancy. In ICRISAT and elsewhere, a very few wild *Cicer* accessions have been used with significant contributions towards chickpea improvement. This is mainly because of a lack of focused research on the use of wild species. Recently, in ICRISAT, systematic pre-breeding activities have been given due emphasis, and efforts are in progress to use wild *Cicer* accessions as donors and cultivated chickpea varieties as recipients to develop large-sized pre-breeding populations with good agronomic background for further use by breeders for chickpea improvement. Recent studies on vernalization and extended photoperiod treatments have proved effective in reducing the vegetative phase of wild *Cicer* species in Patancheru, India. This knowledge would assist in efficient characterization of germplasm accessions of wild *Cicer* species. This would also generate information on important morpho-agronomic and nutrition-related traits as well as on reaction to biotic and abiotic stresses, leading to selection of promising donors for further use in hybridization programs and in generation of pre-breeding populations. Further, scanty information is available for the use of cross-incompatible tertiary genepool species in chickpea improvement programs. Hence, efforts should be made to utilize these cross-incompatible wild *Cicer* species by following tissue culture techniques, as well as by exploiting the potential of the cross-compatible wild *Cicer* species *Cicer reticulatum* and *C. echinospermum* as a bridge to transfer useful traits from cross-incompatible tertiary genepool species into the cultivated genetic background. Besides this, the availability and use of genomic tools, especially the chickpea genome sequence information, large-scale genomic resources and beneficial wild species alleles would increase the efficiency and precision of these pre-breeding activities. This would also enrich the genetic variability available in the primary genepool for further use by breeders in developing new cultivars with a broad genetic base.

References

Abbo, S., C. Molina, N.N.R. Jungma, K.A. Grusa, Z. Berkovitch, R. Reifen, G. Kahl, P. Winter and R. Reifen. 2005. Quantitative trait loci governing carotenoid concentration and weight in seeds of chickpea (*Cicer arietinum* L.). Theor. Appl. Genet. 111: 185-195.

Abbo, S., J. Berger and N.C. Turner. 2003. Evolution of cultivated chickpea: four bottlenecks limit diversity and constrain adaptation. Funct. Plant Biol. 30: 1081-1087.

Abbo, S., S. Lev-Yadun and N. Galwey. 2002. Vernalization response of wild chickpea. New Phytol. 154: 695–701.

Agarwal, G., S. Jhanwar, P. Priya, V.K. Singh, M.S. Saxena, S.K. Parida, R. Garg, A.K. Tyagi and M. Jain. 2012. Comparative analysis of kabuli chickpea transcriptome with desi and wild chickpea provides a rich resource for development of functional markers. PLoS One 7:e52443

Ahmad, F. and A.E. Slinkard. 2004. The extent of embryo and endosperm growth following interspecific hybridization between *Cicer arietinum* L. and related annual wild species. Genet. Resour. Crop Evol. 51: 765-772.

Ahmad, F., A.E. Slinkard and G.J. Scoles. 1988. Investigations into the barrier/s to interspecific hybridization between *Cicer arietinum* L. and eight other annual *Cicer* species. Plant Breed. 100: 193-198.

Badami, P.S., N. Mallikarjuna and J.P. Moss. 1997. Interspecific hybridization between *Cicer arietinum* and *C. pinnatifidum*. Plant Breed. 116: 393-395.

Bajaj, D., S. Das, S. Badoni, V. Kumar, M. Singh, K.C. Bansal, A.K. Tyagi and S.K. Parida. 2015. Genome-wide high-throughput SNP discovery and genotyping for understanding natural (functional) allelic diversity and domestication patterns in wild chickpea. Sci. Rep. 5: 12468.

Berger, J.D., S. Abbo and N.C. Turner. 2003. Ecogeography of annual wild *Cicer* species: The poor state of the world collection. Crop Sci. 43: 1076-1090.

Berger, J.D., R. Buck, J.M. Henzell and N.C. Turner. 2005. Evolution in the genus *Cicer*-vernalization response and low temperature pod set in chickpea (*C. arietinum* L.) and its annual wild relatives. Aust. J. Agric. Res. 56: 1191-1200.

Berger, J.D., S. Kumar, H. Nayyar, K.A. Street, J.S. Sandhu, J.M. Henzell, J. Kaur and H.C. Clarke. 2012. Temperature-stratified screening of chickpea (*Cicer arietinum* L.) genetic resource collections reveals very limited reproductive chilling tolerance compared to its annual wild relatives. Field Crops Res. 126: 119-129.

Canci, H. and C. Toker. 2009. Evaluation of annual wild *Cicer* species for drought and heat resistance under field conditions. Genet. Resour. Crop Evol. 56: 1-6.

Chaturvedi, S.K. and N. Nadarajan. 2010. Genetic enhancement for grain yield in chickpea – accomplishments and resetting research agenda. Electron. J. Plant Breed. 1: 611-615.

Collard, B.C.Y., P.K. Ades, E.C.K. Pang, J.B. Brouwer and P.W.J. Taylor. 2001. Prospecting for sources of resistance to *Ascochyta* blight in wild *Cicer* species. Australas. Plant Pathol. 30: 271-276.

Collard, B.C.Y., E.C.K. Pang and P.W.J. Taylor. 2003b. Selection of wild *Cicer* accessions for generation of mapping populations segregating for resistance to Ascochyta blight. Euphytica 130: 1-9.

Collard, B.C.Y., E.C.K. Pang, P.K. Ades and P.W.J. Taylor. 2003a. Preliminary investigation of QTLs associated with seedling resistance to Ascochyta blight from *Cicer echinospermum*, a wild relative of chickpea. Theor. Appl. Genet. 107: 719-729.

Croser, J.S., F. Ahmad, H.J. Clarke and K.H.M. Siddique. 2003. Utilization of wild *Cicer* in chickpea improvement- progress, constraints and prospects. Aust. J. Agric. Res. 54: 429-444.

Cubero, J.I. 1975. The research on chickpea (*Cicer arietinum*) in Spain. Proceedings of the International Workshop on Grain Legumes. International Crops Research Institute for the Semi-Arid Tropics, Hyderabad, India. 17-122.

Deokar, A.A., L. Ramsay, A.G. Sharpe, M. Diapari, A. Sindhu, K. Bett, T.D. Warkentinand and B. Tar'an. 2014. Genome wide SNP identification in chickpea for use in development of a high density genetic map and improvement of chickpea reference genome assembly. BMC Genomics 15: 708.

FAOSTAT. 2013. Production: Crops. Food and Agriculture Organization of the United Nations. http://faostat.fao.org/site/567/desktopdefault.aspx?pageid=567 (accessed on 11 December 2015).

Harlan, J.R. and J.M.T. de Wet.1971. Toward a Rational Classification of Cultivated Plants. Taxon 20: 509-517.

Hiremath, P.J., A. Kumar, R.V. Penmetsa, A. Farmer, J.A. Schlueter, S.K. Chamarthi, A.M. Whaley, N. Carrasquilla-Garcia, P.M. Gaur, H.D. Upadhyaya, P.B. Kavi Kishor, T.M. Shah, D.R. Cook and R.K. Varshney. 2012. Large-scale development of cost-effective SNP marker assays for diversity assessment and genetic mapping in chickpea and comparative mapping in legumes. Plant Biotechnol. J. 10: 716-732.

ICARDA. Annual Report for 1995. Germplasm Program Legumes, ICARDA, Aleppo, Syria: 210pp.

Infantino, A., A. Porta-Puglia and K.B. Singh. 1996. Screening of wild *Cicer* species for resistance to *Fusarium* wilt. Plant Dis. 80: 42-44.

Jaganathan, D., M. Thudi, S. Kale, S. Azam, M. Roorkiwal, P.M. Gaur, P.B. Kavikishor, H. Nguyen, T. Suttonand and R.K. Varshney. 2015. Genotyping-by-sequencing based intra-specific genetic map refines a "*QTL-hotspot*" region for drought tolerance in chickpea. Mol. Genet. Genomics 290: 559-571.

Jain, M., G. Misra, R.K. Patel, P. Priya, S. Jhanwar, A.W. Khan, Niraj Shah, V.K. Singh, R. Garg, G. Jeena, M. Yadav, C. Kant, P. Sharma, G. Yadav, S. Bhatia, A.K. Tyagi and D. Chattopadhyay. 2013. A draft genome sequence of the pulse crop chickpea (*Cicer arietinum* L.). Plant J. 74: 715-729.

Jaiswal, H.K., B.D. Singh, A.K. Singh and R.M. Singh. 1986. Introgression of genes for yield and yield traits from *C. reticulatum* into *C. arietinum*. Int. Chickpea News. 14: 5-8.

Jukanti, A.K., P.M. Gaur, C.L.L. Gowda and R.N. Chibbar. 2012. Nutritional quality and health benefits of chickpea (*Cicer arietinum* L.): a review. Br. J. Nutr. 108: S11-26.

Kale, S.M., J. Jaganathan, P. Ruperao, C. Chen, R. Punna, H. Kudapa, M. Thudi, M. Roorkiwal, M.A. Katta, D. Doddamani, V. Garg, P.B. Kishor, P.M. Gaur, H.T. Nguyen, J. Batley, D. Edwards, T. Sutton and R.K. Varshney. 2015. Prioritization of candidate genes in QTL-hotspot region for drought tolerance in chickpea (*Cicer arietinum* L.). Sci. Rep. 5: 15296.

Kashiwagi, J., L. Krishnamurthy, H.D. Upadhyaya, H. Krishna, S. Chandra, Vincent Vadez and R. Serraj. 2005. Genetic variability of drought-avoidance root traits in the mini-core germplasm collection of chickpea (*Cicer arietinum* L.). Euphytica 146: 213-222.

Kaur, L., A. Sirari, D. Kumar, J.S. Sandhu, S. Singh, I. Singh, K. Kapoor, C.L.L. Gowda, S. Pande, P. Gaur, M. Sharma, M. Imtiaz and K.H.M. Siddique. 2013. Harnessing ascochyta blight and botrytis grey mould resistance in chickpea through interspecific hybridization. Phytopathol. Mediterr. 52: 157-165.

Knights, E.J., R.J. Southwell, M.W. Schwinghamer and S. Harden. 2008. Resistance to *Phytophthora medicaginis* Hansen and Maxwell in wild *Cicer* species and its use in breeding root rot resistant chickpea (*Cicer arietinum* L.). Aus. J. Agric. Res. 59: 383-387.

Kujur, A., H.D. Upadhyaya, T. Shree, D. Bajaj, S. Das, M. Saxena, S. V. Kumar, S. Tripathi, C.L.L. Gowda, S. Sharma, S. Singh, A.K. Tyagi and S.K. Parida. 2015. Ultra-high density intra-specific genetic linkage maps accelerate identification of functionally relevant molecular tags governing important agronomic traits in chickpea. Nat. Sci. Rep. 5: 9468.

Kumar, J. and H.A. van Rheenen. 2000. A major gene for time of flowering in chickpea. J. Hered. 91: 67-68.

Ladizinsky, G. and A. Adler. 1976a. The origin of chickpea *Cicer arietinum* L. Euphytica 25: 211-217.

Ladizinsky, G. and A. Adler. 1976b. Genetic relationships among annual species of *Cicer* L. Theor. Appl. Genet. 48: 197-203.

Lulsdorf, M., N. Mallikarjuna, H. Clarke and B. Tar'an. 2005. Finding solutions for interspecific hybridization problems in chickpea (*Cicer arietinum* L.). 4th International Food Legumes Research Conference, New Delhi, India. 44 pp.

Madrid, E., D. Rubiales, A. Moral, M.T. Moreno, T. Millan, J. Gil and J. Rubio. 2008. Mechanism and molecular markers associated with rust resistance in a chickpea interspecific cross (*Cicer arietinum x Cicer reticulatum*). Eur. J. Plant. Pathol. 121: 43-53.

Malhotra, R.S., M. Baum, S.M. Udupa, B. Bayaa, S. Kabbabe and G. Khalaf. 2003. *Ascochyta* blight resistance in chickpea: Present status and future prospects. pp. 108-117. *In*: R.N. Sharma, G.K. Shrivastava, A.L. Rathore, M.L. Sharma and M.A. Khan (eds.). Chickpea Research for the Millennium: International Chickpea Conference, 20-22 January 2003, Indira Gandhi Agril. Univ., Raipur, Chhattisgarh, India.

Malhotra, R.S., K.B. Singh, M. Vito, N. Greco and M.C. Saxena. 2002. Registration of ILC 10765 and ILC 10766 chickpea germplasm lines resistant to cyst nematode. Crop Sci. 42: 1756.

Mallikarjuna, N. 1999. Ovule and embryo culture to obtain hybrids from interspecific incompatible pollinations in chickpea. Euphytica 110: 1-6.

Mallikarjuna, N. and D.R. Jadhav. 2008. Techniques to produce hybrids between *Cicer arietinum* L. *x Cicer pinnatifidum* Jaub. Indian J. Genet. 68: 398-405.

Mallikarjuna, N. and F.J. Muehlbauer. 2011. Chickpea hybridization using *in vitro* techniques. Methods Mol. Biol. 710: 93-105.

Muehlbauer, F.J. 1993. Use of wild species as a source of resistance in cool-season food legume crops. pp. 359-372. *In*: K.B. Singh and M.C. Saxena (eds.). Breeding for stress tolerance in cool-season food legumes. John Wiley and Sons, Chichester, UK.

Pande, S., D. Ramgopal, G.K. Kishore, N. Mallikarjuna, M. Sharma, M. Pathakand and J. Narayana Rao. 2006. Evaluation of wild *Cicer* species for resistance to *Ascochyta* blight and *Botrytis* gray mold in controlled environment at ICRISAT, Patancheru, India. SAT eJ., ICRISAT.

Parween. S., K. Nawaz, R. Roy, A.K. Pole, B. Venkata Suresh, G. Misra, M. Jain, G. Yadav, S.K. Parida, A.K. Tyagi, S. Bhatia and D. Chattopadhyay. 2015. An advanced draft genome assembly of a desi type chickpea (*Cicer arietinum* L.). Sci. Rep. 5: 12806.

Pfaff, T. and G. Kahl. 2003. Mapping of gene-specific markers on the genetic map of chickpea (*Cicer arietinum* L.) Mol. Genet. Genomics 269: 243-251.

Pundir, R.P.S. and M.H. Mengesha. 1995. Cross compatibility between chickpea and its wild relative *Cicer echinospermum* Davis. Euphytica 83: 241-245.

Rajesh, P.N., M. Tekeoglu, V.S. Gupta, P.K. Ranjekar and F.J. Muehlbauer. 2002. Molecular mapping and characterization of an RGA locus RGAPtokin 1-2 (171) in chickpea. Euphytica 128: 427-433.

Rao, N.K., L.J. Reddy and P.J. Bramel. 2003. Potential of wild species for genetic enhancement of some semi-arid food crops. Genetic Resour. Crop Evol. 50: 707-721.

Robertson, L.D., B. Ocamppo and K.B. Singh. 1997. Morphological variation in wild annual *Cicer* species in comparison to the cultigen. Euphytica 95: 309-319.

Roorkiwal, M., S.L. Sawargaonkar, A. Chitikineni, M. Thudi, R.K. Saxena, H.D. Upadhyaya, M.I. Vales, O. Riera-Lizarazu and R.K. Varshney. 2013. Single nucleotide polymorphism genotyping for breeding and genetics applications in chickpea and pigeonpea using the BeadXpress platform. Plant Gen. 6. doi: 10.3835/plantgenome2013.05.0017.

Roorkiwal, M., E.J. von Wettberg, H.D. Upadhyaya, E. Warschefsky, A. Rathore and R.K. Varshney. 2014. Exploring germplasm diversity to understand the domestication process in *Cicer* spp. using SNP and DArT markers. PLoS One 9: e102016.

Ruperao, P., C.K. Chan, S. Azam, M. Karafiátová, S. Hayashi, J. Cížková. R.K. Saxena, H. Simková, C. Song, J. Vrána, A. Chitikineni, P. Visendi, P.M. Gaur, T. Millán, K.B. Singh, B. Taran, J. Wang, J. Batley, J. Doležel, R.K. Varshney and D. Edwards. 2014. A chromosomal genomics approach to assess and validate the desi and kabuli draft chickpea genome assemblies. Plant Biotechnol. J. 12: 778-786.

Sandhu, J.S., S.K. Gupta, G. Singh, Y.R. Sharma, T.S. Bains, L. Kaur and A. Kaur. 2006. Interspecific hybridization between *Cicer arietinum* L. and *Cicer pinnatifidum* Jaub. et. Spach for improvement of yield and other traits. 4th International Food Legumes Research Conference, Indian Society of Genetics and Plant Breeding, New Delhi. 192 pp.

Santra, D.K., M. Tekeoglu, M. Ratnaparkhe, W.J. Kaiser and F.J. Muehlbauer. 2000. Identification and mapping of QTLs conferring resistance to ascochyta blight in chickpea. Crop Sci. 40: 1606-1612.

Shah, T.M., M. Hassan, M.A. Haq, B.M. Atta, S.S. Alam and H. Ali. 2005. Evaluation of *Cicer* species for resistance to ascochyta blight. Pak. J. Bot. 37: 431-438.

Sharma, K.D., W. Chen and F.J. Muehlbauer. 2005. Genetics of chickpea resistance to five races of *Fusarium* wilt and a concise set of race differentials for *Fusarium oxysporum* f. sp. *ciceris*. Plant Dis. 89: 385-390.

Sharma, Y.R., G. Singh and L. Kaur. 1995. A rapid technique for Ascochyta blight resistance in chickpea. Int. Chickpea Pigeonpea News. 2: 34-35.

Sharma, S. and H.D. Upadhyaya. 2015. Vernalization and photoperiod response in annual wild *Cicer* species and cultivated chickpea. Crop Sci. 55: 1-8. doi: 10.2135/cropsci2014.09.0598.

Sharma, S., H.D. Upadhyaya, M. Roorkiwal, R.K. Varshney and C.L.L. Gowda. 2013a. Chickpea. pp. 85-112. *In:* M. Singh, H.D. Upadhyaya and I.S. Bisht (eds.). Genetic and Genomic Resources of Grain Legume Improvement. Elsevier; 32 Jamestown Road, London NW1 7BY, UK; 225 Wyman Street, Waltham, MA 02451, USA. ISBN: 978-0-12-397935-3.

Sharma, S., H.D. Upadhyaya, R.K. Varshney and C.L.L. Gowda. 2013b. Pre-breeding for diversification of primary gene pool and genetic enhancement of grain legumes. Front. Plant Sci., 20 August 2013 | doi: 10.3389/fpls.2013.00309.

Simon, C.J. and F.J. Muehlbauer. 1997. Construction of a chickpea linkage map and its comparison with maps of pea and lentil. J. Hered. 88: 115-119.

Singh, B.D., H.K. Jaiswal, R.M. Singh and A.K. Singh. 1984. Isolation of early flowering recombinants from the interspecific cross between *Cicer arietinum* and *C. reticulatum*. Int. Chickpea News. 11: 14.

Singh, I., R.P. Singh, S. Singh and J.S. Sandhu. 2012a. Introgression of productivity genes from wild to cultivated *Cicer*. pp. 155-156. *In:* S.K. Sandhu, N. Sidhu and A. Rang (eds.). International Conference on Sustainable Agriculture for Food and Livelihood Security, Crop Improv. 39 (Special issue), Crop Improvement Society of India, Ludhiana.

Singh, K.B. and B. Ocampo. 1997. Exploitation of wild *Cicer* species for yield improvement in chickpea. Theor. Appl. Genet. 95: 418-423.

Singh, K.B. and M.V. Reddy. 1993. Sources of resistance to *Ascochyta* blight in wild *Cicer* species. Neth. J. Plant Pathol. 99: 163-167.

Singh, K.B., G.C. Hawtin, Y.L. Nene and M.V. Reddy. 1981. Resistance in chickpeas to *Ascochyta rabiei*. Plant Dis. 65: 586-587.

Singh, K.B., R.S. Malhotra and M.C. Saxena. 1990. Sources for tolerance to cold in *Cicer* species. Crop Sci. 30: 1136-1138.

Singh, K.B., R.S. Malhotra and M.C. Saxena. 1995. Additional sources of tolerance to cold in cultivated and wild *Cicer* species. Crop Sci. 35: 1491-1497.
Singh, K.B. and B. Ocampo. 1993. Interspecific hybridization in annual *Cicer* species. J. Genet. Breed. 47: 199-204.
Singh, N.P., A. Singh, A.N. Asthana and A. Singh. 1999. Studies on inter-specific crossability barriers in chickpea. Ind. J. Pulses Res. 12: 13-19.
Singh, R.P. and B.D. Singh. 1989. Recovery of rare interspecific hybrids of gram *Cicer arietinum* × *C. cuneatum* L. through tissue culture. Curr. Sci. 58: 874-876.
Singh, R.P., I. Singh, S. Singh and J.S. Sandhu. 2012b. Assessment of genetic diversity among interspecific derivatives in chickpea. J. Food Legum. 25: 150-152.
Singh, S., R.K. Gumber, N. Joshi and K. Singh. 2005. Introgression from wild *Cicer reticulatum* to cultivated chickpea for productivity and disease resistance. Plant Breeding 124: 477-480.
Singh, U. and R.P.S. Pundir. 1991. Amino acid composition and protein content of chickpea and its wild relatives. Int. Chickpea News. 25: 19-20.
Stamigna, C., P. Crino and F. Saccardo. 2000. Wild relatives of chickpea: Multiple disease resistance and problems to introgression in the cultigen. J. Genet. Breed. 54: 213-219.
Stevenson, P.C. and M.P. Haware. 1999. Maackiain in *Cicer bijugum* Rech. F. associated with resistance to *Botrytis* gray mold. Biochem. System Ecol. 27: 761-767.
Summerfield, R.J., R.H. Ellis and E.H. Roberts. 1989. Vernalization in chickpea (*Cicer arietinum*): Fact or artefact?. Ann. Bot. 64: 599-603.
Tayyar, R.I. and J.G. Waines. 1996. Genetic relationships among anuual species of *Cicer* (Fabaceae) using isozyme variation. Theor. Appl. Genet. 92: 245-254.
Tekeoglu, M., P.N. Rajesh and F.J. Muehlbauer. 2002. Integration of sequence tagged microsatellite sites to the chickpea genetic map. Theor. Appl. Genet. 105: 847-854.
Toker, C. 2005. Preliminary screening and selection for cold tolerance in annual wild *Cicer* species. Genet. Resour. Crop Evol. 52: 1-5.
Toker, C., H. Canci and T. Yildirim. 2007. Evaluation of perennial wild *Cicer* species for drought resistance. Genet. Resour. Crop Evol. 54: 1781-1786.
Upadhyaya, H.D. 2008. Crop Germplasm and wild relatives: a source of novel variation for crop improvement. Korean J. Crop Sci. 53: 12-17.
van der Maesen, L.J.G. 1972. A monograph of the genus, with special references to the chickpea (*Cicer arietinum* L.) its ecology and cultivation. Mendelingen Landbouwhoge School: Wageningen, The Netherlands.
van der Maesen, L.J.G. 1987. Origin, history and taxonomy of chickpea. pp. 11-34. *In*: M.C. Saxena and K.B. Singh (eds.). The Chickpea. CAB International, Wallingford, UK.
van Emden, H.F., S.L. Ball and M.R. Rao. 1988. Pest disease and weed problems in pea lentil and faba bean and chickpea. pp. 519-534. *In:* World Crops: Cool Season Food Legumes. Kluwer Academic Publishers, Dordrecht, The Netherlands. ISBN 90-247-3641-2.
Varshney, R.K., S.M. Mohan, P.M. Gaur, S.K. Chamarthi, V.K. Singh, S. Srinivasan, N. Swapna, M. Sharma, S. Pande, S. Singh and L. Kaur. 2013d. Marker-assisted backcrossing to introgress resistance to Fusarium wilt (FW) race 1 and Ascochyta blight (AB) in C 214, an elite cultivar of chickpea. Plant Gen. doi: 10.3835/plantgenome2013.10.0035.
Varshney, R.K. 2015. Exciting journey of 10 years from genomes to fields and markets: Some success stories of genomics-assisted breeding in chickpea, pigeonpea and groundnut. Plant Sci. doi.org/10.1016/j.plantsci.2015.09.009.
Varshney, R.K. and A. Dubey. 2009. Novel genomic tools and modern genetic and breeding approaches for crop improvement. J. Plant Biochem. Biotechnol. 18: 127-138.
Varshney, R.K., A. Graner and M.E. Sorrells. 2005. Genomics-assisted breeding for crop improvement. Trends Plant Sci. 10: 621-630.
Varshney, R.K., C. Song, R.K. Saxena, S. Azam, S. Yu, A.G. Sharpe, S. Cannon, J. Baek, B.D. Rosen, B. Tar'an, T. Millan, X. Zhang, L.D. Ramsay, A. Iwata, Y. Wang, W. Nelson, A.D. Farmer, P.M. Gaur, C. Soderlund, R.V. Penmetsa, C. Xu, A.K. Bharti, W. He, P. Winter, S. Zhao, J.K. Hane, N. Carrasquilla-Garcia, J.A. Condie, H.D. Upadhyaya, M.C. Luo, M. Thudi, C.L.L. Gowda, N.P. Singh, J. Lichtenzveig, K.K. Gali, J. Rubio, N. Nadarajan,

J. Dolezel, K.C. Bansal, X. Xu, D. Edwards, G. Zhang, G. Kahl, J. Gil, K.B. Singh, S.K. Datta, S.A. Jackson, J.Wang and D.R. Cook. 2013b. Draft genome sequence of chickpea (*Cicer arietinum*) provides a resource for trait improvement. Nat. Biotechnol. 31: 240-246.

Varshney, R.K., H. Kudapa, L. Pazhamala, A. Chitikineni, M. Thudi, A. Bohra, P.M. Gaur, P. Janila, A. Fikre, P. Kimurto and N. Ellis. 2015. Translational genomics in agriculture: some examples in grain legumes. Crit. Rev. Plant Sci. 34: 169-194.

Varshney, R.K., J.M. Ribaut, E.S. Buckler, R. Tuberosa, J.A. Rafalski and P. Langridge. 2012. Can genomics boost productivity of orphan crops?. Nat. Biotechnol. 30: 1172-1176.

Varshney, R.K., P.M. Gaur, S.K. Chamarthi, L. Krishnamurthy, S. Tripathi, J. Kashiwagi, S. Samineni, V.K. Singh, M. Thudi and D. Jaganathan. 2013c. Fast-track introgression of QTL-hotspot for root traits and other drought tolerance traits in JG 11, an elite and leading variety of chickpea. Plant Genome. doi:10.3835/plantgenome2013.07.0022.

Varshney, R.K., R.Terauchi and S.R. McCouch. 2014. Harvesting the promising fruits of genomics: applying genome sequencing technologies to crop breeding. PLoS Biol. 12: e1001883.

Varshney, R.K., S.M. Mohan, P.M. Gaur, N.V.P.R. Gangarao, M.K. Pandey, A. Bohra, S.L. Sawargaonkar, A. Chitikineni, P.K. Kimurto, P. Janila, K.B. Saxena, A. Fikre, M. Sharma, A. Rathore, A. Pratap, S. Tripathi, S. Datta, S.K. Chaturvedi, N. Mallikarjuna, G. Anuradha, A. Babbar, A.K. Choudhary, M.B. Mhase, Ch. Bharadwaj, D.M. Mannur, P.N. Harer, B. Guo, X. Liang, N. Nadarajan and C.L.L. Gowda. 2013a. Achievements and prospects of genomics-assisted breeding in three legume crops of the semi-arid tropics. Biotechnol. Adv. 31: 1120-1134.

Verma, M.M., J.S. Sandhu, H.S. Rrar and J.S. Brar. 1990. Crossability studies in different species of *Cicer* (L.). Crop Improv. 17: 179-181.

Verma, M. M., Ravi and J.S. Sandhu. 1995. Characterization of the interspecific cross *Cicer arietinum* L. × *C. judaicum* (Boiss.). Plant Breed. 114: 549-551.

Winter, P., A.M. Benko-Iseppon, B. Huttel, M. Ratnaparkhe, A. Tullu, G. Sonnante, T. Pfaff, M. Tekeoglu, D. Santra, V.J. Sant, P.N. Rajesh, G. Kahl and F.J. Muehlbauer. 2000. A linkage map of the chickpea (*Cicer arietinum* L.) genome based on recombinant inbred lines from a *C. arietinum* × *C. reticulatum* cross: localization of resistance genes for fusarium wilt races 4 and 5. Theor. Appl. Genet. 101: 1155-1163.

Winter, P., T. Pfaff, S.M. Udupa, B. Huttel, P.C. Sharma, S. Sahi, R. Arreguin-Espinoza, F. Weigand, F.J. Muehlbauer and G. Kahl. 1999. Characterization and mapping of sequence-tagged microsatellite sites in the chickpea (*Cicer arietinum* L.) genome. Mol. Gen. Genet. 262: 90-101.

Index

An environmentally friendly book printed and bound in England by www.printondemand-worldwide.com

PEFC Certified

This product is from sustainably managed forests and controlled sources

www.pefc.org

PEFC/16-33-415

This book is made of chain-of-custody materials; FSC materials for the cover and PEFC materials for the text pages.